Communications and Control Engineering Series

Editors: A. Fettweis · J. L. Massey · M. Thoma

M. S. Mahmoud · M. G. Singh

Discrete Systems

Analysis, Control and Optimization

With 87 Figures

Springer-Verlag
Berlin Heidelberg New York Tokyo 1984

MAGDI S. MAHMOUD
Professor, Electrical and Computer Engineering Dept.
Kuwait University - Kuwait

MADAN G. SINGH
Professor of Control Engineering at U.M.I.S.T.
Manchester, U.K.

ISBN-13:978-3-642-82329-9 e-ISBN-13:978-3-642-82327-5
DOI: 10.1007/978-3-642-82327-5

Library of Congress Cataloging in Publication Data
Mahmoud, Magdi S. Discrete systems, analysis, control, and optimization.
(Communications and control engineering series)
Bibliography: p. 1. Discrete-time systems. 2. Control theory. 3. Mathematical optimization.
I. Singh, Madan G. III. Series. QA 402.M29 1984 621.38 84-13915

Softcover reprint of the hardcover 1st edition 1984

2061/3020-543210

Dedicated to

Medhat, Monda and Salwa
(M.S. MAHMOUD)

Alexandre and Anne-Marie
(M. G. SINGH)

Preface

More and more digital devices are being used for information processing and control purposes in a variety of systems applications, including industrial processes, power networks, biological systems and communication networks. This trend has been helped by the advent of microprocessors and the consequent availability of cheap distributed computing power. For those applications, where digital devices are used, it is reasonable to model the system in discrete-time. In addition there are other application areas, e.g. econometric systems, business systems, certain command and control systems, environmental systems, where the underlying models are in discrete-time and here discrete-time approaches to analysis and control are the most appropriate.

In order to deal with these two situations, there has been a lot of interest in developing techniques which allow us to do analysis, design and control of discrete-time systems.

This book provides a comprehensive treatment of discrete-time dynamical systems. It covers the topics of modelling, optimization techniques and control design. The book is designed to serve as a text for teaching at the first year graduate level.

The material included is organized into eight chapters. In the first chapter, a number of discrete-time models taken from various fields are given to motivate the reader. The rest of the book (seven chapters) is split into three parts: *Analysis* (part I), *Control* (part II) and *Optimization* (part III).

Analysis of discrete-time systems is covered in Chapters 2 and 3. Chapter 2 deals with the representation of discrete dynamical systems using transfer functions, difference equations, discrete state equations and modal decomposition. The simplification of high-order transfer functions is also presented using continued fraction expansions. In Chapter 3, we examine the structural properties of discrete control systems such as controllability (reachability), observability (determinability) and stability. By considering the system modes, other properties are then introduced. Following that, we present Lyapunov analysis of stability and give suitable computational algorithms for solving Lyapunov equations.

Part II on control comprises Chapters 4 and 5. In Chapter 4, we consider the design of feedback controllers for discrete systems using state feedback (based on eigenvalue and eigenstructure assignment algorithms) and output feedback. Feedback control schemes are developed for both low-order systems as well as large-scale systems. In Chapter 5, we undertake parallel developments for systems with some inaccessible states.

Part III on optimization comprises three chapters (6 to 8). State and parameter estimation techniques are considered in Chapter 6. In Chapter 7, we examine adaptive control systems via model reference and self-tuning approaches. The final chapter (8) is concerned with dynamic optimization techniques for discrete dynamical systems. Here again, both the standard techniques as well as their extension to large systems are examined.

Throughout the book many worked examples are provided to illustrate various concepts and methods. We also give problems at the end of each chapter with the exception of Chapter 1. The material presented in the book should prove useful for teaching and research to engineers and practitioners.

We are grateful to Mrs. Vera Butterworth and Mrs. Liz Tongue for typing the final version of this book, to Mr. S. Grace for doing the artwork, and Mrs. Beryl Hooley for preparing the index.

Contents

Chapter 1
Discrete Models in Systems Engineering

1.1 Introduction

In studying physical and engineering systems, one usually starts with a mathematical model which is obtained by considering some physical laws and/or empirical formulae. The behaviour of the system is then described by the evolution of appropriate variables (dependent variables) over time or over frequencies (the independent variable). For a broad class of systems, the values of the dependent variables are only known, or can only be defined, at discrete time instants. Typical examples of this are found in the fields of information processing, digital filters, managerial systems, environmental systems, certain command and control systems, socioeconomic systems, to name but a few. In addition, the rapid growth in computing capabilities and the improved technology of microprocessors has attracted systems analysts and modellers to utilise digital computers extensively in solving their problems. This is the case with many industrial processes where digital devices are often used. In such industrial applications, we have batch information processing in contrast to the continuous information processing which was required when traditional analog equipment was used.

For both categories of systems, it is convenient and meaningful to represent their dynamic models by discrete mathematical structures i.e. by using $\mathcal{Z}$-transform theory or difference equations. The resulting models are commonly termed discrete-time dynamical models.

This book is devoted to the analysis, control and optimisation of discrete-time dynamical systems. Although the analytical development is focused on time-domain characterisations, a modest coverage of the frequency-domain representation methods is also given.

We now start with some illustrative examples drawn from different fields to motivate the reader. It should be pointed out that most of the definitions and concepts are stated in simple terms, leaving all rigorous treatment to subsequent chapters.

1.2 Some illustrative Examples

Our purpose in this section is to provide the reader with some feel about the importance of discrete models and discrete-time dynamical systems. This will be done by presenting some illustrative examples. Each example will be described and analysed in a simple way to stress the main features.

1.2.1 *DIRECT DIGITAL CONTROL OF A THERMAL PROCESS*

A schematic diagram of a typical environmental test chamber is shown in Fig. 1.1. The object to be tested is placed inside the chamber and its temperature is measured with a thermocouple (transducer). Since the electrical signal obtained by the temperature transducer usually has a low voltage level, an amplifier-filter unit is used to raise its level and remove any noise components [1]. The resulting signal is then fed into a digital control system consisting of an A/D (analog-to-digital) converter, a processor unit, and a D/A (digital-to-analog) converter. This system performs the following functions

a) sampling and coding of the electrical analog signal into binary format

b) implementation of a suitable algorithm to generate the discrete control signal

c) conversion of the digital signal back into an electrical voltage

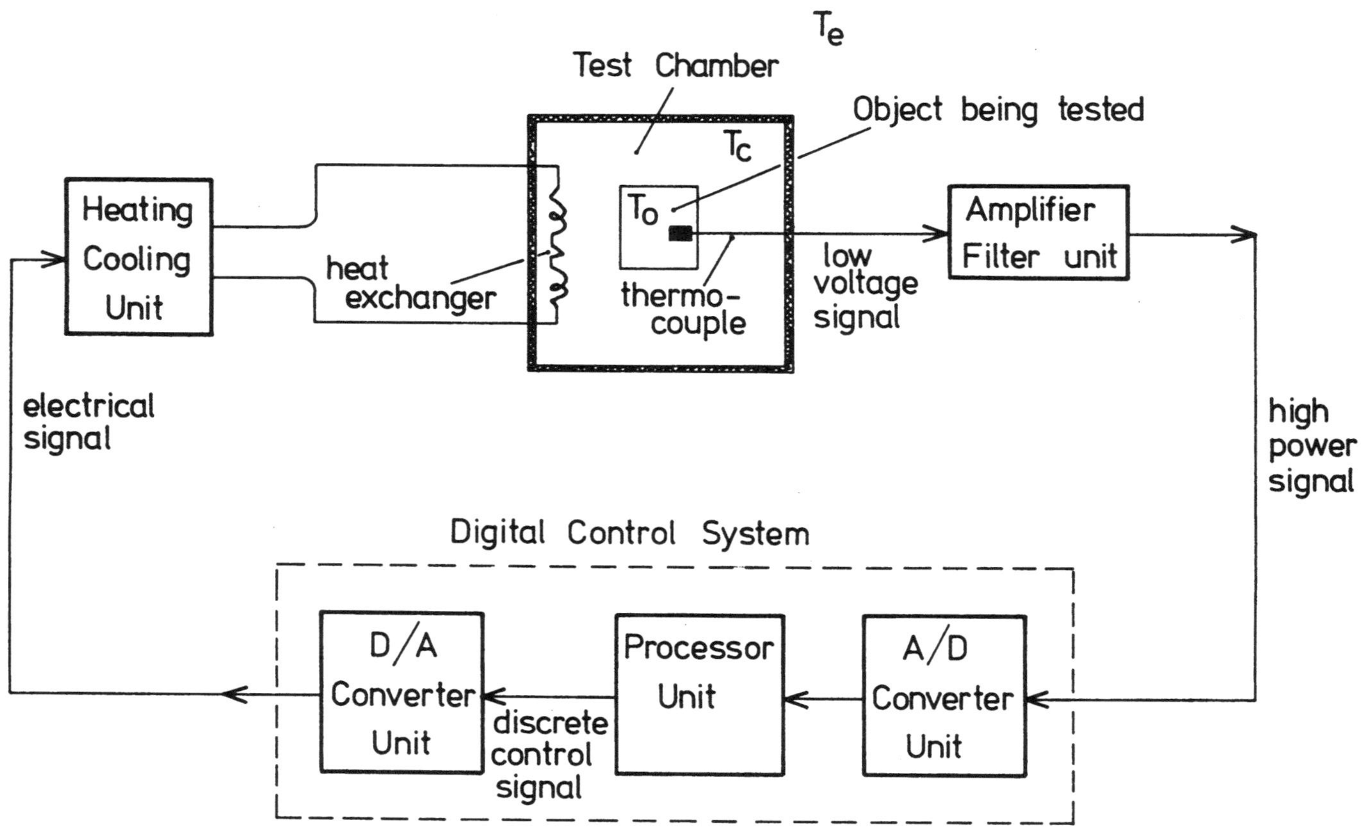

Fig. (1.1) Direct digital control of a thermal process

d) feedback of the electrical signal to the heating/cooling unit.

The above functions are, in general, accomplished in succession. Upon receiving the control signal, the heating/cooling unit responds by providing the appropriate thermal power to the heat exchanger unit. Note that both heating and cooling are achieved by using suitable equipment (for example, one can use electrical resistance for heating and liquid nitrogen for cooling). As to which process is used can be determined by the sign of the control signal (for example, positive for heating and negative for cooling).

To describe the above thermal process by a mathematical model, we assume that

1) both the test object and test chamber can be taken as single, lumped thermal masses m_1 and m_2, respectively *[2]*.
2) the process of heat transfer can be represented by a linear form relating the amount of heat flow to the temperature difference between the two entities
3) the test object can include an internal heat source
4) the sampling period s has been appropriately selected

Let

- p_j be the specific heat of element j,
- T_j be the temperature of element j,
- h_{ij} be the heat transfer coefficient between elements i and j,
- Q_g be the heat generated inside the test object
- Q_s be the heat supplied from the exchanger.

In the thermal system, we have two elements: the test object and test chamber, in addition to the external surrounding.

Direct application of physical laws of heat transfer to our system yields at the n-th sampling instant:

$$m_O P_O [T_O\{(n+1)s\} - T_O\{ns\}] = sh_{Oc}[T_c\{ns\} - T_O\{ns\}] + Q_g\{ns\}$$

$$m_c P_c [T_c\{(n+1)s\} - T_c\{ns\}] = sh_{Oc}[T_O\{ns\} - T_c\{ns\}] - sh_{ce}[T_c\{ns\} - T_e\{ns\}] + Q_s\{ns\}$$

where the subscripts O, c and e stand for object, chamber and external, respectively. The above model can be put into the form:

$$\underline{x}(k+1) = A\underline{x}(k) + \underline{b}u(k) + \underline{d}(k) \tag{1.1}$$

where k represents the nth sampling instant and (k+1) the next instant; (n+1)s. Also we have

$$\underline{x}(.) = \begin{pmatrix} T_O(.) \\ T_c(.) \end{pmatrix}$$

$$u(.) = Q_s(.)$$

$$A = \begin{pmatrix} 1-sh_{Oc}/m_O P_O & sh_{Oc}/m_O P_O \\ sh_{Oc}/m_c P_c & 1-(sh_{Oc}+sh_{ce})/m_c P_c \end{pmatrix}$$

$$\underline{b} = \begin{pmatrix} 0 \\ 1/m_c P_c \end{pmatrix}$$

$$\underline{d}(.) = \begin{pmatrix} Q_g(.)/m_O P_O \\ T_O(.)\, sh_{ce}/m_c P_c \end{pmatrix}$$

It should be stressed that (1.1) can be simulated easily given suitable data for different sampling periods. In this case, we can either consider the control signal u as

i) a known sequence (e.g. step, ramp,.... etc.) or

ii) a known function of $\underline{x}(k)$ e.g. $\{\alpha\underline{x}(k)\}$ (proportional-type), $\{\beta[\underline{x}(k) - \underline{x}(k-1)]\}$ (difference-type) or $\{\theta \sum_{j=0}^{k} \underline{x}(j)$ (summoned-type) or any combination of these [5], or

iii) to be selected subject to a prescribed criteria of performance like minimum settling time [3], minimum consumption of thermal energy [4],... etc.

Despite the simplicity of the above model, the reader should note the following features:

1) the digital control system can be designed to implement a variety of standard control schemes with great flexibility

2) the digital control system can be realised in practise using integrated circuit modules which, these days, are compact in size, cheap and possess high performance capabilities.

3) the digital control system can utilise sophisticated control algorithms in cases where their use would be advantageous.

1.2.2 *AN INVENTORY HOLDING PROBLEM*

We next consider a problem frequently encountered in inventory control and production [6]. Let the decision maker of a multi-product company have a forecast of the demand for the N (>1) products over a time horizon of $k=1,2,...,k_f$ intervals. At successive times, decisions have to be made to replenish inventory $x_j(k)$ of the $j=1,\ldots,N$ products by placing orders $u_j(k)$. In the time interval $(k,k+1)$ sales requirements deplete inventory by $s_j(k)$.

Delivery of orders and purchase may be immediate, in which case the evolution of inventory over the time horizon, from a known initial level $x_j(1) = \beta_j$, can be expressed as

$$j=1,\ldots,N$$

$$x_j(k+1) = x_j(k) + u_j(k) - s_j(k) \tag{1.2}$$

Suppose the delivery today is for π intervals, then (1.2) is replaced by

$$j=1,\ldots,N$$

$$x_j(k+1) = x_j(k) + u_j(k-\Pi) - s_j(k) \tag{1.3}$$

For distributed delays up to a maximum of Θ intervals, the inventory dynamics take the form:

$$j=1,\ldots,N$$

$$x_j(k+1) = x_j(k) + \sum_{m=0}^{m=\Theta} \alpha_j(m) u_j(k-m) - s_j(k) \tag{1.4}$$

where the coefficients $\alpha_j(m)$ are given. In practical situations, both the inventory and ordering levels are bounded, that is:

$$x_j^-(k) \leq x_j(k) \leq x_j^+(k) \tag{1.5}$$

$$u_j^-(k) \leq u_j(k) \leq u_j^+(k) \tag{1.6}$$

where the superscripts +, - represent the upper and lower bounds, respectively. Due to limitations on total production or on total inventory, joint constraints between products may be present. This can be written as

$$\sum_{j=1}^{N} P_j[x_j(k), u_j(k), k] \leq 0 \tag{1.7}$$

The inventory holding problem (1.2), (1.5) - (1.7) is in the form of a constrained discrete-time system. In general, its solution would require the knowledge of a priori ordering levels. A more meaningful situation arises when one considers the selection of both inventory and ordering levels so as to minimise a suitable cost function [7]. Let $q[u_j(k),k]$ be the cost of ordering and purchasing $u_j(k)$ units of the jth product at time k and let $r_j[x_j(k),k]$ be the cost of holding $x_j(k)$ units of the jth product as inventory. Over the time horizon $(1,k_f)$, the total costs are given by

$$J = \sum_{j=1}^{N} \{f_j[x_j(k_f+1), k_f+1] + \sum_{k=1}^{k_f} \left[q_j[u_j(k),k] + r_j[x_j(k),k] \right]\} \tag{1.8}$$

The terminal cost $f_j(\cdot,\cdot)$ has been added in (1.8) to penalise unfulfilled customer requirements or excessive stocks which are not sold. The problem of minimising (1.8) subject to the constraints (1.2), (1.5)-(1.7), falls within the class of constrained dynamic optimisations of discrete systems. It has been solved in [8] using hierarchical computational methods [9] when

1) the holding costs are quadratic and time independent,
2) the ordering costs are linear and time independent implying fixed per unit cost,
3) the demand is either deterministic (or normally distributed with known mean and standard deviation) over a horizon of k_f intervals.
4) no delivery delays are present.

As discussed in [7,8], the use of discrete models would seem to be appropriate tools for describing a wide class of managerial problems like marketing, advertising, workforce and overtime allocation. This is largely due to the nature of the problems and type of decisions involved.

1.2.3 *MEASUREMENT AND CONTROL OF LIQUID LEVEL*

In the following we present an experimental system developed for measurement and control of liquid level using digital techniques [10]. A schematic drawing of the system is shown in Fig. 1.2. The experimental system consists of a plexi-glass tank fed with water from the bottom through a rotameter. There is a control valve located on the tank outlet pipe. To provide a direct digital control, an air-to-electric (A/E) converter is connected to a transducer. On the other hand, an electric-to-air (E/A) converter is connected to the control valve. A computer-based controller, preceeded by an (A/D) converter and succeeded by a (D/A) converter, is then installed to provide the control action.

We note in this experiment that the liquid level is being converted to an air pressure signal (through the transducer), then to an electric signal (through the A/C converter), and finally, to a digital signal through the A/D converter. A convenient way to model this system is to use z-transform theory [11,18, 19]. It is found that the open-loop transfer function relating the liquid level Y to the feed flow rate X is given by [10]:

$$G_p(z) = Y(z)/X(z) = \alpha z/(z-\beta) \qquad (1.9)$$

where α and β are parameters of the systems. The control valve, (E/A) converter (A/E) converter) are modelled by the constants k_v, k_e and k_a, respectively. In designing this experiment, the computer-based controller acts on the error between the discrete forms of the set-point and measured values using an appropriate control algorithm (like a discrete one-,

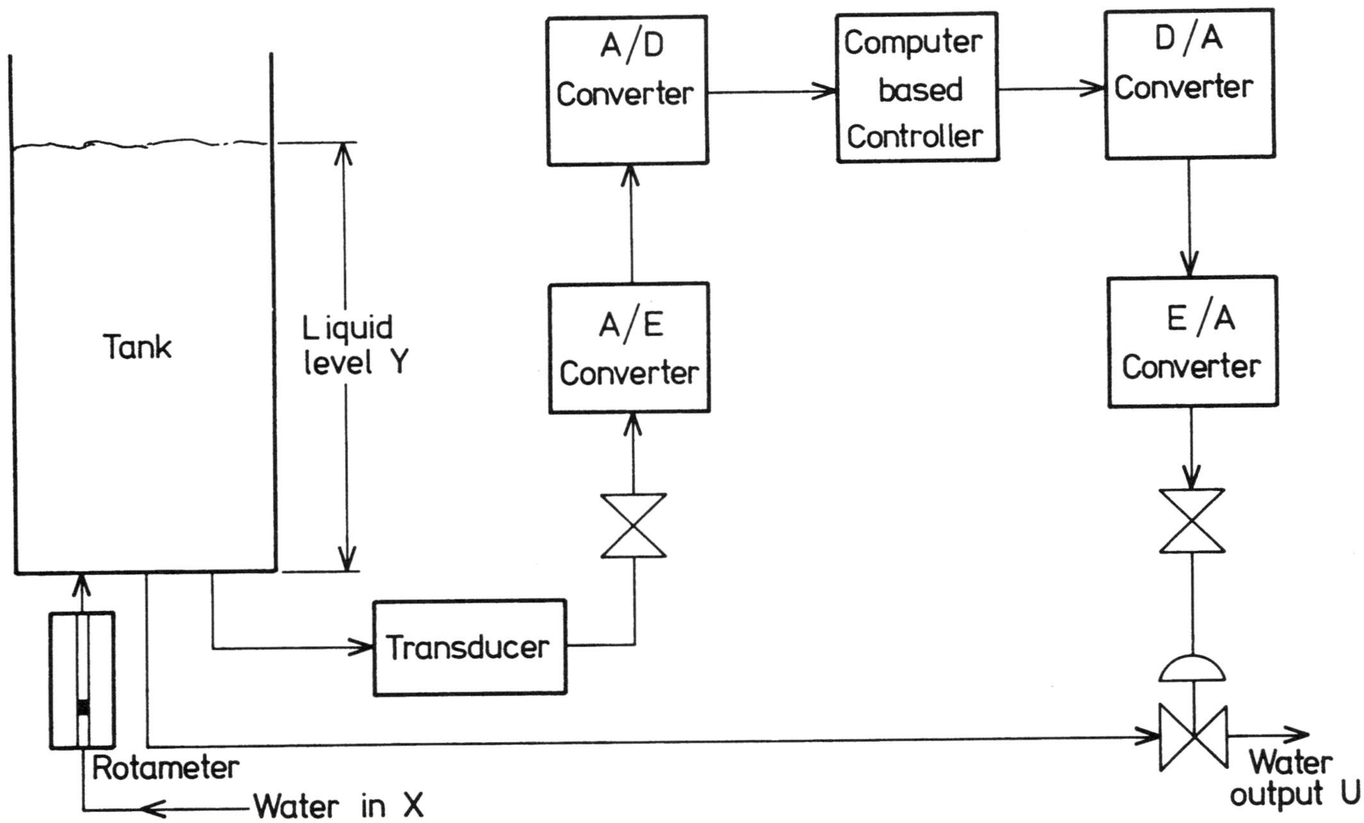

Fig. (1.2) Schematic drawing of the system

two- or three-term controllers) [4,5,19].

The controller delivers the output to the (D/A) converter, which, in effect, acts as a holding device. A complete block diagram representation of the digital control loop is shown in Fig. (1.3). A straightforward analysis will show [10] that the output of the closed loop system, using a proportional-type control algorithm, is given by

$$Y(z) = az/(z^2+bz+c) \tag{1.10}$$

where a, b and c are constants related to the system parameters. Indeed, if another control algorithm is used then (1.10) will be different.

The utility of the above experimental system lies in the ability to select a particular control algorithm so that a desirable profile of the liquid level can be obtained. This is not possible in case a pneumatic analog controller is used.

The discrete models derived in the preceding example are of low order. We now provide a nontrivial example of a discrete-time dynamical system of relatively high order.

1.2.4 *AN AGGREGATE NATIONAL ECONOMETRIC MODEL*

The system we shall consider is the Egyptian economy which is probably quite representative of the economies of developing countries. The model we shall describe is an aggregate, deterministic, linear time-invariant econometric model over the period 1961-1976. It is a yearly model, in the sense that the values of the economic variables are described, or defined, on a one year basis. It is characterised by 13 state variables and 6 control variables. Our main purpose in examining this example is to acquaint the reader with the types of large-scale dynamical systems we could be concerned with. First, we should

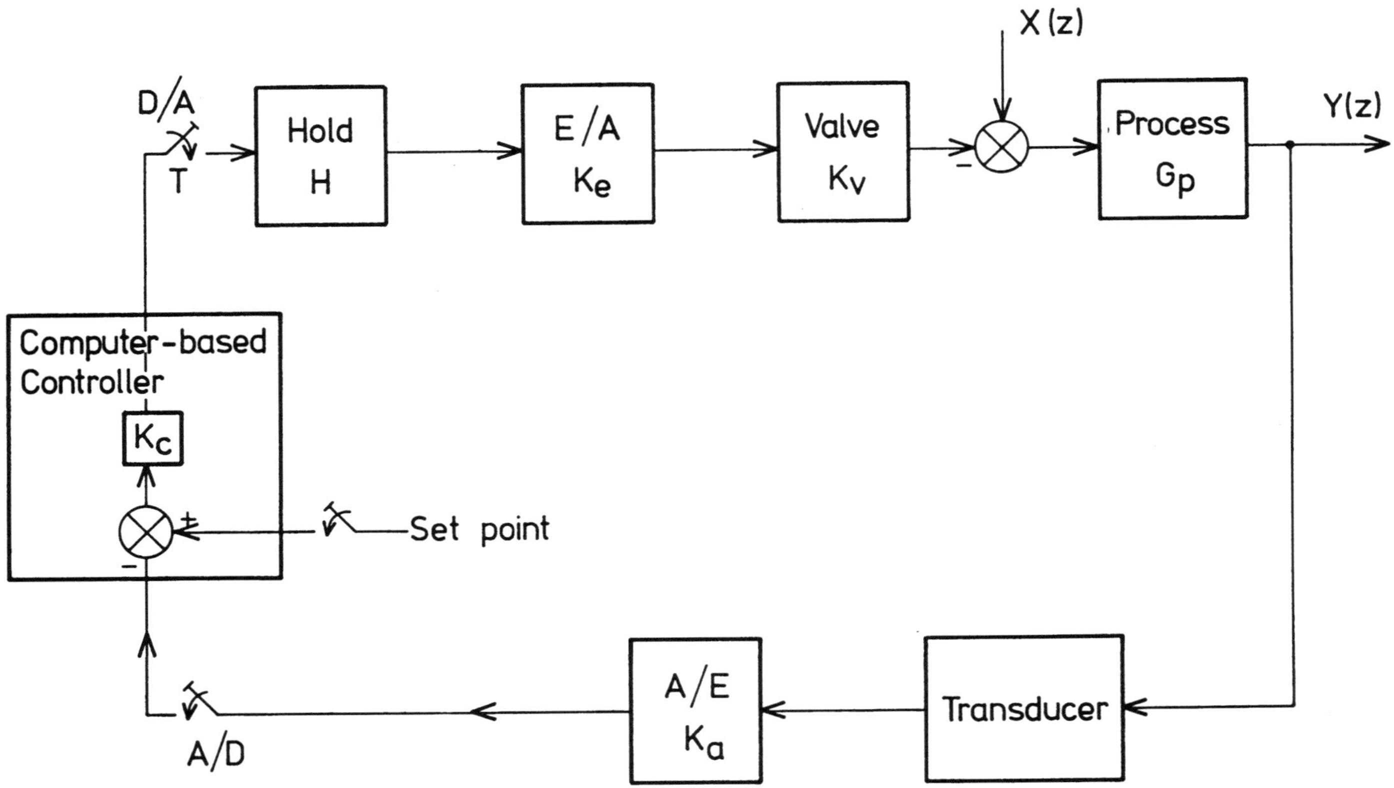

Fig. (1.3) Block diagram representation of the digital contol loop

mention some of the basic features of economic systems and econometrics [12-14]

1) models of economic systems deal with the interrelations of variables of physical interest like gross national product, interest rate, consumption, money supply, wage levels, etc.

2) mathematical models of the economy can be used for forecasting (to deduce the probable outcome of contemplated actions) or for control purposes to adjust frequently a typical policy variable.

3) macroeconomics and economic theory in general attempt to establish some structure in the interrelationship of economic variables using aggregate measures. Econometrics is a body of techniques by which one can transform past records of economic variables into a set of mathematical equations with numerical values for the coefficients. It is based upon statistical theory and time-series analysis [14].

4) In general, macroeconomic models are described by difference equations and may be valid only for short periods. In view of the aggregation used, complexity assumptions, and the structure assumptions, the discrete-time models represent only an approximation to reality.

5) There is no universal agreement on economic laws. Most of the economic relations are based upon empirical observations and logical reasoning.

Before we present the econometric model, it would seem relevant to shed some light on the economies of developing countries. The general features of these economies include low levels of investment, dominance of a traditional sector (usually agriculture), lack of modern technology, much reliance on imports of manufactured products, and a deficit in the balance of payments.

We are interested in the interrelation of economic variables at different instants of time. We shall use the time index

$$k=0,1,2,.... \tag{1.11}$$

to denote discrete time instants. Each k represents the passage of a fiscal year. Thus $k=0$ indexes the start of the econometric model (year 1961). $k=1$ indexes year 1962, $k=2$ year 1963, and so on. The mathematical model [15] consists of two income identities combined with two tax relations and nine structural equations. First, let us define

$x_1(k)$ = disposable income at time k in millions of Egyptian pounds

$x_2(k)$ = value added in the service sector at time k in millions of Egyptian pounds

$x_3(k)$ = value added in the agricultural sector at time k in millions of Egyptian pounds

$x_4(k)$ = value added in the industrial sector at time k in millions of Egyptian pounds

$x_5(k)$ = employed labour force in agriculture at time k in thousands of persons

$x_6(k)$ = employed labour force in industry at time k in thousands of persons

$x_7(k)$ = private consumption at time k in millions of Egyptian pounds

$x_8(k)$ = price level at time k

$x_9(k)$ = imports of capital goods at time k in millions of Egyptian pounds

$x_{10}(k)$ = imports of intermediate goods at time k in millions of Egyptian pounds

$x_{11}(k)$ = imports of consumer goods at time k in millions of Egyptian pounds

$u_1(k)$ = investment in agriculture at time k in millions of Egyptian pounds

$x_{12}(k)$ = $u_1(k-1)$

$u_2(k)$ = investment in industry at time k in millions of Egyptian pounds

$x_{13}(k)$ = $u_2(k-1)$

$u_3(k)$ = investment in services at time k in millions of Egyptian pounds

$u_4(k)$ = exports at time k in millions of Egyptian pounds

$u_5(k)$ = public consumption at time k in millions of Egyptian pounds

$u_6(k)$ = ratio between percentage change in money supply and percentage change in gross domestic product at time k

$z_1(k)$ = 1

$z_2(k)$ = share of total wages in gross domestic product at factor cost at time k

$z_3(k)$ = wage per worker in agriculture at time k in Egyptian pounds

$z_4(k)$ = wage per worker in industry at time k in Egyptian pounds

$z_5(k)$ = total population at time k in millions of persons

$e_1(k)$ = gross domestic product at factor cost at time k in millions of Egyptian pounds

$e_2(k)$ = gross domestic product at market prices at time k in millions of Egyptian pounds

$e_3(k)$ = total investment at time k in millions of Egyptian pounds

$e_4(k)$ = total imports

$e_5(k)$ = net indirect taxes at time k in millions of Egyptian pounds

$e_6(k)$ = direct taxes at time k in millions of Egyptian pounds

where $x(k)$ represents state (endogenous) variables $u(k)$ represents control (policy) variables, $z(k)$ represents exogenous variables and $e(k)$ represents intermediate variables. In terms of these variables, the mathematical model can be simplified into the following form (detailed derivations are found in [15,16])

$$x_1(k) = \alpha_1[x_7(k) + u_1(k) + u_2(k) + u_3(k) + u_4(k) - x_9(k) - x_{10}(k) - x_{11}(k)] + \alpha_2 z_1(k) \tag{1.12}$$

$$x_2(k) = \alpha_3 x_1(k) - x_3(k) - x_4(k) + \alpha_4 z_1(k) \tag{1.13}$$

$$x_3(k) = x_3(k-1) + \beta_1[x_5(k) - x_5(k-1)] + \beta_2 x_{12}(k) + \beta_3 z_1(k) \tag{1.14}$$

$$x_4(k) = x_4(k-1) + \beta_4[x_6(k) - x_6(k-1)] + \beta_5 x_{13}(k) \tag{1.15}$$

$$x_5(k) = x_5(k-1) + \gamma_1[z_3(k-1) - z_4(k-1)] + \gamma_2 z_5(k-1) + \gamma_3 z_1(k) \tag{1.16}$$

$$x_6(k) = x_6(k-1) + \gamma_4 x_{13}(k) + \gamma_5 z_4(k-1) + \gamma_6 z_1(k) \tag{1.17}$$

$$x_7(k) = \pi_1 x_1(k) + \pi_2 x_8(k) + \pi_3 z_2(k-1) + \pi_4 z_5(k-1) + \pi_5 z_1(k) \tag{1.18}$$

$$x_8(k) = \theta_1 u_6(k) + \theta_2 z_6(k-1) + \theta_3 z_1(k) \tag{1.19}$$

$$x_9(k) = \underset{1}{\emptyset} u_1(k) + \underset{2}{\emptyset} u_2(k) + \underset{3}{\emptyset} z_1(k) \tag{1.20}$$

$$x_{10}(k) = x_{10}(k-1) + \psi_1 \, x_3(k) - x_3(k-1)$$

$$+ \psi_2[x_4(k) - x_4(k-1)] + \psi_3 z_1(k) \qquad (1.21)$$

$$x_{11}(k) = \mu_1 \, x_7(k) + \mu_2 \, u_5(k) + \mu_3 \, z_1(k) \qquad (1.22)$$

$$x_{12}(k) = u_1(k-1) \qquad (1.23)$$

$$x_{13}(k) = u_2(k-2) \qquad (1.24)$$

where the α's, β's,...., 's are parameters and structural coefficients to be estimated using statistical records. It is interesting to note that, after some manipulations, the model (1.12)-(1.24) can be put into the vector-matrix form:

$$\underline{x}(k+1) - \underline{x}(k) = A\underline{x}(k) + B\underline{u}(k) + C\underline{z}(k) \qquad (1.25)$$

where $\underline{x}(k) \triangleq \{x_1(k),....,x_{13}(k)\}$, $\underline{u}(k) \triangleq \{u_1(k),....,u_6(k)\}$ and $\underline{z}(k) \triangleq \{z_1(k),....,z_6(k)\}$. The matrices A, B and C can then be obtained in terms of the system parameters. The form (1.25) is called the state-variable form, of which we shall say more in subsequent chapters. Before we can use the econometric model (1.12)-(1.24), the parameters and structural coefficients need to be estimated using past statistical records. This has been performed [15,16] using the two-stage least squares technique [14]. To experiment with the model, a linear-quadratic optimal tracking problem has been formulated by minimising a suitable cost functional (see [15,17] for further details) and utilising different development strategies. Samples of the results are shown in Figs. (1.4)-(1.8) and for an economic interpretation, the reader is referred to [15]. In general, the model seems to be a relatively realistic one given the wide span of time involved and its small dimensionality.

From the above we can see that:

1) More and more digital devices are being used for control purposes in a variety of engineering systems.

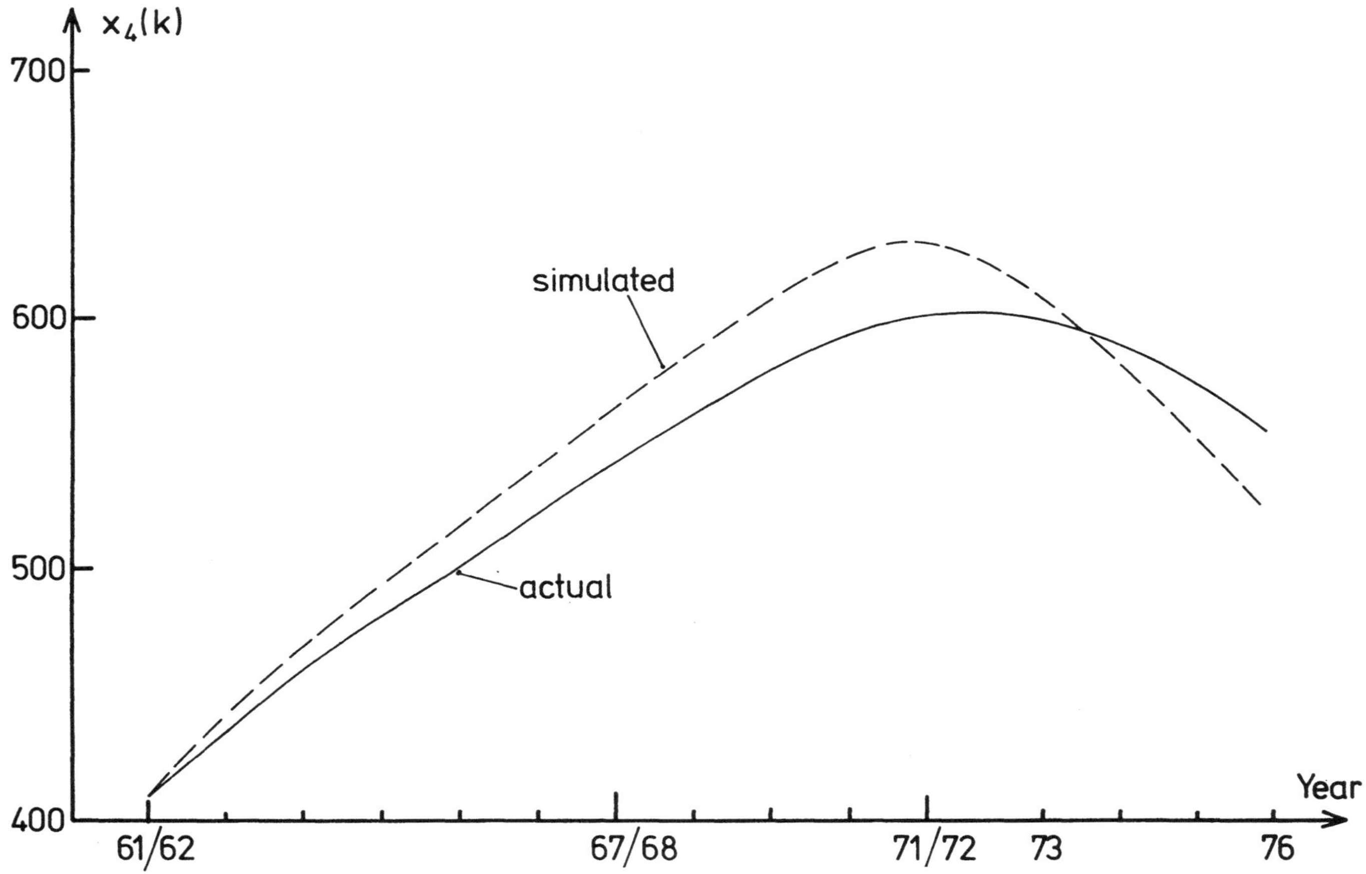

Fig. (1.4) Trajectory of value added in industry

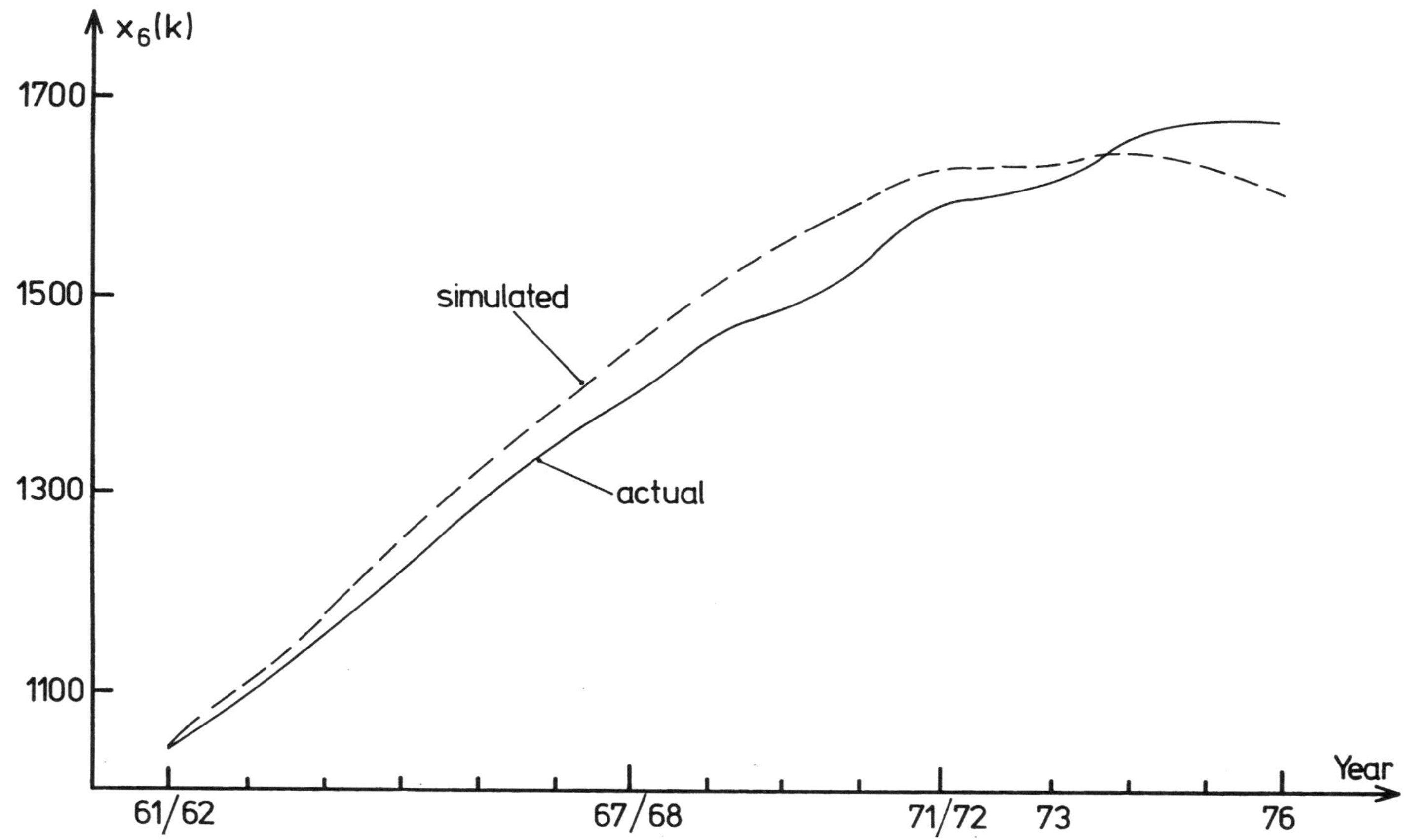

Fig. (1.5) Trajectory of labour in industry

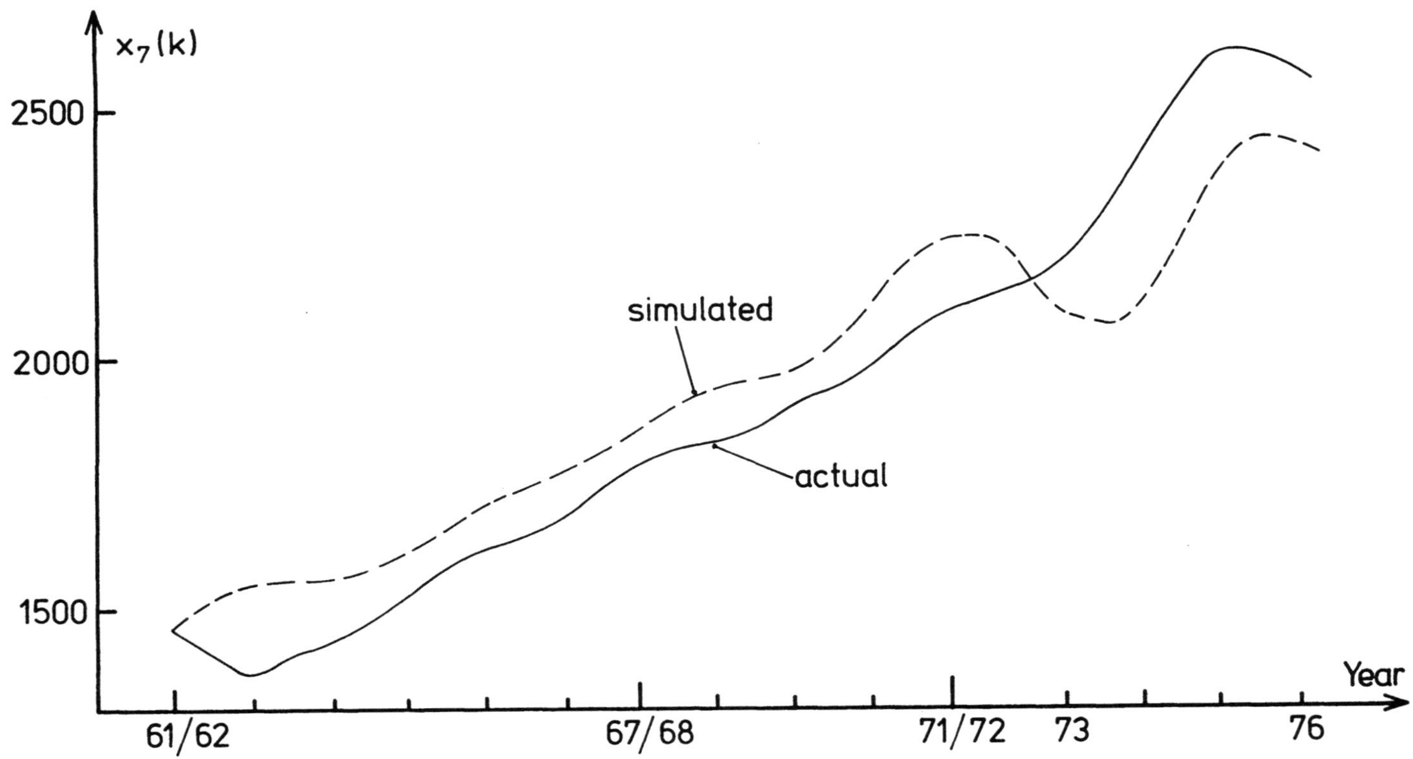

Fig. (1.6) Trajectory of private consumption

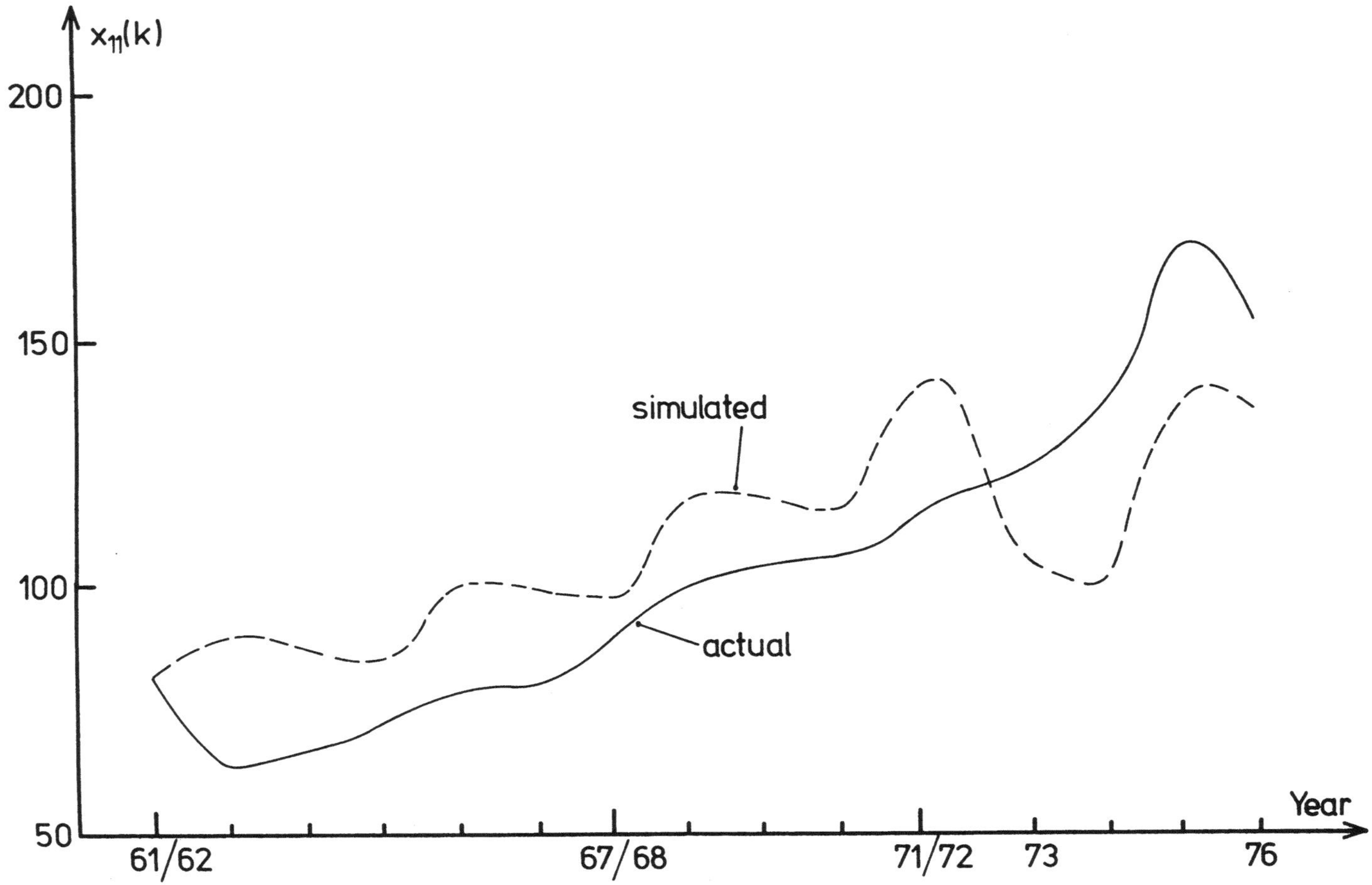

Fig. (1.7) Trajectory of imports of consumer goods

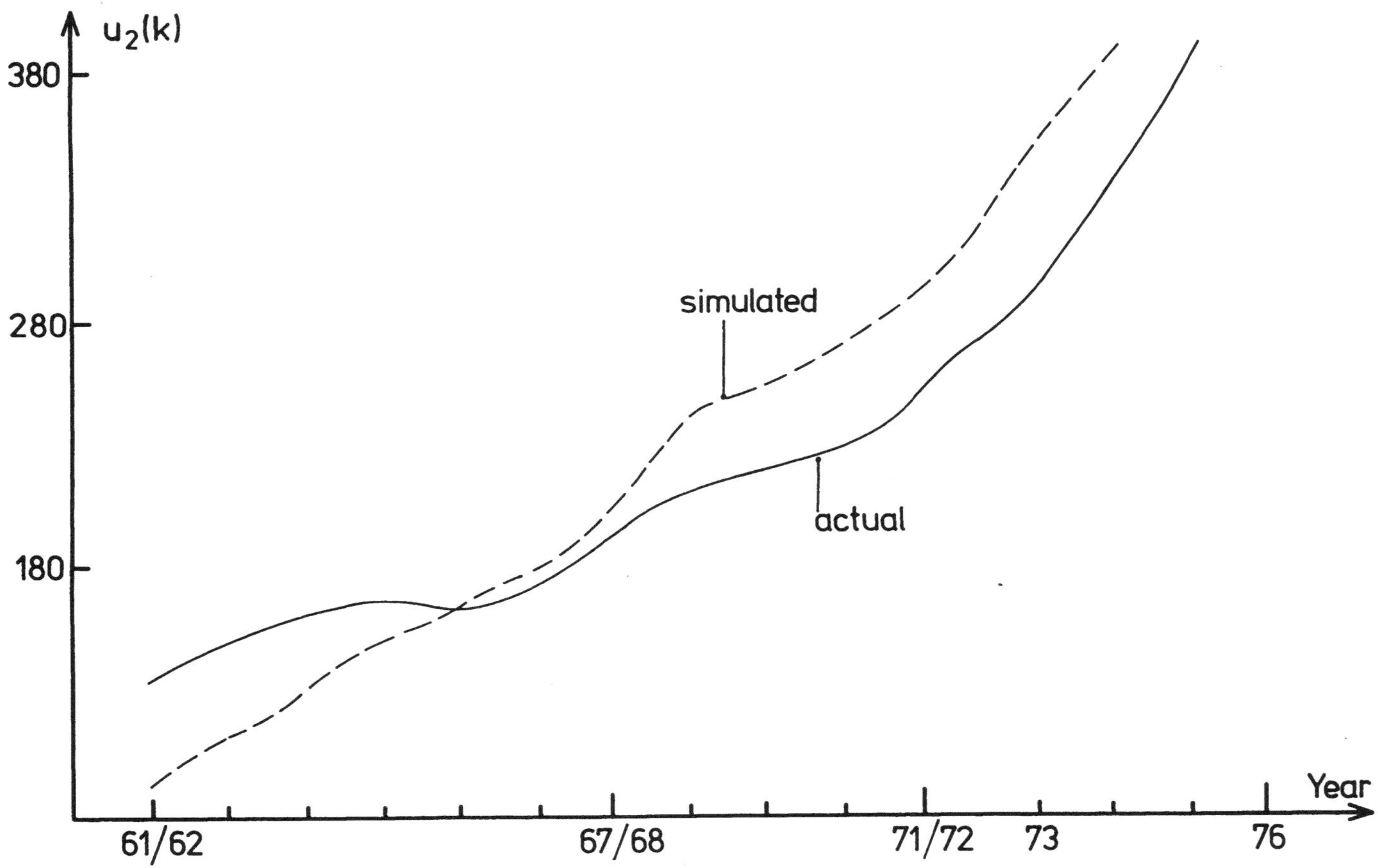

Fig. (1.8) Trajectory of investment in industry

2) there exists a wide-class of nonengineering systems whose dynamic evolution is best described by means of discrete-time models. Two typical examples of these systems have already been described in subsection 1.2.2 and 1.2.4. More examples can be drawn from the fields of socio-economics, business and administration, biology,....etc.

3) in some important engineering applications, the flow of information and type of decisions made eventually lead the system analyst to formulate the entire problem in discrete-time. These applications include signal processing, digital filters, traffic control, river pollution control and modern communication networks, amongst others. We have selected examples pertinent to these applications and these will be discussed throughout the text.

4) from a control system's point of view, the use of digital hardware in implementing control algorithms has been shown to provide great flexibility in design, improved performance and reliability as compared to classical analog hardware.

1.3 Objectives and Outline of the Book

As discussed in the previous sections, our main concern is with the wide class of systems in which the information content is being processed in discrete space; that is, at discrete-time instants or at specified values of complex frequencies. No distinction is made between systems whose natural description is discrete and those obtained using digital devices. The ultimate goal of this book is to acquaint the reader with analytic tools and to provide him with control design algorithms and optimisation methods for discrete-time dynamical systems. Specifically our objectives are:

1) To provide an in-depth treatment of time-domain analysis for discrete dynamical systems. We will also

examine the frequency-domain characterisation of control systems.

2) To deal side by side with small scale (low order) systems and large-scale (high-order or interconnected) systems. We feel that this represents a new approach to tackling systems engineering problems.

3) To cover a wide spectrum of topics on discrete-time dynamical systems including control analysis and design, optimisation techniques, and adaptive control systems. The rationale behind the choice of material is to give the reader a comprehensive pool of knowledge about system-theoretic ideas, concepts and tools.

4) To develop computational algorithms for the solution of different control problems. These algorithms are numerically efficient and easily programmed in practise.

In order to meet our objectives, the book has been organised into eight chapters. In the first chapter, a number of discrete models taken from various fields have been given to motivate the reader. The rest of the book is then split into three parts: *ANALYSIS, CONTROL and OPTIMISATION*.

Part I comprising Chapters 2 and 3, covers the analysis and basic properties of discrete-time dynamical systems. Methods of representing discrete systems are the subject of Chapter 2. These include transfer functions, difference equations, discrete state equations and modal analysis and decomposition. In Chapter 3, we discuss the structural properties of control systems like controllability (reachability, stabilisability), observability (determinability, detectability) and stability.

Part II on control consists of Chapters 4 and 5. In Chapter 4 the design of state and output feedback control schemes for discrete-time systems is considered for both low-order systems as well as for large-scale systems. In Chapter 5 the same is done for control systems with inaccessible states.

Part III on optimisation comprises three chapters. Parameter estimation techniques are considered in Chapter 6, whilst adaptive control systems are examined in Chapter 7. The final chapter (8) is concerned with dynamic optimisation techniques for discrete-time control systems. Here again, both the standard techniques as well as their extension to large systems is examined.

Many worked examples are provided throughout the text and exercises of graded difficulty are given at the end of each chapter.

1.4 References

[1] Ogata, K.
"Modern Control Engineering", Prentice-Hall, N.J., 1970.

[2] Wellstead, P.E.
"Introduction to Physical Systems Modelling", Academic Press, London, 1979.

[3] Cadzow, J.A. and H.R. Martens
"Discrete-time and Computer Control Systems", Prentice-Hall, N.J., 1970.

[4] Kuo, B.C.
"Digital Control Systems", Halt, Rinehart and Winston, N.Y., 1980.

[5] Bishop, A.B.
"Introduction to Discrete Linear Controls", Academic Press, N.Y., 1975.

[6] Mariani, L. and B. Nicoletti
"Optimisation of Deterministic Multiproduct Inventory Model with Joint Replenishment", Management Science, Vol. 20, pp. 349-362, 1973.

[7] Drew, S.A.W.
"Hierarchical Control Methods with Application to Managerial Problems", Ph.D. Thesis, Control Engineering, Cambridge University, 1975.

[8] Drew, S.A.W.
"The Application of Hierarchical Control Methods to a Managerial Problem", Int. J. Systems Science, Vol. 6, pp. 371-395, 1975.

[9] Singh, M.G. and A. Titli
"Systems: Decomposition, Optimisation and Control", Pergamon Press, Oxford, 1978.

[10] Abd El-Bary, M.F.
"Direct Digital Control Liquid Level Experiment", Chemical Engineering Education, Vol. 24, pp. 28-31, 1983.

[11] Deshpande, P.B. and R.H. Ash
"Computer Process Control with Advanced Control Applications", Instrument Society of America, 1981.

[12] Kogiku, K.C.
"An Introduction to Macroeconomic Models", McGraw-Hill, N.Y., 1968.

[13] Allen, R.G.D.
"Macroeconomic Theory", St. Martin's Press, N.Y., 1967.

[14] Intriligator, M.D.
"Econometric Models, Techniques and Applications", Prentice-Hall, N.J., 1978.

[15] Hanna, M.T.
"Control Modelling and Optimisation of the National Economy", M.Sc. Thesis, Cairo University, 1980.

[16] Mahmoud, M.S. and M.T. Hanna
"Linear Optimal Control of National Econometric Models", Int. J. Systems Science, Vol. 13, pp. 1061-1081, 1982.

[17] Pindyck, R.S.
"Optimal Planning for Economic Stabilisation", North-Holland, Amsterdam, 1973.

[18] Steiglitz, K.
"An Introduction to Discrete Systems", J. Wiley, N.Y., 1974.

[19] Franklin G.F. and J.D. Powell
"Digital Control of Dynamic Systems", Addison-Wesley, N.Y., 1980.

Chapter 2

Representation of Discrete Control Systems

2.1 Introduction

The past few years have seen significant advances in computer technology and its incorporation into engineering and other applications. In these applications, the various signals acting in and on systems occur as data sequences occurring at discrete time instants. Some familiar examples of systems controlled by discrete data sequences are:

1. National Economies
2. Radar-tracking systems
3. Data-transmission links

We note that in such systems the information is available only in samples. In this chapter we present methods for the analysis of systems incorporating digital devices, which are called *discrete-time systems*. As used here, the term *discrete system* refers to any system whose operation or output is conveniently described on a discrete time scale even though the system characteristics may in general be given in continuous time.

The analysis of discrete systems depends on the availability of a mathematical representation (or model) of the system that adequately characterises its behaviour. The mathematical model should be sufficiently simple to make the resultant analysis tractable. In this chapter, attention will be focused, most of the time, on systems that are linear. This restriction is justified by the fact that a wide class of systems can be described by linear or approximately linear relationships. Moreover, the mathematical techniques available for the study of discrete-time systems are mainly restricted to linear systems.

The system model can be written in terms of two sets of variables called the *input* and the *output variables*. The objective of system analysis is to determine the nature of this input-output relationship. In Fig. 2.1, a block diagram representation of a multi input-multi output system S is shown in which the set of variables $\{u_1, u_2, \ldots, u_m\}$ is the system input while the set $\{y_1, y_2, \ldots, y_p\}$ is the system output. This is frequently called a multivariable system.

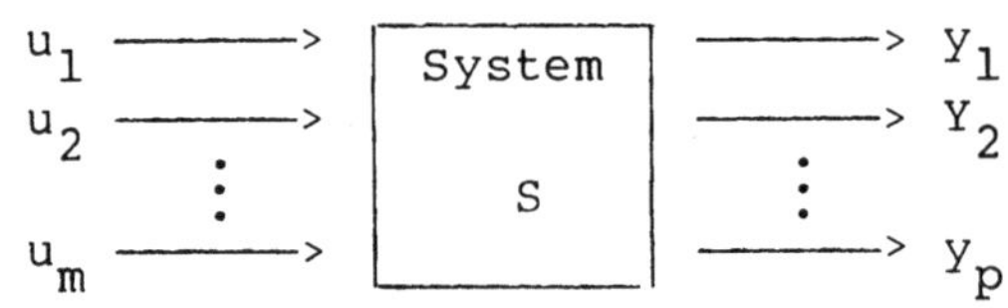

Fig. 2.1 Block diagram representation of a multi input-multi output system.

In many typical situations, there exists only one input (u_1) and one output (y_1). This is often called a single-variable system. The behaviour of multivariable systems can be described either in the time-domain or in the frequency-domain. In the time-domain, the input and output variables are written as functions of the index k; that is, $\{u_1(k), u_2(k), \ldots, u_m(k)\}$, $\{y_1(k), y_2(k), \ldots, y_p(k)\}$. This index k, which is called the discrete-time variable, takes on only a finite or at most an infinitely countable number of values. If the model is obtained by discretising a continuous model, then $k = t/T$ where t is the continuous time and T is the sampling time. A detailed discussion of the process of sampling is given in [1,2].

In the succeeding sections, different methods of representing discrete control systems are given. Since we will be dealing with linear systems, it is important to define the concept of linearity. Let

$$\underline{u} = \{u_1(k), u_2(k), \ldots, u_j(k), \ldots\},$$

$$\underline{y} = \{y_1(k), y_2(k), \ldots, y_j(k), \ldots\},$$

and

$$\underline{y} = S(\underline{u})$$

Here $\underline{u}(k)$ is the input to the system at time k and $\underline{y}(k)$ is the system output at the same time instant. Now suppose that the input $\underline{u}_a$ produces the output $\underline{y}_a$ and that input $\underline{u}_b$ produces the output $\underline{y}_b$; that is

$$\underline{y}_a = S(\underline{u}_a)$$

$$\underline{y}_b = S(\underline{u}_b)$$

Define the composite input $\underline{u}$ as a linear combination of $\underline{u}_a$ and $\underline{u}_b$

$$\underline{u} = \alpha \underline{u}_a + \beta \underline{u}_b$$

where α and β are constants, and define $\underline{y}$ as the output resulting from the input $\underline{u}$. Thus

$$\underline{y} = S(\underline{u}) = S(\alpha \underline{u}_a + \beta \underline{u}_b)$$

We can then say that *the system is linear* if the relationship

$$S(\alpha \underline{u}_a + \beta \underline{u}_b) = \alpha S(\underline{u}_a) + \beta S(\underline{u}_b) \qquad (2.1)$$

is satisfied for any set of real numbers (α, β). We note that (2.1) implies that under the transformation S, the output resulting from a linear combination of inputs is made up of the same linear combination of the individual outputs which result from those inputs alone. In simple terms, linear systems obey the principle of superposition [3]. For example, it is obvious that the input-output equation

$$y(k+3) + 2y(k+2) + 4y(k+1) + y(k) = u(k)$$

represents a linear system with input $u(k)$ and output $y(k)$. The discrete system

$$y(k+1) - y(k) = u^2(k)$$

is not linear since (2.1) is not satisfied.

2.2 Transfer Functions

In linear discrete systems, both the input and output variables are sequences of numerical values. To study the behaviour of these systems, we need to know the mechanism by which the input sequence is converted into the output sequence. There are two fundamental methods for describing this mechanism. One method is in the time-domain where by means of a state-space formulation, we arrive at a set of dynamical relationships. The other method is in the complex frequency-domain where by using the z-transform we obtain the system transfer function [1,2]. In this section, we investigate the second method and examine its basic properties.

2.2.1 *REVIEW OF Z-TRANSFORMS*

Simply stated, the z-transform is a transformation that operates on a sequence of numbers (values of a discrete-time function) to produce a function of the complex variable z. For the case where the sequence of numbers $g(k)$ is defined only for positive integers k, we can define the one-sided z-transform as [3]:

$$\begin{aligned} Z[g(k)] &= G(z) \\ &= \sum_{k=0}^{\infty} g(k) z^{-k} \end{aligned} \tag{2.2}$$

When $g(k)$ is given for all integers, the two-sided z-transform of $g(k)$ takes the form:

$$Z[g(k)] = G(z) = \sum_{k=-\infty}^{\infty} g(k) z^{-k} \tag{2.3}$$

The following are some of the important properties of z-transform of the type (2.2). A detailed treatment of these can be found in [1-4]:

(1) The z-transform is a linear operator, that is:

$$Z[\alpha f(k) + \beta g(k)] = \alpha F(z) + \beta G(z) \tag{2.4}$$

where α, β are given scalars and $F(z)$ and $G(z)$ are the z-transforms of the number sequences $f(k)$ and $g(k)$, respectively.

(2) For shifted data sequences with positive integers, the z-transform is given by:

$$Z[g(k+m)] = z^m G(z) - \sum_{j=0}^{m-1} g(j) z^{m-j} \tag{2.5a}$$

(left shift)

$$Z[g(k-m)] = z^{-m} G(z) \tag{2.5b}$$

(right shift)

We note that in (2.5) $g(k) = 0$ for $k < 0$. A useful application of (2.5) occurs when dealing with the backward difference $[g(k) - g(k-1)]$, whose z-transform is therefore expressed as:

$$Z[g(k) - g(k-1)] = (1 - z^{-1}) G(z) \tag{2.6}$$

Likewise, the use of (2.4) enables us to derive the z-transformation of the forward difference $[g(k+1) - g(k)]$ to obtain:

$$Z[g(k+1) - g(k)] = (z-1) G(z) - z\, g(0) \tag{2.7}$$

(3) Given the one-sided z-transform $G(z)$, the initial and asymptotic values of $g(k)$ can be obtained from the relations:

$$g(0) = \lim_{z\to\infty} G(z) \tag{2.8}$$

$$g(\infty) = \lim_{z\to 1} (z-1)G(z) \tag{2.9}$$

provided the limits exist in the convergence region of z.

(4) In the case of sequences of the summation-type, we have:

$$Z\left[\sum_{j=0}^{k} g(j)\right] = zG(z)/(z-1) \tag{2.10}$$

For weighted-type sequences, the z-transform is given by

$$Z[\alpha^k g(k)] = G(z/\alpha) \tag{2.11}$$

It should be emphasized that in deriving the properties (2.4 - (2.11) it is implicitly assumed that the z-transform has an appropriate convergence region. We now proceed to develop the transfer function of discrete systems. The derivation is based on the observations that

(a) the system S can be viewed as a functional mechanism that assigns different weights to the input signals,

(b) the system is initially at rest,

(c) each input signal is represented by a sequence of impulses.

For simplicity, we first consider the case of single-variable systems. Let the system mechanism be expressed by the sequence $\{g(0), g(1), g(2), \ldots, g(j), \ldots\}$. In terms of the Kronecker delta function $\delta_j(k)$, which has the value 1 for k=j and 0 otherwise, the input sequence can generally be written as:

$$u(k) = u_0\delta_0(k) + u_1\delta_1(k) + \ldots + u_j\delta_j(k) + \ldots \tag{2.12}$$

Since the system is linear, the response $y(k)$ is equal to the sum of the individual responses caused by the inputs $u(0), u(1), \ldots, u(k)$, taken separately, that is

$$y(k) = \sum_{j=0}^{k} y_j(k) \tag{2.13}$$

By virtue of (2.12) and (2.13), it is easy to see that the first impulse $u_0 = u(0)$ produces the response $y_0(k) = u(0)g(k)$; the second impulse $u_1 = u(1)$ superimposes on this the response $y_1(k) = u(1)g(k-1)$; the third impulse $u_2 = u(2)$ adds its contribution $y_2(k) = u(2)g(k-2)$; and so on. The result of repeating this process

$$\begin{aligned} y(k) &= u(0)g(k) + u(1)g(k-1) + \ldots \\ &= \sum_{j=0}^{k} u(j)g(k-j) \end{aligned} \tag{2.14}$$

which describes the output data sequence. (2.14) is often called the convolution summation. Define

$$\left.\begin{aligned} Y(z) &= Z[y(k)] \\ U(z) &= Z[u(k)] \\ G(z) &= Z[g(k)] \end{aligned}\right\} \tag{2.15}$$

then from (2.2) we have

$$Y(z) = \sum_{k=0}^{\infty} \{ \sum_{j=0}^{\infty} g(k-j)u(j) \} z^{-k} \tag{2.16}$$

In the region where (2.2) is uniformly convergent, (2.16) may be written as:

$$Y(z) = \sum_{j=0}^{\infty} u(j) \sum_{k=0}^{\infty} g(k-j) z^{-k} \tag{2.17}$$

Making the change of variables $m = k-j$ in (2.17) it becomes:

$$Y(z) = \sum_{j=0}^{\infty} u(j) \sum_{m=0}^{\infty} g(m) z^{-m-j}$$

$$= \sum_{j=0}^{\infty} u(j) z^{-j} \sum_{m=0}^{\infty} g(m) z^{-m} \qquad (2.18)$$

Using (2.15) yields:

$$Y(z) = U(z)G(z) \qquad (2.19)$$

which is the desired result. In (2.19), $G(z) = Y(z)/U(z)$ is called the *transfer function* of the given discrete system. According to the above analysis $G(z)$ is the z-transform of the system weighting sequence $g(k)$. More importantly, if $u(k) = \delta_0(k)$, then from (2.2), (2.15) and (2.19) we obtain $y(z) = G(z)$. This asserts the fact [2] that the transfer function is the z-transform of the impulse response, which is the system output with a unit impulse applied at the input.

A convenient way to express $G(z)$ is in terms of two polynomials in z, that is

$$G(z) = \frac{N(z)}{D(z)} = \frac{b_0 + b_1 z + b_2 z^2 + \ldots + b_m z^m}{a_0 + a_1 z + a_2 z^2 + \ldots + a_n z^n} \qquad (2.20)$$

$$n \geq m$$

After factoring the numerator $(N(z))$ and denominator $(D(z))$ polynomials of $G(z)$, we have

$$G(z) = H_0 \frac{(z-z_1)(z-z_2)\ldots(z-z_m)}{(z-p_1)(z-p_2)\ldots(z-p_n)} \qquad (2.21)$$

where H_0 is a gain factor.

The numbers z_i are referred to as the zeros of $G(z)$, for the obvious reason that $G(z_i) = 0$, whereas the numbers p_i are called the poles of $G(z)$. It is clear that at a pole, $G(z)$ takes an infinite value. As we shall see later, the zeros and poles play a major role in determining the characteristics of discrete-time systems.

The process of obtaining the sequence of numbers $g(k)$ that generates a given z-transform $G(z)$ is known as the inverse z-transform. There are several methods of doing this, for example:

(a) the power series method
(b) the partial fraction expansion method
(c) the inversion integral method.

In the power series method [3], the value of $g(k)$ can be obtained as the coefficient of z^{-k} in the power series expansion of $G(z)$ as a function of z^{-1}. We seek $G(z)$ of the form

$$G(z) = \sum_{j=0}^{\infty} c_j z^{-j} \tag{2.22}$$

Let $G(z)$ be given as a ratio of two polynomials in z^{-1}:

$$G(z) = \frac{r_0+r_1z^{-1}+r_2z^{-2}+\ldots+r_nz^{-n}}{q_0+q_1z^{-1}+q_2z^{-2}+\ldots+q_nz^{-n}} \tag{2.23}$$

It should be observed that (2.23) can be obtained from (2.20) by simple manipulation. The coefficients $c_0, c_1, c_2, \ldots,$ can be computed by equating (2.22) and (2.23) and matching like-powers in z^{-k} to yield:

$$\left.\begin{aligned} r_0 &= c_0q_0 \\ r_1 &= c_1q_0 + c_0q_1 \\ &\vdots \\ r_n &= c_nq_0 + c_{n-1}q_1 + \ldots + c_0q_n \end{aligned}\right\} \tag{2.24}$$

Having determined the coefficients $c_0, c_1, c_2, \ldots,$ the sequence $g(k)$ is obtained by inverting (2.22) term by term to yield:

$$g(k) = c_0\delta_0(k) + c_1\delta_1(k) + \ldots + c_j\delta_j(k) \qquad (2.25)$$

This method is useful only in determining the first few terms of a sequence. If the general expression for $g(k)$ is desired the inversion integral and the partial fraction expansion methods can be used.

In terms of $G(z)$ in (2.20), let the denominator polynomial $D(z)$ have n_1 roots equal to p_1, n_2 roots equal to $p_2, \ldots$ $\ldots, n_r$ roots equal to p_r where

$$\sum_{j=1}^{r} n_j = n$$

Then we can write $G(z)$ as

$$G(z) = \sum_{j=0}^{m} b_j z^j \Big/ \prod_{j=1}^{r} (z-p_j)^{n_j} \qquad (2.26)$$

The partial fraction expansion representation of $G(z)$ is given by [4,5]:

$$G(z) = b_n + \sum_{j=1}^{r} \sum_{k=1}^{n_j} d_{jk}/(z-p_j)^k \qquad (2.27a)$$

where $b_n = 0$ unless $m = n$. The coefficients of d_{jk} are calculated by

$$d_{jk} = \frac{1}{(n_j-k)!} \frac{d^{n_j-k}}{dz^{n_j-k}} [(z-p_j)^{n_j} z^{-1} G(z)]_{z \to p_j} \qquad (2.27b)$$

The particular coefficients d_{j1}, $j = 1,\ldots,r$ are called the residues of $G(z)$ at p_j, $j = 1,\ldots,r$. If none of the roots are repeated, then

$$G(z) = b_n + \sum_{j=1}^{n} d_{j1}/(z-p_j) \qquad (2.28a)$$

where

$$d_{j1} = (z-p_j)G(z)\Big|_{z \to p_j} \qquad (2.28b)$$

We note that each term in (2.27a) constitutes an elementary function whose inverse z-transform can be found in standard tables. We should mention that comprehensive tables of this kind are available in [3-5].

The most general technique for obtaining the inverse z-transform is the inversion integral method. It is based on the fundamental inversion formula [3]:

$$g(k) = \frac{1}{2\pi i} \oint G(z) z^{k-1} \, dz \tag{2.29}$$

where $i = \sqrt{-1}$ and the integration in (2.29) is taken over any closed contour which contains within it all the singularities of $G(z)$. A detailed discussion of (2.29) and its application is found in [3-6] and in standard texts on complex variable theory [7,8].

2.2.2 *EFFECT OF POLE LOCATIONS*

We will now examine the behaviour of discrete systems in terms of the locations of the poles of $G(z)$. Recall from (2.28a) that a system with a single pole on the real axis can be written as

$$G_1(z) = \beta/(z-\alpha) \tag{2.30a}$$

It can be easily verified that $g_1(k)$, the inverse z-transform of $G_1(z)$, is a data sequence with elements

$$\beta\alpha^{k-1}, \quad k \geq 1 \tag{2.30b}$$

Depending on the location of the pole in the Z-plane, we can consider the following cases:

(a) If $\alpha > 1$, the data sequence (2.30b) diverges; that is $g_1(k) \to \infty$ as $k \to \infty$.

(b) If $\alpha = 1$, all the elements will have the same magnitude b .

(c) If $0 \leq \alpha < 1$, the resulting sequence will contain positive numbers which are monotonically decreasing.

(d) If $0 > \alpha > -1$, we obtain a sequence of decreasing numbers with alternating signs.

(e) If $\alpha = -1$, $g_1(k)$ will be $\{\beta, -\beta, \beta, -\beta \ldots\}$, and

(f) If $\alpha < -1$, we have a divergent sequence with alternating signs.

The above six cases are shown graphically in Fig. 2.2 .

The case of complex poles can be handled in the same way. Consider a pair of complex conjugate poles $z = \alpha \pm i\,\beta$. These poles will give rise to the sequence [2]

$$\lambda r^{k-1} \cos [(k-1)\Theta + \phi] \quad k \geq 1 \tag{2.31a}$$

where λ and ϕ are constants dependent upon the partial fraction expansion coefficients and

$$r = \sqrt{\alpha^2 + \beta^2} \; ; \quad \Theta = \tan^{-1} \frac{\beta}{\alpha} \tag{2.31b}$$

By examining (2.31) for possible values of α and β we find that

(g) If $r > 1$, $\Theta \neq \Pi$; the complex poles lie outside the unit circle. The corresponding data sequence will be oscillatory and increasing in magnitude.

(h) If $r = 1$, $\Theta \neq \Pi$; the poles are located on the unit circle and thus give rise to an oscillating sequence with a constant amplitude.

(i) If $r < 1$, $\Theta \neq \Pi$; now the poles are inside the unit circle and the resulting sequence defines an oscillation with decreasing amplitude.

(j) When $\Theta = \Pi$, we have a double pole on the negative real axis. Again the data sequence is oscillatory and decreasing in magnitude. The four additional cases discussed

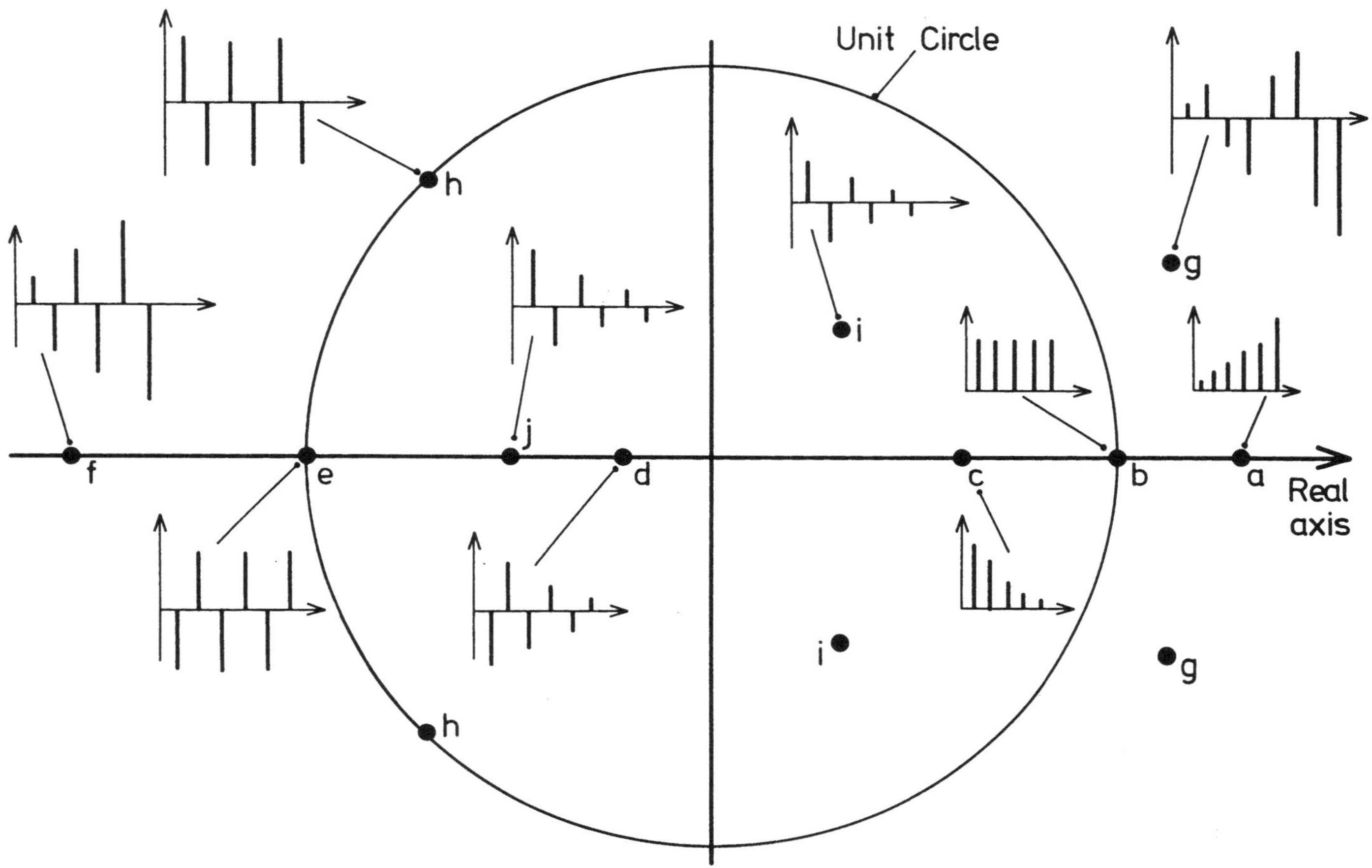

Fig. (2.2) Forms of data sequences associated with Z-plane pole locations

above have been illustrated in Fig. 2.2 for comparison. It should be noted that since the indicated responses of the pole locations (i)-(j), are not affected by the input sequence, they represent transient responses in the time-domain. A typical system transfer function may contain any possible combination of the indicated pole locations. We can now state the following:

(1) When z-plane poles lie within the unit circle, the transient system response decays to zero as $k \to \infty$.

(2) Z-plane poles outside the unit circle correspond to data sequences $g(k)$ for which $|g(k)| \to \infty$ as $k \to \infty$.

(3) Multiple z-plane poles on the unit circle give rise to data sequences $g(k)$ for which $|g(k)| \to \infty$ as $k \to \infty$.

(4) Simple z-plane poles on the unit circle yield data sequences, the terms of which oscillate with k, except for a pole at $z = 1$, which produces a data sequence of constants.

In the light of (1)-(4) above, we may generally conclude that a discrete system is stable if the open-loop transfer function has its poles inside the unit circle in the complex plane.

2.2.3 *STABILITY ANALYSIS*

Examination of the system transient responses due to different pole locations has led us to infer that the open-loop transfer function has all of its poles within the unit circle. We now examine the condition of stability for discrete systems described by the input-output relation

$$Y(z) = G(z)U(z) \tag{2.32}$$

where $G(z)$, the system transfer function, is a rational function of two polynomials in z. We know from Section 2.2.1 that the output sequence is given by

$$y(k) = \sum_{j=0}^{k} y_j(k)\delta_j(k) \tag{2.33a}$$

where

$$y_j(k) = \frac{1}{2\pi i}\oint U(z)G(z)\, z^{j-1}dz \tag{2.33b}$$

It is known [1,3] that the output sequence can be separate for the most part into steady state and transient terms. The steady state part of the output can be written as:

$$\sum_{m=1}^{r} R_m G(p_m) p_m^{j-1} \tag{2.34a}$$

where R_m is the residue of $U(z)$ at a typical pole p_m. If the input sequence is bounded, that is,

$$|p_m| \leq 1 \tag{2.34b}$$

then the steady state part of the output is also bounded. Similarly, the transient part of the output can be written as:

$$\sum_{q=1}^{n} R_n U(p_q) p_q^{j-1} \tag{2.35a}$$

where R_n is the residue of $G(z)$ at a typical pole p_q. It is evident from (2.35a) that for the transient part of the output to remain bounded, the following condition must be satisfied

$$|p_q| < 1 \qquad q = 1,\ldots,n \tag{2.35b}$$

that is, the poles of the system transfer function must be contained within the unit circle. Thus we can state that

> *A linear discrete-time system is considered to be stable if to a bounded input sequence there always corresponds a bounded output sequence.*

The above discussion suggests that it is possible to check the stability of a discrete system without finding the actual numerical values of the poles [6,9]. Stability tests can be broadly categorised into two groups. The first consists of those which are applied directly to $D(z)$, the denominator of the transfer function. The other group includes those tests which first require a bilinear transformation before they can be applied.

The Jury stability test [10,11] is the most efficient method for direct evaluation of absolute stability vs instability of the system transfer function. It involves the construction of the Jury table using the coefficients of the denominator polynomial

$$D(z) = a_n + a_{n-1}z + a_{n-2}z^2 + \ldots + a_0 z^n \qquad (2.36)$$

The Jury table takes the form

row	z^0	z^1	z^2	z^3	z^4	...	z^{n-1}	z^n
1	a_n	a_{n-1}	a_{n-2}	a_{n-3}	a_{n-4}	...	a_1	a_n
2	a_0	a_1	a_2	a_3	a_4	...	a_{n-1}	a_0
3	c_{n-1}	c_{n-2}	c_{n-3}	c_{n-4}	c_{n-5}	...	c_0	
4	c_0	c_1	c_2	c_3	c_4	...	c_{n-1}	
5	d_{n-2}	d_{n-3}	d_{n-4}	d_{n-5}	d_{n-6}	...		
6	d_0	d_1	d_2	d_3	d_4	...		
	⋮	⋮	⋮	⋮	⋮			
2n-5	p_3	p_2	p_1	p_0				
2n-4	p_0	p_1	p_2	p_3				
2n-3	q_2	q_1	q_0					

The first two rows of this table consist of the coefficients in $D(z)$ arranged in ascending order of powers of z in row 1 and in reverse order in row 2. All even-numbered rows are simply the reverse of the immediately preceding odd-numbered row. To calculate the elements for rows 3 through (2n-3), we use the following determinants:

$$c_k = \begin{vmatrix} a_n & a_{n-1-k} \\ a_0 & a_{k+1} \end{vmatrix} \qquad k = 0,1,\ldots,n-1 \qquad (2.37a)$$

$$d_k = \begin{vmatrix} c_{n-1} & c_{n-2-k} \\ c_0 & c_{k+1} \end{vmatrix} \qquad k = 0,1,\ldots,n-2 \quad (2.37b)$$

This process continues until we finally calculate

$$q_k = \begin{vmatrix} p_3 & p_{2-k} \\ p_0 & p_{k+1} \end{vmatrix} \qquad k = 0,1,2 \qquad (2.38)$$

which means that the (2n-3)rd row is reached such that it contains exactly three elements. According to the Jury stability criterion, a system is stable if *all* of the following conditions hold:

(1) When 1 is substituted for z in $D(z)$, $D(z)$ takes on a positive value, that is $D(1) > 0$.

(2) $D(-1) \begin{cases} > 0 & \text{for } n \text{ even} \\ < 0 & \text{for } n \text{ odd} \end{cases}$

(3) The coefficients in the Jury table meet the (n-1) constraints:

$$\begin{aligned} |a_n| &< |a_0| \,, \\ |c_{n-1}| &> |c_0| \,, \\ |d_{n-2}| &> |d_0| \,, \\ &\vdots \\ |q_2| &> |q_0| \end{aligned}$$

We note that the first (n-2) constraints of condition (3) involve elements of the first column only. All the determinants in (2.37) and (2.38) have dimensions (2x2), a factor which simplifies the amount of calculations required to form the Jury table. It should be emphasized that failure to meet either of the first two conditions (1) and (2), implies that there is no further need to construct the Jury table since the system is definitely unstable. We observe that the Jury table is directly amenable to machine calculations [10]. Finally, from the theory of polynomials,

$$a_n/a_0 = (-)^n \prod_{j=1}^{n} p_j \tag{2.39a}$$

where p_j is a root of (2.36). For all roots to be less than one in magnitude it is necessary, but not sufficient, that

$$a_n/a_0 < 1 \tag{2.39b}$$

which obviously corresponds to the first constraint in condition (3) of the Jury criterion.

Now, we direct attention to another analytic method for testing the stability of system transfer functions. It is based on the use of the bilinear transformation [1,3]

$$w = (z+1)/(z-1) \tag{2.40a}$$

$$z = (w+1)/(w-1) \tag{2.40b}$$

which maps the unit circle in the z-plane into the left half of the w-plane. Fig. 2.3 illustrates the effect of such a transformation. Let (2.36) describe the denominator polynomial in the z-plane. When mapped into the w-plane by using the transformation (2.40a), the zeros of the corresponding w-functions are given by:

$$F(w) = f_0 w^n + f_1 w^{n-1} + \ldots + f_n = 0 \tag{2.41}$$

The relationship between the coefficients a_k in (2.36) and f_j in (2.41) is given by [1]:

$$f_j = \sum_{k=0}^{n} a_k \Big[\binom{n-k}{j} - \binom{k}{1}\binom{n-k}{j-1} + \binom{k}{2}\binom{n-k}{j-2} - \ldots \\ + (-1)^{j-1}\binom{k}{j-1}\binom{n-k}{1} + (-1)^j \binom{k}{j} \Big] \tag{2.42a}$$

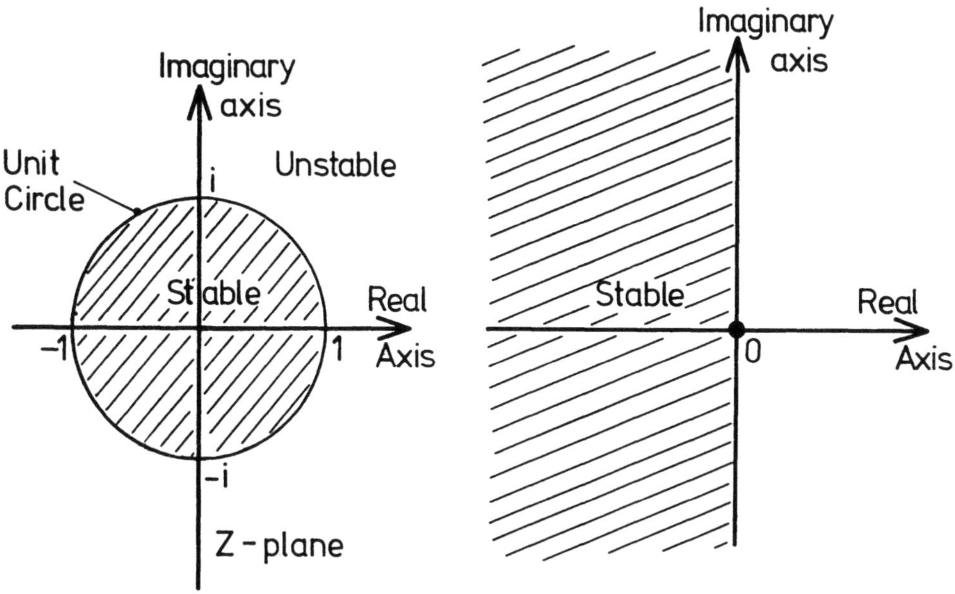

Fig. (2.3) Bilinear transformation of the complex Z-plane into the W plane

where

$$\binom{\alpha}{\beta} = \alpha!/\beta!\,(\alpha-\beta)! \tag{2.42b}$$

$$|f_n| = |f_{n-j}| \tag{2.42c}$$

One of the standard methods to test the stability of (2.41), which implies that of (2.36), is the Routh-Hurwitz array *[9]*. It states that for a system to be stable, all of the following conditions must be met by $F(w)$, the w-domain transform of $D(z)$:

(1) All powers of w in descending order from the highest to the constant terms in (2.41) must exist.

(2) The coefficients of all powers of w must have the same algebraic sign.

Once conditions (1) and (2) are satisfied, implying that no indication of instability is present, we can proceed to examine the final condition:

(3) There should be no changes of sign in the left-hand column of the following Routh array

$$\begin{array}{lllll} f_0 & f_2 & f_4 & f_6 & \cdots \\ f_1 & f_3 & f_5 & f_7 & \cdots \\ e_0 & e_2 & e_4 & e_6 & \cdots \\ e_1 & e_3 & e_5 & e_7 & \cdots \\ h_0 & h_2 & h_4 & h_6 & \cdots \\ h_1 & h_3 & h_5 & h_7 & \end{array}$$

where

$$e_0 = \frac{f_1 f_2 - f_0 f_3}{f_1}, \quad e_2 = \frac{f_1 f_4 - f_0 f_5}{f_1}, \quad \ldots$$

$$e_1 = \frac{e_0 f_3 - e_2 f_1}{e_0}, \quad e_3 = \frac{e_0 f_4 - e_4 f_1}{e_0}, \quad \ldots$$

$$h_0 = \frac{e_1 e_2 - e_0 e_3}{e_1}, \quad h_2 = \frac{e_1 e_4 - e_0 e_5}{e_1}, \quad \ldots$$

The array continues to the right and downward as long as the values last. There will be (n+1) rows in the array. We will demonstrate the application of both stability tests by examples in Section 2.2.5 .

2.2.4 *SIMPLIFICATION BY CONTINUED FRACTION EXPANSIONS*

Frequently, one is given a transfer function $G(z)$ in the form (2.20) with n being large and it is desired to develop another transfer function $G_s(z)$ whose denominator is of order $s < n$ such that $g_s(k)$, the inverse z-transform of $G_s(z)$, is a reasonable approximation to the data sequence $g(k)$. This problem is called model simplification and has been extensively treated in the literature *[12-14]*. Here, we will be mainly concerned with the development of model simplification methods using continued fraction expansion *[15,16]*. The basic idea behind these methods is to expand $G(z)$ in a power series about an arbitrary point and then retain some terms only. There are two types of power series which have a well defined physical interpretation. The first type is the power series expansion about infinity $(z = \infty)$ and has the form:

$$G(z) = \sum_{j=0}^{\infty} M_j z^{-j} \tag{2.43}$$

where the quantities M_j, $j = 0,1,\ldots,$ are called the Markov parameters. By comparing (2.2) and (2.43), it is readily seen that the Markov parameters correspond to the values of the data sequence $g(k)$ at different discrete instants. In this sense, the Markov parameters define the impulse response of discrete systems.

Another type of power series is the expansion about $z = 1$ and it can be expressed as:

$$G(z) = \sum_{m=0}^{\infty} \frac{(-1)^m}{m!} T_m (z-1)^m \tag{2.44a}$$

where the quantities

$$T_m = (-1)^m \left[\frac{d^m}{dz^m} z^m G(z) \right]_{z \to 1} , \quad m = 0,1,\ldots \tag{2.44b}$$

are the time-moments of the discrete system. In the light of the final value theorem (2.9), the time-moments represent the coefficients of the data sequence $g(k)$ for asymptotic values of k. Thus, in terms of the time-domain behaviour of discrete systems, the Markov parameters describe the initial transients whereas the time-moments account for the steady-state operation.

We now develop a model simplification procedure using the continued fraction expansion about $z = 1$. Let the system transfer function $G(z)$ be of the form:

$$G(z) = \frac{a_{21}+a_{22}z+a_{23}z^2 +\ldots+ a_{2n}z^{n-1}}{a_{11}+a_{12}z+a_{13}z^2 +\ldots+ a_{1,n+1}z^n} \tag{2.45}$$

where we have used double subscripts for convenience. Define

$$z = p+1 \tag{2.46}$$

Then substituting for z in (2.46) gives

$$G(p) = \frac{b_{21}+b_{22}p+b_{23}p +\ldots+b_{2n}p^{n-1}}{b_{11}+b_{12}p+b_{13}p +\ldots+ b_{1,n+1}p^n} \tag{2.47}$$

Now (2.47) is expanded into a Cauer-type continued fraction about $p = 0$ to yield *[12]*:

$$G(p) = \cfrac{1}{h_1 + \cfrac{p}{h_2 + \cdots + \cfrac{p}{h_{2n-1} + \cfrac{p}{h_{2n}}}}} \tag{2.48}$$

where the coefficients h_i, $i = 1,2,\ldots,2n$, may be derived by applying the Routh algorithm. Given the coefficients b_{ij} in (2.47), we construct the Routh array

$$\begin{array}{lllll} b_{11} & b_{12} & b_{13} & b_{14} & \cdots \\ b_{21} & b_{22} & b_{23} & b_{24} & \cdots \\ b_{31} & b_{32} & b_{33} & \cdots & \\ b_{41} & b_{42} & b_{43} & \cdots & \\ b_{51} & b_{52} & & & \end{array}$$

where

$$b_{j,k} = b_{j-2,k+1} - b_{j-2,1}b_{j-1,k+1}/b_{j-1,1} \; ; \quad j = 3,4,\ldots,2n+1, \; k = 1,\ldots,n \tag{2.49a}$$

then we compute the quotients h_i using the formula

$$h_i = b_{i,1}/b_{i+1,1} \; ; \quad i = 1,2,\ldots,2n \tag{2.49b}$$

Simplification is effected by truncating (2.48) after a certain number of terms, say 2s, and using (2.46) to yield:

$$G_s(z) = \cfrac{1}{h_1 + \cfrac{1}{\cfrac{h_2}{z-1} + \cfrac{1}{h_3 + \ddots + \cfrac{1}{\cfrac{h_{2s}}{z-1}}}}} \tag{2.50}$$

and by inverting (2.50), the simplified transfer function of order s is obtained. It has been shown *[12]* that the power series of $G_s(z)$ about $z = 1$ agrees with that of $G(z)$ up to and including the term in $(z-1)^{2s-1}$. In the light of (2.44), it is readily seen that using the continued fraction expansion about $z = 1$ to retain 2s terms implies matching the first 2s time-moments. Consequently, the simplified model (2.50) provides a good approximation to the system response at large time instants.

In a similar manner, it can be shown *[15]* that the expansion of $G(z)$ in powers of $(1/z)$ and truncating after 2s terms yields a simplified transfer function of order s which matches the first 2s Markov-parameters.

To improve the accuracy of approximation at small time instants, one can use the double series expansions, first about $z = \infty$ and then about $z = 1$ or vice versa. Applying this procedure and truncating after $(r+p)$ quotients, one arrives at:

$$G_{r,p}(z) = G_s(z) = \cfrac{1/z}{h_1 + \cfrac{1/z}{h_2 + \ddots + \cfrac{1/z}{h_r + \cfrac{1}{h_{r+1} + \cfrac{z-1}{h_{r+2} + \ddots + \cfrac{z-1}{h_{r+p}}}}}}} \tag{2.51}$$

and by inverting (2.51), a simplified transfer function of order $s = (r+p)/2$ is obtained. We can visualise that the power series expansion of $G_s(z)$ derived from (2.51) about $z = 1$ and $z = \infty$, agrees with that of $G(z)$ up to and including the terms in $(z-1)^{r-1}$ and z^{-p} respectively. This corresponds to the matching of the first r time-moments and the initial p Markov-parameters.

To conclude this section, we provide some examples to illustrate the various concepts and methods presented so far.

2.2.5 *EXAMPLES*

The first example given here is meant to demonstrate the utility of z-transform theory.

Example 1

Let a discrete system be described by

$$y(k) - \alpha y(k-1) = u(k) \; ; \quad |\alpha| < 1$$

Application of (2.2) and (2.5) to the above system results in

$$Y(z) - \alpha z^{-1} Y(z) = U(z)$$

Solving for $Y(z)/U(z)$ we obtain the transfer function

$$G(z) = \frac{Y(z)}{U(z)} = \frac{1}{1-\alpha z^{-1}}$$

Now let the input be a discrete step, that is

$$u(k) = \begin{cases} 1 & \text{for } k \geq 0 \\ 0 & \text{for } k < 0 \end{cases}$$

It is easy to see from (2.2) that $U(z)$, the z-transform of $u(k)$, is given by

$$U(z) = 1/(1-z^{-1})$$

Hence, the z-transform of the system response is expressed as

$$Y(z) = \frac{1}{1-\alpha z^{-1}} \cdot \frac{1}{1-z^{-1}}$$

$$= \frac{1}{(1-\alpha)} \left(\frac{1}{1-z^{-1}} - \frac{\alpha}{1-\alpha z^{-1}} \right)$$

The right hand side of $Y(z)$ is in standard form and therefore we can write directly $y(k) = (1-\alpha^{k+1})/(1-\alpha)$ with

$$y(0) = \frac{1-\alpha}{1-\alpha} = 1$$

$$y(\infty) = \frac{1-0}{1-\alpha} = \frac{1}{1-\alpha} \qquad \text{(note that } |\alpha| < 1)$$

In order to validate these results, let us consider

$$Y(z) = \frac{1}{1-z^{-1}} \cdot \frac{1}{1-\alpha z^{-1}}$$

Now, from (2.8) we have

$$y(0) = \lim_{z\to\infty} Y(z)$$

$$= \lim_{z\to\infty} \frac{1}{(1-z^{-1})(1-\alpha z^{-1})} = 1$$

The application of (2.9) results in

$$y(\infty) = \lim_{z\to 1} (z-1)Y(z)$$

$$= \lim_{z\to 1} \frac{(z-1)}{(1-z^{-1})(1-\alpha z^{-1})}$$

$$= \lim_{z\to 1} \frac{z}{1-\alpha z^{-1}} = \frac{1}{1-\alpha}$$

Example 2

The purpose of this example is to illustrate the application of stability criteria. Consider the denominator of a transfer function

$$D(z) = z^3 - .2z^2 - .25z + .05$$

In comparison with (2.36) we find that

$$a_0 = 1, \quad a_1 = -.2, \quad a_2 = -.25, \quad a_3 = .05$$

and $n = 3$. First we examine

$$D(1) = 1 - .2 - .25 + .05 = .6 > 0$$

and

$$D(-1) = -1 - .2 + .25 + .05 = -.9 < 0$$

Since $D(z)$ is an odd-order polynomial, the first two conditions of the Jury criterion are met. We now proceed to construct the Jury table as follows:

Row	z^0	z^1	z^2	z^3
1	.05	-.25	-.2	1
2	1	-.2	-.25	.05
3	-.9975	.1875	.24	

The third-row elements were computed from (2.37a) to be

$$c_2 = q_2 = \begin{vmatrix} .05 & 1 \\ 1 & .05 \end{vmatrix} = -.9975$$

$$c_1 = q_1 = \begin{vmatrix} .05 & -.2 \\ 1 & -.25 \end{vmatrix} = .1875$$

$$c_0 = q_0 = \begin{vmatrix} .05 & -.25 \\ 1 & -.2 \end{vmatrix} = .24$$

Note that c_j and q_j are equivalent since there are only three rows in the table. Applying the third set of stability conditions we have:

$$|a_n| = |a_3| = .05 \quad , \quad |a_0| = 1$$

$$\text{thus} \quad |a_n| < |a_0|$$

and

$$|c_{n-1}| = |c_2| = .9975 \ , \quad |c_0| = .24$$

$$\text{thus} \quad |c_{n-1}| > |c_0|$$

Therefore the constraint conditions on the coefficients of $D(z)$ are also met, and we conclude that the system is stable. It should be observed that $D(z)$ can be factored into

$$D(z) = (z - .5)(z + .5)(z - .2)$$

which means that the roots of $D(z)$ are .5, -.5 and .2 and since all of them are less in absolute value than unity, the system is stable.

In order to examine the system stability in the w-domain, we use (2.40a) to yield

$$F(w) = \left(\frac{w+1}{w-1}\right)^3 - .2\left(\frac{w+1}{w-1}\right)^2 - .25\left(\frac{w+1}{w-1}\right) + .05$$

which upon manipulation reduces to

$$F(w) = .6w^3 + 2.9w^2 + 3.6w + .9$$

We first note that all powers of w exist and all coefficients have the same algebraic sign. Then we construct the Routh array

.6	3.6
2.9	.9
19.33	0
.9	

Since there are no changes of sign in the left-hand column, the system is stable.

Example 3

This example was chosen to illustrate the use of continued fraction expansions in developing simplified transfer functions. Consider the following transfer function

$$G(z) = \frac{8z^2 - 15.283z + 7.313}{z^3 - 2.628z^2 + 2.3z - .67}$$

Using long division, the Markov parameters are computed to be

$$M_1 = 8, \; M_2 = 5.744, \; M_3 = 4.01, \; M_4 = 2.685, \ldots$$

Application of (2.44) yields the time-moments

$$T_0 = 18.418, \quad T = -51.831, \quad T_2 = 4158.878,$$

$$T_3 = 109316.925, \ldots$$

We consider the design of two simplified models. The first simplified model is desired to fit the first three time-moments and the initial Markov parameter of the system. Therefore, we expand $G(z)$ about $z = \infty$ for one term and about $z = 1$ for another three terms to give

$$G_{3,1}(z) = \cfrac{1/z}{.125 - \cfrac{1/z}{14.144 + \cfrac{z-1}{.18 + \cfrac{z-1}{.027}}}}$$

which when inverted results in

$$G_{3,1}(z) = \frac{8z - 7.961}{z^2 - 1.56z + .562}$$

As a matter of emphasis, we compute the first Markov parameter and the three time-moments of $G_{3,1}(z)$

$$M_1 = 8, \quad T_0 = 18.4175, \quad T_1 = -51.831, \quad T_2 = 4158.878$$

which are obviously in very close agreement with those of $G(z)$ as expected.

A second simplified model is sought to match the first time-moment and the initial three Markov parameters of the system. This means that $G(z)$ is expanded into a continued fraction about $z = \infty$ for three terms and then about $z = 1$ for one term only. After inverting, the result is

$$G_{1,3}(z) = \frac{8z - 7.239}{z^2 - 1.623z + .664}$$

By long division, we can easily check that $G_{1,3}(z)$ has

$$M_1 = 8, \quad M_2 = 5.744, \quad M_3 = 4.009$$

and

$$T_0 = 18.4176$$

which are identical with those of G(z) within the computational accuracy. It should be emphasized that the simplified models $G_{3,1}(z)$ and $G_{1,3}(z)$ match only some of the time-moments and/or Markov parameters of the original transfer function G(z). It is in this sense that they are "approximate" models since the other time-moments and/or Markov parameters are not generally matched.

2.3 Difference Equations

As mentioned earlier, one can describe the mechanism by which the input sequence is transformed into the output sequence, either in the time domain or in the complex frequency-domain. In Section 2.2, we presented the concept of system transfer function in order to characterise the system's behaviour in the z-plane. Here, we examine the mechanism in the time-domain by means of linear difference equations. In the sequel, we will point out the relationship between the time- and frequency-domain characterisations.

Difference equations provide the basic structure of almost all discrete-time control systems. In these equations, the independent variable is the discrete time index (or an index related to discrete time). The dependent variable is the one whose explicit functional relationship with the independent variable is sought as the solution of a difference equation. With reference to Fig. 2.1, the independent variable is usually taken as the output. A known or assumed function of the independent variable which influences the functional relationship between the dependent and independent variables is called the forcing function. Again with respect to Fig. 2.1, the forcing function includes the input sequence and any external perturbations.

The equation

$$a_0 y(k) + a_1 y(k-1) + \ldots + a_n y(k-n) = f(k) \qquad (2.52)$$

where k is the independent variable, $y(k)$ the dependent variable, $f(k)$ the forcing function and a_j a constant real number the coefficient of the dependent variable delayed j steps, $j = 0, \ldots, n$ is called a linear difference equation with constant coefficients. We note that the linearity of (2.52) in the sense of (2.1) implies that the forcing function $f(k)$ is a linear combination of input sequences. An important class of forcing functions is given by

$$f(k) = b_0 u(k) + b_1 u(k-1) + \ldots + b_m u(k-m) \; ; \quad m \le n \qquad (2.53)$$

which represents a spectrum of the input sequence. An alternative form of (2.52) and (2.53) is given by

$$a_0 y(k+n) + a_1 y(k+n-1) + \ldots + a_n y(k)$$
$$= b_0 u(k+m) + b_1 u(k+m-1) + \ldots + b_m u(k) \qquad (2.54)$$

In both (2.52) and (2.54), the order of the difference equation is n which represents the difference between the largest and smallest arguments of the dependent variable. Finally, we observe that (2.52) and (2.54) designate right-shifted and left-shifted data sequences, respectively.

In the remainder of this section we present the classical methods for solving linear difference equations with constant coefficients of the type (2.52) or (2.54). We first discuss the general nature of the solutions of difference equations.

2.3.1 *THE NATURE OF SOLUTIONS*

From the calculus of difference equations *[17]*, it is known that the solutions of linear difference equations with constant coefficients comprise two main parts: the homogeneous solution and the particular solution. Given the expanded form (2.54), $y_h(k)$ is defined as the solution to the homogeneous equation

$$a_0 y(k+n) + a_1 y(k+n-1) + \ldots + a_n y(k) = 0 \tag{2.55}$$

where $a_0 \neq 0$, otherwise the order of the difference equation will be decreased by one. The particular solution $y_p(k)$ is the one which results from solving the original equation (2.54) with the particular forcing function $f(k)$. Note that similar arguments apply equally to (2.52) and (2.53). In fact, any $y(k)$ which satisfies (2.55) will not affect the particular solution. More importantly, the complete solution is

$$y(k) = y_h(k) + y_p(k) \tag{2.56}$$

for which $y_h(k)$ is a linear combination of all those $y(k)$ which satisfy the homogeneous equation (2.55). To elaborate on this point, let us insert $y(k)$ given by (2.56) as the dependent variable in the compact form of (2.54) to obtain

$$\sum_{j=0}^{n} a_j [y_h(k+n-j) + y_p(k+n-j)] = \sum_{i=0}^{m} b_i u(k+m-i) \tag{2.57}$$

By definition, the first term in the summation on the left-hand side of (2.57) vanishes leaving

$$\sum_{j=0}^{n} a_j y_p(k+n-j) = \sum_{i=0}^{m} b_i u(k+m-i) \tag{2.58}$$

which corresponds to the definition of $y_p(k)$. Now, the complete solution (2.56) is determined in two steps: first, the homogeneous part and then the particular part.

2.3.2 *The Free Response*

We now consider the solution of the homogeneous equation (2.55).

In terms of linear discrete control systems, this solution corresponds to the system response without application of the input sequence and it is thus labelled as the free response. It is well-known that the homogeneous solution is generally assumed to be of the form *[17]*:

$$y(k) = h^k \tag{2.59a}$$

For all k, $0 \leq k \leq n$, it follows that

$$y(k+n-j) = h^{k+n-j} \tag{2.59b}$$

Then the substitution of (2.59) into (2.55) after rearrangement yields

$$h^k [a_0 h^n + a_1 h^{n-1} + \ldots + a_{n-1} h + a_n] = 0 \tag{2.60}$$

From the theory of linear equations, a nontrivial solution, that is where $h \neq 0$, exists for all k by solving

$$a_0 h^n + a_1 h^{n-1} + \ldots + a_{n-1} h + a_n = 0 \tag{2.61}$$

which is frequently known as the characteristic (indicial) equation. Since (2.61) is an nth-order ordinary polynomial, it must have exactly n roots subject to complex pairing. Consequently, there will be n functions $y(k)$, a linear combination of which will eventually constitute the homogeneous solution. In Table 2.1 we provide the various possible types of roots and their contribution to the homogeneous solution. Then, by imposing the physical boundary conditions, the free response is immediately obtained. For illustrative purposes only, let us consider an eleventh-order equation which has the following roots:

single real root at $-.5$
single complex pair at $.05 \pm i\ .12$
fourth-order real root at $.5$
double complex pair at $-.3 \pm i\ .4$

Type of Root	Contributions to the Homogeneous Solution
1) single real root at h	ch^k c is a combinatorial coefficient
2) single complex pair at $h_1 \pm i\, h_2$	$R^k[c_1 \cos(k\Theta) + c_2 \sin(k\Theta)]$; $R = \sqrt{h_1^2 + h_2^2}$ $\Theta = \tan^{-1}(h_2/h_1)$
3) p-th order real root at h	$h^k[c_1 k^{p-1} + c_2 k^{p-2} + \ldots$ $\ldots + c_{p-1}k + c_p]$
4) p-th order complex pair at $h_1 \pm ih_2$	$R^k\{[c_1 k^{p-1} + c_2 k^{p-2} + \ldots + c_p]$ $\cos(k\Theta) + [d_1 k^{p-1} + d_2 k^{p-2} +$ $\ldots + d_p] \sin(k\Theta)\}$; $R = \sqrt{h_1^2 + h_2^2}$ $\Theta = \tan^{-1}(h_2/h_1)$

TABLE 2.1 Summary of Roots

Simple calculations show that for the single complex pair $R = .13$ and $\Theta = 1.186$ rad, whereas for the double complex pair $R = .5$ and $\Theta = 2.217$ rad. By applying the rules of Table 2.1, the homogeneous solution takes the form:

$$
\begin{aligned}
y_h(k) \;=\; & c_1(-.5)^k + (.13)^k[C_2 \cos(1.186k) \\
& + C_3 \sin(1.186k)] + (.5)^k[C_4K^3 + C_5K^2 \\
& + C_6K + C_7] + (.5)^k\{[C_8K + C_9]\cos(2.217k) \\
& + [C_{10}K + C_{11}]\sin(2.217k)\}
\end{aligned}
$$

which is a linear combination of 11 clearly distinct and identifiable terms. To determine the free response, one obviously needs 11 boundary conditions so that the unknown coefficient

C_1 up to C_{11} can be evaluated. Next, we discuss the particular solution.

2.3.3 *THE FORCED RESPONSE*

Two procedures are presented for finding the particular solution to linear difference equations with constant coefficients. Both require the prior determination of the homogeneous solution $y_h(k)$. The method of undetermined coefficients is restricted to equations with a limited but fairly common set of forcing functions. The method of variation of parameters is applicable to general forcing functions; however, the amount of calculations is quite large.

Undetermined Coefficients

The method of undetermined coefficients is useful when $f(k)$ in (2.52) consists of terms having certain special forms. Corresponding to each such term which is present in $f(k)$ we consider a *trial solution* containing a number of unknown constant coefficients which are to be determined by substitution into the difference equation. The trial solutions to be used in each case are shown in the following table where the letters $A, B, A_0, A_1, \ldots$ represent the unknown constant coefficients to be determined.

TABLE 2.2 Common Trial Solutions

Terms in f(k)	Trial Solution
β^K	$A\beta^K$
$\sin(\alpha K)$ or $\cos(\alpha K)$	$A\cos(\alpha K) + B\sin(\alpha K)$
$\sum_{j=0}^{m} \alpha_j K^j$	$\sum_{j=0}^{m} A_j K^j$
$\beta^K \sum_{j=0}^{m} \alpha_j K^j$	$\beta^K \sum_{j=0}^{m} A_j K^j$
$\beta^K \sin(\alpha K)$ or $\beta^K \cos(\alpha K)$	$\beta^K [A\cos(\alpha K) + B\sin(\alpha K)]$

The only requirement which must be met to guarantee success of the method is that no term of the trial solution can appear in the homogeneous solution. If any terms of the trial solution does happen to be in the homogeneous solution then the entire trial solution corresponding to this term must be multiplied by a positive integer power of K which is just large enough so that no term of the new trial solution will appear in the homogeneous solution. The following represent the main steps involved:

(1) Given (2.52) or (2.54), determine the homogeneous solution $y_h(k)$.

(2) Construct the trial solution in accordance with Table 2.2.

(3) In case that common terms between the trial and homogeneous solutions exist, we multiply the trial solution by an appropriate positive integer.

(4) Form $y_p(k)$ as a linear combination of all trial solutions

(5) Determine the unknown coefficients by treating $y_p(k)$ as $y(k)$ and substituting into the original difference equation.

Variation of Parameters

This method requires the prior determination of the roots of the characteristic equation (2.61) and the resulting homogeneous solution $y_h(k)$. All roots must be given in their original forms. Let $y_h(k)$ be expressed as

$$y_h(k) = \sum_{j=0}^{n} c_j v_j(k) \tag{2.62}$$

where $v_j(k)$ includes the appropriate root h raised to the kth power and any power of k by which h^k must be multiplied in the case of multiple roots. The c_j are combinatorial coefficients. We now assume a complete solution of the form

$$y(k) = \sum_{j=1}^{n} A_j(k) v_j(k) \tag{2.63}$$

where the $A(k)$ are functions of k to be determined. Since n conditions are needed to find $A_1, A_2, \ldots, A_n$ and one of these is that (2.63) be satisfied, there remain $(n-1)$ arbitrary conditions which can be imposed. These are usually chosen so as to simplify as much as possible the expressions for successive forward differences *[18]*. The solutions obtained in this way are

$$V.\Delta A = F \tag{2.64a}$$

where

$$V = \begin{bmatrix} v_1(k+1) & v_2(k+1) & \ldots & v_n(k+1) \\ v_1(k+2) & v_2(k+2) & \ldots & v_n(k+2) \\ \vdots & \vdots & & \\ v_1(k+n) & v_2(k+n) & \ldots & v_n(k+n) \end{bmatrix} \tag{2.64b}$$

$$\Delta A = \begin{bmatrix} \Delta A_1(k) \\ \Delta A_2(k) \\ \vdots \\ \Delta A_n(k) \end{bmatrix} ; \qquad F = \begin{bmatrix} 0 \\ 0 \\ \vdots \\ f(k)/a_0 \end{bmatrix} \tag{2.64c}$$

In the light of (2.64a)-(2.64c), the solution for ΔA_j is thus

$$\Delta A_j(k) = v_{jn} f(k)/a_0 \tag{2.65}$$

where v_{jn} is the element in row j and column n of V^{-1}. Then the $A_j(k)$ are calculated from

$$A_j(k) = A_j(0) + \sum_{m=0}^{k-1} \Delta A_j(m) \; ; \quad j = 1,\ldots,m \tag{2.66}$$

Substitution of (2.66) into (2.63) yields the complete solution $y(k)$. We note that

(1) The boundary coefficients are included in the form of the $A_j(0)$ and therefore they must be evaluated from boundary conditions.

(2) It is not necessary to calculate $\Delta A_j(k)$ by direct matrix inversion. Convenient methods for doing this would be the Gauss-Jordan complete-elimination procedure *[19]* or the repeated use of determinants. In the latter method, one should observe that the denominator of all $\Delta A_j(k) = |V|$, the determinant of V. The numerator of $\Delta A_j(k)$ is the determinant of the matrix formed by replacing column j in V by the constant vector F.

(3) The method of variation of parameters can solve linear difference equations with general forms of the forcing function. This is apparent from (2.64c) since the nth entry contains only $f(k)/a_0$.

(4) By considering the extended form of (2.65), we find that the particular solution part (or forced response) contains terms which are of the convolution-summation type *[2]*. A typical term would be

$$\sum_{j=0}^{n} f(k-j)g(j)$$

or

$$\sum_{j=0}^{n} f(j)g(k-j)$$

where $g(j)$ is an appropriate function of the roots.

We now illustrate the above methods on two examples.

2.3.4 *EXAMPLES*

The first example shows the application of the method of undetermined coefficients.

Example 1

Consider the difference equation of the type (2.54), with $n = 2$

$$y(k+2) - 6y(k+1) + 8y(k) = 3k^2 + 2 - 5.3^k$$

From (2.61), we have the characteristic equation

$$h^2 - 6h + 8 = 0$$

which has two real distinct roots at 2 and 4. Using Table 2.1, the homogeneous solution is given by

$$y_h(k) = c_1 2^k + c_2 4^k$$

According to the method of undetermined coefficients, the trial solution corresponding to the polynomial $3k^2 + 2$ would be

$$A_1 k^2 + A_2 k + A_3$$

since none of these terms occur in the homogeneous solution $y_h(k)$. We assume the trial solution

$$A_4 3^k \quad \text{corresponding to the term } -5.3^k$$

Thus corresponding to the given forcing function we consider the trial solution (or particular solution)

$$y(k) = A_1 k^2 + A_2 k + A_3 + A_4 3^k$$

Substituting in the original difference equation and manipulating we arrive at:

$$y(k+2) - 6y(k+1) + 8y(k) = 3A_1 k^2 + (3A_2 - 8A_1)k + 3A_3 - 4A_2 - 2A_1 - A_4 3^k$$

By comparing the above equation with the given difference equation and equating coefficients of like terms we obtain

$$A_1 = 1, \quad A_2 = 8/3, \quad A_3 = 44/9, \quad A_4 = 5$$

Thus the particular solution is

$$y_p(k) = k^2 + 8k/3 + 44/9 + 5.3^k$$

Adding this to the homogeneous solution $y_h(k)$ we find

$$y(k) = c_1 2^k + c_2 4^k + k^2 + 8k/3 + 44/9 + 5.3^k$$

It is interesting to note that if the forcing function was

$$3k^2 + 2 - 5.2^k$$

Then the trial solution would be

$$A_1 k^2 + A_2 k + A_3 + A_4 k.2^k$$

in order to make it distinct from the homogeneous solution $y_h(k)$.

Example 2

Consider the second-order difference equation

$$y(k+2) - 5y(k+1) + 6y(k) = k^2$$

which we would like to solve using the method of variation of parameters. It is simple to note that the characteristic equation (2.61) has two real roots at 2 and 3. According to (2.63), the general solution of the complete difference equation becomes

$$y(k) = A_1 2^k + A_2 3^k$$

where $v_1(k) = 2^k$ and $v_2 = 3^k$. The application of (2.64) results in

$$\begin{pmatrix} 2^{k+1} & 3^{k+1} \\ 2^{k+2} & 3^{k+1} \end{pmatrix} \begin{pmatrix} \Delta A_1 k \\ \Delta A_2 (k) \end{pmatrix} = \begin{pmatrix} 0 \\ k^2 \end{pmatrix}$$

which when solved for $\Delta A_1(k)$, $\Delta A_2(k)$ gives

$$\Delta A_1(k) = -\frac{1}{2}k^2(\frac{1}{2})^k$$

$$\Delta A_2(k) = \frac{1}{3}k^2(\frac{1}{3})^k$$

Now, from (2.66) we have

$$A_1(k) = A_1(0) + \sum_{m=0}^{k-1} -\frac{1}{2}m^2(\frac{1}{2})^m$$

whose expansion takes the form *[18]:*

$$A_1(k) = A_1(0) + 4[\frac{1}{4}k^2(\frac{1}{2})^k + \frac{k}{2}(\frac{1}{2})^k + \frac{1}{4}(\frac{1}{2})^k - \frac{3}{4} + \frac{1}{2}(\frac{1}{2})^k]$$

$$= [A_1(0) - 3] + (\frac{1}{2})^k[k^2 + 2k + 3]$$

$$= c_1 + (\frac{1}{2})^k[k^2 + 2k + 3]$$

Similarly, we can show that

$$A_2(k) = -\frac{1}{2}(\frac{1}{3})^k[k^2 + k+1] + c_2$$

The complete solution thus becomes:

$$y(k) = c_1 2^k + c_2 3^k + \frac{1}{2}k^2 + \frac{3}{2}k + 5/2$$

We conclude this section by pointing out the relationship between linear difference equations and transfer functions.

2.3.5 *RELATIONSHIP TO TRANSFER FUNCTIONS*

Consider a linear, time-invariant discrete system characterised by the linear difference equation

$$y(k) + a_1y(k-1) + \ldots + a_ny(k-n) = b_0u(k) + b_1u(k-1) + \ldots + b_nu(k-n) \quad (2.67)$$

which is of the type (2.52) with $a_0 = 1$ and $f(k)$ is a linear combination of data sequences. Bearing in mind that the right and left sides of (2.67) are equal for all values of k, we can write

$$g(k) = y(k) + a_1 y(k-1) + \ldots + a_n y(k-n)$$

$$= b_0 u(k) + b_1 u(k-1) + \ldots + b_n u(k-n) \quad (2.68)$$

which defines a data sequence representing the input or output information. Taking the z-transform of $g(k)$ using (2.2), (2.4) and (2.5), and dropping out the initial conditions, we arrive at:

$$Y(z) + a_1 z^{-1} Y(z) + \ldots + a_n z^{-n} Y(z)$$
$$= b_0 U(z) + b_1 z^{-1} U(z) + \ldots + b_n z^{-n} U(z) \quad (2.69)$$

where $Y(z) = Z[y(k)]$ and $U(z) = Z[u(k)]$.
Now, we can write (2.69) in a convenient way as

$$Y(z) = \frac{b_0 + b_1 z^{-1} + \ldots + b_n z^{-n}}{1 + a_1 z^{-1} + \ldots + a_n z^{-n}} U(z)$$

$$= G(z) U(z) \quad (2.70)$$

from which we recognise the following

(1) The transfer function $G(z)$ can be constructed directly from the difference equation. It should be observed that if we start with a difference equation of the type (2.54) we will arrive at

$$Y(z) = \frac{b_o + b_1 z + \ldots + b_n z^n}{1 + a_1 z + \ldots + a_n z^n} U(z)$$

$$= G(z) U(z) \quad (2.71)$$

(2) Given $U(z)$ as a rational function of z, we can derive $y(k)$ as the inverse z-transform of $Y(z)$ by using any of

the techniques discussed before. This gives another method of obtaining linear difference equations with constant coefficients.

It should perhaps be mentioned that what we have presented in this section is only a part of the calculus of difference equations. Nevertheless, it provides a reasonable coverage of the material related to the classical methods of linear discrete systems. We now turn our attention to the development of modern techniques of representing discrete control systems.

2.4 Discrete State Equations

2.4.1 *INTRODUCTION*

With reference to Fig. 2.1, the last two sections have been concerned with the characterisation of the mechanism which converts the input sequence into output sequence via a transfer function in the complex frequency domain and difference equations in the time domain. Although these techniques are well-developed and are useful in certain applications, they suffer from some limitations, the most important of which are:

(1) In control engineering practice, one is often interested in recording the time response of the system as measured by the output sequence. Using the concept of transfer function, it is not always an easy job to invert z-transform functions. Sometimes it requires a great deal of complex integration.

(2) The solution of difference equations generally needs a lot of calculation effort. This is particularly so when analysing higher-order systems.

(3) Both the transfer function concept and the difference equation representation specify only the input-output relationship. Nothing can be said about the internal dynamics of the systems which is of particular importance in the analysis of control engineering problems.

In seeking a total description of a dynamic system, the key concept of *State* was introduced *[19]* through which a complete summary of the status of the system at a particular instant in time could be given. Knowledge of the state at some initial discrete instant k_0, plus knowledge of the input sequence starting from that instant, allows the determination of the state for the present and all future discrete instants. In this sense, the state at k_0 constitutes a complete history of the system behaviour prior to k_0, insofar as that history affects future behaviour. Thus, knowledge of the present state allows a sharp separation between the past and the future.

At any given time instant, the state of a system can be described by the values of a set of variables x_i, called *state variables* *[20]*. To emphasise this point, we draw in Fig. 2.4 a block diagram of the discrete system using a state variable representation. We note that the input and output sequences appear to be external to the system, whereas the state sequences are internal. This means that to model dynamic systems we utilise three types of sequences: the input, the output and the state sequences.

The input sequences u_i serve as external inputs that influence the dynamical behaviour of the system. The output sequences y_j are the characteristic quantities that can be directly recorded. The state sequences x_r characterise the internal dynamics of the system. We note that the state variable representation of a dynamic system is not unique and entirely depends on the particular choice of the states that one has made. Our objective now is to develop methods for deriving the state equations.

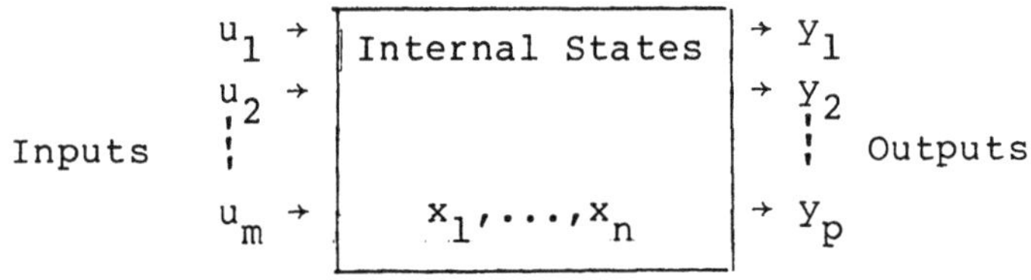

Fig. 2.4 Block diagram of state variable representation.

2.4.2 *OBTAINING THE STATE EQUATIONS*

The methods of selecting state variables of a discrete-time system depend upon the form in which the given information is available. From now onwards we will consider that the system is described by difference equations or transfer functions with forcing functions of the linear type. The various methods of obtaining the state equation are based on the use of the *ideal delay element* as a memory device whose function is illustrated in Figure 2.5 . In terms of data sequences, the input of the ideal delay element corresponds to its output advanced by one discrete step. Application of (2.5a) with $m = 1$ gives another way of achieving the ideal delay in the complex domain. Whenever convenient, the symbol D will be used to represent the delay operation.

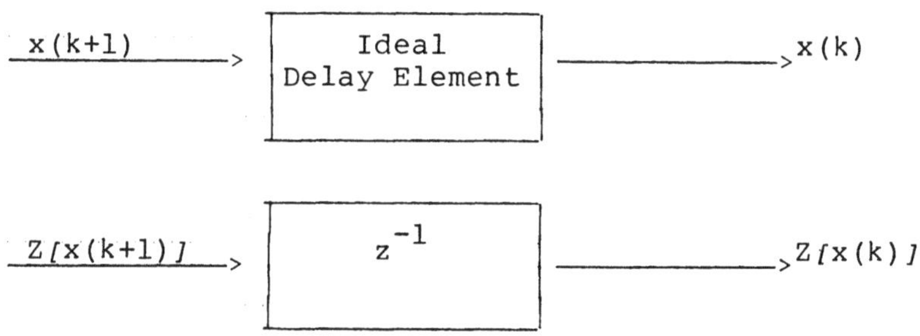

Fig. 2.5 Ideal delay element (Shift Register)

A. From Difference Equations

Consider the n-th order difference equation

$$\begin{aligned} & y(k+n) + a_1 y(k+n-1) + \ldots + a_n y(k) \\ & = b_0 u(k+n) + b_1 u(k+n-1) + \ldots + b_n u(k) \end{aligned} \tag{2.72}$$

which is of the type (2.54) with $a_0 = 1$ and $n = m$. If $m < n$, then some coefficients b_j can be set to zero. It is impossible that $m > n$ for physically realizable systems. We first solve (2.72) for the most advanced output term:

$$y(k+n) = -a_1 y(k+n-1) - \ldots - a_n y(k)$$

$$+ b_0 u(k+n) + \ldots + b_n u(k) \qquad (2.73)$$

Then we delay every term in (2.73) n times so that $y(k)$ is obtained on the left-hand side in the form:

$$y(k) = -a_1 \mathcal{D}(y(k)) - \ldots - a_{n-1} \mathcal{D}^{n-1}(y(k))$$

$$- a_n \mathcal{D}^n (y(k)) + b_n \mathcal{D}^n (u(k))$$

$$+ b_{n-1} \mathcal{D}^{n-1}(u(k)) + \ldots + b_0 u(k) \qquad (2.74)$$

By rearranging this expression as a nested sequence of delayed terms, we arrive at:

$$y(k) = b_0 u(k) + \mathcal{D}\{-a_1 y(k) + b_1 u(k) +$$

$$\mathcal{D} - a_2 y(k) + b_2 u(k) + \mathcal{D}(\ldots + \mathcal{D}\{-a_n y(k)$$

$$+ b_n u(k)\})]\} \qquad (2.75)$$

We can now draw the simulation diagram of Fig. 2.6 to represent (2.75) by noting that everything which is operated upon by the delay operator $\mathcal{D}$ forms the input to that delay. From the simulation diagram, the state equation can be written directly by using the fact that if the output of a delay is $x_j(k)$, then its input sequence must be $x_j(k+1)$. Hence,

$$\left.\begin{aligned} x_1(k+1) &= -a_1 x_1(k) + x_2(k) + [b_1 - a_1 b_0] u(k) \\ x_2(k+1) &= -a_2 x_1(k) + x_3(k) + [b_2 - a_2 b_0] u(k) \\ &\vdots \\ x_{n-1}(k+1) &= -a_{n-1} x_1(k) + x_n(k) + [b_{n-1} - a_{n-1} b_0] u(k) \\ x_n(k+1) &= -a_n x_1(k) + [b_n - a_n b_0] u(k) \end{aligned}\right\} \qquad (2.76a)$$

with $$y(k) = x_1(k) + b_0 u(k) \qquad (2.76b)$$

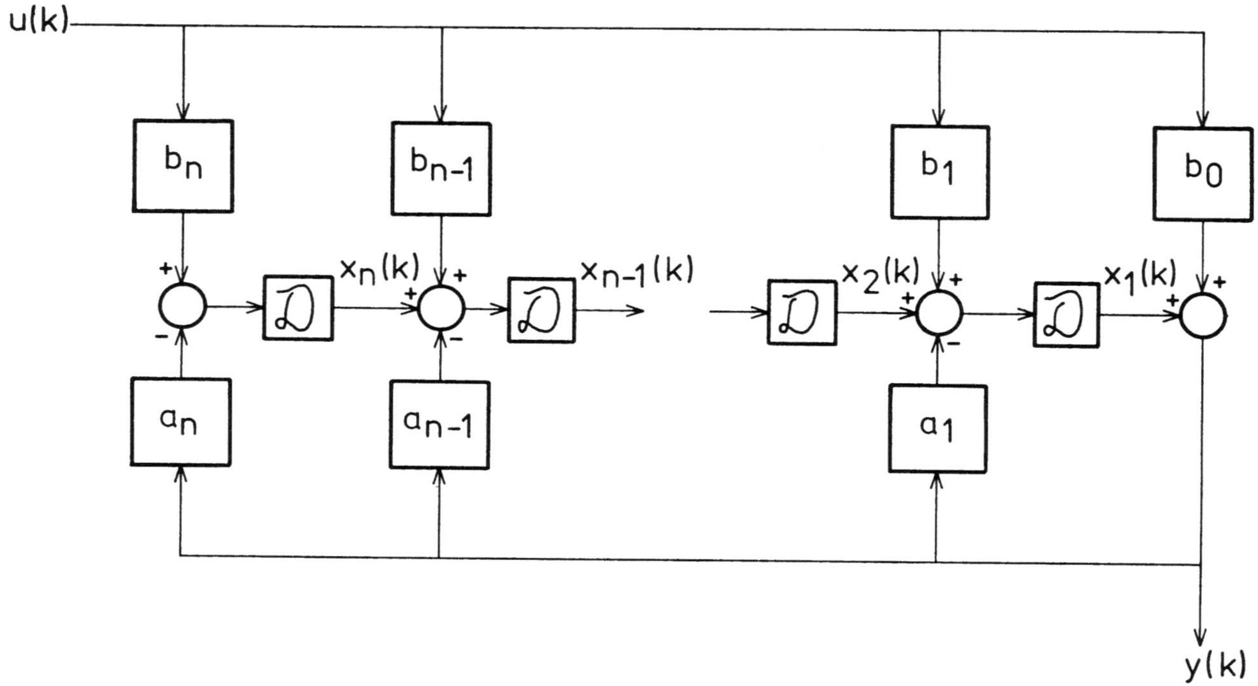

Fig. (2.6) Simulation diagram of n-th order difference equation using state variables

Define

$$\underline{x}(k) = \begin{bmatrix} x_1(k) \\ x_2(k) \\ \vdots \\ x_n(k) \end{bmatrix}$$

as the state vector, we can write (2.76) in the standard form

$$\underline{x}(k+1) = A\underline{x}(k) + Bu(k) \qquad (2.77a)$$

$$y(k) = C\underline{x}(k) + Du(k) \qquad (2.77b)$$

where

$$A = \begin{bmatrix} -a_1 & 1 & 0 & \dots & 0 \\ -a_2 & 0 & 1 & \dots & 0 \\ \vdots & & & & \\ -a_{n-1} & 0 & 0 & \dots & 1 \\ -a_n & 0 & 0 & \dots & 0 \end{bmatrix}$$

is the (n×n) system matrix,

$$B = \begin{bmatrix} b_1 - a_1 b_0 \\ b_2 - a_2 b_0 \\ \vdots \\ b_n - a_n b_0 \end{bmatrix}$$

is the (n×1) input matrix,

$$C = [1 \quad 0 \quad 0 \quad \dots \quad 0]$$

is the (1×n) output matrix and

$$D = b_0$$

is the (1×1) feedforward matrix.

It should be noted that:

(1) The preceding method is immediately applicable to sets of coupled linear constant coefficient difference equations,

which correspond to multivariable systems. In this case, we obtain the general state equations in vector-matrix form as:

$$\underline{x}(k+1) = A\underline{x}(k) + B\underline{u}(k) \qquad (2.78a)$$

$$\underline{y}(k) = C\underline{x}(k) + D\underline{u}(k) \qquad (2.78b)$$

where $\underline{u}(k)$, $\underline{y}(k)$ denote the m input and p output vectors, respectively.

(2) In the special case of (2.72), in which $b_0 = b_1 = \ldots = b_{n-1} = 0$, an alternative way of defining the state variables would be:

$$x_1(k) = y(k)$$
$$x_2(k) = y(k+1)$$
$$\vdots$$
$$x_n(k) = y(k+n-1)$$

In terms of (2.77), the associated matrices would be

$$A = \begin{bmatrix} 0 & 1 & 0 & 0 & \ldots & 0 \\ 0 & 0 & 1 & 0 & \ldots & 0 \\ \vdots & \vdots & \vdots & \vdots & & \vdots \\ -a_n & -a_{n-1} & -a_{n-2} & -a_{n-3} & \ldots & -a_1 \end{bmatrix}$$

$$B = \begin{bmatrix} 0 \\ 0 \\ \vdots \\ b_n \end{bmatrix} ; \quad C = [1 \quad 0 \quad \ldots \quad 0] \; ; \quad D = [0]$$

This situation is often encountered in systems where there is no direct feedthrough from the input to the output.

B. From Transfer Functions

The discrete system to be investigated is characterised by the transfer function

$$G(z) = \frac{Y(z)}{U(z)} = \frac{b_0 z^m + b_1 z^{m-1} + \ldots + b_m}{z^n + a_1 z^{n-1} + \ldots + a_n} \; ; \; m \leq n \qquad (2.79)$$

which is obviously the z-transform of (2.54) with $a_0 = 1$. There are four principal methods by which one can obtain the state equations (2.77). The first method is based on the partial-fraction expansion of (2.79). In Section 2.2.1 we have presented the main expressions (2.26), (2.27) to calculate this expansion. For simplicity in exposition, we will only treat the case where the denominator has a multiple root of order r at $p = p_n$ with all its other roots being distinct. Thus, we can write (2.79) in the light of (2.27a) as:

$$G(z) = \frac{d_1}{(z-p_1)} + \frac{d_2}{(z-p_2)} + \ldots + \frac{d_{n-r}}{(z-p_{n-r})} + \frac{d_{11}}{(z-p_n)} + \frac{d_{12}}{(z-p_n)^2} + \ldots + \frac{d_{1r}}{(z-p_n)^r} \tag{2.80}$$

where the coefficients $d_1, d_2, \ldots, d_{n-r}$ are calculated in the manner of (2.28b) and the coefficients $d_{11}, \ldots, d_{1r}$ are obtained through the use of (2.27b). Since all terms in (2.80) are multiplied by the same input $U(z)$, therefore their contribution to the output will be additive and can be generally represented by parallel branches in the simulation diagram as shown in Fig. 2.7. Selecting the state variables as indicated and applying the inverse z-transform, it is readily seen that the associated matrices of the state equations (2.77) are given by:

$$A = \left[\begin{array}{cccc|cccc} p_1 & 0 & \cdots & & & & \cdots & 0 \\ 0 & p_2 & & & & & & 0 \\ \vdots & & \ddots & & & & & \vdots \\ & & & p_{n-r} & 0 & & & 0 \\ \hline \vdots & & & & p_n & 0 & \cdots & 0 \\ & & & & 1 & p_n & & 0 \\ & & & & 0 & 1 & & \\ & & & & & & \ddots & \\ 0 & \cdots & & & 0 & 0 & \cdots & p_n \end{array}\right] \quad \begin{array}{c} \\ \\ \\ \\ \uparrow \\ r \text{ rows} \\ \\ \\ \downarrow \end{array}$$

$$\qquad\qquad |\leftarrow \quad r \text{ columns} \quad \rightarrow|$$

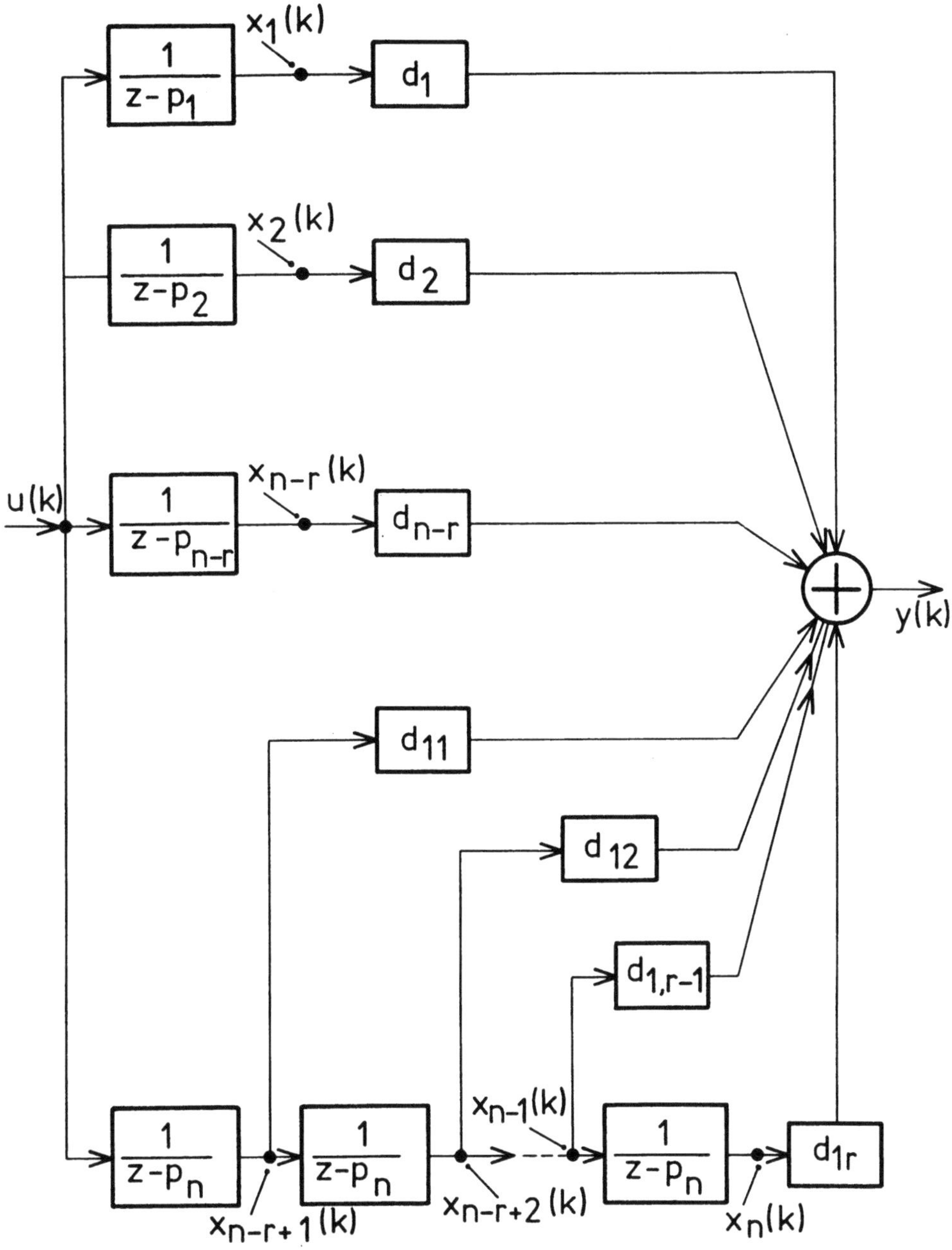

Fig. (2.7) State variable representation of partial-fraction expansion method

$$B = \begin{bmatrix} 1 \\ 1 \\ \cdot \\ \cdot \\ \cdot \\ 1 \\ 0 \\ \cdot \\ \cdot \\ \cdot \\ 0 \end{bmatrix} \begin{matrix} \\ \\ \\ \\ \\ \\ \overline{\downarrow} \\ \\ (r-1) \text{ entries} \\ \\ \underline{\downarrow} \end{matrix} \quad ; \quad C^t = \begin{bmatrix} d_1 \\ d_2 \\ \cdot \\ \cdot \\ \cdot \\ d_{n-r} \\ d_{11} \\ \cdot \\ \cdot \\ \cdot \\ d_{1r} \end{bmatrix} \quad ; \quad D = [0]$$

The system matrix A, which is obtained by a partial-fraction expansion of $G(z)$ when it has multiple poles, is said to be in a Jordan canonical form *[2]*. In the special case where all poles are distinct, the system matrix A becomes diagonal. By comparing the choice of state variables made when dealing with the difference equation and partial-fraction expansion we find that the use of the delay device is a fundamental step.

Now suppose that the transfer function $G(z)$ can be put in the factored form (2.20), that is:

$$G(z) = \frac{Y(z)}{U(z)}$$

$$= H_0 \frac{(z-z_1)(z-z_2)\ldots(z-z_m)}{(z-p_1)(z-p_2)\ldots(z-p_n)} \quad ; \; m \leq n \qquad (2.81)$$

We can arrange (2.81) as a tandem connection of blocks, some of which will have both numerator and denominator dynamics and the others will have only denominator dynamics. This means that we will have a combination of the two blocks in Fig. (2.8). The identity $H_0 = h_1 \, h_2 \ldots h_n$ is used.

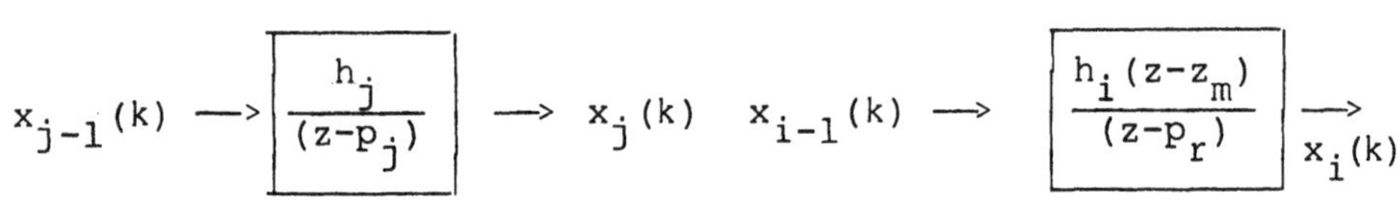

(a) (b)

Fig. 2.8 Basic blocks

Note that the arrangement process is quite arbitrary and the order in which the zeros and poles of G(z) appear is obviously not important. The first block will have u(k) as its input and the last block will produce $x_n(k) = y(k)$ as its output. From Fig. 2.8, the two representative first-order equations will be

$$\left.\begin{aligned} x_{j-1}(k+1) &= p_j\, x_{j-1}(k) + h_j\, x_j(k) \\ x_{i-1}(k+1) - p_r x_{i-1}(k) &= h_i[x_i(k+1)-x_i(k)] \end{aligned}\right\} \qquad (2.82)$$

Upon manipulating the first-order equations of the type (2.82), we can obtain (2.77) as the state equations. In general, the system matrix A tends to be lower-triangular. The above method is sometimes called iterative programming [2]. It should be emphasized that the usefulness of the partial-fraction expansion and iterative programming methods hinges upon the availability of G(z) in a factored form.

Direct methods of obtaining state equations from transfer functions depend on appropriate manipulation of G(z) of the form (2.79). We will discuss two such methods for the general case where m = n-1. According to the first method, we multiply the numerator and denominator of (2.79) by z^{-n} and arrange to yield:

$$\frac{Y(z)}{b_1 z^{-1}+b_2 z^{-2}+\ldots+b_n z^{-n}} = \frac{U(z)}{1+a_1 z^{-1}+\ldots+a_n z^{-n}}$$

$$= P(z) \qquad (2.83)$$

Alternatively, we have:

$$P(z) = -a_1 z^{-1}P(z) -\ldots- a_n z^{-n}P(z) + U(z) \qquad (2.84a)$$

and

$$Y(z) = b_1 z^{-1}P(z)+b_2 z^{-2}P(z)+\ldots+b_n z^{-n}P(z) \qquad (2.84b)$$

By virtue of (2.84b), a suitable choice of the state variables would be

$$X_j(z) = z^{-j}P(z) \quad ; \quad j = 1,\dots,n \tag{2.85a}$$

or

$$zX_{i+1}(z) = X_i(z) \quad ; \quad i = 1,\dots,n-1 \tag{2.85b}$$

in which case the output would be a linear combination of the states. By applying the inverse z-transform to (2.85b), to $zX_1(z) = P(z)$, using (2.84a), and to (2.84b), we obtain the standard state form (2.77) whose matrices are given by:

$$A = \begin{bmatrix} -a_1 & -a_2 & \cdots & -a_{n-1} & -a_n \\ 1 & 0 & \cdots & 0 & 0 \\ 0 & 1 & \cdots & 0 & 0 \\ \vdots & \vdots & & \vdots & \vdots \\ 0 & 0 & & 1 & 0 \end{bmatrix}$$

$$B = \begin{bmatrix} 1 \\ 0 \\ \vdots \\ 0 \end{bmatrix} \quad ; \quad C^t = \begin{bmatrix} b_1 \\ b_2 \\ \vdots \\ b_n \end{bmatrix} \quad ; \quad D = [0]$$

Note that if $m < n-1$, the same procedure applies but where some of the coefficients b_j are set to zero. In this case A and B remain unchanged. The only change will be in C, the output matrix. It is interesting to observe the similarity between the above choice of state variables and the one made in difference equations using the alternative method.

By adopting a different way of manipulating (2.79), we obtain a fourth method to derive the state equations. First, multiply the numerator and denominator by z^{-n}, then cross multiply both sides and arrange in equal powers of z^{-1} to obtain:

$$[Y(z) - b_0U(z)] + z^{-1}[a_1Y(z) - b_1U(z)] + z^{-2}[a_2Y(z) - b_2U(z)] +\dots+ z^{-n}[a_nY(z) - b_nU(z)] = 0 \tag{2.86}$$

Let the state variables be defined as:

$$
\begin{aligned}
X_1(z) &= z^{-1}[b_1U(z) - a_1Y(z) + X_2(z)] \\
&\vdots \\
X_j(z) &= z^{-1}[b_jU(z) - a_jY(z) + X_{j+1}(z)] \\
&\vdots \\
X_n(z) &= z^{-1}[b_nU(z) - a_nY(z)]
\end{aligned}
\tag{2.87}
$$

The substitution of (2.87) into (2.86) results in

$$
Y(z) = b_0U(z) + X_1(z) \tag{2.88}
$$

By inverting (2.87) using (2.88), we obtain n first-order equations which can be put in the form (2.77) with the matrices

$$
A = \begin{bmatrix} -a_1 & 1 & 0 & 0 & \dots & 0 \\ -a_2 & 0 & 1 & 0 & \dots & 0 \\ \vdots & & & & & \vdots \\ -a_{n-1} & 0 & 0 & \dots & \dots & 1 \\ -a_n & 0 & 0 & & & 0 \end{bmatrix}
$$

$$
B = \begin{bmatrix} b_1 - a_1b_0 \\ b_2 - a_2b_0 \\ \vdots \\ b_n - a_nb_0 \end{bmatrix} \; ; \quad C^t = \begin{bmatrix} 1 \\ 0 \\ \vdots \\ 0 \end{bmatrix} \; ; \quad D = [b_0]
$$

Careful examination of the above method in the light of other methods discussed this far leads to:

(1) The above method is equivalent to the one developed for difference equations, as both yield the same structural matrices. This is obviously due to the fact that the choice of state variables in both cases has been the same.

(2) The above method yields a system matrix which is the transpose of the one developed using the iterative programming method.

To conclude this section, it should be clear that one can use any method which seems appropriate and in particular one can utilise either the transfer function or the difference equations interchangeably.

2.4.3 *SOLUTION PROCEDURE*

Consider the state equations

$$\underline{x}(k+1) = A\underline{x}(k) + B\underline{u}(k) \tag{2.90a}$$

$$\underline{y}(k) = C\underline{x}(k) + D\underline{u}(k) \tag{2.90b}$$

which can be obtained from the description of a multi-input multi-output discrete system in the manner of the methods discussed before. Note that the frequency-domain form will be a matrix transfer-function. We wish to determine the state (or output) sequence for all k, given the input sequence $\underline{u}(0)$, $\underline{u}(1), \ldots$. The initial conditions, as summarised by $\underline{x}(0)$, must be known. Consider first the homogeneous part of (2.90a), that is:

$$\underline{x}(k+1) = A\underline{x}(k) \tag{2.91}$$

Starting from $\underline{x}(0)$ at $k = 0$ and iterating we get

$$\begin{aligned}
\underline{x}(1) &= A\underline{x}(0) \\
\underline{x}(2) &= A\underline{x}(1) \\
&= A^2\underline{x}(0) \\
\underline{x}(3) &= A\underline{x}(2) \\
&= A^3\underline{x}(0)
\end{aligned}$$

and in general,

$$\underline{x}(k) = A^k \underline{x}(0) \tag{2.92}$$

The non-homogeneous case is now considered. Given the sequence of input vectors $\underline{u}(0)$, $\underline{u}(1),\ldots,$ as well as the initial conditions $\underline{x}(0)$, then

$$\begin{aligned} \underline{x}(1) &= A\underline{x}(0) + B\underline{u}(0) \\ \underline{x}(2) &= A\underline{x}(1) + B\underline{u}(1) \\ &= A^2\underline{x}(0) + AB\underline{u}(0) + B\underline{u}(1) \end{aligned}$$

and continuing this process to a general discrete instant k, we arrive at:

$$\underline{x}(k) = A^k\underline{x}(0) + \sum_{j=0}^{k-1} A^{k-j-1}B\underline{u}(j) \tag{2.93}$$

A change in the dummy summation index allows this result to be written in the alternative form:

$$\underline{x}(k) = A^k\underline{x}(0) + \sum_{j=1}^{k} A^{k-j}B\underline{u}(j-1) \tag{2.94}$$

For time-invariant systems, we designate

$$\Phi(k,j) = A^{k-j} \tag{2.95}$$

as the discrete state transition matrix. Thus,

$$\underline{x}(k) = \Phi(k,0)x(0) + \sum_{j=0}^{k-1} \Phi(k-1,j)B\underline{u}(j) \tag{2.96}$$

which represents the general solution of the state equations. Note that the output sequence can be directly obtained from (2.96) by using (2.90b):

$$y(k) = C\,\Phi(k,0)\underline{x}(0) + \sum_{j=0}^{k-1} C\,\Phi(k-1,j)B\underline{u}(j) + D\underline{u}(k) \tag{2.97}$$

It is seen from (2.96) that the state sequence $\underline{x}(k)$ is composed of a factor dependent only on the initial state $\underline{x}(0)$ and another factor dependent solely on the input sequence $\underline{u}(0)$, $\underline{u}(1), \ldots, \underline{u}(k-1)$. In fact, through the second factor one has the ability to specify and thereby to influence the resultant state vector in some manner. An alternative solution procedure can be obtained via the application of Z-transform methods. Since we deal with arrays of scalars, vectors or matrices, we define

$$Z[\underline{x}(k)] = Z\begin{bmatrix} x_1(k) \\ \cdot \\ \cdot \\ \cdot \\ x_n(k) \end{bmatrix}$$

$$= \begin{bmatrix} X_1(z) \\ \cdot \\ \cdot \\ \cdot \\ X_n(z) \end{bmatrix} = \underline{X}(z) \tag{2.98}$$

where $Z[x_j(k)] = X_j(z)$. It is easy to see that $Z[A\underline{x}(k)] = A\underline{X}(z)$. Now, taking the z-transform of (2.90), we get

$$z\underline{X}(z) - z\underline{x}(0) = A\underline{X}(z) + B\underline{U}(z) \tag{2.99a}$$

$$\underline{Y}(z) = C\underline{X}(z) + D\underline{U}(z) \tag{2.99b}$$

where $\underline{U}(z) = Z[\underline{u}(k)]$ and $\underline{Y}(z) = Z[\underline{y}(k)]$ in the manner of (2.98). From (2.99a) the solution of $\underline{X}(z)$ is given by:

$$\underline{X}(z) = [zI-A]^{-1} z\underline{x}(0) + [zI-A]^{-1} B\underline{U}(z) \tag{2.100}$$

where I is the $n\times n$ identity matrix. By applying the inverse z-transform to (2.100) we can obtain $\underline{x}(k)$. A comparison of (2.96) with (2.100) indicates that

$$\Phi(k,0) = Z^{-1}\{z[zI-A]^{-1}\} \tag{2.101}$$

which provides another way of defining the state transition matrix. We observe that (2.95) requires successive multiplication of the system matrix, whereas (2.101) requires the inversion of a matrix followed by the inverse transformation of a z-transform. An algorithm for determining the inverse of $[zI-A]$ will now be outlined. Compute the coefficients d_j and the matrices L_j from the sequential relationships:

$$
\begin{aligned}
d_1 &= \mathrm{Tr}(A) \quad &, \quad L_1 &= A - d_1 I \\
d_2 &= \tfrac{1}{2}\mathrm{Tr}(AL_1) \quad &, \quad L_2 &= AL_1 - d_2 I \\
&\vdots & &\vdots \\
d_j &= \tfrac{1}{j}\mathrm{Tr}(AL_{j-1}) \quad &, \quad L_j &= AL_{j-1} - d_j I \\
&\vdots & &\vdots \\
d_n &= \tfrac{1}{n}\mathrm{Tr}(AL_{n-1}) \quad &, \quad L_n &= AL_{n-1} - d_n I
\end{aligned}
$$

where $\mathrm{Tr}(H)$ is the trace of H defined as the sum of the diagonal elements of H. Then, construct the polynomials

$$R(z) = z^n - d_1 z^{n-1} - d_2 z^{n-2} -\ldots d_n \tag{2.102}$$

$$T(z) = Iz^{n-1} + L_1 z^{n-2} + L_2 z^{n-3} +\ldots+ L_{n-1} \tag{2.103}$$

and thus derive $[zI-A]^{-1}$ as:

$$[zI-A]^{-1} = T(z)/R(z) \tag{2.104}$$

where $T(z)$ is the adjoint matrix of $[zI-A]$ and $R(z)$ is its determinant.

2.4.4 *EXAMPLES*

We now consider four examples to illustrate the methods of obtaining state equations.

Example 1

Consider the transfer function

$$G(z) = \frac{Y(z)}{U(z)} = \frac{1}{z^3-7z^2+14z-8}$$

The poles of the system are located at 1, 2, 4 and by a partial fraction expansion we can express $G(z)$ as:

$$G(z) = \frac{1/3}{z-1} + \frac{-1/2}{z-2} + \frac{1/6}{z-4}$$

Let the state variables be defined as

$$X_1(z) = \frac{1}{z-1} U(z)$$
$$X_2(z) = \frac{1}{z-2} U(z)$$
$$X_3(z) = \frac{1}{z-4} U(z)$$

Thus the output can be written as

$$y(z) = 1/3\ X_1(z) - 1/2\ X_2(z) + 1/6\ X_3(z)$$

From the above relationships, we see that the state model (2.77) is given by:

$$\begin{bmatrix} X_1(k+1) \\ X_2(k+1) \\ X_3(k+1) \end{bmatrix} = \begin{bmatrix} 1 & 0 & 0 \\ 0 & 2 & 0 \\ 0 & 0 & 4 \end{bmatrix} \begin{bmatrix} X_1(k) \\ X_2(k) \\ X_3(k) \end{bmatrix} + \begin{bmatrix} 1 \\ 1 \\ 1 \end{bmatrix} u(k)$$

$$y(k) = [1/3 \quad -1/2 \quad 1/6]\ \underline{x}(k)$$

Since the poles of the transfer function are distinct, the system matrix is diagonal as expected.

Example 2

The purpose of this example is to show how we can obtain the state model for a multivariable discrete system. Let the system be described by the following set of coupled difference equations:

$$y_1(k+3) + .6[y_1(k+2) - y_3(k+2)] + .2\, y_1(k+1)$$
$$+ y_2(k+1) + y_1(k) - .3\, y_3(k) = u_1(k) + u_2(k+1) \, , \tag{2.105a}$$

$$y_2(k+2) + .4\, y_2(k+1) - y_1(k+1) + .5\, y_2(k) + y_3(k)$$
$$= u_1(k) + u_2(k) + u_3(k) \, , \tag{2.105b}$$

$$y_3(k+1) + .1\, y_3(k) - y_2(k) = u_3(k) - u_2(k)$$
$$+ .7\, u_3(k+1) \tag{2.105c}$$

The system has three inputs $u_1(k)$, $u_2(k)$, $u_3(k)$ and three outputs $y_1(k)$, $y_2(k)$, $y_3(k)$. For (2.105a) we delay each term three times to obtain

$$y_1(k) = D\{-.6[y_1(k)-y_3(k)] + D[u_2(k) - .2\, y_1(k)$$
$$- y_2(k) + D\{u_1(k) - y_1(k) + .3\, y_3(k)\}]\} \tag{2.106a}$$

Delaying each term in (2.105b) twice gives

$$y_2(k) = D\{-.4\, y_2(k) + y_1(k) + D[u_1(k) + u_2(k)$$
$$+ u_3(k) - .5\, y_2(k) - y_3(k)]\} \tag{2.106b}$$

and from (2.105c), delayed once,

$$y_3(k) = .7\, u_3(k) + D\{u_3(k) - u_2(k)$$
$$- .1\, y_3(k) + y_2(k)\} \tag{2.106c}$$

The simulation diagram of (2.106a) - (2.106c) can be represented as shown in Fig. (2.9). Since the order of the overall system is the sum of the orders of (2.105a) - (2.105c), which is six, we have six delay elements. Labelling the output of these delays as $x_1, \ldots, x_6$, we obtain the state model (2.78) with the associated matrices:

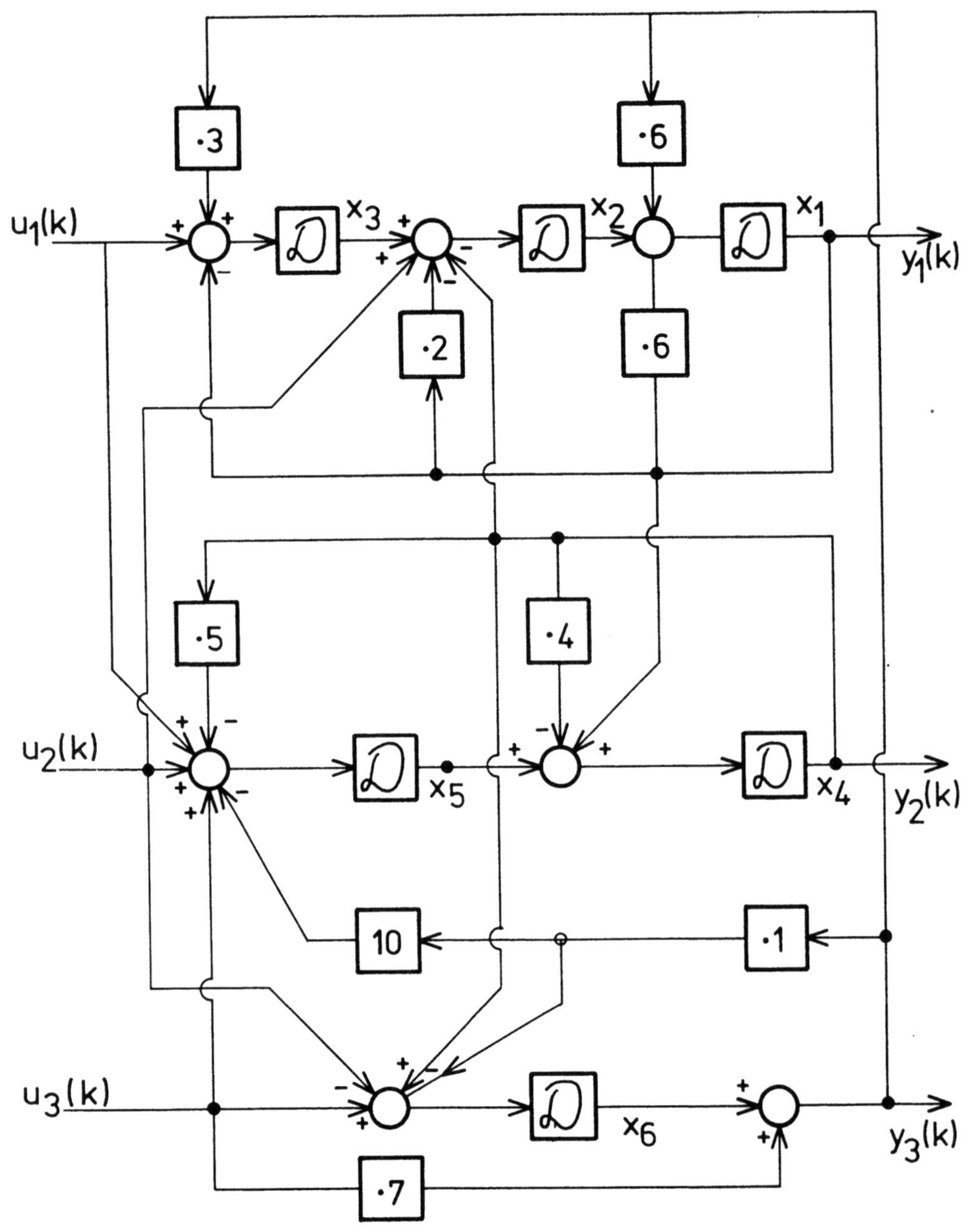

Fig. (2.9) Simulation diagram of example 2

$$A = \begin{bmatrix} -.6 & 1 & 0 & 0 & 0 & .6 \\ -.2 & 0 & 1 & -1 & 0 & 0 \\ -1 & 0 & 0 & 0 & 0 & .3 \\ 1 & 0 & 0 & -.4 & 1 & 0 \\ 0 & 0 & 0 & -.5 & 0 & -1 \\ 0 & 0 & 0 & 1 & 0 & -.1 \end{bmatrix}$$

$$B = \begin{bmatrix} 0 & 0 & .42 \\ 0 & 1 & 0 \\ 1 & 0 & .21 \\ 0 & 0 & 0 \\ 1 & 1 & -.3 \\ 0 & -1 & .93 \end{bmatrix}$$

$$C = \begin{bmatrix} 1 & 0 & 0 & 0 & 0 & 0 \\ 0 & 0 & 0 & 1 & 0 & 0 \\ 0 & 0 & 0 & 0 & 0 & 1 \end{bmatrix}$$

$$D = \begin{bmatrix} 0 & 0 & 0 \\ 0 & 0 & 0 \\ 0 & 0 & .7 \end{bmatrix}$$

We note that the feedforward matrix D has one element since $u_3(k)$ affects $y_3(k)$ directly, see (2.105c), whilst there is a delay between the inputs and outputs otherwise.

Example 3

It is required to put the transfer function

$$G(z) = \frac{Y(z)}{U(z)}$$

$$= \frac{z+1}{(z+2)(z+3)}$$

in state variable form using different methods. First, by partial-fraction expansion, G(z) becomes:

$$G(z) = \frac{-1}{(z+2)} + \frac{2}{z+3}$$

Let $$X_1(z) = \frac{1}{(z+2)} U(z)$$

$$X_2(z) = \frac{1}{(z+3)} U(z)$$

then the first state model has the matrices

$$A = \begin{pmatrix} -2 & 0 \\ 0 & -3 \end{pmatrix} \quad ; \quad B = \begin{pmatrix} 1 \\ 1 \end{pmatrix}$$

$$C = [-1 \quad 2]$$

Using the iterative programming method we arrange $G(z)$ as:

$$G(z) = \left(\frac{1}{(z+2)} \right) \cdot \left(\frac{(z+1)}{(z+3)} \right)$$

Let $$X_1(z) = \frac{1}{(z+2)} U(z)$$

$$X_2(z) = \frac{(z+1)}{(z+2)} X_1(z) \quad ; \quad X_2(z) = Y(z)$$

Applying the inverse z-transform and manipulating we obtain the second state model whose matrices are given by:

$$A = \begin{pmatrix} -2 & 0 \\ -1 & -3 \end{pmatrix} \quad ; \quad B = \begin{pmatrix} 1 \\ 1 \end{pmatrix}$$

$$C = [0 \quad 1]$$

Multiplying the numerator and denominator of $G(z)$ by z^{-2} and cross-multiplying gives:

$$\frac{Y(z)}{z^{-1}+z^{-2}} = \frac{U(z)}{1+5z^{-1}+6z^{-2}} = P(z)$$

Direct application of (2.85) results in the third state model with the associated matrices:

$$A = \begin{pmatrix} -5 & -6 \\ 1 & 0 \end{pmatrix} \quad ; \quad B = \begin{pmatrix} 1 \\ 0 \end{pmatrix}$$

$$C = [1 \quad 1]$$

Another way to manipulate G(z) yields:

$$Y(z) = z^{-1}\{[U(z) - 5Y(z)] + z^{-1}[U(z) - 6Y(z)]\}$$

Now, define the state and output variables by:

$$Y(z) = X_1(z)$$

$$X_1(z) = z^{-1}[U(z) - 5Y(z) + X_2(z)]$$

$$X_2(z) = z^{-1}[U(z) - 6Y(z)]$$

from which we obtain the associated matrices of the fourth state model:

$$A = \begin{pmatrix} -5 & 1 \\ -6 & 0 \end{pmatrix} \quad ; \quad B = \begin{pmatrix} 1 \\ 1 \end{pmatrix}$$

$$C = [1 \quad 0]$$

By examining the system matrices of the four state models, it is readily evident that the first two methods give the values of the poles at the main diagonal whereas the last two methods provide the coefficients of the denominator along the first row or the first column.

Example 4

Solve the following homogeneous state model

$$x(k+1) = A\underline{x}(k) = \begin{pmatrix} .5 & -.5 & 1 \\ 0 & .5 & 2 \\ 0 & 0 & .5 \end{pmatrix} \underline{x}(k)$$

with initial state $\underline{x}(0) = [2 \quad 4 \quad 6]^t$

From (2.93) the solution is given by

$$x(k) = A^k \underline{x}(0)$$

we can use (2.101) to compute the state transition matrix. Instead, we shall use another procedure based on the Cayley-Hamilton theorem *[21]*. According to this theorem analytic functions of the (nxn) square matrix can be expanded into a power series up to the (n-1)th term. Thus we can write

$$A^k = \alpha_0 I + \alpha_1 A + \alpha_2 A^2$$

since A is of order 3. To compute the α coefficients, the eigenvalues of A are required. In this example, A is an upper triangular matrix, therefore its eigenvalues are located along the main diagonal and are .5, .5, .5 . Then, the coefficients α_0, α_1, α_2 satisfy the equations *[21]*:

$$(.5)^k = \alpha_0 + .5\alpha_1 + .25\alpha_2$$

$$k(.5)^{k-1} = \alpha_1 + \alpha_2$$

$$k(k-1)(.5)^{k-2} = 2\alpha_2$$

The solution of these equations yields

$$\alpha_0 = (1 - 1.5k + .5k^2)(.5)^k$$

$$\alpha_1 = -k(k-2)(.5)^{k-1}$$

$$\alpha_2 = k(k-2)(.5)^{k-1}$$

from which we obtain

$$A^k = \begin{pmatrix} (.5)^k & -k(.5)^k & -2k(k-2)(.5)^k \\ 0 & (.5)^k & 4k(.5)^k \\ 0 & 0 & (.5)^k \end{pmatrix}$$

and thus $\underline{x}(k)$ is given by

$$\begin{pmatrix} x_1(k) \\ x_2(k) \\ x_3(k) \end{pmatrix} = \begin{pmatrix} 2(.5)^k - 4k(.5)^k - 6k(k-2)(.5)^{k-1} \\ 4(.5)^k + 6k(.5)^{k-2} \\ 6(.5)^k \end{pmatrix}$$

Indeed, for higher-order systems the above procedure requires a computer program for its implementation. The same is true for all the algorithms presented so far.

2.5 Modal Decomposition

In the last section, it has been observed that the system matrix A contains information about the poles of the discrete system. When the partial-fraction expansion is used, we found that the poles are located along the main diagonal. Depending on the position of these poles in the complex-plane, different time-sequences can be produced. In linear system theory *[20]*, the poles of the transfer function are the eigenvalues of the system matrix. It is the purpose of this section to examine the properties and role of eigenvalues in the analysis of linear discrete systems using a state variable representation.

2.5.1 *EIGEN-STRUCTURE*

We first define eigenvalues and eigenvectors of the system matrix. Those particular vectors $\underline{v}_j$ and the particular scalars λ_j which satisfy

$$A\underline{v}_j = \lambda_j \underline{v}_j \quad , \quad j = 1,\ldots,n \tag{2.107}$$

are called eigenvalues and eigenvectors, respectively *[20]*.

Note that the trivial case $\underline{x} = \underline{0}$ is explicitly excluded. The set of all scalars λ for which the transformation $(A-I\lambda)$ has no inverse is called the spectrum of A. The set of vectors $\{\underline{v}_1, \underline{v}_2, \ldots, \underline{v}_n\}$ accordingly constitutes a basis for an n-dimensional vector space R_n, which can therefore be used as a state space *[20,32]* for the system model (2.78). A necessary condition for the existence of nontrivial solutions to the set of n homogeneous equations (2.107) is that rank $[A-I\lambda_j]$ $< n$. This is equivalent to requiring

$$\det[A-\lambda_j I] = 0 \tag{2.108}$$

which is called the characteristic equation. The roots of (2.108) are the n eigenvalues of A. The eigenvectors $\underline{v}$ can be obtained by solving (2.107) for each of $\underline{v}_j$, after substitution of the corresponding eigenvalue λ_j into the appropriate equation. It should be noted that the $\underline{v}_j$ are thus determined to within a scalar multiplier.

In addition to the eigenproperties of A, the corresponding properties of A^t, the transposed system matrix, play an important role in the analysis of system modes. Let μ and $\underline{w}$ be respectively the eigenvalues and eigenvectors of A^t, thus:

$$A^t \underline{w}_m = \mu_m \underline{w}_m \quad ; \quad m = 1, \ldots, n \tag{2.109}$$

which is obviously analogous to (2.107). Following similar arguments, the n eigenvalues of A^t are the n roots of

$$\det[A^t - \mu_m I] = 0 \tag{2.110}$$

where I is the (nxn) identity matrix. For any square matrix H, $\det[H] = \det[H^t]$ and since I is symmetric, it is easy to see that (2.110) implies that

$$\det[A-\mu_m I] = 0 \tag{2.111}$$

A comparison of (2.108) and (2.111) reveals that A and A^t have the same eigenvalues

$$\mu_j = \lambda_j \quad , \quad j = 1,\ldots,n \tag{2.112}$$

The fact that A and A^t each has the same set of eigenvalues does not imply that the corresponding sets of eigenvectors will in general be equal.

To develop the relationship between the sets of vectors $\{\underline{v}\}$ and $\{\underline{w}\}$, we first use (2.112) in (2.109) to yield:

$$A^t \underline{w}_m = \lambda_m \underline{w}_m \quad , \quad m = 1,\ldots,n \tag{2.113}$$

Upon transposing (2.113) and then post-multiplying by $\underline{v}_j$ for $j \neq m$, it follows that

$$\underline{w}_m^t A \underline{v}_j = \lambda_m \underline{w}_m^t \underline{v}_j \quad , \quad (j \neq m,\ m = 1,\ldots,n) \tag{2.114}$$

On the other hand, the pre-multiplication of (2.107) by $\underline{w}_m^t$, $m \neq j$, results in:

$$\underline{w}_m^t A \underline{v}_j = \lambda_j \underline{w}_m^t \underline{v}_j \quad , \quad (m \neq j,\ j = 1,\ldots,n) \tag{2.115}$$

Now, subtracting (2.114) from (2.115) gives:

$$(\lambda_j - \lambda_m) \underline{w}_m^t \underline{v}_j = 0 \quad , \quad (m \neq j,\ j,m = 1,\ldots,n) \tag{2.116}$$

For distinct eigenvalues, (2.116) implies that:

$$\underline{w}_m^t \underline{v}_j = 0 \tag{2.117}$$

whose interpretation is that eigenvectors of A and A^t corresponding to different eigenvalues are orthogonal. For nondistinct eigenvalues, (2.116) implies that:

$$\underline{w}_j^t \underline{v}_j = \gamma_j , \quad j = 1,\dots,n \qquad (2.118)$$

where the γ_j are non-zero constants. By virtue of the fact that eigenvectors are determined only to within a scalar multiplier, we can normalise them such that $\gamma_j = 1$ and under this condition, we can combine (2.117) and (2.118) into one form:

$$\underline{w}_m^t \underline{v}_j = \underline{v}_j^t \underline{w}_m$$

$$= \delta_{jm} \qquad (j,m = 1,\dots,n) \qquad (2.119)$$

where δ_{jm} is the Kronecker delta. In view of (2.119), the set of vectors $\{\underline{w}_1,\underline{w}_2,\dots,\underline{w}_n\}$ is said to constitute a reciprocal basis *[20]* for the vector space R_n. The vectors $\underline{v}$ and $\underline{w}^t$ are sometimes called the right and left eigenvectors respectively of the matrix A. For the system matrix A, we define

$$V = [\underline{v}_1,\underline{v}_2,\dots,\underline{v}_n] \qquad (2.120a)$$

as its modal matrix, and similarly

$$W = [\underline{w}_1,\underline{w}_2,\dots,\underline{w}_n] \qquad (2.120b)$$

is defined as the modal matrix of A^t. The (nxn) matrix

$$\Lambda = \text{diag}\{\lambda_1,\dots,\lambda_n\} \qquad (2.121)$$

is defined as the eigenvalue matrix of A and A^t. In view of (2.120) and (2.121), we can write (2.107), (2.109) and (2.119) in the following compact forms:

$$AV = V\Lambda \qquad (2.122a)$$

$$A^tW = W\Lambda \qquad (2.122b)$$

$$W^tV = I \qquad (2.122c)$$

Simple manipulations of (2.122) give us some important relations:

$$W^t = V^{-1} \tag{2.123a}$$

$$W = V^{-t} \tag{2.123b}$$

and

$$V^{-1}AV = VAV^{-1} = \Lambda \tag{2.123c}$$

The usefulness of (2.123) depends on the availability of the eigenvalues and eigenvectors (eigen-structure) of the system matrix. Having determined this structure, (2.123c) tells us that any system matrix with distinct eigenvalues can be put in a diagonal form by a suitable linear transformation, frequently called a similarity transformation *[21]*. In the case of repeated eigenvalues, V becomes the generalised modal matrix and (2.122a) takes the form:

$$AV = VJ \tag{2.124}$$

where J is an (nxn) block diagonal matrix, called the Jordan form *[21]*,

$$J = \mathrm{diag}[J_1, J_2, \ldots, J_n] \tag{2.125a}$$

Each submatrix J_i of dimension $(n_i \times n_i)$ has the form:

$$J_i = \begin{bmatrix} \lambda_j & 1 & & \text{\huge 0} \\ & \lambda_j & 1 & \\ & & \ddots & \ddots \quad 1 \\ \text{\huge 0} & & & \lambda_j \end{bmatrix} \tag{2.125b}$$

We emphasize that this situation corresponds to the partial-fraction expansion of transfer functions with multiple poles.

The determination of the eigen-structure has been extensively studied in the literature on numerical analysis and linear algebra [22-24]. Computer packages are now available and widely used [25]. We will restrict attention here to the basic features of the eigen-structure.

When the determinant (2.108) is expanded, it yields an nth degree polynomial in the scalar λ, that is:

$$\det(A-I) = (-1)^n\lambda^n + c_{n-1}\lambda^{n-1} + \ldots + c_0 = \Delta(\lambda) \tag{2.126}$$

Some of the well-known facts about $\Delta(\lambda)$ are [22]:

(1) It has exactly n roots, subject to complex pairing.

(2) In factored form,

$$\Delta(\lambda) = (-1)^n(\lambda-\lambda_1)^{r_1}(\lambda-\lambda_2)^{r_2} \ldots (\lambda-\lambda_p)^{r_p}$$

which indicates that $\lambda = \lambda_1$ is an r_1-th order root, $\lambda = \lambda_2$ is an r_2-th order root, ..., etc. The integer r_j is called the algebraic multiplicity of the eigenvalue λ_j such that

$$\sum_{j=1}^{p} m_j = n$$

(3) $$T_r(A) = \sum_{j=1}^{n} \lambda_j = (-)^{n+1} c_{n-1}$$

(4) $$\det[A] = \prod_{j=1}^{n} \lambda_j = c_0$$

Let q_j denote the geometric multiplicity of λ_j (degeneracy) and is defined by

$$q_j = n - \text{rank}[A-\lambda_j I] \; ; \quad 1 \leq q_j \leq r_j \tag{2.127}$$

The following table summarises the different classes of eigen-structure.

Type of Eigenvalues	Corresponding Eigenvectors
distinct ($r_j = 1$ for all j)	1) n linearly independent eigenvectors, each is given by any nonzero column of the adjoint matrix $Ad_j[A-\lambda_j I]$.
repeated ($r_j > 1$ for some j)	2) $q_j = r_j$ (full degeneracy) a full set of r_j eigenvectors associated with λ_j is obtained. They can be found from the nonzero columns of $\frac{1}{(r_j-1)!}\left[\frac{d^{r_j-1}}{d\lambda^{r_j-1}}\{Ad_j[A-\lambda_j I]\}\right]_{\lambda\to\lambda_j}$ 3) $q_j = 1$ (simple degeneracy) there will be one eigenvector and (r_j-1) generalised eigenvectors*. 4) $1 \le q_j \le m_j$ there will be q_j eigenvectors and (r_j-q_j) generalised eigenvectors*.

*A generalised eigenvector of rank K is defined as a nonzero vector satisfying $[A-\lambda_j I]^k \underline{v}_k = \underline{0}$ and $[A-\lambda_j I]^{k-1}\underline{v}_k \neq \underline{0}$ [23].

TABLE 2.3 Classes of Eigen-Structure

In connection with (2.125), we emphasize that the number of Jordan blocks is equivalent to the sum of geometric multiplicities of the eigen-values. As a general rule, every (nxn) matrix has n eigenvalues and n linearly independent vectors either eigenvectors or generalised eigenvectors. The modal matrix will be unique in the case of distinct eigenvalues. However, it will not be unique for repeated eigenvalues although all generalised modal matrices will give the same Jordan canonical form. We now examine the use of eigenstructure in linear discrete control systems.

2.5.2 *SYSTEM MODES*

Consider first the case where the system matrix has n distinct eigenvalues. By virtue of the fact that the corresponding eigenvectors are linearly independent, we introduce the new vector $\theta(k)$ defined by:

$$\underline{x}(k) = V\underline{\theta}(k) \tag{2.128}$$

Now, (2.128) transforms the free state model

$$\underline{x}(k+1) = A\underline{x}(k) \tag{2.129}$$

into the form

$$V\underline{\theta}(k+1) = AV\underline{\theta}(k)$$

or equivalently

$$\begin{aligned}\underline{\theta}(k+1) &= V^{-1}AV\underline{\theta}(k) \\ &= \Lambda\underline{\theta}(k) \end{aligned} \tag{2.130}$$

The importance of (2.130) as compared with (2.129) is that Λ is a diagonal matrix, whereas A is, in general, non-diagonal. An alternative form of (2.130) would be

$$\theta_j(k+1) = \lambda_j\theta_j(k) \; ; \quad j = 1,\ldots,n \tag{2.131}$$

which represents n uncoupled, first order difference equations whose solutions in the manner of Table 2.1 are given by:

$$\theta_j(k) = \lambda_j^k \, \theta_j(0) \tag{2.132}$$

where $\theta_j(0)$, $j = 1,\ldots,n$, are the initial values of $\underline{\theta}(k)$.

In the light of (2.123a), it follows that, by putting $k = 0$ in (2.128),

$$\underline{\theta}(0) = W^t \underline{x}(0) \tag{2.133a}$$

or in component form

$$\theta_j(0) = \underline{w}_j^t \underline{x}(0) \quad ; \quad j = 1,\ldots,n \tag{2.133b}$$

Substituting (2.132) and (2.133b) in (2.128) and manipulating, we obtain

$$\underline{x}(k) = \underline{v}_1\underline{w}_1^t\underline{x}(0)\lambda_1^k + \underline{v}_2\underline{w}_2^t\underline{x}(0)\lambda_2^k + \ldots + \underline{v}_n\underline{w}_n^t\underline{x}(0)\lambda_n^k$$

$$\underline{x}(k) = \left(\sum_{j=1}^{n} \underline{v}_j\lambda_j^k\underline{w}_j^t \right)\underline{x}(0) \tag{2.134}$$

which shows clearly that the free motion of the discrete state model (2.129) is a linear combination of n functions of the form $\lambda_j^k\underline{v}_j$ $(j = 1,\ldots,n)$ which are said to describe the n dynamical modes of the system. Thus, the shape of a mode is described by its associated eigenvector $\underline{v}_j$, and its characteristics in the discrete time-domain by its associated eigenvalue λ_j. In particular, it is clear from (2.134) that the equilibrium state $\underline{x} = \underline{0}$ of the system (2.129) will be asymptotically stable in the sense that $\underline{x}(k) \to \underline{0}$ as $k \to \infty$ if and only if $\lambda_j < 1$ for all $j = 1,\ldots,n$. This is in agreement with the discussion in Section 2.2.3. In view of (2.123a) we can rewrite (2.134) in the compact form:

$$\underline{x}(k) = V\Lambda^k V^{-1}\underline{x}(0) \tag{2.135}$$

which when compared with (2.92) using (2.95) shows that

$$\begin{aligned} \Phi(k,0) &= A^k \\ &= V\Lambda^k V^{-1} \end{aligned} \tag{2.136}$$

The above expression provides a means of computing the state transition matrix, when A has a full set of eigenvectors.

Consider now the general state model

$$\underline{x}(k+1) = A\underline{x}(k) + B\underline{u}(k) \tag{2.137a}$$

$$\underline{y}(k) = C\underline{x}(k) + D\underline{u}(k) \tag{2.137b}$$

in which A may have repeated eigenvalues. Then, if the change of basis $\underline{x}(k) = V\underline{\theta}(k)$, with V being the generalised modal matrix, is used, the state model (2.137) reduces to

$$\underline{\theta}(k+1) = J\underline{\theta}(k) + B_m\underline{u}(k) \tag{2.138a}$$

$$y(k) = C_m\ \theta(k) + D\underline{u}(k) \tag{2.138b}$$

where

$$J = V^{-1}AV \tag{2.139a}$$

$$B_m = V^{-1}B \tag{2.139b}$$

$$C_m = CV \tag{2.139c}$$

are the Jordan canonical, modal input and modal output matrices respectively. We note that the modal equations (1.38) are as nearly uncoupled as possible and provide the same advantages as for the case of distinct eigenvalues.

2.5.3 *SOME IMPORTANT PROPERTIES*

Our purpose here is to present some important properties of the modal expansion (2.134) which have recently been found to be useful in control system's studies *[26-30]*. Further examination of the eigen-structure of linear, shift-invariant systems reveals *[26,27]* that there exist associations between groups of state variables and groups of natural modes of the system matrix. These associations are termed "dynamic pattern of behaviour" and can be displayed by means of the "participation matrix"

$$P = [p_{m_j}]$$

$$= [w_{mj} v_{mj}]$$

$$= \begin{bmatrix} w_{11}v_{11} & w_{12}v_{12} & \cdots\cdots & w_{1n}v_{1n} \\ w_{21}v_{21} & w_{22}v_{22} & & \vdots \\ \vdots & & \ddots & \vdots \\ w_{n1}v_{n1} & \cdots\cdots & & w_{nn}v_{nn} \end{bmatrix} \tag{2.140}$$

where v_{mj} and w_{mj} are the m-th entry of the j-th right and left eigenvectors of A, respectively; see (2.107), (2.113).

In the case of normalized eigenvectors, that is (2.119) holds, the sum of the values of all the entries of a single row or column of P is always equal to 1. This motivates defining the entries of P as "participation factors" since they are dimensionless. Now, it is easy to see from (2.134) that the quantity $[\underline{w}_j^t \, \underline{x}(0)]$ gives the contribution of the initial condition $\underline{x}(0)$ to the j-th mode and $\underline{v}_j$ describes the activity of each state variable in the j-th mode [26,27]. To clarify this point, let $\underline{x}(0) = \underline{e}_m$ (the m-th elementary unit vector), then from (2.134) and (2.140) we obtain

$$x_m(k) = \sum_{j=1}^{n} p_{mj} \lambda_j^k \tag{2.141}$$

from which it is clear that p_{mj} provides the relative participation of the j-th mode in the m-th state at $k = 0$.

Suppose that only the j-th mode is excited, that is $\underline{x}(0) = \underline{v}_j$ then the substitution in (2.134) yields:

$$\begin{aligned} \underline{x}(k) &= \underline{v}_j \lambda_j^k (\underline{w}_j^t \underline{v}_j) \\ &= (\underline{w}_j^t \underline{v}_j) \underline{v}_j \lambda_j^k \\ &= \left(\sum_{m=1}^{n} p_{mj} \right) \underline{v}_j \lambda_j^k \end{aligned} \tag{2.142}$$

with p_{mj} measuring the relative participation of the m-th state in building the dynamic response of the j-th mode.

Let the system matrix A in the free model (2.129) be

$$A = [a_{ms}]$$

and consider that the generic element a_{ms} is perturbed due to a change in the system parameters. Partial differentiation of (2.107) with respect to a_{ms} results in

$$(\partial A/\partial a_{ms})\underline{v}_j + A(\partial \underline{v}_j/\partial a_{ms})$$

$$= (\partial \lambda_j/\partial a_{ms})\underline{v}_j + \lambda_j(\partial \underline{v}_j/\partial a_{ms})$$

which upon pre-multiplication by $\underline{w}_j^t$ reduces to

$$\underline{w}_j^t(\partial A/\partial a_{ms})\underline{v}_j + \lambda_j \underline{w}_j^t(\partial \underline{v}_j/\partial a_{ms})$$

$$= \underline{w}_j^t(\partial \lambda_j/\partial a_{ms})\underline{v}_j + \lambda_j \underline{w}_j^t(\partial \underline{v}_j/\partial a_{ms}) \qquad (2.143)$$

In view of (2.119) and the fact that

$$(\partial A/\partial a_{ms}) = \delta_{im}\delta_{js} = [\alpha_{ij}]$$

(2.143) simplifies to

$$[\partial \lambda_j/\partial a_{ms}] = w_{jm}v_{js} \; ; \qquad (j,m,s = 1,\ldots,n) \qquad (2.144)$$

By virtue of (2.140) and (2.144), it is readily evident that $p_{mj} = [\partial \lambda_j/\partial a_{mm}]$, which show that the participation factors are the first-order eigenvalue sensitivities with respect to the diagonal elements of the system matrix. Looked at in this light, we can rewrite (2.140) as:

$$P = \begin{bmatrix} \partial\lambda_1/\partial a_{11} & \cdots\cdots & \partial\lambda_n/\partial a_{11} \\ \vdots & & \vdots \\ \partial\lambda_1/\partial a_{nn} & \cdots\cdots & \partial\lambda_n/\partial a_{nn} \end{bmatrix} \qquad (2.145)$$

Using the basic concept of participation factors, the "selective modal analysis" method has been developed and utilized in order-reduction *[26]*, dynamic stability *[27]* and model approximation *[28]*.

A final remark concerns the interpretation of the dynamics of the system eigenvectors. Consider the free system (2.129) whose right eigenvectors are defined by (2.107). Suppose that the initial state vector is set equal to a scalar multiple of the j-th eigenvector $\underline{v}_j$, that is, $\underline{x}(0) = \rho\underline{v}_j$. Then from (2.107) and (2.129) it is readily seen that

$$\underline{x}(1) = A\underline{x}(0) = \rho A\underline{v}_j = \rho\lambda_j\underline{v}_j$$

$$= \lambda_j\ \underline{x}(0) \qquad (2.146a)$$

and recursively

$$\underline{x}(k) = \lambda_j^k\ \underline{x}(0) \qquad (2.146b)$$

This means that, in this particular case, the next state is just λ_j times the initial state and all successive states are also various scalar multiples of the initial state. The above simple analysis leads us to the observation that the *right eigenvector is most naturally regarded as a vector in the state space.*

If we consider (2.128) in the light of (2.123) we see that a left eigenvector $\underline{w}^t$ defines the scalar function

$$\theta(k) = \underline{w}^t\ \underline{x}(k) \qquad (2.147)$$

It simply associates a scalar with each value of the state vector. As an example, let $\underline{w}^t = [0 \;\; 1 \;\; 0 \; \ldots]$, then from (2.147) we have $\theta(k) = x_2(k)$, that is, the second component of $\underline{x}(k)$. If, as another example, $w^t = [1 \;\; 1 \;\; \ldots \;\; 1]$, then (2.147) yields $\theta(k) = x_1(k) + \ldots + x_n(k)$, that is, the scalar function would be the sum of the components of $\underline{x}(k)$.

In general, *a left eigenvector defines a certain linear combination of the components of the state vector.* Of course, as the state vector evolves in time, the associated value of the linear combination also evolves.

This section shows that modal decomposition is useful in the analysis of linear discrete systems because of the insight it gives regarding the intrinsic properties of the system. The structural concepts of controllability and observability, which will be examined in Chapter 3, are more easily understood and evaluated in this framework. Modal decomposition provides a simple geometrical picture for the behaviour of the state vector in the discrete-time domain. By retaining only the dominant modes, a higher-order system can be approximated by a lower-order system, as will be shown in Chapter 4.

2.5.4 *EXAMPLES*

Example 1

Given the discrete state model

$$\underline{x}(k+1) = \begin{bmatrix} .5 & .5 & 0 \\ 0 & 1 & 0 \\ .833 & -2.167 & -.333 \end{bmatrix} \underline{x}(k) + \begin{bmatrix} 3 & 1 \\ 2 & 0 \\ -1 & 1 \end{bmatrix} \underline{u}(k)$$

$$\underline{y}(k) = \begin{bmatrix} -1 & 3 & 1 \\ 0 & 1 & 1 \end{bmatrix} \underline{x}(k)$$

which is of the type (2.137) with $D = 0$. We wish to express it in the normal form (2.138). First the eigenvalues are determined:

$$\det[A-\lambda I] = \begin{bmatrix} .5-\lambda & .5 & 0 \\ 0 & 1-\lambda & 0 \\ .833 & -2.167 & -.333-\lambda \end{bmatrix}$$

$$= (1-\lambda)(.5-\lambda)(-.333-\lambda) = 0$$

which gives $\lambda_1 = 1$, $\lambda_2 = .5$, $\lambda_3 = -.333$. Since the eigenvalues are distinct, then case 1 of Table 2.3 is applied. Compute

$$\mathrm{Ad}_j[A-\lambda I] = \begin{bmatrix} -(1-\lambda)(\lambda+.333) & .5(.333+\lambda) & 0 \\ 0 & -(.5-\lambda)(\lambda+.333) & 0 \\ -.833(1-\lambda) & -2.167(.5-\lambda)-.4165 & (.5-\lambda)(1-\lambda) \end{bmatrix}$$

from which we obtain

$$\underline{v}_1 = \begin{bmatrix} 1 \\ 1 \\ -1 \end{bmatrix}, \quad \underline{v}_2 = \begin{bmatrix} 1 \\ 0 \\ 1 \end{bmatrix}, \quad \underline{v}_3 = \begin{bmatrix} 0 \\ 0 \\ 1 \end{bmatrix}$$

so that

$$V = \begin{bmatrix} 1 & 1 & 0 \\ 1 & 0 & 0 \\ -1 & 1 & 1 \end{bmatrix}$$

and

$$V^{-1} = \begin{bmatrix} 0 & 1 & 0 \\ 1 & -1 & 0 \\ -1 & 2 & 1 \end{bmatrix}$$

From (2.139), the modal matrices are

$$J = \Lambda = V^{-1}AV$$

$$= \begin{bmatrix} 1 & 0 & 0 \\ 0 & .5 & 0 \\ 0 & 0 & -.333 \end{bmatrix}$$

$$B_n = V^{-1}B = \begin{bmatrix} 2 & 0 \\ 1 & 1 \\ 0 & 0 \end{bmatrix}$$

$$C_n = CV = \begin{pmatrix} 1 & 0 & 1 \\ 0 & 1 & 1 \end{pmatrix}$$

Example 2

This example is a discrete version of the linearized model of an advanced turbofan engine. The data for the continuous-time model *[31]* is discretized at sampling intervals of 0.1 sec to yield the system matrix of order 16x16 of the form:

$$A = \begin{bmatrix} A_1 & A_2 \\ A_3 & A_4 \end{bmatrix}$$

where

$$A_1 = \begin{bmatrix} .62 & .1956 & .4764 & .3633 & -16.35 & -1.174 & -.0315 & .2860 \\ -.0186 & .7217 & .3453 & -.3213 & -5.902 & .2449 & .0346 & .2156 \\ -.0052 & .0256 & .0164 & -.0086 & 2.158 & -.0355 & -.0023 & .0165 \\ -.0009 & .0054 & .0037 & -.0015 & .5072 & -.0108 & -.0011 & .0036 \\ .0006 & -.0001 & .0014 & .0017 & .4829 & -.0027 & -.0007 & -.0006 \\ .0279 & .0043 & .0203 & .0208 & .5940 & .0929 & -.0050 & .0040 \\ .0232 & .0032 & .0195 & .0208 & 1.798 & -.0411 & .1269 & .0016 \\ .0230 & .0535 & .0569 & -.0040 & .2391 & .2952 & -.0001 & .1621 \end{bmatrix}$$

$$A_2 = \begin{bmatrix} -.0004 & -.0203 & .1502 & .1102 & .7679 & -.0355 & .003 & .0044 \\ .1820 & 1.087 & .1096 & -.0025 & .0582 & .0003 & .0033 & .0030 \\ .0064 & .0197 & .0034 & -.0009 & -.0211 & .0042 & .0015 & -.0005 \\ .0014 & -.0035 & .0006 & -.0001 & -.0071 & .0008 & .0002 & -.0001 \\ -.0001 & -.0046 & -.0001 & .0001 & -.0053 & .0007 & .0001 & -.0001 \\ -.0007 & -.0134 & .0039 & .0048 & .0045 & -.0006 & -.0001 & -.0003 \\ -.0005 & -.0209 & .0028 & .0040 & -.0089 & .0005 & .0011 & -.0007 \\ -.0108 & .0458 & .0111 & .0031 & .0159 & .0001 & .0013 & 0 \end{bmatrix}$$

$$A_3 = \begin{bmatrix} .0287 & -.0408 & .0063 & .0133 & -9.532 & .7375 & .0362 & .2424 \\ .0005 & -.0008 & .0002 & .0002 & -.1909 & .0151 & .0008 & .0053 \\ .0423 & -.0511 & -.0029 & .0245 & -10.55 & .6976 & .0334 & .1984 \\ .0282 & -.0409 & .0014 & .0196 & -7.812 & .5886 & .0252 & .1907 \\ .0014 & -.0024 & .0002 & .0018 & -.435 & .0334 & .0016 & .0114 \\ .0204 & -.0589 & -.0173 & .0271 & -4.925 & .6697 & .0301 & .1836 \\ .0164 & .0023 & .0136 & .0151 & 1.62 & -.2493 & .2608 & .0016 \\ -.0244 & -.0011 & -.0039 & -.0078 & 2.953 & .2651 & .1257 & .11 \end{bmatrix}$$

$$A_4 = \begin{bmatrix} -.0045 & .0306 & .0373 & .0028 & .1088 & -.0084 & -.0003 & .0030 \\ -.0128 & .9361 & .0016 & .0001 & .0020 & -.0001 & 0 & .0001 \\ -.0156 & .0289 & .0135 & .0039 & .1547 & -.0128 & -.0003 & .0031 \\ -.0161 & -.1950 & .0193 & .0101 & .0840 & -.0086 & -.0015 & .0023 \\ -.0013 & -.0106 & .0025 & -.0337 & .8229 & -.0004 & 0 & .0001 \\ -.0276 & -.1798 & .0458 & -.0071 & -.0679 & .1348 & -.0028 & .0015 \\ -.0001 & -.0120 & .0013 & .0023 & -.0086 & -.0003 & .1364 & -.0007 \\ -.0099 & -.1010 & .0397 & -.0133 & -.1337 & .1396 & .0840 & .0060 \end{bmatrix}$$

A computer program has been written for the UNIVAC Computer Centre at Kuwait University, to determine the eigenvalues $(\lambda_1, \ldots, \lambda_{16})$, right modal matrix V and matrix of reciprocal basis W. Then it uses (2.140) to determine the participation matrix P.

In Table 2.4 we present the eigenvalues and the absolute value of the participation matrix. Entries smaller than .001 have been omitted and other entries have been rounded to four decimals. Careful examination of $|P|$ in Table 2.4 shows the main dynamic patterns of the turbofan model; these are presented in Table 2.5 .

Although these results were obtained in a routine way, they have interesting interpretations. On the one hand, any feedback control scheme based on certain states implies that the associated modes in Table 2.5 will be more greatly affected than other modes. For example, using x_{10} in the feedback loop means that the first mode will be subsequently changed.

|P| :

$\lvert\underline{P}_1\rvert$	$\lvert\underline{P}_2\rvert$	$\lvert\underline{P}_3\rvert$	$\lvert\underline{P}_{4,5}\rvert$	$\lvert\underline{P}_{6,7}\rvert$	$\lvert\underline{P}_8\rvert$	$\lvert\underline{P}_{9,10}\rvert$	$\lvert\underline{P}_{11}\rvert$	$\lvert\underline{P}_{12}\rvert$	$\lvert\underline{P}_{13}\rvert$	$\lvert\underline{P}_{14}\rvert$	$\lvert\underline{P}_{15}\rvert$	$\lvert\underline{P}_{16}\rvert$
.0050	.0259	.0668	.5613	.1046	.0042	.0299	.0027	.0279	.0062	.0160	.0016	.0055
.0035	.0248	.7345	.1260	.0428	.0438	.0259	.1821	.0459	.0018	.0396	-	.0145
-	-	.0350	.0046	.0085	.0479	.0049	-	-	.0028	.0146	.0313	.9738
-	-	.0024	.0044	.0026	-	.0015	-	-	-	-	1.0269	.0308
.0013	.0025	.0417	.5514	.0723	.0065	.0219	.0282	.0357	.0268	.0099	.0027	.0193
.0014	.0065	.0374	.1436	.5592	.0428	.0208	.0026	.0343	.0027	.0161	-	.0033
-	-	.0035	.0117	.0229	.3981	.2901	.0024	.0036	-	.0012	-	-
.0028	.0086	.0672	.0216	.3018	.0027	.2120	.0810	.0503	.0069	.0155	-	.0027
.0039	.0015	.0220	.0374	.0759	.0016	.5225	.5070	.2507	.0181	.3650	-	-
.9893	.0021	.0081	.0014	.0016	-	-	.0019	-	-	.0011	-	-
-	.0032	.0081	.0068	.0803	-	.0090	.7026	.0703	.0013	.3570	.0013	.0109
-	.0266	.0132	.0350	.0660	.0037	.0182	.0152	.7123	.0553	.2205	-	-
-	.951	.0510	.0062	.0029	-	-	-	.0029	-	.0047	-	-
-	-	.0012	.0384	.2167	.0746	.2356	.0040	.0132	.0164	.0038	-	-
-	-	-	-	.0088	.4903	.2503	-	-	-	-	-	-
-	-	-	.0039	.0128	.0017	.0034	-	.0562	.996	.0459	-	-

Eigenvalues: $\lambda_1 = .9373$, $\lambda_2 = .8266$, $\lambda_3 = .7695$, $\lambda_{4,5} = .5066 \pm j.0669$, $\lambda_{6,7} = .1497 \pm j.0775$,
$\lambda_8 = .1557$, $\lambda_{9,10} = .1181 \pm j.0097$, $\lambda_{11} = .0209$, $\lambda_{12} = .8998 \times 10^{-2}$,
$\lambda_{13} = .6306 \times 10^{-2}$, $\lambda_{14} = .2697 \times 10^{-2}$, $\lambda_{15} = -.3662 \times 10^{-5}$, $\lambda_{16} = -.6549 \times 10^{-6}$.

TABLE 2.4 Absolute value of participation matrix and eigenvalues of Example 2.

State Variable	Associated Modes
x_1 = fan speed	λ_4 and λ_5
x_2 = compressor speed	λ_3
x_3 = compressor discharge pressure	λ_{16}
x_4 = interturbine volume pressure	λ_{15}
x_5 = augmentor pressure	λ_4 and λ_5
x_6 = fan inside diameter discharge temperature	λ_6 and λ_7
x_7 = duct temperature	λ_9 and λ_{10}
x_8 = compressor discharge temperature	λ_{14}
x_9 = burner exit fast response temperature	λ_9 and λ_{10}
x_{10} = burner exit slow response temperature	λ_1
x_{11} = burner exit total temperature	λ_{11}
x_{12} = fan turbine inlet fast response temperature	λ_{12}
x_{13} = fan turbine inlet slow response temperature	λ_2
x_{14} = fan turbine exit temperature	λ_6 and λ_7
x_{15} = duct inlet temperature	λ_8
x_{16} = duct exit temperature	λ_{13}

TABLE 2.5 State variables and associated modes of Example 2.

On the other hand, for any model which retains certain modes, the corresponding states will appear in the new model. For example, a fifth-order model preserving $\lambda_1,\lambda_2,\lambda_3,\lambda_4$ and λ_5 will have $x_{10},x_{13},x_2,x_1,x_5$ (or a combination of these) as the state vector of the new model.

2.6 Concluding Remarks

The objective of this chapter has been the development of different methods of modelling the behaviour of linear, discrete systems. In terms of the input-output relationship, we have presented the concept of transfer function to express the system mechanism in the complex z-domain. When the system is described by difference equations which relate the output sequence to the input sequence, we have discussed the nature of the solutions and the relationship to transfer functions has been pointed out. We then moved on to developing discrete state variable models, starting from either transfer functions or difference equations. Finally, we discussed modal decomposition techniques and examined the system modes. The material covered in this chapter provides the basic starting step for all subsequent chapters.

2.7 Problems

1. Indicate whether the discrete systems represented by the following characteristic polynomials are stable, unstable or marginally stable.
   (a) $D(z) = z^4-.9z$
   (b) $D(z) = z^4-1$
   (c) $D(z) = z^3+1.2z^2-.6z-.8$
   (d) $D(z) = z^4-.3z^3-.15z^2-.075z-.1$
   (e) $D(z) = z^3-.5z^2+.4z-.2$

2. Apply the Jury test to examine the stability of the discrete systems whose $D(z)$ are given below:
   (a) $D(z) = z^7-.8z^6+.4z^5-.32z^4-.0625z^3+.05z^2-.025z+.02$
   (b) $D(z) = z^6-.4z^5+.35z^4-.65z^3+.1z^2-.02z+.01$

3. Use the Routh-Hurwitz test to evaluate the stability of the systems described by the transformed characteristic polynomials in the w-plane:
   (a) $F(w) = w^9+w^8+4w^7+8w^6+5w^5+5w^4+7w^2+2w+17$
   (b) $F(w) = w^6+9w^5+36w^4+86w^3+125w^2+97w+30$
   (c) $F(w) = w^5+3w^4+2w^3+6w^2+w+3$

4. Determine the z-transform of the following sequences:
   (a) $g(k) = k^3$ (b) $g(k) = |k-4|$
   (c) $g(k) = \sum_{n=1}^{k} n^2$ (d) $g(k) = (k+2)(k+1)k$

5. Find the inverse z-transforms of the following complex functions:
   (a) $G(z) = \dfrac{2z+3}{(z-2)^3}$ (d) $G(z) = \dfrac{4}{z^3(2z-1)}$
   (c) $G(z) = \dfrac{2z^2-2z}{(z-3)(z-5)^2}$ (d) $G(z) = \dfrac{30z^2-12z}{6z^2-5z+1}$

6. Consider the transfer function

$$G(z) = \frac{z^2-1.77z+.782}{z^3-2.65z^2+2.535z-.684}$$

Obtain a second-order model that approximates the first four time moments. Validate the results by comparing the step responses of the original and approximate transfer functions.

7. The transfer function

$$G(z) = \frac{8z^2-15.283z+7.313}{z^3-2.628z^2+2.3z-.67}$$

was treated in Example 3, Section 2.2, by model simplification techniques. Develop three second-order approximate models with the following specifications:

(i) to fit the first two time-moments and the first two Markov parameters;

(ii) to fit the first four time-moments;

(iii) to fit the first four Markov parameters.

Are all three models stable? Plot the step response of the stable ones in comparison with the original model.

8. A direct digital control system is described by

$$G(z) = \frac{1.682z^8+1.116z^7-.21z^6+.152z^5-.516z^4-.262z^3-.044z^2-.006z}{z^9-.159z^8-.283z^7-.663z^6+.181z^5-.334z^4-.127z^3-.153z^2-.025z+.003}$$

Using $w = (z-1)/(z+1)$, obtain $G(w)$. Expand $G(w)$ by a continued fraction expansion into the second Cauer form. Truncate the expansion after four and six terms to yield second-order and third-order approximate models, respectively. Plot the step response of $G(z), G_2(z)$ and $G_3(z)$ and evaluate your results.

9. Solve the following difference equations:

(a) $y(k+2)-2y(k+1)+y(k) = k+4$

(b) $y(k+3)-6y(k+2)+11y(k+1)-6y(k) = 3^k+3$

(c) $y(k+3)-3y(k+2)+3y(k+1)-y(k) \;=\; k^3+k2^k$

(Hint: consider that all boundary conditions are zero.)

10. Draw a simulation diagram for the third-order difference equation:

$$y(k+3) + 3y(k+2) + 4y(k+1) + y(k)$$
$$= 2u(k+3) + 3u(k+2) + u(k+1) + 2u(k)$$

11. Find the transfer function matrix of the system:

$$y_1(k+3) + 6y_1(k+2) + 11y_1(k+1) + 6y_1(k)$$
$$= u_1(k+1) + u_1(k) + u_2(k)$$

$$y_2(k+2) + 5y_2(k+1) + 6y_2(k)$$
$$= u_2(k+1) + u_2(k)$$

using the Z-transform method.

12. Compute the transition matrix of the free systems described by:

(a) $A = \begin{pmatrix} 0 & 1 \\ -6 & -5 \end{pmatrix}$ (b) $A = \begin{pmatrix} 0 & 1 \\ -1 & -2 \end{pmatrix}$

(c) $A = \begin{pmatrix} 0 & 1 \\ -2 & 2 \end{pmatrix}$ (d) $A = \begin{pmatrix} .5 & 0 \\ 0 & .8 \end{pmatrix}$

13. Determine the solution to the difference equation:

$$y(k+2) + 5y(k+1) + 6y(k) = 1$$

by various methods.

14. Set up the state equations for the system shown below in different ways.

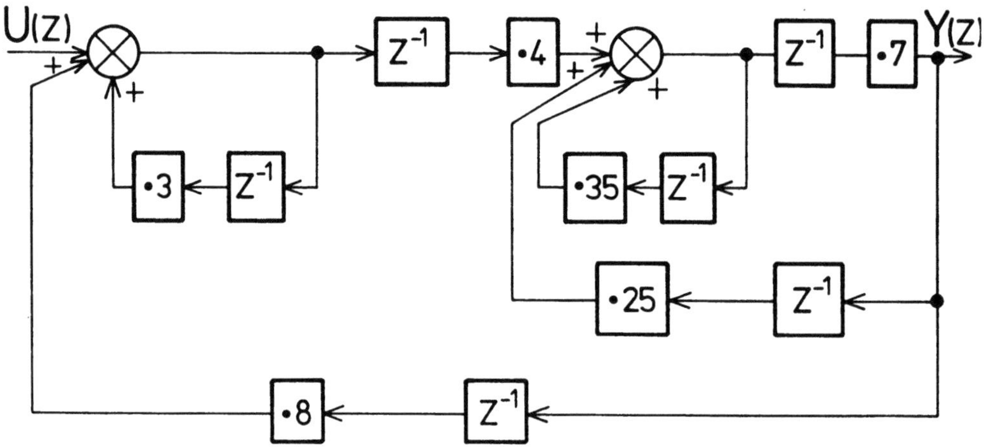

15. Solve the third-order model

$$\underline{x}(k+1) = \begin{pmatrix} .5 & .5 & 0 \\ .5 & .25 & .25 \\ 0 & .5 & .5 \end{pmatrix} \underline{X}(k)$$

for $k \geq 0$. What will be the value of $\underline{x}(k)$ as $k \to \infty$?

16. Find the Jordan form for the matrix

$$A = \begin{pmatrix} 0 & 1 & 0 \\ 0 & 0 & 1 \\ -6 & -11 & -6 \end{pmatrix}$$

17. Find the eigenvalues, the modal matrix and the Jordan form of the matrix

$$A = \begin{pmatrix} 5 & 19 & 9 \\ 0 & 0 & 1 \\ -1 & -4 & -2 \end{pmatrix}$$

18. An insect population model is described by:

$$\begin{pmatrix} n_1(k+1) \\ n_2(k+1) \\ n_3(k+1) \end{pmatrix} = \begin{pmatrix} 0 & 0 & b \\ s_1 & 0 & 0 \\ 0 & s_2 & 0 \end{pmatrix} \begin{pmatrix} n_1(k) \\ n_2(k) \\ n_3(k) \end{pmatrix}$$

where $n_j(m)$ is the number of insects between ages j and j+1 at the m-th period. The parameters s_1, s_2, b represent the survival rate of the second group, the survival of the third group and the reproduction rate of the insect population, respectively.

(a) Determine the eigenvalues and discuss the modes of this system (note $s_1, s_2, b < 1$).

(b) What will be the size of the population groups for large values of k? Consider that $n_1(0)$, $n_2(0)$, $n_3(0)$ are nonzero constants.

19. A steam power system model has the form

$$\underline{x}(k+1) = \begin{pmatrix} .915 & .204 & .152 & .30 & .152 \\ -.0075 & .889 & -.0005 & .023 & .111 \\ -.0015 & .468 & .247 & .007 & .948 \\ -.3575 & -.044 & -.042 & .24 & -.048 \\ -.037 & -.003 & -.004 & .045 & -.026 \end{pmatrix} \underline{x}(k)$$

Determine the dynamic pattern of association of this model.

2.8 References

[1] Jury, E.I.
"Sampled-Data Control Systems",
Wiley, New York, 1958.

[2] Cadzow, J.S. and H.R. Martens
"Discrete-Time and Computer Control Systems",
Prentice-Hall, New Jersey, 1970.

[3] Jury, E.I.
"Theory and Application of the Z-Transform Method",
Wiley, New York, 1964.

[4] Ragazzini, J.R. and G.F. Franklin
"Sampled-Data Control Systems",
McGraw-Hill, New York, 1958.

[5] Tou, J.T.
"Digital and Sampled-Data Control Systems",
McGraw-Hill, New York, 1959.

[6] Churchill, R.V.
"Introduction to Complex Variables and Applications",
McGraw-Hill, New York, 1948.

[7] Kuo, B.J.
"Analysis and Synthesis of Sampled-Data Control Systems", Prentice-Hall, New Jersey, 1963.

[8] Kaplan, W.
"Advanced Calculus",
Addison-Wesley, Cambridge, 1952.

[9] Shinners, S.M.
"Control System Design",
Wiley, New York, 1964.

[10] Jury, E.I.
"A Simplified Stability Criterion for Linear Discrete Systems",
Proc. IRE, Vol. 50, pp. 1493-1500, 1962.

[11] Jury, E.I. and J. Blanchard,
"A Stability Test for Linear Discrete Systems in Table Form",
Proc. IRE, Vol. 49, pp. 1947-1948, 1961.

[12] Mahmoud, M.S. and M.G. Singh
"Large-Scale Systems Modelling",
Pergamon Press, Oxford, 1981.

[13] Bosley, M.J. and F.P. Lee
"A Survey of Simple Transfer-Function Derivations from Higher-Order State-Variable Models", Automatica, Vol. 8, pp. 765-775, 1972.

[14] Hickin, J. and N.K. Sinha
"Model Reduction for Linear Multivariable Systems", IEEE Trans. Automat. Contr., Vol. AC-25, pp. 1121-1127, 1980.

[15] Shamash, Y.
"Continued Fraction Methods for the Reduction of Discrete-Time Dynamic Systems", Int. J. Control, Vol. 20, pp. 267-275, 1974.

[16] Shih, Y.P. and W.T. Wu
"Simplification of Z-Transfer Functions by Continued Fractions", Int. J. Control, vol. 17, pp. 1089-1094, 1973.

[17] Miller, K.S.
"An Introduction to the Calculus of Finite Differences and Difference Equations", Holt, Rinehart and Winston, New York, 1960.

[18] Bishop, A.B.
"Introduction to Discrete Linear Controls", Academic Press, New York, 1975.

[19] Kalman, R.E.
"Mathematical Description of Linear Dynamical Systems", J. SIAM Control, Vol. 1, pp. 152-192, 1963.

[20] Zadeh, L.A. and C.A. Desoer
"Linear System Theory : The State Space Approach", McGraw-Hill, New York, 1963.

[21] Chen, C.T.
"Introduction to Linear System Theory", Holt, Rinehart and Winston, Inc., New York, 1970.

[22] Noble, B.
"Applied Linear Algebra", Prentice-Hall, New Jersey, 1969.

[23] Wilkinson, J.H.
"The Algebraic Eigenvalue Problem", Clarendon Press, Oxford, 1965.

[24] Wilkinson, J.H. and C. Reinsch
"Handbook for Automatic Computation", Vol. 2 (Linear Algebra), edited by F.L. Bauer, Springer-Verlag, New York, 1974.

[25] Smith, B.T., J.M. Boyle, B.S. Garbow, Y. Ikebe, V.C. Klema and C.B. Moler "Matrix Eigensystem Routines", Springer-Verlag, New York, 1974.

[26] Perez-Arriaga, I.J., G.C. Verghese and F.C. Schweppe "Selective Modal Analysis with Applications to Electric Power Systems, Part I : Heuristic Introduction", IEEE Trans. Power Appar. Systems, Vol. PAS-101, 1982, pp. 3117-3125.

[27] Verghese, G.C., I.J. Perez-Arriaga and F.C. Schweppe "Selective Modal Analysis with Applications to Electric Power Systems, Part II : The Dynamic Stability Problem", IEEE Trans. Power Appar. Systems, Vol. PAS-101, 1982, pp. 3126-3134.

[28] Verghese, G.C., I.J. Perez-Arriaga and F.C. Schweppe "Rational Approximation via Selective Modal Analysis" IEEE Trans. Circuits, Systems and Signal Processing, Vol. CSS-17, 1983.

[29] Perez-Arriaga, I.J., F.C. Schweppe and G.C.Verghese "Selective Modal Analysis : Basic Results", Proc. IEEE Conference on Circuits and Computers, N.Y., 1980, pp. 649-656.

[30] Perez-Arriaga, I.J., G.C. Verghese and F.C. Schweppe "Determination of Relevant State Variables for Selective Modal Analysis", Proc. JACC, VA, 1980, No. TA-4F.

[31] Cook, P.A. and M.M.M. Hassan "The Use of Model Following Methods to Simplify Linear Systems", Large Scale Systems, Vol. 2, 1981, pp. 123-142.

[32] Luenberger, D.G. "Introduction to Dynamic Systems", J. Wiley & Sons, N.Y., 1979.

Chapter 3

Structural Properties

3.1 Introduction

A fundamental prerequisite for the design of feedback control systems is the availability of full information about the structural properties of the discrete system under consideration. These properties are closely related to the concepts of controllability, observability and stability. In this chapter we study such properties in detail, develop methods for their evaluation and then examine their role in determining the behaviour of linear feedback control systems. We shall restrict our discussions to the time-domain description using state-space analysis.

It has been shown previously that the description of a linear discrete system depends upon four matrices: the system matrix A, the input B, the output matrix C, and the feedforward matrix D. We have also seen that depending on the choice of state variables, or alternatively on the choice of the basis for the state space R_n, different matrices can be used to describe the same system (for instance, see example 3 in section 2.4.4). A particular set {A,B,C,D} is often called a *system representation* or *realisation*. It is readily seen, following the terminology of linear system theory [1], that the four matrices are actually representations of the transformations on the n-dimensional state space R_n, the m-dimensional input space R_m and the p-dimensional output R_p; that is

$$A : R_n \to R_n \qquad D : R_m \to R_p$$

$$B : R_m \to R_n$$

$$C : R_n \to R_p$$

The above relationships are illustrated in Fig. (3.1)

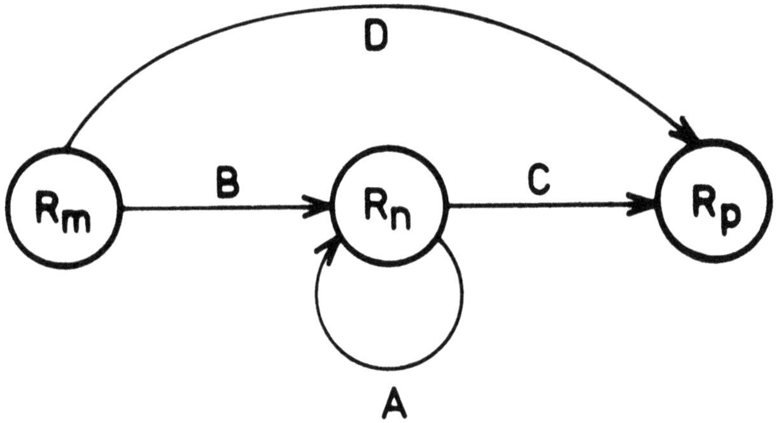

Fig. (3.1) System representation as transformation

Generally speaking, controllability is a property of the coupling between the input and state, and thus involves the matrices A and B and some functions of these. In a similar way, observability expresses the coupling between the state and the output, and is thus described in terms of the matrices A and C and some functions of these. Stability is a structural feature of the free dynamic system and is thus related to the system matrix A. The subsequent sections contain detailed discussions of these structural concepts.

3.2 Controllability

For the solution of linear control problems, it is important to know whether or not a given dynamic system has the property that it may be driven from any given state to any other given state. This leads to one of the key concepts in modern control theory;

the concept of *controllability* which was originally developed by Kalman [2].

3.2.1 BASIC DEFINITIONS

Consider the linear, time-invariant discrete system

$$\underline{x}(k+1) = A\underline{x}(k) + B\underline{u}(k) \tag{3.1}$$

where $\underline{x}(k)$ is the n-dimensional state vector, $\underline{u}(k)$ is the m-dimensional input vector and for simplicity we assume that the input matrix B is of full rank $m < n$. One should note that this is not a restrictive assumption since any linearly dependent columns of B corresponding to redundant inputs can always be eliminated. Let U represent a given control sequence that drives (3.1) from an initial state $\underline{x}_s$ to a desired one $\underline{x}_d$ where both $\underline{x}_s$ and $\underline{x}_d$ are fixed but arbitrary. We shall distinguish between two different situations:

1) The initial state $\underline{x}_s$ is the origin $\underline{O}$ of the state-space and U_1 is given by the sequence $\{\underline{u}(k-\theta+1),\underline{u}(k-\theta+2)\ldots,\underline{u}(k)\}$. This situation corresponds to *controllability from the origin* or *reachability*.

2) The desired state $\underline{x}_d$ is the origin $\underline{O}$ of the state-space and U_2 is given by the sequence $\{\underline{u}(k),\underline{u}(k+1),\ldots,\underline{u}(k+\theta-1)\}$ This situation corresponds to *controllability to the origin* or *controllability*.

In 1) and 2) above, θ is a given index which will be explained below. The following definitions clarify both situations.

Definition 3.1: (Weiss [3])

The linear system (3.1) is completely θ-step reachable if there exists a control sequence $U_1 = \{\underline{u}(k-\theta+1),\underline{u}(k-\theta+2),$

$\dots, \underline{u}(k)\}$ such that the state of (3.1) is transferred from the origin $\underline{O}$ to $\underline{x}(k)$ under the action of U_1 in θ time steps.

In order to translate this definition into a workable criterion, we recall that the solution of (3.1) is given by (2.93) in the form:

$$\underline{x}(k) = A^k \underline{x}(k_0) + \sum_{j=k_0}^{k-1} A^{k-j-1} B\underline{u}(j) \tag{3.2}$$

Let the initial discrete instant k_0 be equal to $k-\theta$ and set $\underline{x}(k_0) = \underline{x}(k-\theta) = \underline{O}$; then (3.2) after expansion and arrangement reduces to:

$$\underline{x}(k) = W_r \, \underline{U}_r(k) \tag{3.3}$$

where the reachability matrix W_r is defined by:

$$W_r = [B, \quad AB, \quad \dots, \quad A^{\theta-1}B] \tag{3.4a}$$

and

$$\underline{U}_r(k) = \begin{bmatrix} \underline{u}(k-1) \\ \underline{u}(k-2) \\ \cdot \\ \cdot \\ \cdot \\ \cdot \\ \underline{u}(k-\theta) \end{bmatrix} \tag{3.4b}$$

It has been established [3] that a necessary and sufficient condition for (3.1) to be completely θ-step reachable is that

$$\text{rank } [W_r] = n \quad \text{for some } \theta \leqslant n \tag{3.5}$$

To show this, let us first assume that (3.5) holds. Now, define an n-vector $\underline{d}(k)$ by the relation

$$\underline{U}_r(k) = W_r^t \underline{d}(k) \tag{3.6}$$

Then, from (3.3) and (3.6) we get

$$\underline{d}(k) = [W_r W_r^t]^{-1} \underline{x}(k) \tag{3.7}$$

and so, we can solve for $\underline{d}(k)$ and hence obtain, from (3.6), the appropriate sequence of controls needed to reach any given $\underline{x}(k)$, in θ-steps.

Alternatively, suppose rank $[W_r] < n$, but the system (3.1) is completely θ-step reachable. Then there exists a nonzero n-vector λ such that $\underline{\lambda}^t W_r = \underline{O}$. Upon premultiplying (3.3) by $\underline{\lambda}^t$ we arrive at

$$\underline{\lambda}^t \underline{x}(k) = 0$$

regardless of $\underline{u}(k)$. By virtue of the fact that (3.1) is completely θ-step reachable we choose $\underline{U}_r(k)$ such that $\underline{x}(k) = \underline{\lambda}(k)$. Hence, $\underline{\lambda}^t\underline{\lambda} = 0$ which contradicts the assumption that $\underline{\lambda} \neq \underline{O}$, and completes the demonstration of validity of (3.5) as a condition of reachability. The quantity θ is called the reachability index whose minimum value is $(n-m+1)$ where n and m are the number of state and control variables, respectively [3]. A useful interpretation of (3.3) is that the columns of W_r span the n-dimensional state space R_n, which in turn leads to the necessity of (3.5).

Definition 3.2: (Weiss [3])

> The linear system (3.1) is completely θ-step controllable if there exists a control sequence $U_2 = \{\underline{u}(k,\underline{u}(k+1),\ldots .,\underline{u}(k+ -1)\}$ such that the state of (3.1) is transferred from $\underline{x}(k)$ to the origin $\underline{O}$ under the action of U_2 in θ time steps.

It has been pointed out [3] that criterion (3.5) is also a

sufficient condition for complete θ-step controllability. However, it is not a necessary condition unless A is invertible. To demonstrate this fact, we consider the solution to (3.2) at time $(k+\theta)$ starting from the initial state $\underline{x}(k)$. Thus

$$\underline{x}(k+\theta) = A^{k+\theta}\underline{x}(k) + \sum_{j=k}^{k+\theta-1} A^{k+\theta-j-1} B\underline{u}(j)$$

$$= A^{k+\theta}\underline{x}(k) + A^{\theta-1} B\underline{u}(k) + A^{\theta-2} B\underline{u}(k+1)$$

$$+\dots+ B\underline{u}(k+\theta-1)$$

$$= A^{k+\theta}\underline{x}(k) + [A^{\theta-1}B, A^{\theta-2}B,\dots,B] \begin{bmatrix} \underline{u}(k) \\ \underline{u}(k+1) \\ \vdots \\ \underline{u}(k+\theta-1) \end{bmatrix} \tag{3.8}$$

It is clear from (3.8) that a necessary and sufficient condition for complete θ-step controllability would be

$$A^{k+\theta}\underline{x}(k) \in R[A^{\theta-1}B, A^{\theta-2}B,\dots.B] \tag{3.9}$$

where $R[\dots]$ denotes the range space, or the collection of all linear combinations, of the columns of $[A^{\theta-1}B, A^{\theta-2}B,\dots,B]$. In the case where A is nonsingular, we can write (3.8) with $\underline{x}(k+\theta) = \underline{O}$, as:

$$-\underline{x}(k) = A^{-k} W_c \underline{U}_c(k) \tag{3.10}$$

where W_c is the controllability matrix defined by

$$W_c = [A^{-1}B, A^{-2}B,\dots,A^{-\theta}B] \tag{3.11a}$$

and

$$\underline{U}_c(k) = \begin{bmatrix} \underline{u}(k) \\ \underline{u}(k+1) \\ \vdots \\ \underline{u}(k+\theta-1) \end{bmatrix} \tag{3.11b}$$

It is now easy to see, from the way we developed the notion of reachability, that a necessary and sufficient condition for (3.1) to be completely θ-step controllable is that

$$\text{rank } [W_c] = n \qquad \text{for some } \theta \leqq n \tag{3.12}$$

In this case, θ is called the controllability index. The comparison of (3.5) and (3.12) reveals that controllability and reachability are equivalent when the system matrix is nonsingular. The importance of the controllability concept is crucial when studying systems for which it is required to determine the control sequences that bring the states to a dead stop in at most n steps. This is known [4] as the "dead beat control problem" and will be examined in Chapter 4. On the other hand, the concept of reachability is important being the dual of another structural concept called the observability about which we shall say more in section 3.3. From now onwards if (3.12) is satisfied we will call (A,B) a controllable pair. Similarly, (A,B) is a reachable pair if (3.5) is satisfied.

In terms of the complex frequency-domain, the transfer function matrix of system (3.1) is $[zI-A]^{-1}B$ and we can therefore state that

The linear system (3.1) is completely reachable ($\theta=n$) if and only if the row vectors of $[zI-A]^{-1}B$ are linearly independent.

The above statement can be illustrated by showing its equivalence to condition (3.5) as follows. Consider the expansion of the transfer function

$$\begin{aligned} [zI-A]^{-1}B &= z^{-1}[I-z^{-1}A]^{-1}B \\ &= z^{-1}B + z^{-2}AB + z^{-3}A^2B + \dots \end{aligned} \tag{3.13}$$

It is clear from (3.13) that if all the rows of $[B\ AB\ A^2B\ \dots\ A^{n-1}B]$ are linearly independent, so are the rows of $[zI-A]^{-1}B$. Now if rank $[B\ AB\ \dots\ A^{n-1}B] < n$, by definition there exists a nonzero constant vector $\underline{\lambda}$ such that $\underline{\lambda}B = \underline{O}$, $\underline{\lambda}AB = \underline{O},\ldots,$ $\underline{\lambda}A^{n-1}B = \underline{O}$. From the Cayley-Hamilton theorem [8], it follows that $\underline{\lambda}A^kB = \underline{O}$ for $k = n, n+1, \ldots$. Hence from (3.13), we have $\underline{\lambda}[zI-A]^{-1}B = \underline{O}$. That is if $\text{rank}[B \;\; AB \;\ldots\; A^{n-1}B] < n$, then the rows of $[zI-A]^{-1}B$ are not linearly independent and this completes the demonstration.

When defining the concepts of controllability and reachability, the system matrix A has been assumed to be of the general form. We will now examine the related definitions of these basic concepts by using the eigenstructure of the system matrix.

3.2.2 *MODE-CONTROLLABILITY STRUCTURE*

In the sequel, we will use, unless otherwise stated, the common term controllability to mean both controllability from and to the origin. Recall from section 2.5.2 that the system (3.1) can be brought under the similarity transformation $\underline{x}(k) = V\underline{\alpha}(k)$ into the modal form:

$$\underline{\alpha}(k+1) = J\underline{\alpha}(k) + B_m\underline{u}(k) \tag{3.14}$$

where V is the generalised modal matrix, and

$$\begin{aligned} B_m &= V^{-1}B \\ &= W^tB \end{aligned} \tag{3.15}$$

is the (nxm) modal input matrix. For reasons to be explained shortly, B_m is called the mode-controllability matrix [5].
In order to investigate the controllability properties of (3.14) we consider different cases of eigenvalue distributions. In the first case when the system matrix A has n distinct eigenvalues $\{\lambda_1, \lambda_2, \ldots, \lambda_n\}$ then

$$J = \text{diag}\,\{\lambda_1, \lambda_2, \ldots, \lambda_n\} \tag{3.16}$$

and (3.13) can be written as n uncoupled scalar equations of the form

$$\alpha_i(k+1) = \lambda_i \alpha_i(k) + \sum_{j=1}^{m} b_{m,ij} u_j(k); \qquad i=1,\dots,n \tag{3.17}$$

It is evident from (3.16) that the jth control input $u_j(k)$ can influence the mode $\alpha_i(k)$ if and only if

$$b_{m,ij} = \underline{w}_i^t \, \underline{b}_j \neq 0 \tag{3.18}$$

where $\underline{w}_i$ is the ith column of W and $\underline{b}_j$ is the jth column of B. This means that the ith mode is controllable if and only if it is controllable through at least one control input, which leads us to state that

The linear discrete system (3.1) is controllable if and only if each of the n modes is controllable.

For the general case when the system matrix A has repeated eigenvalues, the Jordan canonical matrix J is formed by appropriate Jordan blocks. Depending on the eigenvalue distribution and the associated eigen-structure, this case may be broken into one of the following versions:

1) single eigenvalue with algebraic multiplicity n and associated with single Jordan block,

2) repeated eigenvalues with algebraic multiplicities $n_1,\dots,n_\mu$ and associated with μ distinct Jordan blocks, and

3) repeated eigenvalues with algebraic multiplicities $n_1,\dots,n_\mu$ and associated with μ nondistinct Jordan blocks. The main results of the three versions are summarised in Table 3.1, from which we draw the following

TABLE 3.1 MODE-CONTROLLABILITY STRUCTURE FOR REPEATED EIGENVALUES

	Algebraic Multiplicity	Jordan Canonical Form	Conditions for Mode Excitation	Controllability Conditions
1)	n for eigenvalue λ_1	$J = \begin{pmatrix} \lambda_1 & 1 & & 0 \\ & \ddots & \ddots & \\ & & \ddots & 1 \\ 0 & & & \lambda_1 \end{pmatrix}$ dimension nxn	$b_{m,rj} = \underline{w}_r^t \underline{b}_j \neq 0;$ $r < n$ $b_{m,kj} = \underline{w}_k^t \underline{b}_j = 0;$ $k = r+1, \ldots, n$	The Jordan matrix J is controllable if and only if $b_{m,nj} \neq 0$ (at least one input). The system (3.1) is controllable if J is controllable
2)	n_1 for λ_1, n_2 for λ_2, ..., n_μ for λ_μ; $\sum_j^{\mu} n_j = n$	$J = J_{n_1}(\lambda_1) \oplus J_{n_2}(\lambda_2) \oplus \ldots \oplus J_{n_\mu}(\lambda_\mu)$ $J_{n_j}(\lambda_j) = \begin{pmatrix} \lambda_j & 1 & & 0 \\ & \ddots & \ddots & \\ & & \ddots & 1 \\ 0 & & & \lambda_j \end{pmatrix}$ dimension $n_j x n_j$	For $i = 1, \ldots, \mu$ $b_{m,rj}^{(i)} = \underline{w}_r^{(i)t} \underline{b}_j^{(i)} \neq 0;$ $r \leq n_i$ $b_{m,kj}^{(i)} = \underline{w}_k^{(i)t} \underline{b}_j^{(i)} \neq 0;$ $k = r+1, \ldots, n_i$	The Jordan matrix J_i is controllable if and only if $b_{m,n_i j}^{(i)} \neq 0$ (at least one input). The system (3.1) is controllable if and only if each of the μ Jordan blocks is controllable.
3)	n_1 for λ_1, n_2 for λ_2, ..., n_μ for λ_μ; $\sum_j^{\mu} n_j = n$	$J = J(\lambda_1) \oplus J(\lambda_2) \oplus \ldots \oplus J(\lambda_\mu)$ $J(\lambda_j) \overset{\mu}{=} J_{n_1}(\lambda_j) \oplus J_{n_2}(\lambda_j) \ldots \oplus J_{n_\nu}(\lambda_j)$ $J_{n_j}(\lambda_j) = \begin{pmatrix} \lambda_j & 1 & & 0 \\ & \ddots & \ddots & \\ & & \ddots & 1 \\ 0 & & & \lambda_j \end{pmatrix}$ dimension $n_j x n_j$	For $J(\lambda_j)$; Same conditions as above are applied.	The Jordan matrix $J(\lambda_j)$ is controllable if and only if all the rows of the mode-controllability matrix which correspond to the last rows of its constituent Jordan blocks are linearly independent. The system (3.1) is controllable if and only if all constituent Jordan blocks are controllable.

conclusion

The linear discrete system (3.1) is controllable if and only if all the rows of the mode-controllability matrix which correspond to the last rows of Jordan blocks containing the same eigenvalue are linearly independent.

It is interesting to note [6] that the above condition implies that it is necessary for the controllability of the system (3.1) that the number of input variables, m, must satisfy the inequalities

$$m \geq e_j \quad ; \quad j=1,\ldots,\mu \tag{3.19}$$

where e_j is the number of Jordan blocks associated with the eigenvalue λ_j. In particular, for single-variable systems, condition (3.19) can be satisfied only if $e_j = 1$, $j=1,\ldots,\mu$. That is, if all the Jordan blocks of the Jordan canonical form of the system matrix A are distinct.

We emphasise that the usefulness of the mode-controllability structure lies in its ability to provide information about the modes which are accessible for excitation. Next, we examine the application of modal analysis to the reachability matrix in order to deduce the mode excitation pattern.

3.2.3 *MODAL ANALYSIS OF STATE REACHABILITY*

We recall that the application of condition (3.5) tells us whether the state of a linear discrete model is reachable or unreachable. Sometimes, this is called a state-controllability test. We note however that it does not provide any information about the system modes which are excited by control inputs. In this section, we shed some light on the properties of the state-reachability matrix W_r (the same applies for W_c when A is invertible) in terms of the eigen-structure of the system. For this purpose, we consider the input matrix B in the partitioned form

$$B = [\underline{b}_1 \quad \underline{b}_2 \quad \ldots \quad \underline{b}_m] \tag{3.20}$$

in accordance with the m input variables.

Using (3.20), we can write (3.1) as:

$$\underline{x}(k+1) = A\underline{x}(k) + \sum_{j=1}^{m} \underline{b}_j \, u_j(k) \tag{3.21}$$

where $u_j(k)$ is the jth component of the input vector $\underline{u}(k)$. Now, to examine the reachability characteristics of a given element $u_j(k)$, it is only necessary to study the properties of the (nxn) matrix

$$W_{r_j} = [\underline{b}_j \quad A\underline{b}_j \quad A^2\underline{b}_j \quad \ldots \quad A^{n-1}\underline{b}_j] \tag{3.22}$$

associated with the input variable $u_j(k)$. From section 2.5.2, we know that the system matrix A can be put in the modal form

$$\begin{aligned} J &= V^{-1}AV \\ &= W^tAV \end{aligned} \tag{3.23}$$

where V is the generalised modal matrix comprising the n linearly independent generalised eigenvectors. Therefore, we can express the jth column $\underline{b}_j$ in the form:

$$\underline{b}_j = \sum_{i=1}^{n} b_{m,ij}\underline{v}_i \tag{3.24}$$

where $b_{m,ij}$ (i,j=1,....,m) are the elements of the mode-controllability matrix B_m given by (3.15). Direct application of the theory of modal analysis [5] shows that

$$W_{r_j} = VM_jE \tag{3.25}$$

where E is the (nxn) matrix of the Vandemonde type and M_j

is the (nxn) matrix derived from the elements of B_m, the mode-controllability matrix. Explicit forms of M_j and E depend upon the eigen-structure of the system (3.21) and are summarised in Table 3.2. It should be noted that:

1) The matrix E is upper-triangular for which a typical non-zero element is given by

$$e_{ij} = \frac{(j-1)!\ \lambda_1^{j-i}}{(j-i)!\ (i-1)!} \quad (i \leqslant j, i, j=1, \ldots, n) \quad (3.26)$$

In general, (3.26) applies to the sub-matrix E_j in a truncated form.

2) In the case where A has distinct eigenvalues, both V and E have rank n. It thus follows from (3.25) that the state-reachability condition is given by

$$\text{rank } [W_{r_j}] = \text{rank } [M_j] \quad (3.27)$$

Since M_j is a diagonal matrix, (3.27) implies that rank $[W_{r_j}]$ is equal to the number of non-zero elements of M_j. By virtue of condition (3.18), we can thus state that:

The rank of the state-reachability matrix associated with the input variable $u_j(k)$ is equal to the number of modes of the system (3.18) which can be reached by that input.

3) When A has a single eigenvalue λ_1 with algebraic multiplicity n, the same condition (3.27) is applied with M_j being obtained from the second row of Table 3.2. In this case rank $[W_{r_j}]$ is equal to s_j where

$$b_{m,s_j j} \neq 0 \quad (1 \leqslant s_j \leqslant n) \quad (3.28a)$$

and

$$b_{m,k_j} = 0 \quad (k = s_j + 1, s_j + 2, \ldots, n) \quad (3.28b)$$

TABLE 3.2 DIFFERENT FORMS OF MATRICES RELATED TO THE MODAL FORM OF THE STATE-REACHABILITY MATRIX

Eigenvalue Distribution	Jordan Canonical Form	Matrix M_j	Matrix E
1) Distinct	$J = \begin{bmatrix} \lambda_1 & & 0 \\ & \ddots & \\ 0 & & \lambda_n \end{bmatrix}$	$M_j = \begin{bmatrix} b_{m,ij} & & & 0 \\ & b_{m,2j} & & \\ & & \ddots & \\ 0 & & & b_{m,nj} \end{bmatrix}$	$E = \begin{bmatrix} 1 & \lambda_1 & \lambda_1^2 \cdots \lambda_1^{n-1} \\ 1 & \lambda_2 & \lambda_2^2 \cdots \lambda_2^{n-1} \\ \cdot & \cdot & \cdots \\ \cdot & \cdot & \cdots \\ 1 & \lambda_n & \lambda_n^2 \quad \lambda_n^{n-1} \end{bmatrix}$
2) Repeated with algebraic multiplicity n for λ_1	$J = \begin{bmatrix} \lambda_1 & 1 & & 0 \\ & \ddots & \ddots & \\ & & \ddots & 1 \\ 0 & & & \lambda_1 \end{bmatrix}$	$M_j = \begin{bmatrix} b_{m,1j} & b_{m,2j} & \cdots & b_{m,nj} \\ b_{m,2j} & b_{m,3j} & \cdots & 0 \\ \cdot & \cdot & \cdots & \cdot \\ \cdot & \cdot & \cdots & \cdot \\ b_{m,nj} & 0 & & 0 \end{bmatrix}$	$E = \begin{bmatrix} 1 & 1 & 1^2 \cdots & 1^{n-1} \\ 0 & 1 & 2 \; 1 \cdots (n-1) & 1^{n-2} \\ 0 & 0 & 1 \cdots (n-1)(n-2) & 1^{n-3} \\ \vdots & \vdots & \vdots & \\ 0 & 0 & 0 & 1 \end{bmatrix}$
3) Repeated with algebraic mmultiplicity n_j for λ_2; $j=1,\ldots,\mu$	$J = J_{n_1}(\lambda_1) \oplus \ldots \oplus J_{n_\mu}(\lambda_\mu)$ $J_{n_j}(\lambda_j) = \begin{bmatrix} \lambda_j & 1 & & 0 \\ & \ddots & \ddots & \\ & & \ddots & 1 \\ 0 & & & \lambda_j \end{bmatrix}$	$M_j = M_j^{(1)} \oplus \ldots \oplus M_j^{(\mu)}$ $M_j^{(i)} = \begin{bmatrix} b_{m,1j}^{(i)} & b_{m,2j}^{(i)} & \cdots & b_{m,nj}^{(i)} \\ b_{m,2j}^{(i)} & b_{m,3j}^{(i)} & \cdots & 0 \\ \cdot & \cdot & \cdots & \cdot \\ \cdot & \cdot & \cdots & \cdot \\ b_{m,nj}^{(i)} & 0 & & 0 \end{bmatrix}$	$E = \begin{bmatrix} E_1 \\ E_2 \\ \vdots \\ E \end{bmatrix}$ $E_j = \begin{bmatrix} 1 & \lambda_j & \lambda_j^2 \cdots \lambda_j^{n-1} \\ 0 & 1 & 2\lambda_j \cdots (n-1)\lambda_j^{n-2} \\ 0 & 0 & 1 \cdots (n-1)(n-2)\lambda_j^{n-3} \\ \vdots & \vdots & \vdots \\ 0 & 0 & 0 \; \dfrac{(n-1)!\,\lambda_j^{n-n_j}}{(n-n_j)!\,(n_j-1)!} \end{bmatrix}$

4) By inspecting the third row of Table 3.2, it can be readily verified that the state-reachability condition takes the form

$$\text{rank } [W_{r_j}] = \sum_{k=1}^{\mu} M_j^{(k)} \tag{3.29}$$

which implies that

$$\text{rank } [W_{r_j}] = \sum_{k=1}^{\mu} s_{kj} \qquad (j=1,\ldots,m) \tag{3.30}$$

where

$$b^{(k)}_{m_{s_{kj},j}} \neq 0 \qquad (1 \leq s_{kj} \leq n_j; \quad k=1,2,\ldots,\mu) \tag{3.31a}$$

and

$$b^{(k)}_{m_{rj}} = 0 \qquad (r = s_{kj} + 1,\ s_{kj} + 2,\ldots,n_j; \quad k=1,2,\ldots,\mu) \tag{3.31b}$$

The above analysis provides an alternative way of checking the reachability of dynamic systems through the use of the properties of the mode-controllability matrix.

3.2.4 *SOME GEOMETRICAL ASPECTS*

It has been shown in section 3.2.1 that the pair (A,B) is completely controllable from the origin (reachable) if condition (3.5) is satisfied, which implies that the column vectors of W_r span the n-dimensional state space. As suggested by the mode-controllability structure, some of the system modes may not be reached by some inputs which would then imply that the system is not completely controllable. The purpose of this section is to analyse in some detail the structure of linear time-invariant systems that are not completely controllable. In this regard, it would be of interest to determine what part of the state space can be reached. This motivates the following definition.

Definition 3.3:

> The controllable subspace of the linear system (3.1) is the linear subspace consisting of the states that can be transferred from the origin to a desired state within a finite number of steps.

In the light of definition 3.1 and condition (3.5), it is obvious that the controllable subspace is the one spanned by the column vectors of the reachability matrix. A basic property of this subspace is that it is invariant under A, that is, if a vector $\underline{h}(k)$ is in the controllable subspace then $A\underline{h}(k)$ is also in this subspace. This fact can be demonstrated as follows. By definition, the controllable subspace is spanned by the column vectors of W_r. Thus $A\underline{h}(k)$, where $\underline{h}(k)$ is in the controllable subspace, that is the linear subspace spanned by the column vectors of $[AB, A^2B, \ldots A^{\theta}B]$. Since the column vectors of $A^{\theta}B$ depend linearly upon the column vectors of W_r, therefore $A\underline{h}(k)$ is in the subspace spanned by the column vectors of W_r. This means that $A\underline{h}(k)$ is in the controllable subspace and in turn this subspace is invariant under A.

Let us suppose that W_r has rank $s \leqslant n$, that is, W_r possesses s linearly independent column vectors. This means that the controllable subspace of the system (3.1) has dimension s. Choose the set of vectors $\{\underline{a}_1, \underline{a}_2, \ldots, \underline{a}_s\}$ as a basis for the controllable subspace, and let $\{\underline{a}_{s+1}, \underline{a}_{s+2}, \ldots, \underline{a}_n\}$ be a set of (n-s) linearly independent vectors. Thus the set $\{\underline{a}_1, \underline{a}_2, \ldots, \underline{a}_n\}$ spans the entire n-dimensional space. Now, the coordinate transformation

$$\begin{aligned} \underline{x}(k) &= M\underline{\alpha}(k) \\ &= [M_1 \quad M_2]\ \underline{\alpha}(k) \end{aligned} \tag{3.32}$$

where

$$M_1 = [\underline{a}_1 \quad \underline{a}_2 \; \cdots \; \underline{a}_s] \tag{3.33a}$$

$$M_2 = [\underline{a}_{s+1} \quad \underline{a}_{s+2} \; \cdots \; \underline{a}_n] \tag{3.33b}$$

converts the system (3.1) into

$$\underline{\alpha}(k+1) = M^{-1}AM\underline{\alpha}(k) + M^{-1}B\underline{u}(k) \tag{3.34}$$

We partition M^{-1} as follows

$$M^{-1} = \begin{pmatrix} N_1 \\ N_2 \end{pmatrix} \tag{3.35}$$

where the partitioning corresponds to that of M in the sense that N_1 has s rows and N_2 has $(n-s)$ rows. It thus follows that

$$M^{-1}M = \begin{pmatrix} N_1 \\ N_2 \end{pmatrix} [M_1 \quad M_2]$$

$$= \begin{pmatrix} N_1M_1 & N_1M_2 \\ N_2M_1 & N_2M_2 \end{pmatrix}$$

$$= \begin{pmatrix} I_s & 0 \\ 0 & I_{n-s} \end{pmatrix} \tag{3.36}$$

where I_j is the (jxj) identity matrix, and we conclude that

$$N_2M_1 = 0 \tag{3.37}$$

The importance of (3.37) lies in the fact that since the controllable subspace is invariant under A and contains all columns of M_1, thus:

$$N_2AM_1 = 0 \tag{3.38}$$

Moreover, since B is a part of the reachability matrix, all of its columns are in the controllable subspace and consequently we have

$$N_2B = O \tag{3.39}$$

The relevance of (3.38) and (3.39) is now clear. The matrices of the transformed system (3.34) can be written using (3.32) and (3.35) as:

$$M^{-1}AM = \begin{pmatrix} N_1 \\ N_2 \end{pmatrix} A \; [M_1 \quad M_2]$$

$$= \begin{pmatrix} N_1AM_1 & N_1AM_2 \\ N_2AM_1 & N_2AM_2 \end{pmatrix}$$

which with the aid of (3.38) reduces to

$$M^{-1}AM = \begin{pmatrix} N_1AM_1 & N_1AM_2 \\ O & N_2AM_2 \end{pmatrix} \tag{3.40}$$

and

$$M^{-1}B = \begin{pmatrix} N_1 \\ N_2 \end{pmatrix} B = \begin{pmatrix} N_1B \\ N_2B \end{pmatrix}$$

using (3.39). this becomes

$$M^{-1}B = \begin{pmatrix} N_1B \\ O \end{pmatrix} \tag{3.41}$$

In view of (3.40) and (3.41) we can partition the transformed state vector $\underline{\alpha}(k)$ as

$$M^{-1}B = \begin{pmatrix} \underline{\alpha}_1(k) \\ \underline{\alpha}_2(k) \end{pmatrix} \tag{3.42}$$

where $\underline{\alpha}_1(k)$ has dimension s and $\underline{\alpha}_2(k)$ has dimension (n-s). The fact that $\underline{\alpha}_2(k)$ behaves completely independently while $\underline{\alpha}_1(k)$ is influenced by both $\underline{\alpha}_2(k)$ and $\underline{u}(k)$ means that any state of the form $(\underline{\alpha}_0,\underline{0})$ lies in the controllable subspace of the system

$$\underline{\alpha}(k+1) = \begin{pmatrix} A_1 & A_2 \\ 0 & A_4 \end{pmatrix} \underline{\alpha}(k) + \begin{pmatrix} B_1 \\ 0 \end{pmatrix} \underline{u}(k) \qquad (3.43)$$

where

$$A_1 = N_1 A M_1 \qquad (3.44a)$$

$$A_2 = N_1 A M_2 \qquad (3.44b)$$

$$A_4 = N_2 A M_2 \qquad (3.44c)$$

$$B_1 = N_1 B \qquad (3.44d)$$

Based on this fact, it has been proved *[7]* that:

The pair (A_1,B_1) is completely reachable

Discrete models of the type (3.43) are frequently called the reachability canonical forms and in general, due to the arbitrary choice of the coordinate transformation (3.32), they are not unique. However, it can be easily verified that no matter how the coordinate transformation is chosen, the eigenvalues of both A_1 and A_4 are always the same *[8]*. For this purpose, we refer to the eigenvalues of A_1 as the reachable modes of the system and to the eigenvalues of A_4 as the unreachable modes. It should be emphasised that this agrees with the results of section 3.2.3. Consequently we can state that:

The controllable subspace of the system (3.43) is spanned by the characteristic vectors corresponding to the reachable modes of the system.

It is easy to see that the uncontrollable subspace of the system (3.1) is spanned by the characteristic vectors corresponding to the unreachable modes. We now examine further the role of the unreachable modes. In section 2.2.3, we found that a discrete system is asymptotically stable if all of its characteristic values are in the unit disc in the complex plane. For systems that are not asymptotically stable, it is convenient to refer to those eigenvalues with magnitude less than one as the stable eigenvalues and to the remaining ones as the unstable eigenvalues. Let the system matrix A have μ eigenvalues $\lambda_1,\ldots,\lambda_\mu$ with algebraic multiplicities $n_1,\ldots,n_\mu$; respectively. Express the initial state $\underline{x}(0)$ of the system (3.1) as

$$\underline{x}(0) = \sum_{j=1}^{\mu} \underline{m}_j \tag{3.45a}$$

with

$$\underline{m}_j \in N_j = N\{[A-\lambda_j I]^{n_j}\} \tag{3.45b}$$

where $N\{H\}$ is the null space of H. It is easy to see that

$$R_n = N_1 \oplus \cdots \oplus N_\mu \tag{3.45c}$$

From section 2.5.2 and using (3.45a), the system response can be written as:

$$\underline{x}(k) = \sum_{i=1}^{\mu} \underline{v}_i J_i^k \underline{w}_i^k \underline{m}_i \tag{3.46}$$

where J_i is the Jordan block associated with the eigenvalue λ_i. In the light of the discussion of section 2.5.2, the behaviour of the factor J_i^k is determined by the eigenvalue λ_i; only if $|\lambda_i| < 1$ does the corresponding component of the state approach the origin. This leads to the following definition:

Definition 3.4:

The stable subspace for the system (3.1) is the real subspace of the direct sum of those null spaces that correspond to eigenvalues with magnitudes less than one. Similarly, the unstable space is the real subspace of the direct sum of those null spaces that correspond to eigenvalues with magnitudes greater than or equal to one.

The importance of this definition arises from the fact that any initial state $\underline{x}(0)$ can now be uniquely written as

$$\underline{x}(0) = \underline{x}_s(0) + \underline{x}_u(0) \tag{3.47}$$

where $\underline{x}_s(0)$ lies in the stable subspace and $\underline{x}_u(0)$ is in the unstable subspace. For control design, we require that the unstable component be completely controlled which is the case if $\underline{x}_u(0)$ is in the controllable subspace. This leads to the notion of stabilisability *[9]* and we thus state that

The linear time-invariant system (3.1) is stabilisable if its unstable subspace is contained in its controllable subspace.

The above statement implies that any vector $\underline{h}(k)$ in the unstable subspace is also in the controllable subspace. It is obvious that any completely reachable system is stabilisable. Now if the system (3.1) is transformed into the form (3.43) and all the qualifications hold, then we can state that:

The linear time-invariant system (3.1) is stabilisable if and only if the pair (A_1, A_2) is completely reachable and all the eigenvalues of the matrix A_4 have moduli strictly less than one.

The various concepts developed in this section will now be illustrated on some examples.

3.2.5 *EXAMPLES*

Three examples are worked out and each one serves a particular purpose:

Example 1

It is required to investigate the reachability and mode controllability properties of the third order system

$$\underline{x}(k+1) = \begin{bmatrix} .5 & .5 & 0 \\ 0 & 1 & 0 \\ .833 & -2.167 & -.333 \end{bmatrix} \underline{x}(k) + \begin{bmatrix} 3 & 1 \\ 2 & 0 \\ -1 & 1 \end{bmatrix} \underline{u}(k)$$

The reachability matrix W_r is

$$W_r = \begin{bmatrix} 3 & 1 & \vdots & 2.5 & .5 & \vdots & 2.25 & .25 \\ 2 & 0 & \vdots & 2 & 0 & \vdots & 2 & 0 \\ -1 & 1 & \vdots & -1.5 & .5 & \vdots & -1.75 & .25 \end{bmatrix}$$

It is readily seen that the rank of W_r is not 3, since subtracting row 3 from row 1 yields twice row 2. This means that condition (3.5) is not satisfied and therefore the system is not completely reachable.

Now, to examine the mode-controllability matrix, the eigenvalues and eigenvectors must be determined:

$$\det[A-\lambda I] = (1-\lambda)(\tfrac{1}{2}-\lambda)(\tfrac{1}{3}-\lambda)$$

yields distinct eigenvalues. The eigenvectors are

$$\underline{v}_1 = \begin{bmatrix} 1 \\ 1 \\ -1 \end{bmatrix} \qquad \underline{v}_2 = \begin{bmatrix} 1 \\ 0 \\ 1 \end{bmatrix} \qquad \underline{v}_3 = \begin{bmatrix} 0 \\ 0 \\ 1 \end{bmatrix}$$

so that

$$V = \begin{bmatrix} 1 & 1 & 0 \\ 1 & 0 & 0 \\ -1 & 1 & 1 \end{bmatrix} \quad \text{and} \quad V^{-1} = \begin{bmatrix} 0 & 1 & 0 \\ 1 & -1 & 0 \\ -1 & 2 & 1 \end{bmatrix}$$

From (3.15), the mode-controllability matrix is given by

$$B_m = \begin{bmatrix} 2 & 0 \\ 1 & 1 \\ 0 & 0 \end{bmatrix}$$

Since A has distinct eigenvalues and the third row of B_m contains only zeros, no control can affect the third mode of this system. We thus conclude that the system is not completely controllable from the origin.

Example 2

Consider the second-order system

$$\underline{x}(k+1) = \begin{bmatrix} 1 & .632 \\ 0 & .368 \end{bmatrix} \underline{x}(k) + \begin{bmatrix} .368 \\ .632 \end{bmatrix} \underline{u}(k)$$

We wish to determine the control sequence that forces the system to the origin from the initial condition $[1 \quad 1]^t$.

First, we need to check the controllability of the system. It is easy to see that

$$A^{-1} = \begin{bmatrix} 1 & -1.718 \\ 0 & 2.718 \end{bmatrix}$$

hence the controllability matrix W_c is given by

$$W_c = \begin{bmatrix} .718 & 3.671 \\ -1.718 & -4.671 \end{bmatrix}$$

which has rank 2 and the system is completely controllable. Thus it can be forced to the origin in at most two-steps. Note that from (3.2) by setting $k_o = 0$, $k = 2$, $\underline{x}(2) = \underline{0}$ and multiplying both sides by A^{-2}, we can write

$$\begin{bmatrix} x_1(0) \\ x_2(0) \end{bmatrix} = W_c \begin{bmatrix} u(0) \\ u(1) \end{bmatrix}$$

from which we obtain

$$\begin{bmatrix} u(0) \\ u(1) \end{bmatrix} = W_c^{-1} \begin{bmatrix} 1 \\ 1 \end{bmatrix}$$

$$= \begin{bmatrix} -1.582 & -1.243 \\ .582 & .243 \end{bmatrix} \begin{bmatrix} 1 \\ 1 \end{bmatrix}$$

$$= \begin{bmatrix} -2.825 \\ .825 \end{bmatrix}$$

The above result shows that if the control input level is made -2.825 during the first period and then reduced to .825 during the second period, the system can be brought to the origin in two-steps.

Example 3

A second-order system is described by

$$\underline{x}(k+1) = \begin{bmatrix} .8 & -.2 \\ .3 & .3 \end{bmatrix} \underline{x}(k) + \begin{bmatrix} 1 \\ 1 \end{bmatrix} \underline{u}(k)$$

By constructing the reachability matrix

$$W_r = \begin{bmatrix} B & \vdots & AB \end{bmatrix}$$

$$= \begin{bmatrix} 1 & .6 \\ 1 & .6 \end{bmatrix}$$

it is readily seen that the system is not completely reachable since rank $[W_r] = 1$. To examine the stabilisability, we choose the coordinate transformation M as:

$$M = \begin{bmatrix} 1 & 1 \\ 1 & 0 \end{bmatrix} \quad \text{and } M^{-1} = \begin{bmatrix} 0 & 1 \\ 1 & -1 \end{bmatrix}$$

where the first column of M is the linearly independent column of W_r; the second column of M is chosen arbitrarily to make M nonsingular. From (3.40), (3.41) we obtain:

$$M^{-1}AM = \begin{bmatrix} .6 & .3 \\ 0 & .5 \end{bmatrix}$$

$$M^{-1}B = \begin{bmatrix} 1 \\ 0 \end{bmatrix}$$

By inspection, we see that the eigen-mode .6 is controllable whereas the eigen-mode .5 is not. However, this mode is stable and hence the system is stabilisable.

3.3 Observability

In this section, we study another structural problem in linear control systems. This problem is concerned with the conditions under which some information about output records can lead to a specification of the system state. According to Kalman [2], this problem reveals the concept of *observability* as a basic property of linear systems. Sometimes, the problem is called state-determination [10]. We shall see later the correspondence between the concepts of observability and controllability as established by the principle of duality [2].

3.3.1 *BASIC DEFINITIONS*

The appropriate model for our study of observability is:

$$\underline{x}(k+) = A\underline{x}(k) + B\underline{u}(k)$$
$$\underline{y}(k) = C\underline{x}(k) \qquad (3.48)$$

where $\underline{y}(k)$ is the p-dimensional output vector, $\underline{x}(k)$ and $\underline{u}(k)$ are as in (3.1) and the different matrices are defined by the transformations in section 3.1. Depending on the nature of the output records and the states to be determined, we have the following two definitions:

Definition 3.5: (Weiss [3])

The linear system (3.48) is completely N-step observable at time k_0 if and only if there exist integer numbers N such that knowledge of the output sequence $Y_a = \{\underline{y}(k_0), \underline{y}(k_0+1), \ldots, \underline{y}(k_0+N-1)\}$ and the input sequence $U_a = \{\underline{u}(k_0), \underline{u}(k_0+1), \ldots, \underline{u}(k_0+N-2)\}$ is sufficient to determine $\underline{x}(k_0)$.

It is readily seen from the observability definition that a past state (sometimes the initial state) can be specified from future records.

Definition 3.6: (Weiss [3])

The linear system (3.48) is completely N-step determinable at time k_0 if and only if there exist integer numbers N such that any state at time k_0 can be determined from knowledge of the output sequence $Y_b = \{\underline{y}(k_0-N+1), \ldots, \underline{y}(k_0)\}$ and input sequence $U_b = \{\underline{u}(k_0-N+1), \ldots, \underline{u}(k_0-1)\}$.

The definition of determinability above implies that the present state can be specified from past records. In principle, the definitions of determinability and observability are different. If the system (3.48) is observable (determinable) without any time designation, then it is completely observable (determinable). We now proceed to develop a mathematical criterion of observability that can be tested in practical situations.

Let the initial time and initial state of system (3.48) be k_o and $\underline{x}(k_o)$; respectively. The solution to (3.48) at the mth instant is then given by:

$$\underline{y}(m) = CA^{m-k_o}\,\underline{x}(k_o) + \sum_{j=k_o}^{m=1} CA^{m-j-1}\,B\underline{u}(j) \tag{3.49}$$

In terms of the modified output sequence

$$Y_N(k_o) = \begin{bmatrix} \tilde{\underline{y}}(k_o) \\ \tilde{\underline{y}}(k_o+1) \\ \vdots \\ \tilde{\underline{y}}(k_o+N-1) \end{bmatrix} \tag{3.50}$$

we write (3.49) in the form:

$$Y_N(k_o) = W_o^t \underline{x}(k_o) \tag{3.51}$$

where

$$\tilde{\underline{y}}(k) = \underline{y}(k) - \sum_{j=k_o}^{k-1} CA^{k-j-1}\,B\underline{u}(j) \tag{3.52}$$

and W_o is the observability matrix defined by:

$$W_o = [C^t, A^tC^t, \ldots, (A^t)^{N-1} C^t] \tag{3.53}$$

It has been shown [3] that a necessary and sufficient condition for the solvability of (3.51), which implies that (3.48) is N-step observable, is that

$$\text{rank } [W_o] = n \tag{3.54}$$

We note that the validity of (3.54) enables us to recover $\underline{x}(k_o)$

from the output sequence $\underline{Y}_N(k_o)$. This implies the sufficiency part of the observability condition. Now suppose that rank $[W_o]$ $<$ n but the system (3.48) is completely N-step observable at time k_o. Then there exists a nonzero n-vector $\underline{h}$ such that $\underline{h}^t W_o = 0$. Using (3.51) with $\underline{x}(k_o) = \underline{h}$ implies that $W_o = 0$ which contradicts the complete N-step observability assumption since the output will be identically zero. This shows the necessity part of the observability condition.

In view of definitions 3.5 and 3.6, we would expect the criterion of determinability to be similar to (3.54). This is partially true, but with an important difference. The criterion (3.54) is only sufficient for complete N-step determinability at k_o unless the matrix A is nonsingular. In the same way as in section 3.2, we can say that the criterion of determinability requires that

$$\text{rank } [W_d] = n \tag{3.55}$$

where the determinability matrix is defined by:

$$W_d = [A^{-t}C^t, \ldots, (A^t)^{-N} C^t] \tag{3.56}$$

At this stage, we should stress that both the controllability and determinability matrices, as given by (3.11a) and (3.56) respectively, require the nonsingularity of the matrix A. However, the reachability matrix (3.4a) and the observability matrix (3.53) do not require this condition.

3.3.2 *PRINCIPLE OF DUALITY*

We shall now discuss the relations between the concept of controllability (to or from the origin) and observability (using past or future output records). We shall introduce the principle of duality, due to Kalman [2], to clarify apparent analogies between the two structural concepts. Consider the system S_1

described by (3.48) for which the controllability, reachability, observability and determinability matrices are given by (3.11a), (3.4a), (3.53) and (3.56); respectively. Let another system S_2 be defined by

$$\underline{z}(k+1) = A^t\underline{z}(k) + C^t\underline{v}(k)$$
$$\underline{w}(k) = B^t\underline{z}(k) \tag{3.57}$$

where $\underline{z}(k)$, $\underline{v}(k)$, $\underline{w}(k)$ are n-, p- and m- dimensional vectors, representing the state, control and output respectively. In the light of section 3.2, we can see that the controllability conditions of system S_2 are:

1) The rank of the (nxnp) controllability matrix

$$\bar{W}_c = [A^{-t}C^t,\ldots\ldots,(A^t)^{-N}C^t] \tag{3.58}$$

must be n.

2) The rank of the (nxnp) reachability matrix

$$\bar{W}_r = [C^t,A^tC^t,\ldots\ldots,(A^t)^{N-1}C^t] \tag{3.59}$$

must be n.

From section 3.3.1, we can also see that the observability conditions of system S_2 are:

3) The rank of the (nxnm) observability matrix

$$\bar{W}_o = [B,AB,\ldots\ldots,A^{N-1}B] \tag{3.60}$$

must be n.

4) The rank of the (nxnm) determinability matrix

$$\bar{W}_d = [A^{-1}B,A^{-2}B,\ldots\ldots,A^{-N}B] \tag{3.61}$$

must be n.

A simple comparison of the conditions for systems S_1 and S_2 indicates that the pairings [(3.4a),(3.60)], [(3.11a),(3.61)], [(3.53(,(3.59)] and [(3.56,(3.58)] are identical. This verifies the principle of duality which can be stated as follows:

The system S_1 is completely controllable (determinable) if and only if system S_2 is completely determinable (controllable). And the system S_1 is completely reachable (observable) if and only if system S_2 is completely observable (reachable).

In the case where the principle of duality is valid for the two systems, they are termed *dual* to each other. The usefulness of this principle lies in the simplicity of checking the rank condition for one system or its dual. It should be emphasised that, a direct consequence of the duality principle, the pairing of reachability-observability and controllability-determinability as dual properties is very attractive, see Fig. (3.2). It allows us to use the results developed in section 3.2 for controllability via duality to cover determinability. This will be examined in the next section.

3.3.3 *MODE-OBSERVABILITY STRUCTURE*

In section 2.5.2, it was shown that the similarity transformation $\underline{x}(k) = V\_(k)$, where V is the generalised modal matrix, converts linear systems of the form (3.48) into:

$$\underline{\alpha}(k+1) = J\underline{\alpha}(k) + B_m\underline{u}(k) \qquad (3.62a)$$

$$\underline{y}(k) = C_m\underline{\alpha}(k) \qquad (3.62b)$$

where J is the Jordan canonical form,

B_m is the mode-controllability matrix, (see 3.15)

$C_m = CV$ is the mode-observability matrix

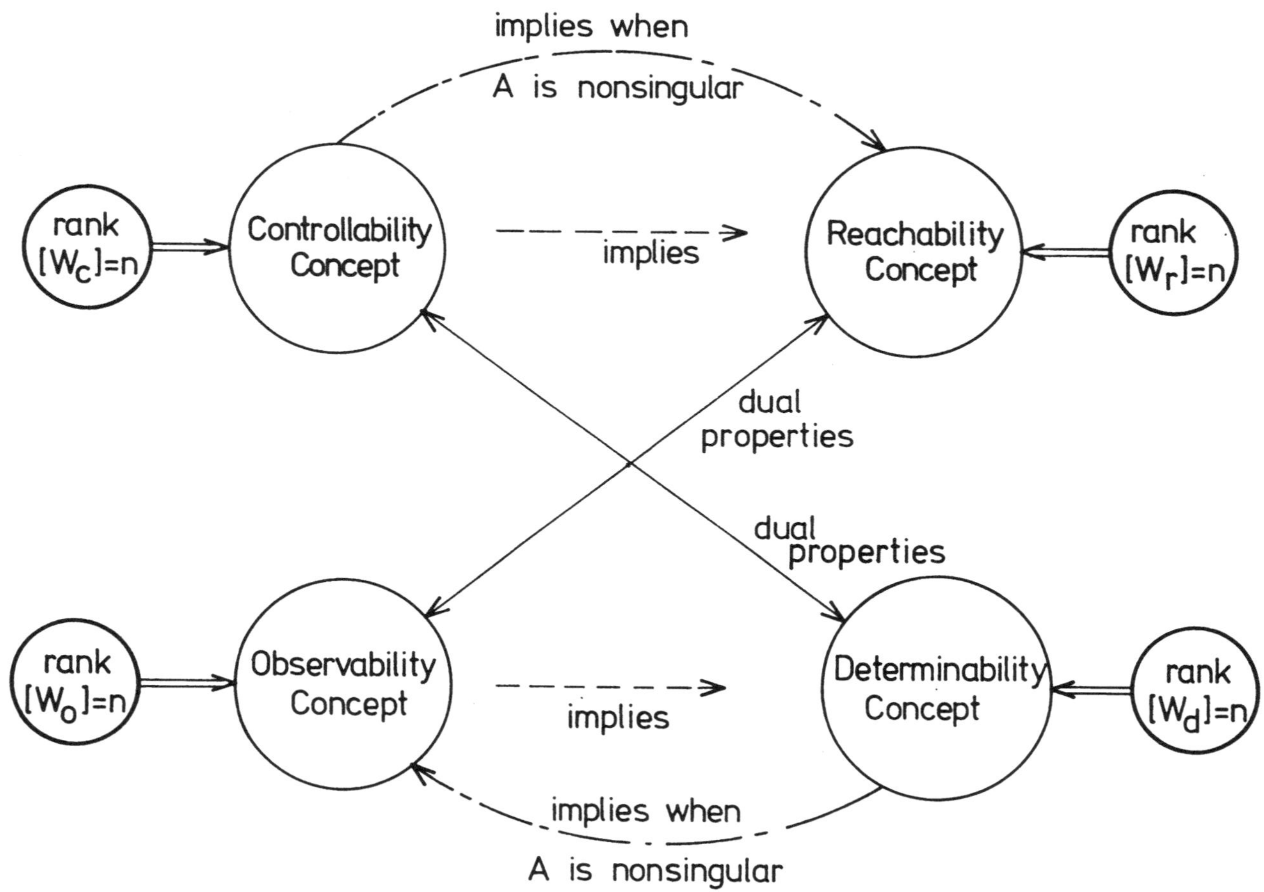

Fig. (3.2) Relationship between the structural concepts

It follows from (3.62b) that

$$y_j(k) = \sum_{i=1}^{n} C_{m,ij}\, \alpha_i(k) \quad ; \qquad j=1,\ldots,p \tag{3.63}$$

with

$$C_{m,ij} = \underline{c}_i^t \underline{v}_j \tag{3.64}$$

where $\underline{c}_i^t$ is the ith row vector of the output matrix and $\underline{v}_j$ is the jth column of the modal matrix. When the matrix A has distinct eigenvalues, then $\alpha_i(k)$ is decoupled from $\alpha_j(k)$. It is evident from (3.63) that the ith component $\alpha_i(k)$ of the modal state vector will contribute to the jth output variable $y_j(k)$ if and only if

$$C_{m,ij} \neq 0 \tag{3.65}$$

This means that the ith mode of (3.48) can be observed if condition (3.65) is satisfied by at least one output variable. Thus we have

The system (3.48) is observable if and only if each of the n modes is observable.

Now to relate this statement to the condition (3.54), we write the output matrix in the partitioned form

$$C^t = [\underline{c}_1^t, \underline{c}_2^t, \ldots, \underline{c}_p^t] \tag{3.66}$$

The observability characteristics of the jth output $y_j(k) = \underline{c}_j \underline{x}(k)$ can be determined by evaluating the rank of the (nxn) matrix

$$W_{O,j} = [\underline{c}_j^t, \underline{A}^t \underline{c}_j^t, \ldots, (A^t)^{N-1} \underline{c}_j^t] \tag{3.67}$$

But since $C^t = W^t C_m^t$, the manipulation of (3.67) using (3.64) and (3.66) results in:

$$W_{O,j} = W^t \overline{C}_{m,j} P \qquad (3.68)$$

where $\overline{C}_{m,j} = \mathrm{diag}\,[C_{m,1j}, C_{m,2j}, \ldots, C_{m,nj}]$

and P is the (nxn) Vandermonde matrix

$$P = \begin{bmatrix} 1 & \lambda_1 & \lambda_1^2 & \cdots & \lambda_1^{n-1} \\ 1 & \lambda_2 & \lambda_2^2 & \cdots & \lambda_2^{n-1} \\ \cdot & \cdot & \cdot & \cdots & \cdot \\ \cdot & \cdot & \cdot & \cdots & \cdot \\ 1 & \lambda_n & \lambda_n^2 & & \lambda_n^{n-1} \end{bmatrix} \qquad (3.69)$$

By virtue of the fact that W^t and P have rank n when A has distinct eigenvalues, it follows that

$$\mathrm{rank}\,[W_{O,j}] = \mathrm{rank}\,[\overline{C}_{m,j}] \qquad (3.70)$$

which clearly implies that rank $[W_{Oj}]$ is equal to the number of non-zero elements $c_{m,ij}$. Thus rank $[W_{Oj}]$ is equal to the number of modes of (3.48) which can be observed by the jth output. Let the number of modes be n_j, then from $C_m = CV$ and using (3.66) we have:

$$\underline{c}_j^t = \sum_{i=1}^{n_j} \underline{w}_i^t \, c_{m,ij} \qquad (3.71)$$

$$(A^t)^r \underline{c}_j^t = \sum_{i=1}^{n_j} c_{m,ij} \, \lambda_i^r \, \underline{w}_i^t \qquad (3.72)$$

Note that in this case rank $[W_{O,j}] = n_j$, thus from the Cayley-Hamilton theorem [8] we can write:

$$(A^t)^{n_j} \underline{c}_j^t = \sum_{k=0}^{n_j-1} \psi_{kj} (A^t)^k \, \underline{c}_j^t \qquad (3.73)$$

The manipulation of (3.72) and (3.73) yields

$$C_{m,ij}\lambda_i^{n_j} = \sum_{k=0}^{n_j-1} \psi_{kj} C_{m,ji}\lambda_i^k$$

But $C_{m,ji} \neq 0$, for $1 \leqslant i \leqslant n_j$, thus the above relation reduces to

$$\lambda^{n_j} = \sum_{k=0}^{n_j-1} \psi_{kj}\lambda^k \tag{3.74}$$

which means that the observable modes can be identified by determining the associated eigenvalues from the solution of (3.74).

The case of repeated eigenvalues can be readily dealt with by dualising the results of section 3.2.2, particularly tables 3.1 and 3.2.

3.3.4 *CONCEPT OF DETECTABILITY*

We have seen in section 3.3.2 that the state of linear discrete systems can be specified (reconstructed) from output records provided a rank condition on the pair (A, C^t) is satisfied. In the last section, it has been pointed out that not all the system modes can be observed from the output. Similarly to the analysis of section 3.2, it can be easily seen that the observability condition (3.54) implies that the row vectors of the observability matrix W_O span the n-dimensional space. Therefore if a system is not completely observable, it is never possible to completely determine the state from output records. We can thus have only partial information about the state.

Definition 3.7:

> The unobservable subspace of the linear discrete system (3.48) is the null space of the observability matrix W_O.

This definition suggests that linear systems can be split into two parts: the observable part and the unobservable part. Let us assume that W_O has rank $s \leqslant n$ meaning that it possesses only s linearly independent vectors. Define the (sxn) matrix H, such that its rows $[\underline{h}_1^t,....,\underline{h}_s^t]$ are the basis for the observable subspace, spanned by the rows of W_O. Let $H_2 = [\underline{h}_{s+1}^t,....,\underline{h}_n^t]$ be a set of linearly independent vectors which together with H_1 span the whole n-dimensional space. The application of the coordinate transformation

$$\underline{\alpha}(k) = \begin{bmatrix} H_1 \\ H_2 \end{bmatrix} \underline{x}(k) = H\underline{x}(k) \qquad (3.75)$$

to the system (3.48) converts it to:

$$\begin{aligned} \underline{\alpha}(k+1) &= H\ A\ H^{-1}\underline{\alpha}(k) + H\ B\underline{\alpha}(k) \\ \underline{y}(k) &= C\ H^{-1}\underline{\alpha}(k) \end{aligned} \qquad (3.76)$$

Let H^{-1} be partitioned into

$$H^{-1} = [F_1 \quad F_2] = F \qquad (3.77)$$

where F_1 and F_2 have dimensions (nxs) and (nxn-s); respectively. Since $HF=I_n$ we obtain:

$$H_1F_2 = O \qquad (3.78)$$

Recall that $H_1 = [\underline{h}_1^t,....,\underline{h}_s^t]$ are made up of linear combinations of the rows of W_O. Thus if $H_1\underline{z} = \underline{O}$, for any $\underline{z}$, then $W_O\underline{z} = 0$ implying that $\underline{z}$ is in the null (unobservable) subspace. In view of this fact, the interpretation of (3.78) is that all column vectors of F_2 must be in the unobservable subspace, and they form a basis for the subspace. In the manner of section 3.2.4, we can establish that the unobservable subspace is invariant under A. We then conclude that

$$H_1A\ F_2 = O \qquad (3.79a)$$

$$C\ F_2 = O \qquad (3.79b)$$

Using (3.75), (3.77) and (3.79) in (3.76), it becomes:

$$\begin{bmatrix} \underline{\alpha}_1(k+1) \\ \underline{\alpha}_2(k+1) \end{bmatrix} = \begin{bmatrix} A_1 & 0 \\ A_3 & A_4 \end{bmatrix} \begin{bmatrix} \underline{\alpha}_1(k) \\ \underline{\alpha}_2(k) \end{bmatrix} + \begin{bmatrix} B_1 \\ B_2 \end{bmatrix} \underline{u}(k)$$

$$\underline{y}(k) = [C_1 \quad 0] \begin{bmatrix} \underline{\alpha}_1(k) \\ \underline{\alpha}_2(k) \end{bmatrix} \qquad (3.80)$$

where

$A_1 = H_1AF_1$ of dimension (sxs)

$A_3 = H_2AF_1$ of dimension (n-sxs)

$A_4 = H_2AF_2$ of dimension (n-sxn-s)

$B_1 = H_1B$ of dimension (sxm)

$B_2 = H_2B$ of dimension (n-sxm)

$C_1 = CF_1$ of dimension (pxs)

It should be noted that the pair (A_1, C_1^t) is now completely observable. From (3.80) we see that the output records would not be of help in specifying the component $\underline{\alpha}_2(k)$. Moreover, the system (3.80) is lower triangular and consequently its eigen-values, which corresponds exactly to those of (3.48), are those of A_1 (observable modes) and A_4 (unobservable modes). To suppress the effects of unobservable modes, it is required that their zero-input response converge to zero. This requires in turn that any state in the unobservable subspace should lie also in the stable subspace of the system, and we have the following definition [8]:

The linear discrete system (3.48) is detectable if its unobservable subspace is contained within its stable subspace.

The importance of the concept of detectability lies in the ability to have accurate information about the system state provided that in the transformed structure (3.80), the characteristic value of the matrix A_4 have moduli strictly less than one.

We now illustrate the preceding analysis on some examples.

3.3.5 *EXAMPLES*

Example 1

Consider the third-order example of the form (3.48) with the following matrices:

$$A = \begin{bmatrix} .5 & .5 & 0 \\ 0 & 1 & 0 \\ .833 & -2.167 & -.333 \end{bmatrix}$$

$$B = \begin{bmatrix} 3 & 1 \\ 2 & 0 \\ -1 & 1 \end{bmatrix}; \quad C = \begin{bmatrix} -1 & 3 & 1 \\ 0 & 1 & 1 \end{bmatrix}$$

The same example was treated in section 3.2.5, where it was shown that $\lambda_1 = 1, \quad \lambda_2 = .5, \quad \lambda_3 = -.333$

$$V = \begin{bmatrix} 1 & 1 & 0 \\ 1 & 0 & 0 \\ -1 & 1 & 1 \end{bmatrix} \quad ; \quad W = V^{-1} = \begin{bmatrix} 0 & 1 & 0 \\ 1 & -1 & 0 \\ -1 & 2 & 1 \end{bmatrix}$$

The mode-observability matrix C_m is

$$C_m = CV$$

$$= \begin{bmatrix} 1 & 0 & 1 \\ 0 & 1 & 1 \end{bmatrix}$$

from which it can be deduced that $c_{m,11} \neq 0, c_{m,12} = 0,$

$C_{m,13} \neq 0$ which implies that the first and third modes are observable in the first output. Moreover, since $C_{m,21} = 0$ and $C_{m,22}$, $C_{m,23} \neq 0$ meaning that the first mode is unobservable by the second output. Simple calculation gives the mode-observability matrix (3.53) as:

$$W_O = \begin{bmatrix} -1 & 0 & \vdots & .333 & .833 & \vdots & -.111 & .139 \\ 3 & 1 & \vdots & .333 & -1.167 & \vdots & -.222 & -.028 \\ 1 & 1 & \vdots & -.333 & -.333 & \vdots & .111 & .111 \end{bmatrix}$$

which can easily be shown to have rank =3. This means that the system is completely observable.

Recall from example 1, section 3.2.5, that the same system was not completely controllable. By comparing the mode controllability and mode-observability matrices, B_m and C_m, it is evident that all modes are sensed at the output but some of them (the third mode) cannot be excited by the input.

Example 2

A discrete-system is described by the model

$$\underline{x}(k+1) = \begin{bmatrix} 0 & 1 & 0 \\ 0 & 0 & 1 \\ -1/4 & 1/4 & 1 \end{bmatrix} \underline{x}(k) + \begin{bmatrix} 1 & 3 \\ 0 & 2 \\ -1 & 0 \end{bmatrix} \underline{u}(k)$$

$$\underline{y}(k) = \begin{bmatrix} 2 & -3 & -2 \\ 2 & 3 & 1 \end{bmatrix} \underline{x}(k)$$

The characteristic equation is

$$\lambda^3 - \lambda^2 - 1/4\ \lambda + 1/4 = 0$$

which gives $\lambda_1 = 1$, $\lambda_2 = 1/2$ and $\lambda_3 = -1/3$.

For this system,

$$A^t C^T = \begin{bmatrix} -1/2 & 1/4 \\ 5/2 & 7/2 \\ -1 & 2 \end{bmatrix} \quad \text{and} \quad (A^t)^2 C^t = \begin{bmatrix} -1/4 & 1/2 \\ -1/4 & -1/4 \\ 7/2 & -1/4 \end{bmatrix}$$

so that it may be deduced from (3.67) that:

$$W_{O,1} = [\underline{c}_1^t, \ A^t \underline{c}_1^t, \ (A^t)^2 \underline{c}_2^t]$$

$$= \begin{bmatrix} 2 & -1/2 & -1/4 \\ -3 & 5/2 & -1/4 \\ -2 & -1 & 7/2 \end{bmatrix}$$

$$W_{O,2} = [\underline{c}_2^t, \ A^t \underline{c}_2^t, \ (A^t)^2 \underline{c}_2^t]$$

$$= \begin{bmatrix} 2 & 1/4 & 1/2 \\ 3 & 7/4 & 2 \\ 1 & 2 & -1/4 \end{bmatrix}$$

Since rank $[W_{O,1}]$ = rank $[W_{O,2}] = 3$, it follows that all the three modes are observed by the outputs and hence the system is completely observable.

3.4 Stability

3.4.1 *INTRODUCTION*

Stability of single-input, single-output shift-invariant systems was discussed in section 2.2.3. There the conditions for stability were given in terms of the pole locations of the input-output transfer functions. Classical methods of stability analysis were also presented.

The purpose of this section is to extend the previous stability concepts to multivariable systems described in terms of state variables. In particular we are interested in the overall time behaviour of discrete systems. Consider the general nonlinear state difference equation

$$\underline{x}(k+1) = \underline{g}[\underline{x}(k),\underline{u}(k),k] \tag{3.81}$$

An important property of the system is whether or not the solutions of the state difference equation diverge or tend to grow indefinitely as $k \to \infty$. There are many different definitions of "stability". A few of the more common definitions are given next, along with methods of investigating these. A small sample of the many works on stability may be found in [11-14].

3.4.2 *DEFINITIONS OF STABILITY*

For simplicity, we assume that we are dealing with an autonomous system, that is, a system without an input $\underline{u}(k)$ or, equivalently, a system where $\underline{u}(k)$ is a fixed time function. Thus we focus our attention on the system

$$\underline{x}(k+1) = \underline{f}\,[\underline{x}(k),k] \tag{3.82}$$

A particular point $\underline{x}_e$ is an *equilibrium state* of the system (3.82), if

$$\underline{x}_e = \underline{f}\,[\underline{x}_e,k] \quad \text{for all } k \tag{3.83}$$

The origin of the state space is always an equilibrium state for linear systems, although it need not be the only one. This can be observed from the linearised form of (3.83), $\underline{x}_e = A\underline{x}_e$, which means that a unity eigenvalue may be associated with an infinity of vectors. Without loss of generality, we assume in the sequel that $\underline{x}_e = \underline{0}$.

Stability deals with the following question: If at time k_0 the state is perturbed from its equilibrium position, does the state return to $\underline{x}_e$, or remain close to $\underline{x}_e$, or diverge from it? For discrete systems of the type (3.82) with initial state $\underline{x}(k_0) = \underline{x}_0$, we provide the following definitions [8,11,14]:

Definition 3.8:

The origin is a *Stable* equilibrium state if for any given value $\varepsilon > 0$ there exists a number $\delta(\varepsilon, k_0) > 0$ such that if $||\underline{x}_0|| < \delta$, then the resulting trajectory $\underline{x}(k)$ satisfies $||\underline{x}(k)|| < \varepsilon$ for all $k > k_0$.

This definition of stability is sometimes called *stability in the sense of Lyapunov [11]*. If a system possesses this type of stability, then we know that the state can be kept within ε, in the norm, of the origin by restricting the initial perturbation to be less than $\delta \leqslant \varepsilon$, in norm.

Definition 3.9:

The origin is an *asymptotically stable* equilibrium state if

(a) it is stable, and if in addition,

(b) there exists a number $\delta_1(k_0) > 0$ such that whenever $||\underline{x}(k_0)|| < \delta_1(k_0)$ the resulting trajectory satisfies $\lim_{k\to\infty} ||\underline{x}(k)|| = \underline{0}$.

If δ and δ_1 are not functions of k_0, then the origin is said to be *uniformly stable* and *uniformly asumptotically stable*, respectively. By making $\delta_1(k_0)$ arbitrarily large, meaning that all $\underline{x}_0$ converge to $\underline{0}$, the origin is said to be *globally* asymptotically stable *in the large*. In Fig. (3.3), we give a graphical representation of the foregoing definitions to clarify them. We use $S(\varepsilon)$ and $S(\delta)$ to represent spherical regions of radii ε and δ about the equilibrium state $\underline{x}_e$.

When the general system (3.81) is considered, two additional types of stability are often used.

Definition 3.10: (Bounded input-bounded state stability)

If there is a fixed, finite constant M such that $||\underline{u}(k)|| \leq M$ for all k, then the input is said to be bounded. If for every bounded input, and for arbitrary initial state $\underline{x}_0$, there exists a scalar $\delta(M,k_0,\underline{x}_0) > 0$ such that the resulting state satisfies $||\underline{x}|| \leq \delta$, then the system is *bounded input, bounded state* stable (BIBS stable). We note that the above three definitions of stability deal with the behaviour of the state vector relative to an equilibrium state. If the system output behaviour is of interest, we have the final stability definition.

Definition 3.11: (Bounded input, bounded output stability)

Let $\underline{u}(k)$ be a bounded input with N as the least upper bound. If there exists a scalar α such that for every k, the output satisfies $||\underline{y}(k)|| \leq \alpha N$, then the system is *bounded input, bounded output* stable (BIBO stable).

3.4.3 *LINEAR SYSTEM STABILITY*

We now specialise the preceding analysis to linear systems of the form

$$\underline{x}(k+1) = A(k)\,\underline{x}(k) + B(k)\,\underline{u}(k)$$

$$\underline{y}(k) = C(x)\underline{x}(k) \tag{3.84}$$

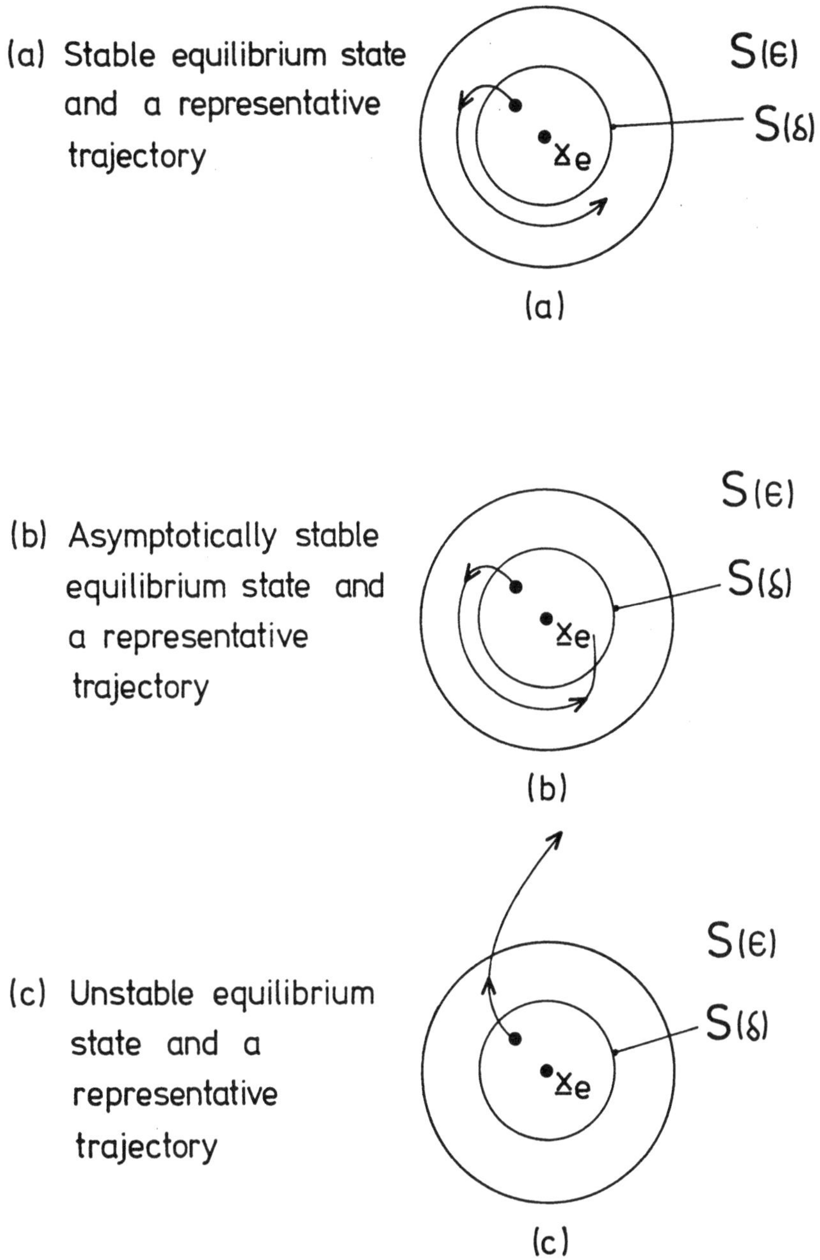
(a) Stable equilibrium state and a representative trajectory
S(ϵ)
S(δ)
$\underline{x}_e$
(a)
(b) Asymptotically stable equilibrium state and a representative trajectory
S(ϵ)
S(δ)
$\underline{x}_e$
(b)
(c) Unstable equilibrium state and a representative trajectory
S(ϵ)
S(δ)
$\underline{x}_e$
(c)

Fig. (3.3)

The free case is treated first. With $\underline{u}(k) = \underline{0}$, the state vector is given by

$$\underline{x}(k) = \Omega(k-1,k_0)\underline{x}_0 \tag{3.85}$$

where the real (nxn) matrix-valued function Ω is defined by

$$\begin{aligned} &\Omega(m,j) = A(m)A(m-1)\ldots.A(j+1)A(j) \\ &\Omega(m,m+1) \triangleq I_n \\ &\Omega(m,j) \quad \text{undefined for } j > m+1 \end{aligned} \tag{3.86}$$

Taking the norm of (3.85), we get:

$$\begin{aligned} ||\underline{x}(k)|| &= ||\Omega(k-1,k_0)\underline{x}_0|| \\ &\leqslant ||\Omega(k-1,k_0)||\,||\underline{x}_0|| \end{aligned} \tag{3.87}$$

Let $S(k_0)$ be defined by

$$||\Omega(k-1,k_0)|| \leqslant S(k_0) \qquad \text{for all } k \geqslant k_0 \tag{3.88}$$

It is easy to see that the conditions of Definition 3.8 can be satisfied for any $\varepsilon > 0$ by letting $\delta(k_0,\varepsilon) = \varepsilon/S(k_0)$. Note however that (3.88) is only sufficient to ensure that the origin is stable in the sense of Lyapunov. If (3.88) is valid and moreover let $||\Omega(k-1,k_0)|| \to 0$ for $k \to \infty$; then the origin is asymptotically stable.

For the forced case, the state vector is described by:

$$\underline{x}(k) = \Omega(k-1,k_0)\underline{x}_0 + \sum_{m=k_0}^{K-1} \Omega(k-1,m+1)B(m)\underline{u}(m) \tag{3.89}$$

In view of the fact that $\underline{u}(k) = \underline{0}$, is bounded it is clear that stability in the sense of Lyapunov is a necessary condition for BIBS stability. By taking the norm of (3.89)

$$||\underline{x}(k)|| \leq ||\Omega(k-1,k_0)\quad \underline{x}_0 + \sum_{m=k_0}^{K-1} ||\Omega(k-1,m+1)B(m)||\,||\underline{u}(m)|| \quad (3.90)$$

it is readily evident that $||\underline{x}(k)||$ remains bounded, and thus the origin is BIBS stable, if (3.88) holds and if, in addition. there exists a number $N_1(k_0)$ such that

$$\sum_{m=k_0}^{k-1} ||\Omega(k-1,m+1)B(m)|| \leq N_1(k_0) \quad \text{for all } k \geq k_0 \quad (3.91)$$

Similar results can be derived from BIBO stability. In particular, the output behaviour of system (3.84) is given in terms of the pulse response matrix $W(k,m)$ as

$$\underline{y}(k) = \sum_{j=-\infty}^{k} W(k,j)\underline{u}(j)$$

$$= \sum_{j=-\infty}^{k} C(k)\Omega(k-1,m+1)B(m)\underline{u}(j) \quad (3.91)$$

where we view the initial state $\underline{x}_0$ as having arisen because of a bounded input over the interval $(-\infty,k_0)$. A necessary and sufficient condition for BIBO stability is that

$$\sum_{j=-\infty}^{k} ||W(k,j)|| \leq M > 0 \quad \text{for all } k \quad (3.92)$$

In the case where the system (3.89) has a constant system matrix A, then the previous stability conditions can be expressed in terms of the eigenvalues of the system. The resulting conditions are summarised in Table 3.3.

TABLE 3.3 Stability Results
for Linear Constant Systems*

Stability Results	Patterns of Eigenvalues**
a) Unstable	1) if $\lvert\lambda_j\rvert > 1$ for any simple eigenvalue 2) if $\lvert\lambda_j\rvert \geq 1$ for any repeated eigenvalue
b) Stable in the sense of Lyapunov	1) if $\lvert\lambda\ \rvert \leq 1$ for all simple eigenvalues 2) if $\lvert\lambda_j\rvert < 1$ for all repeated eigenvalues
c) Asymptotically stable	$\lvert\lambda_j\rvert < 1$ for all eigen-values

* The model is $\underline{x}(k+1) = A\underline{x}(k)$

** eigenvalues of A are

$$\lambda_j = \alpha_j \pm i\beta_j \quad \text{where } i = \sqrt{-1}$$

3.4.4 *LYAPUNOV ANALYSIS*

From the classical theory of mechanics, we know that a vibratory system is stable if its total energy is continually decreasing until an equilibrium state is reached. The second method of Lyapunov is based on generalisation of this fact for determining the stability of dynamic systems [8]. Before discussing this method, we first recall [14,15] that:

1) a scalar function $V(x)$ is said to be positive definite in a region Π (including the origin) is $V(\underline{x}) > 0$ for all $\underline{x} \neq \underline{0}$ and $V(\underline{0}) = 0$.

2) a time-varying function $V(\underline{x},k)$ is said to be positive definite in a region Π (including the origin) if it is bounded from below by a time-invariant positive-definite function, that is, if there exists a positive definite function $V(\underline{x})$ such that

$$V(\underline{x},k) > V(\underline{x}) \qquad \text{for all} \quad k \geq k_0$$
$$V(0,k) = 0 \qquad \text{for all} \quad k \geq k_0$$

Lyapunov analysis of stability [8,11,14] leads to the following result:

The nonlinear system (3.82) is stable if and only if there exists a real-valued function $V(\underline{x},k)$ such that

(i) $V(\underline{0},k) = 0$ *for all k*

(ii) $V(\underline{x},k)$ *is positive definite; that is*

$$0 < \alpha(||\underline{x}||) \leq V(\underline{x},k) < \beta\ ||\underline{x}||)$$

where $\alpha(\cdot)$ and $\beta(\cdot)$ are real-valued, continuous monotone-increasing, positive-definite functions with $\alpha(0) = 0$ and $\beta(0) = 0$

(iii) $\Delta V(k,\underline{x},k_0)$ *is negative definite; that is*

$$\Delta V(k,\underline{x},k_0) \leqq -\gamma(||\underline{x}||)$$

where Δ *is the forward difference operator and* $\gamma(\cdot)$ *is a scalar-valued, continuous function such that* $\gamma(0) = 0$.

$V(\underline{x},k)$ *is called a Lyapunov function.*

To utilise this result successfully, considerable ingenuity is required in determining the appropriate Lyapunov function. Nevertheless, it has been the most generally used stability method [14,15].

One of the basic results of Lyapunov analysis can be obtained by considering the linear, shift-invariant system:

$$\underline{x}(k+1) = A\underline{x}(k) \tag{3.93}$$

where the origin $\underline{x} = \underline{0}$ is the equilibrium state. To investigate the stability of this state by use of Lyapunov analysis, we choose a possible Lyapunov function

$$V(\underline{x}) = \underline{x}^t(k)P\underline{x}(k) \tag{3.94}$$

where P is a real symmetric, positive-definite matrix. This ensures that $V(\underline{x})$ is positive-definite. Using (3.93), we calculate the first-forward difference of $V(\underline{x})$ as:

$$\begin{aligned}\Delta V(\underline{x}) &= V(\underline{x}(k+1) - V(\underline{x}(k)) \\ &= \underline{x}^t(k+1)P\underline{x}(k+1) - \underline{x}^t(k)P\underline{x}(k) \\ &= \underline{x}^t(k)[A^tPA - P]\underline{x}(k)\end{aligned} \tag{3.95}$$

For asymptotic stability, we require that $\Delta V(\underline{x})$ be negative definite and therefore let

$$\Delta V(\underline{x}) = -\underline{x}^t(k)Q\underline{x}(k) \tag{3.96}$$

where Q is a positive-definite matrix. The substitution of (3.96) in (3.95) yields

$$A^t PA - P = -Q \tag{3.97}$$

which is often called the Lyapunov equation. It is convenient to specify Q and then see whether or not the P matrix determined from the solution of (3.97) is positive definite. We thus conclude that:

A necessary and sufficient condition for the equilibrium state $\underline{x} = \underline{0}$ *to be asymptotically stable is that, given any positive-definite, real symmetric matrix* Q, *there exists a positive-definite real symmetric matrix* P *such that (3.97) is satisfied.*

It is important to note that the solution of (3.97) is generally non-unique. The condition for uniqueness of the solution of (3.97) is [16] that the eigenvalues λ of A are such that $\lambda_k \lambda_j \neq 1$.

Application of Lyapunov theory to the analysis of nonlinear systems of the form (3.82) requires construction of Lyapunov functions. We only mention methods like variable-gradient [14,16], Zubov's technique [16] and Krasovskii's theorem [14] for doing this. Further details of these methods can be found in [17].

3.4.5 *SOLUTION AND PROPERTIES OF THE LYAPUNOV EQUATION*

The results of Lyapunov analysis discussed in the last section indicate that the positive definiteness of the Lyapunov matrix P is a necessary and sufficient condition for asymptotic stability of linearised models. This requires that the solution of (3.97) be determined first and then by an application of Sylvester's criterion [10,16], we check the positive definiteness of the P matrix. The criterion is that all the principal minor determinants must be positive; that is, if

$$P = \begin{pmatrix} P_{11} & P_{12} \cdots\cdots P_{1n} \\ P_{12} & P_{22} \quad\quad \vdots \\ \vdots & \vdots \\ P_{1n} & P_{nn} \end{pmatrix} \tag{3.98}$$

then we must have:

$$P_{11} > 0; \quad \det \begin{pmatrix} P_{11} & P_{12} \\ P_{12} & P_{22} \end{pmatrix} > 0; \ldots; \det[P] > 0 \tag{3.99}$$

For convenience, we use $Q = I_n$ in (3.97). Since P is symmetric, the number of unknown elements is $n+\frac{1}{2}(n^2-n) = \frac{1}{2}n(n+1)$. A direct solution of (3.97) would thus require the inversion of a matrix of order $\frac{1}{2}n(n+1)$. We illustrate this procedure for the case of the second-order system:

$$\begin{pmatrix} a_{11} & a_{21} \\ a_{12} & a_{22} \end{pmatrix} \begin{pmatrix} P_{11} & P_{12} \\ P_{12} & P_{22} \end{pmatrix} \begin{pmatrix} a_{11} & a_{12} \\ a_{21} & a_{22} \end{pmatrix} - \begin{pmatrix} P_{11} & P_{12} \\ P_{12} & P_{22} \end{pmatrix} = \begin{pmatrix} -1 & 0 \\ 0 & -1 \end{pmatrix} \tag{3.100}$$

Upon expansion, we obtain:

$$(a_{11}^2-1)P_{11} + 2a_{11}a_{21}P_{12} + a_{21}^2P_{22} = -1$$
$$a_{11}a_{12}P_{11} + (a_{11}a_{22} + a_{12}a_{12}-1)P_{12} + a_{21}a_{22}P_{22} = 0$$
$$a_{12}^2P_{11} + 2a_{12}a_{22}P_{12} + (a_{22}^2-1)P_{22} = -1$$

which can be written in the matrix form:

$$L\,\underline{p} = -\underline{q} \tag{3.101}$$

where

$$L = \begin{pmatrix} (a_{11}^2-1) & 2a_{11}a_{21} & a_{21}^2 \\ a_{11}a_{12} & (a_{11}a_{22}+a_{12}a_{21}-1) & a_{21}a_{22} \\ a_2^2 2 & 2a_{12}a_{22} & (a_{22}^2-1) \end{pmatrix}$$

$$\underline{p} = [P_{11} \quad P_{12} \quad P_{22}]^t$$

$$\underline{q} = [-1 \quad 0 \quad -1]$$

In the general (nxn) case the matrices $\underline{p}$ and $\underline{q}$ in (3.101) are given by

$$\underline{p} = [P_{11} \quad P_{12} \quad P_{22} \quad P_{13} \quad P_{23} \quad P_{33} \ldots . P_{nn}]^t \qquad (3.102)$$

$$\underline{q} = [+1 \quad 0 \quad -1 \quad 0 \quad 0 \quad -1 \ldots . -1] \qquad (3.103)$$

and L is formed from A as

$$L = \begin{bmatrix} (a_{11}^2-1) & 2a_{11}a_{12} & a_{21}^2 & 2a_{11}a_{31} & 2a_{21}a_{31} & a_{31}^2 \ldots\ldots\ldots \\ a_{11}a_{12} & a_{12}a_{21}+a_{11}a_{22}-1 & a_{21}a_{22} & a_{12}a_{31}+a_{11}a_{32} & a_{22}a_{31}+a_{21}a_{32} & a_{31}a_{32} \ldots\ldots\ldots \\ a_{12}^2 & 2a_{12}a_{22} & (a_{22}^2-1) & 2a_{12}a_{32} & 2a_{22}a_{32} & a_{32}^2 \ldots\ldots\ldots \\ a_{11}a_{13} & a_{21}a_{13}+a_{11}a_{23} & a_{21}a_{23} & a_{11}a_{33}+a_{31}a_{13}-1 & a_{31}a_{23}+a_{21}a_{33} & a_{31}a_{33} \ldots\ldots\ldots \\ a_{12}a_{13} & a_{12}a_{22}+a_{12}a_{23} & a_{22}a_{23} & a_{32}a_{13}+a_{12}a_{33} & a_{32}a_{23}+a_{22}a_{23}-1 & a_{32}a_{33} \ldots\ldots\ldots \\ a_{13}^2 & 2a_{13}a_{23} & a_{23}^2 & 2a_{13}a_{33} & 2a_{23}a_{33} & (a_{33}^2-1) \ldots\ldots\ldots \\ \vdots & \vdots & \vdots & \vdots & \vdots & \vdots \\ a_{1n}^2 & 2a_{1n}a_{2n} & & & & (a_{nn}^2-1) \end{bmatrix}$$

It is interesting to note that we have a systematic way of forming L from A, which can be easily programmed for use on a digital machine. From (3.101) the unique solution is given by:

$$\underline{p} = -L^{-1}\underline{q} \tag{3.104}$$

For large n, the inversion in (3.104) may be computationally prohibitive. When A is in the canonical form [10], it is readily seen that L will be sparce and hence the computational effort is reduced. Specific results are repcrted in [20,21] for single-input single-output systems:

(1) If $Q = BB^t$ where

$$B = \begin{bmatrix} 1-a_n^2 \\ a_1-a_n a_{n-1} \\ a_2-a_n a_{n-2} \\ \vdots \\ a_{n-1}-a_n a_1 \end{bmatrix} \tag{3.105}$$

$$A = \begin{bmatrix} 0 & 0 & \cdots & 0 & -a_n \\ 1 & 0 & \cdots & 0 & -a_{n-1} \\ 0 & 1 & & & \\ \vdots & \vdots & & & \\ 0 & 0 & \cdots & 1 & -a_1 \end{bmatrix} \tag{3.106}$$

then the solution of (3.97) is the Schur-Cohn matrix whose (ij)th element is given by [21]

$$p_{ij} = \sum_{k=1}^{\min(i,j)} [\, a_{i-k}a_{j-k} - a_{n-i+k}a_{n-j+k} \,] \tag{3.107}$$

The matrix A in (3.106) is often called the reachability canonical form [10]

(2) Suppose that the linear system is in the controllable canonical form [10]; that is

$$A = \begin{bmatrix} 0 & 1 & 0 & \cdots & 0 \\ 0 & 0 & 1 & \cdots & 0 \\ \vdots & \vdots & \vdots & & \vdots \\ \vdots & \vdots & \vdots & & 1 \\ -a_n & -a_{n-1} & -a_{n-2} & & -a_1 \end{bmatrix} \tag{3.108}$$

$$B = [0 \quad 0 \quad \cdots \cdots \quad 1]^t \tag{3.109}$$

then the solution of (3.97) with $Q = BB^t$ is shown to be the inverse of the Schur-Cohn matrix. That is, the P matrix is the inverse of that given by (3.107).

Due to the close connection between (3.97)

$$F^tP + PB = -Q_0 \tag{3.110}$$

frequently known as the continuous matrix Riccati equation [10, 19], it has been shown by Barnett [22] that the bilinear transformation

$$F = (A-I_n)(A+I_n)^{-1} \tag{3.111}$$

converts (3.97) into (3.110) such that

$$Q_0 = -2(I_nAF^t)^{-1}Q(I_n+F)^{-1} \tag{3.112}$$

This suggests that instead of solving (3.97) we employ (3.111) and (3.112) and solve (3.110) by any of the available methods [23-26].

In some cases, one can iterate on (3.97) directly by the sequence:

$$P_{j+1} = A^t P_j A + Q \tag{3.113}$$

$$P_{j+1} \leftarrow \alpha P_{j+1} + \beta P_j; \quad \alpha+\beta = 1 \tag{3.114}$$

If A is a stable matrix, then the scheme (3.113), (3.114) will approach the solution of (3.97) with a linear rate of convergence.

By virtue of the fact that the eigenvalues of a matrix provide valuable information about the system dynamics, it is sometimes important to have *a priori* bounds on the external eigenvalues of the positive definite solution P of (3.97). For reference, we define:

$\lambda_m(R)$ as the minimum eigenvalue of R

$\lambda_M(R)$ as the maximum eigenvalue of R

$\lambda_m^+(R)$, $\lambda_m^-(R)$ as the upper and lower bound on the minimum eigenvalue of R_i respectively

$\lambda_M^+(R)$, $\lambda_M^-(R)$ as the upper and lower bound on the maximum eigenvalue of R; respectively.

Results on the extremal properties of the eigenvalues of P are derived based on information about A and Q. The recent results developed in *[27]* indicates that

$$\left.\begin{aligned} \lambda_m^-(P) &= \lambda_m(Q)/[1-\lambda_m(A^tA)] \\ \lambda_m^+(P) &= \lambda_M(Q)/[1-\sigma_m^2] \\ \lambda_M^-(P) &= \lambda_m(Q)/[1-\sigma_m^2] \\ \lambda_M^+(P) &= \lambda_M(Q)/[1-\sigma_M(A^tA)] \end{aligned}\right\} \tag{3.115}$$

where

$$\left.\begin{aligned} \sigma_m &= \min_j \theta_j; \quad \theta_j = |\lambda_j(A)| \\ \sigma_M &= \max_j \theta_j \end{aligned}\right\} \tag{3.116}$$

Other bounds are developed in [28]; however, those of (3.115) give better estimates. It should be emphasised that $[1-\lambda_M(A^tA)]$ in (3.115) should be positive; otherwise $\lambda_M^+(P)$ becomes meaningless.

This concludes the material that we would like to cover on Lyapunov stability analysis.

3.4.6 *EXAMPLES*

We clarify the various notions introduced in this section by some examples.

Example 1

Consider the system

$$\underline{x}(k+1) \begin{bmatrix} .5 & .5 & 1 \\ 0 & .5 & 2 \\ 0 & 0 & .5 \end{bmatrix} \underline{x}(k)$$

which was treated as example 4 in section 2.4. Since the A matrix is upper-triangular, its eigenvalues are found by inspection to be $\lambda_1 = \lambda_2 = \lambda_3 = .5$. Since all the $|\lambda\ | < 1$, then according to Table 3.3 the system is asymptotically stable. This is verified by the explicit solution found in section 2.4, which shows that $\underline{x}(k) \to \underline{0}$ as $k \to \infty$.

Example 2

Recall from example 1, section 3.3, that the eigenvalues of the system are $\lambda_1 = 1$, $\lambda_2 = .5$, $\lambda_3 = .333$. This corresponds to case (b) in Table 3.3 and hence the system is stable in the sense of Lyapunov, but not asymptotically stable.

Example 3

It is required to determine the stability of the equilibrium state of the following system

$$x_1(k+1) = x_1(k) + 3\, x_2(k)$$

$$x_2(k+1) = -3\, x_1(k) - 2\, x_2(k) - 3\, x_3(k)$$

$$x_3(k+1) = x_1(k)$$

To do this, we put the system in the form (3.93) with

$$A = \begin{pmatrix} 1 & 3 & 0 \\ -3 & -2 & -3 \\ 1 & 0 & 0 \end{pmatrix}$$

then select the Q matrix as

$$Q = \begin{pmatrix} 1 & 0 & 0 \\ 0 & 1 & 0 \\ 0 & 0 & 1 \end{pmatrix}$$

and attempt to solve (3.97).

Direct application of any of the numerical algorithms mentioned in section (3.4.5) leads to

$$P = \begin{pmatrix} -.24359 & -.25641 & -.5 \\ -.25641 & -.62821 & -.86538 \\ -.5 & -.86538 & -4.65385 \end{pmatrix}$$

To check the positive-definiteness of P, we have to make sure that all the minor determinants are positive. By inspection, we find that $P_{11} < 0$, which violates Sylvester's criterion. Thus we conclude that the equilibrium state of the third-order system is unstable.

By computing the eigenvalues of A, we get

$$\sigma(A) = \{-1.2346, \quad .11731 \pm i2.6974\}$$

Since some of the eigenvalues have magnitude greater than 1, hence the system is unstable.

Example 4

Consider the following problem faced by a cattle rancher, who has a particular breed:

$x_1(k)$ = average number of 1-year-old cattle (young)

$x_2(k)$ = average number of 2-year-old cattle (mature)

$x_3(k)$ = average number of 3-year-old cattle (old)

In the absence of slaughtering, the net growth of cattle is described by

$$\begin{pmatrix} x_1(k+1) \\ x_2(k+1) \\ x_3(k+1) \end{pmatrix} = \begin{pmatrix} 0 & .8 & .4 \\ 1 & 0 & 0 \\ 0 & 1 & .7 \end{pmatrix} \begin{pmatrix} x_1(k) \\ x_2(k) \\ x_3(k) \end{pmatrix}$$

This model was presented as a part of an exercise in [18], and no solution was given. We shall analyse this problem from a stability point of view.

In the above model, the young cattle (1-year-olds) do not multiply since $a_{11} = 0$. Each 2-year-old produces on the average .8 young cattle per year, and old cattle produce on the average .4 young cattle per year. The natural death rate of old cattle is

.3 per year.

It is easy to evaluate the eigenvalues of the model as

$$\lambda_1 = -.71993, \quad \lambda_2 \doteq 1.24082, \quad \lambda_3 = .17911$$

Since $\lambda_2 > 1$ the model is unstable; meaning that under natural conditions, the numbers would grow without limit.

Evidently a controlled slaughter policy needs to be enforced. For this purpose, we consider that the cattle rancher can slaughter only mature and old cattle at rates S_2, S_3; respectively. The rancher will gain money from selling the cattle he slaughters. It is obvious that S_2, S_3 are non-negative and for simplicity we assume that they are constants. Since the slaughtered cattle eventually reduce the old ones, the dynamic model becomes:

$$\begin{bmatrix} x_1(k+1) \\ x_2(k+1) \\ x_2(k+1) \end{bmatrix} = \begin{bmatrix} 0 & .8 & .4 \\ 1 & 0 & 0 \\ 0 & 1-S_2 & .7-S_3 \end{bmatrix} \begin{bmatrix} x_1(k) \\ x_2(k) \\ x_3(k) \end{bmatrix}$$

There are indeed several ways to determine S_2 and S_3. Here, we concentrate on the stability aspect. By varying S_2 and S_3 and checking the eigenvalues in each case, we obtain the graph given in Fig. (3.4), from which we conclude the following:

1) In general, slaughtering cattle for food keeps cattle population finite.

2) The slaughter rate of mature cattle has to be more than or equal to 50%. This should take place with slaughtering rate of old cattle being below 100%. Otherwise we obtain imaginary eigenvalues corresponding to oscillatory behaviour.

3) The upper bound of slaughtering mature cattle is 100%.

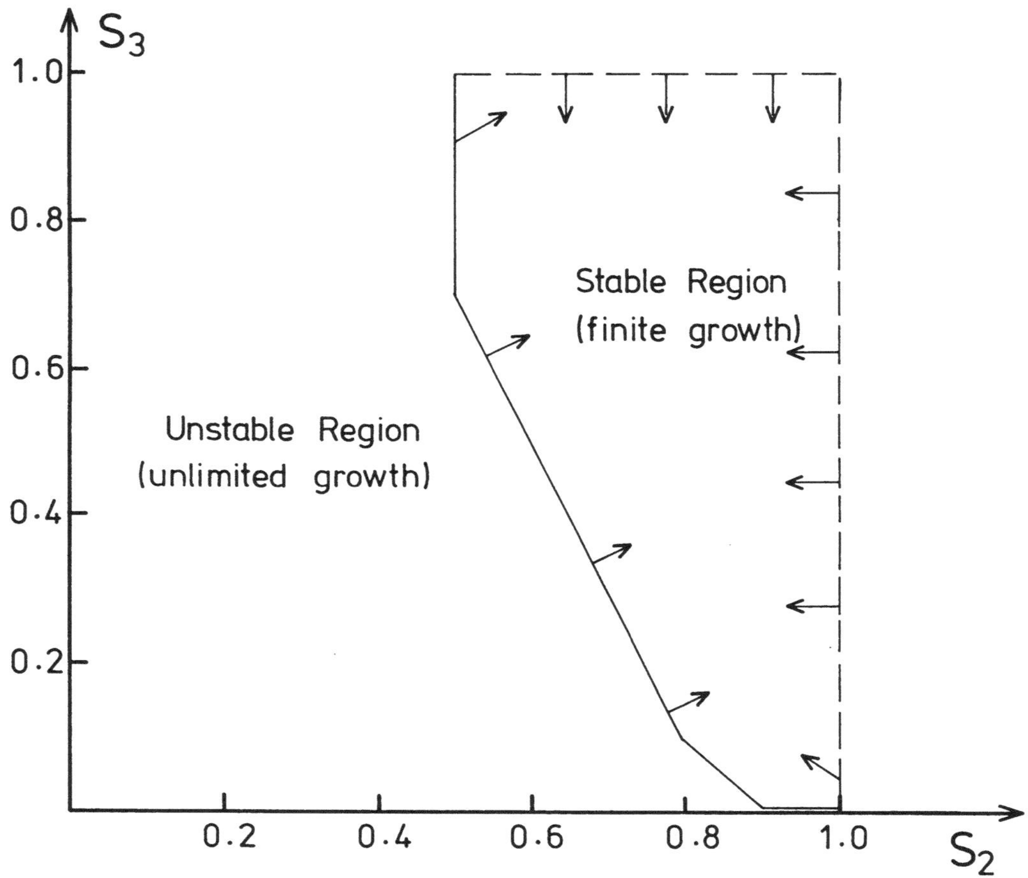

Fig. (3.4) Stable region of the cattle model

Again, beyond that we obtain unrealistic results.

4) Any point in the stable region gives finite growth of the cattle breed. If one takes into consideration the prices of the slaughtered cattle, optimal slaughtering rates corresponding to maximum profit can be obtained.

Example 5

Consider the second-order system

$$\underline{x}(k+1) = \begin{pmatrix} 0 & 1 \\ -.03 & -.4 \end{pmatrix} \underline{x}$$

which is asymptotically stable and has $\lambda_1 = -.1$ and $\lambda_2 = -.3$. Let

$$Q = \text{diag } [2 \quad .6]$$

We wish to evaluate the extremal eigenvalues of the P matrix. To do this, we first compute

$$A^tA = \begin{pmatrix} 0 & -.03 \\ 1 & -.4 \end{pmatrix} \begin{pmatrix} 0 & 1 \\ -.03 & -.4 \end{pmatrix}$$

$$= \begin{pmatrix} .0009 & .012 \\ .012 & 1.16 \end{pmatrix}$$

$$\sigma(A^tA) = \{.00078, \quad 1.16012\}$$

In terms of (3.115), we thus have:

$$\lambda_m(Q) = .2 \quad ; \quad \sigma_M(Q) = .6$$

$$\lambda_m(A^tA) = .00078; \ \lambda_M(A^tA) = 1.16012$$

$$\sigma_m = .1 \quad ; \quad \sigma_M = .3$$

Direct application of (3.115) results in:

$$\lambda\ [P] = .2/\ 1-.00078$$

$$= .20016$$

$$\lambda_m^+ [P] = .6/\ 1-(.1)^2$$

$$\lambda_M^- [P] = .2/\ 1-(.3)^2$$

$$= .21978$$

$\lambda_M^+ [P]$ cannot be obtained since $\lambda_M(A^tA) > 1$.

To examine these results, we apply (3.101) with

$$P = \begin{pmatrix} P_1 & P_2 \\ P_2 & P_3 \end{pmatrix}$$

to yield

$$\begin{pmatrix} -1 & 0 & .0009 \\ 0 & -1.03 & .012 \\ 1 & -.8 & -.84 \end{pmatrix} \begin{pmatrix} P_1 \\ P_2 \\ P_3 \end{pmatrix} = \begin{pmatrix} -.2 \\ 0 \\ -.6 \end{pmatrix}$$

and upon inverting, we get

$$P = \begin{matrix} .20085 & .01099 \\ .01099 & .94293 \end{matrix}$$

which is positive-definite $(p_{11} > 0$ and $p_{11}p_{22} > p_{12}^2)$ and has

$$\sigma(P) = \{.2007,\quad .9431\} \equiv \{\lambda_m[P], \lambda_M[P]\}$$

It is readily evident that

$$\lambda_m^-[P] < \lambda_m[P] < \lambda_m^+[P]$$

$$\lambda_M^-[P] < \lambda_M[P]$$

as expected.

3.5 Remarks

In this chapter, we have presented a detailed discussion about the basic notions of controllability, observability and stability as important structural properties of discrete-time systems. Our approach in treating these properties has been systematic although we hope that the intuitive skills of the reader have been enhanced. Generally, we started with a definition, interpreted its meaning and then, by simple analysis, we arrived at the desired result. It has been shown that the singularity (or nonsingularity) of the system matrix A plays a key role in determining the controllability and observability behaviour of the system. We have clarified the difference between controllability from the origin (reachability) and controllability to the origin. Also, the distinction between specifying the state of the system from past or future output records. We have explained the principle of duality due to Kalman and indicated that (controllability-determinability) and reachability-observability) are dual properties. The utilisation of modal analysis has led us to define two additional notions; stabilisability and detectability. Finally, we have discussed different definitions pertaining to the stability of discrete systems and then introduced Lyapunov analysis.

Most of the discussions were given in the time domain. Several examples have been worked out to demonstrate the various concepts introduced in this chapter.

3.6 Problems

1. Verify that the complete reachability of the pair (A,B) is achieved if either of the following conditions holds

   a) $\underline{z}^t A^j B = \underline{0}$ for $j=0,1,...,n-1$ implies $\underline{z} = \underline{0}$

b) $\underline{z}^t B = \underline{0}$ and $\underline{z}^t A = \mu\, \underline{z}^t$ for some constant μ implies $\underline{z} = \underline{0}$.

(Note: A is an nxn matrix, B is an nxm matrix, $\underline{z}$ is an n-vector)

2) Verify that the complete controllability of the pair (A,B) is achieved if either of the following conditions holds

a) $\underline{z}^t A^j B = \underline{0}$ for $j=0,1,\ldots,n-1$ implies $\underline{z}^t A^n = \underline{0}$

b) $\underline{z}^t B = \underline{0}$ and $\underline{z}^t A = \mu\, \underline{z}^t$ implies $\mu = 0$ or $\underline{z} = \underline{0}$

3) Verify that the pair (A,B) is completely stabilisable if $\underline{z}^t B = \underline{0}$ and $\underline{z}^t A = \mu\, \underline{z}^t$ for some constant μ implies $|\xi| < 1$ or $\underline{z} = \underline{0}$

4) Apply the duality principle to the results of problems 1, 2 and 3 to determine the corresponding conditions for complete observability, determinability and detectability.

5) Discuss the mode-controllability and mode-observability characteristics of the systems described by their normal forms:

i) $J = \begin{pmatrix} .5 & 0 \\ 0 & -.7 \end{pmatrix}, \quad B_m = \begin{pmatrix} 1 \\ 2 \end{pmatrix}, \quad C_m = [1 \quad 0]$

ii) $J = \begin{pmatrix} .8 & 0 & 0 \\ 0 & 0 & 0 \\ 0 & 0 & -.5 \end{pmatrix}; \quad B_m = \begin{pmatrix} 1 & 0 \\ -3 & 1 \\ 2 & -.5 \end{pmatrix};$

$C_m = \begin{pmatrix} 2 & 1 \\ 1 & -3 \end{pmatrix}$

iii) $J = \begin{pmatrix} .2 & 1 & 0 \\ 0 & .2 & 1 \\ 0 & 0 & .2 \end{pmatrix}; \quad B_m = \begin{pmatrix} 0 \\ 0 \\ 1 \end{pmatrix};$

$$C_m = [2 \quad 0 \quad 0]$$

iv) $$J = \begin{pmatrix} 0 & 1 \\ 0 & 0 \end{pmatrix}; \quad B_m = \begin{pmatrix} .5 & .7 \\ 0 & .3 \end{pmatrix}$$

$$C_m = \begin{pmatrix} 2 & 1 \\ -1 & 0 \end{pmatrix}$$

v) $$J = \begin{pmatrix} .4 & 1 & 0 & 0 \\ 0 & .4 & 0 & 0 \\ 0 & 0 & -.6 & 1 \\ 0 & 0 & 0 & -.6 \end{pmatrix}; B_m = \begin{pmatrix} .8 & .4 & 0 \\ -1 & 3 & 0 \\ .5 & -2 & .6 \\ 0 & .1 & .2 \end{pmatrix}$$

6) Determine the mode-controllability and mode-observability structure of the systems described by the following matrces

i) $$A = \begin{pmatrix} .5 & .1 \\ .2 & .3 \end{pmatrix}; \quad B = \begin{pmatrix} 1 & 2 \\ -1 & 1 \end{pmatrix}$$

$$\begin{pmatrix} 0 & 2 \\ -1 & 1 \end{pmatrix}$$

ii) $$A = \begin{pmatrix} 1 & 3 & -2 \\ 0 & -.4 & .5 \\ 0 & .6 & -.5 \end{pmatrix}; \quad B = \begin{pmatrix} 0 & 0 \\ 1 & 1 \\ 0 & 1 \end{pmatrix}$$

$$C = \begin{pmatrix} 0 & 0 & 1 \\ 1 & 0 & -1 \end{pmatrix}$$

iii) $$A = \begin{pmatrix} 0 & 1 & 0 \\ 0 & 0 & 1 \\ .024 & .26 & -.1 \end{pmatrix}; \quad B = \begin{pmatrix} 0 \\ 0 \\ 1 \end{pmatrix}$$

$$C = [1 \quad 0 \quad 0]$$

7) Consider the system

$$\underline{x}(k+1) = \begin{pmatrix} \lambda & 1 & 0 \\ 0 & \lambda & 1 \\ 0 & 0 & \lambda \end{pmatrix} \underline{x}(k) + \begin{pmatrix} a \\ b \\ c \end{pmatrix} u(k)$$

Determine the conditions on (a,b,c) for complete state controllability.

8) Consider the following 'doubling' algorithm for solving the Lyapunov matrix equation (3.

$$R_{j+1} = (R_j)^2 \quad ; \quad R_1 = A$$
$$S_{j+=} = R_j^t S_j R_j + S_j \quad ; \quad S_1 = Q$$

Verify that $R_{j+1} = A^k$ and $S_{j+=} = P_k$ where $k = 2^j$. Study the convergence behaviour of this algorithm and show that

$$P_a = \lim_{\to\infty} N_j \quad \text{where} \quad P_a = \sum_{j=0}^{\infty} (A^t)^j QA^j.$$

9) A simple model of a production and inventory control system is shown below. The input $u_1(k)$ represents the scheduled production rate, $x_1(k)$ represents the actual production rate, $u_2(k)$ represents the sales rate, and $x_2(k)$ represents the current inventory level.

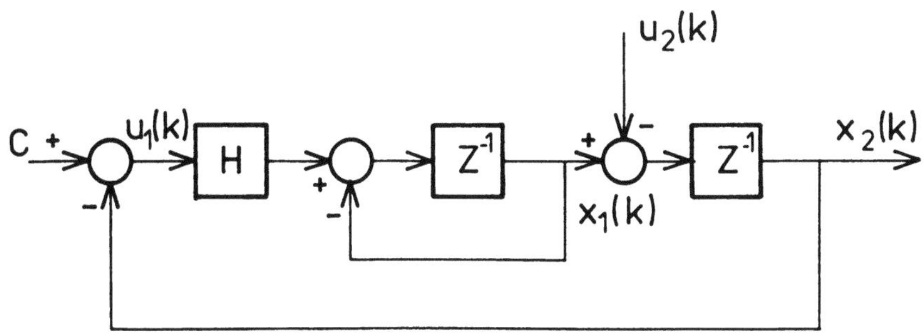

Let c be the desired inventory level. A suitable feedback control policy is to choose $u_1(k) = C - x_2(k)$. Suppose that the system is originally in equilibr um with $x_1(0)$ = sales rate and $x_2(0) = c$.

a) Write the state equations of the system.

b) Examine the stability behaviour of the system when $H = 0$, $H < 0$ and $H > 0$.

c) Suppose that at $k = 0$, the sales rate suddenly increases by $\alpha\%$. Discuss the structural properties when $H = .4$.

10) Show that the discrete Lyapunov matrix equation can be put into the form

$$[I_n - A \oplus A^t]\underline{p} = \underline{q}$$

where $\oplus$ denotes the Kronecker product and $\underline{p}, \underline{q}$ designate column vectors obtained in a one to one manner from the matrices P and Q. Hence show that a unique solution exists if and only if $(1 - \lambda_i \lambda_j) \neq)$, for all i,j where $\{\lambda_i\}$ are the eigenvalues of A.

3.7 References

[1] Zadeh, L.A. and E. Polak
"Systems Theory", McGraw-Hill, N.Y, 1969.

[2] Kalman, R.E.
"On the General Theory of Control Systems", Proc. First IFAC Congress on Automatic Control, Butterworth, London, pp. 481-492, 1960.

[3] Weiss, L.
"Controllability, Realisation and Stability of Discrete-Time Systems", SIAM J. Control, Vol. 10, pp.230-251, 1972.

[4] O"Reilly, J.
"The Discrete Linear Time Invariant Time-Optimal Control Problem - An Overview", Automatica, Vol. 17, pp. 363-370, 1981.

[5] Porter B. and R. Crossley
"Modal Control- Theory and Applications", Taylor & Francis Ltd., London, 1972.

[6] Simon, J.D. and S.K. Mitter
"A Theory of Modal Control", Information and Control, Vol. 13, pp. 316-353, 1968.

[7] Gopinath, B.
"On the Control of Linear Multiple Input-Output Systems", Bell Systems Technical J., Vol. 50, pp. 1063-1081, 1971.

[8] Brogan, W.L.
"Modern Control Theory", Quantum Publishing, Inc., New York, 1974.

[9] Wonham, W.M.
"On a Matrix Riccati Equation of Stochastic Control", SIAM J. Control, Vol. 6, pp. 681-697, 1968.

[10] Kailath, T.
"Linear Systems", Prentice-Hall, Inc., N.J., 1980.

[11] Kalman, R.E. and J.E. Betram
"Control System Analysis and Design via the 'Second Method' of Lyapunov, II. Discrete-time Systems", Trans. ASME J. Basic Engineering, Vol. 82D, pp. 394-400, 1960.

[12] Anderson, B.D.O.
"External and Internal Stability of a Linear System - A New Connection", IEEE Trans. Autom. Control, Vol. AC-17, pp. 107-111, 1972.

[13] Brockett, R.W.
"The Status of Stability Theory for Deterministic Systems" IEEE Trans. Autom. Control, Vol. AC-11, pp. 596-606, 1966.

[14] Ogata, K.
"State Space Analysis of Control Systems", Prentice-Hall, N.J., 1967.

[15] Goldberg, S.
"Introduction to Difference Equations", John Wiley, N.Y., 1958.

[16] Barnett, S.
"Introduction to Mathematical Control Theory", Clarendon Press, Oxford, 1975.

[17] LaSalle, J.P. and S. Lefschetz
"Stability by Lyapunov's Direct Method with Applications", Academic Press, N.Y., 1961.

[18] Athans, M., M.L. Dertouzos, R.N. Spann and S.J. Mason
"Systems, Networks and Computation: Multivariable Methods", McGraw-Hill, N.Y., 1974.

[19] Barnett, S. and C. Storey
"Matrix Methods in Stability Theory", Nelson Ltd., London, 1970.

[20] Bitmead, R.R. and H. Weiss
"On the Solution of the Discrete-Time Lyapunov Matrix Equation in Controllable Canonical Form", IEEE Trans. Autom. Control. Vol. AC-24, pp. 481-482, 1979.

[21] Parks, P.C.
"Lyapunov and the Schur-Cohen Stability Criterion", IEEE Trans. Autom. Control, Vol. AC-9, p. 121, 1964.

[22] Barnett, S.
"Matrices in Control Theory", Van Nostrand Reinhold, London, 1971.

[23] Belanger, P.R. and T.P. McGillivray
"Computational Experience with the Solution of the Matrix Lyapunov Equation", IEEE Trans. Autom. Control, Vol. AC-21, pp. 799-780, 1976.

[24] Hagander, P.
"Numerical Solution $A^tS+SA+Q$", Inform. Science, Vol. 4, pp. 35-50, 1972.

[25] Rothschild, D. and A. Jameson
"Comparison of Four Numerical Algorithms for Solving the Lyapunov Matrix Equation", Int. J. Control, pp. 181-198, 1970.

[26] Bartels, R.H. and G.W. Stewart
"Solution of the Equation AX+XB=C (Algorithm 432)", Common. Ass. Comput. Mach., Vol. 15, pp. 820-826, 1972.

[27] Yasuda, K. and K. Hirai
"Upper and Lower Bounds on the Solution of the Algebraic Riccati Equation", IEEE Trans. Autom. Control, Vol. AC-24, pp. 483-487, 1979.

[28] Patel, R.V. and M. Toda
"On Norm Bounds for Algebraic Riccati and Lyapunov Equations", IEEE Trans. Autom. Control, Vol. AC-23, pp. 87-88, 1978.

Chapter 4
Design of Feedback Systems

4.1 Introduction

Our main objective in this chapter is to introduce some of the central concepts involved in the design of linear feedback controllers for discrete time systems. On the basis of these concepts, we will develop different techniques to compute the feedback gain matrix. The material covered is divided into five major sections. In section 4.2, we discuss the concept of feedback in linear control problems and illustrate its utility by means of typical systems applications. Next, we examine the effect of state and output feedback on the structural properties of the system. Algorithms for eigenvalue assignment are then given. An important class of discrete-type controllers, that is, deadbeat controllers is next investigated in section 4.3. The properties of these controllers are examined and the different approaches are brought together to show their interrelationships.

Section 4.4 is devoted to the development of simplification techniques that approximate the behaviour of the higher-order models. Two basic approaches are used. The first is based on modal analysis whilst the second utilises the concept of output modelling. Controller design is then examined in order to illustrate the main differences between the two approaches.

Discrete two-time-scale systems is the subject of section 4.5. Both modal analysis and feedback control design are presented using the block-diagonalisation and quasi-steady state approaches. Several examples are worked out throughout the chapter to illustrate the various techniques.

We begin our analysis by considering the notion of feedback.

4.2 The Concept of Linear Feedback

The first phase in the analysis of most problems involves the development of a mathematical model of the phenomena under consideration. On the one hand, the models must be an adequate representation of reality whilst on the other, the models should not be so complex that they are difficult to manipulate. There are two distinct approaches to the development of system models. The first method, called physical systems modelling, involves applying the basic laws of science to the system components and to their interconnections. The second, called experimental modelling, corresponds to empirical fitting of recorded data amongst the system inputs and outputs to a set of mathematical relationships. Both methods are used in this book. Since we are concerned throughout this book with discrete-time control systems, we will assume that we have a discrete dynamic model which is given in terms of either a discrete transfer function (frequency-domain) or in the form of difference equations [1-3] (in the time-domain). Such models have become particularly important in the last few years due to the advances that have been made in micro-electronic technology and the consequent availability of cheap distributed computing facilities. Since such facilities are digital, it becomes more meaningful to analyse and design the control structure directly from the discrete model as opposed to doing the design in continuous time and then implementing it in discrete-time using a digital computer. There is thus a real need for the development of new analytical tools for handling discrete time systems. We note that such tools will also be applicable to switching networks and digital process layout [4].

For continuous time systems, the notion of feedback has been paramount in the design of successful control schemes [5,6]. *"Feedback"* essentially means that we utilise some information about the behaviour of the system in a closed loopway in order to obtain a more desirable response from the system. Thus the difference between the actual and desired behaviour of the system is fed back in order to exercise control. This allows the designer to improve amongst others the transient and/or the steady state response, track specified input functions, reduce parametric variations, reject disturbances, etc. A number of excellent text books exist which deal with these issues [5,8].

There are obviously advantages in using feedback for discrete time systems. In the design of the discrete control strategies, a number of questions are of importance, e.g. what are the variables which can be directly measured? can all the variables be reconstructed? are the measurements instantaneous or is there a delay? how will the information be processed? what are appropriate measures of system performance that can be used in feedback design?. These questions are not given here in any order of preference. In this chapter we attempt to provide some answers to such questions, and develop techniques for the implementation of the design. Before embarking on the mathematical analysis, we will give some examples of typical discrete feedback control systems drawn from different fields of application.

Example 1

This example comes from the field of stream quality control in rivers [9]. In multi-reach, non-tidal river systems, the water quality is regulated by controlling the effluent discharges from sewage treatment facilities. The interaction between dissolved oxygen (DO) and biochemical oxygen demand (BDO) provides the basis for a simplified discrete dynamic model. This model is described by two difference equations per reach [10]. One of the main quality control purposes is to keep the in-stream DO concentration close to a desired reference level. A feedback control system can thus be constructed, for a single reach, as shown in Fig. (4.1).

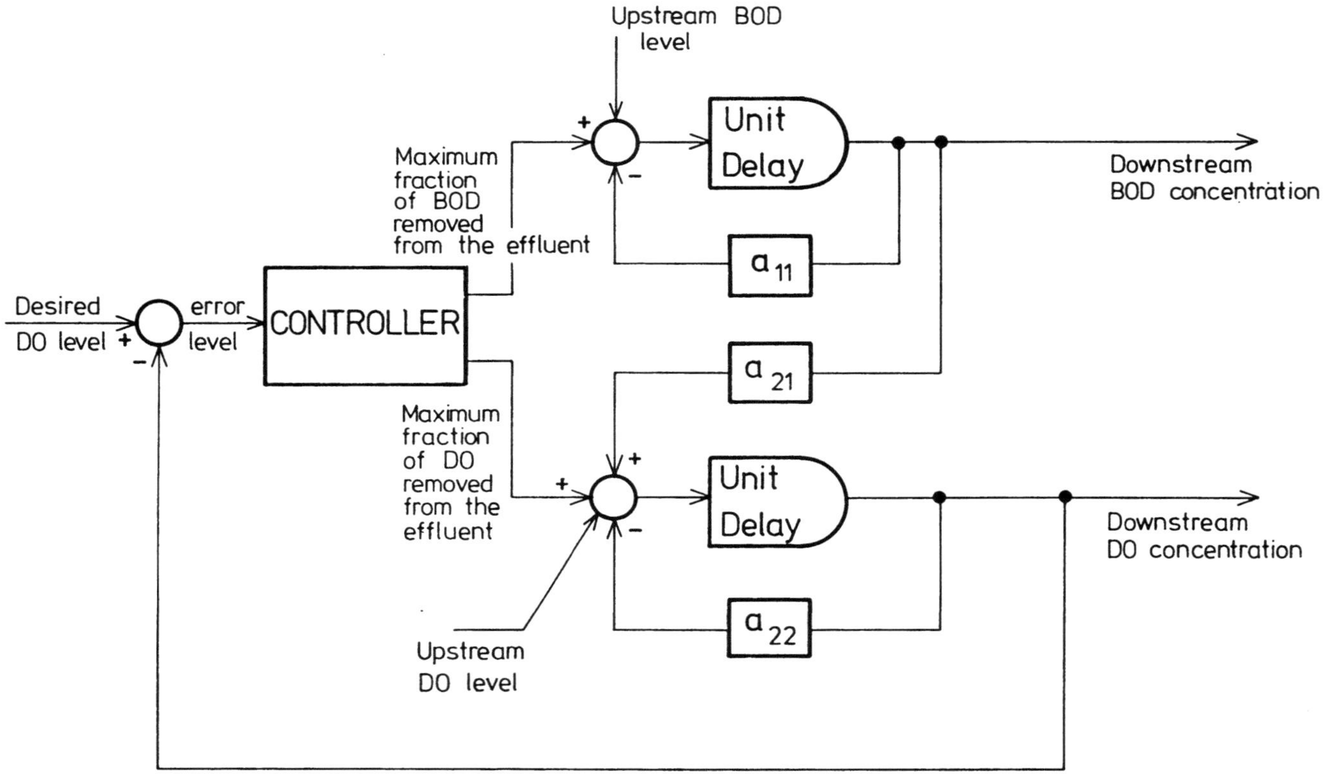

Fig. (4.1) Simplified feedback system to control DO in a single reach

We should point out that the controller can be designed by minimising a certain cost function or by specifying a set of closed-loop poles which correspond to the decay times of DO and BDO. An important question arises here as to whether we can use full information records about DO and BOD concentrations. It turns out [9] that although the DO concentration can be recorded periodically by using in-stream probes, the BOD concentration measurement requires some laboratory manipulation. The control design needs to take this into consideration. In multi-reach river systems, we can visualise the feedback control system as a distributed recording and computing facility for preserving stream quality.

Example 2

The second example is of a dynamic macroeconomic model [11]. This linear, fourth-order, discrete-time model combines an inventory sector and a capital goods sector and exhibits some basic features of macroeconomic theory [13]. A schematic diagram of the model is given in Fig. (4.2), in which the consumer goods production (consumption), consumer goods inventory, plant and equipment (capital stock) and plant and equipment production rate due to induced investment demand are chosen as the state (endogenous) variables. In this model the government expenditure is taken as the control variable. As in the case of most macroeconomic models, there are input quantities, frequently called exogenous variables, whose values are known. These quantities resemble constant disturbance in control engineering terminology.

Utilising typical values to represent the macroeconomic behaviour, it has been shown [12] that this discrete model exhibits oscillations which means that without exercising any control the system is unstable. A stabilisation scheme can be synthesised by using a feedback control which is a linear function of the states. Such a scheme has the economic interpretation that the government expenditure is generated from proportions of the out-

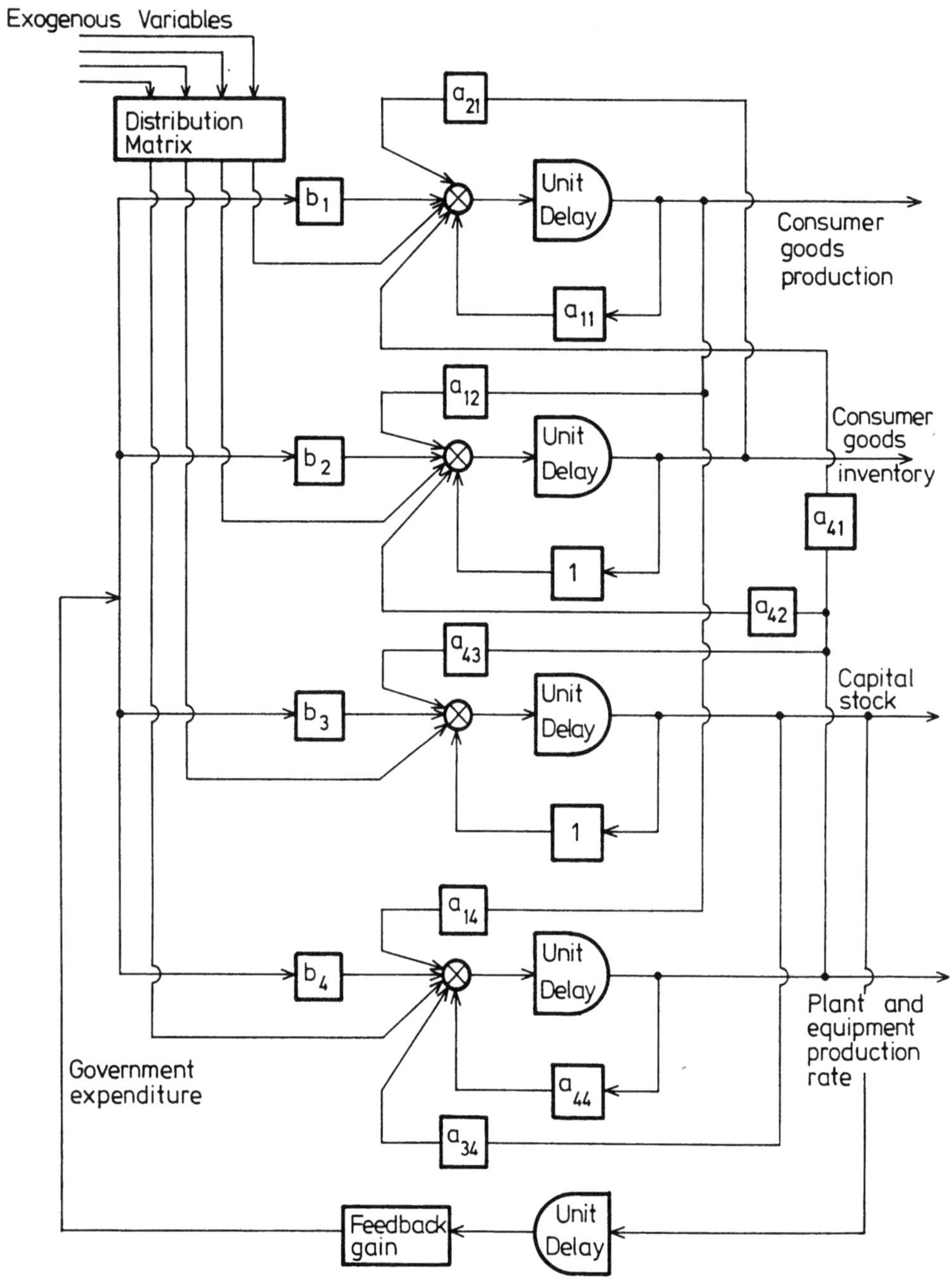

Fig. (4.2) Feeback control of macroeconomic model

puts of the economic sectors. The selection of the feedback coefficients can be based on a set of economic trends specified a priori [12]. Up to now, the feedback design problem would seem to be a standard control one. Strictly speaking, this does not reflect the economic reality in which time is used in order to determine the economic variables. One way to get around this is to generate the control input from delayed information about the variables that can be recorded. As shown in Fig. (4.2), the value of the capital stock at the last discrete instant can be used to determine the government expenditure at the present time. An alternative method [13] would be to design a state reconstructor that provides estimates of the economic variables and then utilises the estimated records to generate the present value of government expenditure. In practice, this means that the government needs to do some economic forecasts using previous records before implementing a particular economic policy.

Example 3

Urban road traffic networks consist of junctions and interconnecting roads. During the rush house, the number of vehicles arriving at any intersection exceeds those leaving it, and this propogates to the adjacent junctions thereby causing oversaturation of the traffic network. The dynamic behaviour of oversaturated networks can be modelled by treating the queues at junction approaches as the state variables [14]. In terms of the simple one-way no-turn intersection shown in Fig. (4.3), we define the following variables:

$q_i(k)$ is the arrival rate of vehicles in the direction i

$s_i(k)$ denotes the saturation flow rate of vehicles in the direction i

C is the duration of the cycle time

ℓ is the loss time due to the amber phase

G_i is the duration of the green in the direction i.

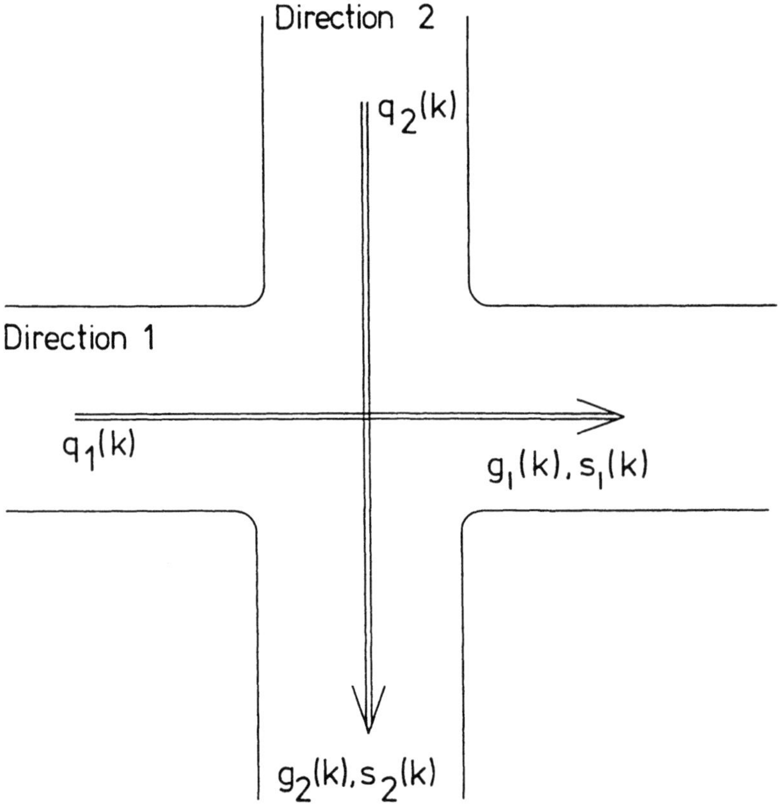

Fig. (4.3) An oversaturated intersection

Let $i=1$ denote the horizontal traffic direction and $i=2$ denote the vertical direction. If the average departure rate is $\bar{g}_i(k)$, then the control variable $u_i(k)$ can be taken as

$$u_i(k) = \bar{g}_i(k)/s_i(k) = G_i/C \qquad i=1,2$$

Since C is assumed constant [14], it is necessary to have only one control $u = u_i$ and the other control is $C - u_i - 1$. In practice, we know that too short a green is wasteful. On the other hand, if the green is too long, drivers tend to believe that the traffic signals have broken down. Hence the controls u must be subject to inequality constraints

$$u_2 \leqq u \leqq u_M$$

Taking the states $x_i(k)$ as the instantaneous queue length in the i-th direction, then from one cycle to the next, the evolution of the queues can be described by

$$x_i(k+1) = x_i(k) + q_i(k) - s_i u_i(k)$$

This simple model can be extended to deal with a traffic network by treating the interconnecting roads as pure delay elements. In the general case, we will have

$$\underline{x}(k+1) = A\underline{x}(k) + \sum_{j=0}^{m} B_j \underline{u}(k-j) + \underline{v}$$

as the dynamic model of the network with a queues $\underline{x}(k)$ and r traffic signal controls $\underline{u}(k)$. The states and controls are constrained by

$$0 \leqq \underline{x}(k) \leqq \underline{x}_M$$

$$\underline{u}_m \leqq \underline{u}(k) \leqq \underline{u}_M$$

The determination of the best control settings can be obtained by minimising delays through penalisation of longer queues [14].

Implementation of the optimal traffic signalling scheme requires multiprocessor arrays which are distributed along the road network and which function on a discrete-time basis.

The three examples presented so far are representative practical applications of computer-oriented control techniques. Indeed, the scope of applications is wide enough to include others like computer control of blast furnaces in the steel industry, modern digital control of machine components and load scheduling in electric power networks, etc. We should emphasise that these applications can be described in a natural way by discrete-time control models. To meet the objective of this part of the book, we have concentrated on linearised, deterministic system models.

In the rest of this chapter we will assume that the open-loop system is modelled in the time-domain by

$$\underline{x}(k+1) = A\underline{x}(k) + B\underline{u}(k) \qquad (4.1a)$$

$$\underline{y}(k) = C\underline{x}(k) + D\underline{u}(k) \qquad (4.1b)$$

where $\underline{x}(k) = \{x_1(k),\ldots,x_n(k)\}^t$ is the state vector, $\underline{u}(k) = \{u_1(k),\ldots,u_m(k)\}^t$ is the input vector, and $\underline{y}(k) = \{y_1(k),\ldots,y_p(k)\}^t$ is the output vector. The matrices A,B,C,D are real constants with dimensions (nxn), (nxm), (pxn) and (pxm), respectively. It is assumed that the model (4.1) characterises a plant to be controlled. A block diagram of this model is depicted in Fig. (4.4a).

In the frequency-domain, the plant is described by the input-output relationship

$$Y(z) = \{C(zI_n-A)^{-1}B + D\}U(z) \qquad (4.2)$$

where I_n is the (nxn) identity matrix, z is the complex variable in the Z-plane and Y(z), U(z) are the Z-transform of the sequences $\underline{y}(k)$, $\underline{u}(k)$, respectively. The z-transfer matrix expressed by (4.2) is shown in Fig. (4.4b). It should be observed that the matrix D provides a direct feedthrough from the input $\underline{u}(k)$, or U(z), to the output $\underline{y}(k)$, or Y(z). For a wide class of control engineering problems, this coupling link is absent which means that D=0. This case will receive much of our attention.

Our aim is to investigate the properties of system (4.1), or correspondingly (4.2), when a static feedback compensator is used in a closed-loop configuration.

4.2.1 *STATE FEEDBACK*

We will now study the problem of designing state feedback compensators for linear discrete-time systems of the type (4.1). We assume that all state variables are available for generating feedback signals. In practise, this would require the installation of many sensors and measuring devices. Occasionally, this is costly or it may be extremely difficult to implement for other reasons. A number of arguments could be fairly raised here concerning the utility and practical interest of linear state feedback control. In our opinion, the usefulness of state feedback lies in the following:

(1) By definition, the state variables provide a complete picture of the internal dynamics of the system. Hence, state feedback would provide the best possible control.

(2) As we shall see in Chapter 5, the reconstruction of state variables from input and output records can be achieved using efficient techniques.

(3) There are some cases where all the state variables can be recorded directly.

In view of the above, it is of interest to determine what can be accomplished by using linear state feedback. For this pur-

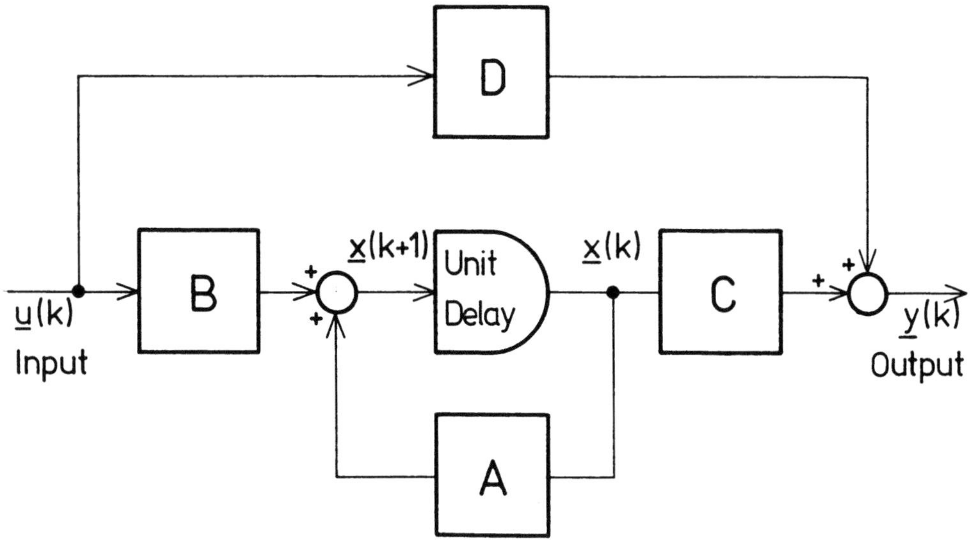

Fig. (4.4a) Discrete system in the time-domaine

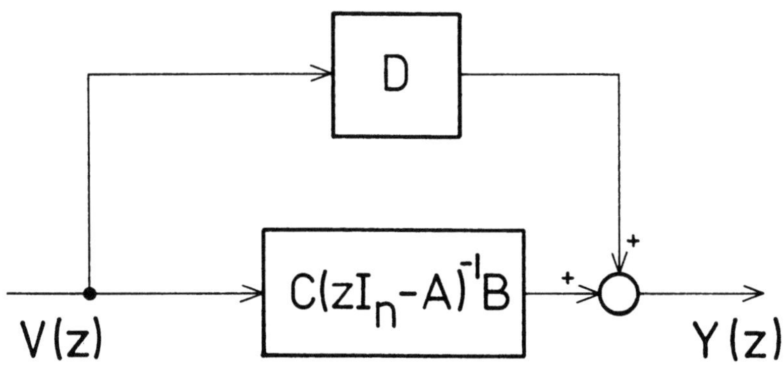

Fig. (4.4b) Discrete system in the frequency-domaine

pose, we analyse system (4.1) when the control input takes the form:

$$\underline{u}(k) = F\underline{v}(k) - K\underline{x}(k) \tag{4.3}$$

where K is the (nxn) matrix of design parameters, $\underline{v}(k)$ is an (mxl) new constant input vector and F is an (nxm) feedforward matrix. Since there is no systematic way to select F, we shall assume from now onwards that $F=I_m$, the m-dimensional identity matrix.

The state feedback control system, shown in Fig. (4.5), is obtained by using (4.3) in (4.1):

$$\underline{x}(k+1) = [A-BK]\ \underline{x}(k) + B\underline{v}(k) \tag{4.4a}$$

$$\underline{y}(k) = [C-DK]\ \underline{x}(k) + D\underline{v}(k) \tag{4.4b}$$

We observe that system (4.4) is similar to (4.1). It is therefore of interest to examine whether we can change the overall system characteristics by choice of the gain matrix K. We recall that

(i) the stability of the closed-loop system (4.4a) depends on the eigenvalues of the matrix $[A-BK]$.

(ii) from the pair $\{[A-BK],B\}$ one obtains both the controllability and reachability matrices.

(iii) observability (reconstructability) properties of (4.4b) can be assessed by use of the pair $\{[A-BK], [C-DK]\}$.

Thus, it is useful to understand the effect of state feedback on the system's structural properties.

We know that the reachability matrix of system (4.1) is

$$W_r = [B \ \vdots\ AB \ \vdots\ A^2B \ \vdots\ \ldots.A^{(n-1)}B]$$

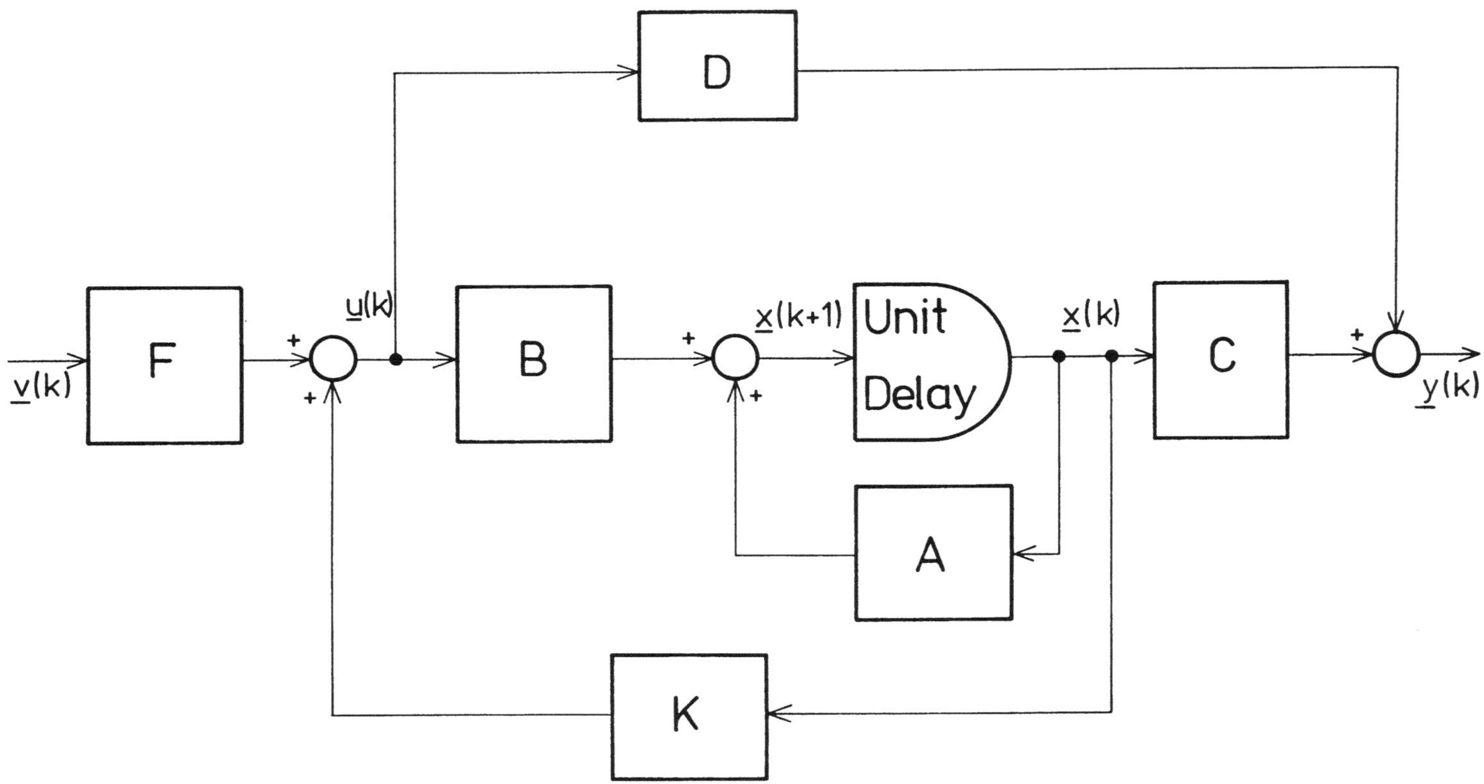

Fig. (4.5) State feedback discrete control system

whereas for system (4.4), the reachability matrix takes the form

$$\tilde{W}_r = [B \vdots (A-BK)B \vdots (A-BK)^2 B \vdots \dots \vdots (A-BK)^{(n-1)} B]$$

If system (4.1) is completely reachable [16,17] then the column vectors of W_r span the n-dimensional space which corresponds to rank $[W_r] = n$. It can be readily seen that a series of elementary column operations reduces the extra terms in $\tilde{W}_r$ to 0. This means that

$$\text{rank } [\tilde{W}_r] = \text{rank } [W_r] = n$$

for any gain matrix K, and we can therefore assert that: *"State Feedback does not alter the reachability of the open-loop system"*. Similar results can be developed for the controllability properties.

The reconstructability (observability) matrix is given by:

$$W_o = [C^t \vdots A^t C^t \vdots \dots A^{(n-1)^t} C^t]$$

For system (4.3), the reconstructability matrix is:

$$\tilde{W}_o = [(C-DK)^t \vdots (A-BK)^t (C-DK) \dots \vdots (A-BK)^{(n-1)^t} (C-DK)^t]$$

We now consider the following situations:

(a) The feedback matrix D is the null matrix. Unlike the reachability matrix, there is no guarantee that rank $[\tilde{W}_o]$ = rank $[W_o]$. This is because the columns of $K^t B^t C^t$ need not be linearly related to those of C^t.

(b) The gain matrix K is selected such that

$$C = DK$$

Obviously in this case observability is lost since rank $[\tilde{W}_o] \neq n$. One is therefore led to assert that:

"State Feedback can cause a loss of observability (reconstructability)".

In the light of the first assertion, control design based on state feedback is useful and robust. We know that the stability of linear discrete systems is closely related to the location of the eigenvalues in the complex plan *[3,6]*. Specifically, if the eigenvalues lie within the unit circle then the system is asymptotically stable. And more importantly, depending on the relative positions of the eigenvalues, the system response may have various forms as shown in the previous chapters. It is thus clear that by applying a feedback gain matrix K, the resulting closed-loop eigenvalues would be, in a broad sense, different from the open-loop eigenvalues. Therefore, we can assert that:

"State Feedback affects the stability of the discrete control system".

In the context of discrete-time systems, it turns out *[15]* that the state feedback gain matrix K plays a major role in all eigenvalue assignment algorithms. We now have the rigorous result *[15]*:

"The closed-loop characteristic values, that is, the eigenvalues of the matrix [A-BK], can be arbitrarily located in the complex plane subject to symmetry condition by choosing the gain K if and only if system (4.1) is completely reachable."

In order to demonstrate this fundamental result we consider, without loss of generality, single-input systems in phase-variable canonical form *[3]*:

$$\underline{x}(k+1) = \begin{bmatrix} 0 & 1 & 0 & \cdots & 0 \\ 0 & 0 & 1 & \cdots & 0 \\ \cdot & \cdot & \cdot & \cdot & \\ 0 & 0 & \cdots & 0 & 1 \\ -\alpha_0 & -\alpha_1 & \alpha_2 & & -\alpha_{n-1} \end{bmatrix} \underline{x}(k) + \begin{bmatrix} 0 \\ 0 \\ \vdots \\ \vdots \\ 1 \end{bmatrix} \underline{u}(k) \qquad (4.5)$$

where the scalars α_i; $i=0,1,\dots,n-1$ are the coefficients of the characteristic polynomial of the system (4.5), that is

$$\det(zI_n - A) = z^n + \alpha_{n-1} z^{n-1} + \dots + \alpha_n z + \alpha_0.$$

Let the scalar state feedback control

$$u(k) = v(k) - \underline{g}^t \underline{x}(k)$$

where the column gain $\underline{g}$ is given by:

$$\underline{g}^t = [\beta_1 \ \beta_2 \dots \beta_n]$$

be applied to (4.5), then the resulting closed-loop system becomes:

$$\underline{x}(k+1) = \begin{bmatrix} 0 & 1 & 0 & \dots & \dots & 0 \\ 0 & 0 & 1 & 0 & \dots & 0 \\ \dots & \dots & \dots & \dots & \dots & \dots \\ 0 & 0 & 0 & 0 & & 1 \\ -(\alpha_0+\beta_1) & -(\alpha_1+\beta_2) & \dots & \dots & \dots & -(\alpha_{n-1}+\beta_n) \end{bmatrix} \underline{x}(k)$$

$$+ \begin{bmatrix} 0 \\ 0 \\ \cdot \\ \cdot \\ \cdot \\ 1 \end{bmatrix} v(k) \qquad (4.6)$$

By the similarity to the system (4.5), the characteristic polynomial of the system (4.6) has the coefficients $(\alpha_j+\beta_{j+1})$; $j=0,1,\dots,n-1$. It should be noted that the scalars β_j; $j=1,\dots,n$ can be arbitrarily selected. Thus, it is clear that the coefficients of the closed-loop characteristic polynomial can be assigned preselected values. Obviously, complex eigenvalues should occur in conjugate pairs (symmetry condition). This demonstrates the general result obtained in [15].

We now examine the effects of output feedback control on the system properties.

4.2.2 *OUTPUT FEEDBACK*

If a finite number of state variables or linear combinations of these can only be used for forming feedback signals, then the control law can be expressed as

$$\underline{u}(k) = F_O\underline{s}(k) - K_O\underline{y}(k) \tag{4.7}$$

where K_O is the (mxp) matrix of design parameters, $\underline{s}(k)$ is an (mxl) new constant input vector and F_O is an (nxm) feed-forward matrix. The closed-loop output feedback is shown in Fig. (4.6).

From (4.1b) and (4.7) we obtain:

$$\underline{y}(k) = C\underline{x}(k) + D\{F_O\underline{s}(k) - K_O\underline{y}(k)\}$$

or

$$\underline{y}(k) = (I_p + DK_O)^{-1} \{C\underline{x}(k) + DF_O\underline{s}(k)\} \tag{4.8}$$

The substitution of (4.8) in (4.1a) results in:

$$\begin{aligned}\underline{x}(k+1) = {} & \{A - BK_O(I_p + DK_O)^{-1}C\}\underline{x}(k) \\ & + B\{F_O - K_O(I_p + DK_O)^{-1}DF_O\}\underline{s}(k)\end{aligned} \tag{4.9}$$

Making use of the matrix inversion identity

$$(I_m + K_OD)^{-1} = I_m - K_O(I_p + DK_O)^{-1}D$$

simplifies (4.9) to

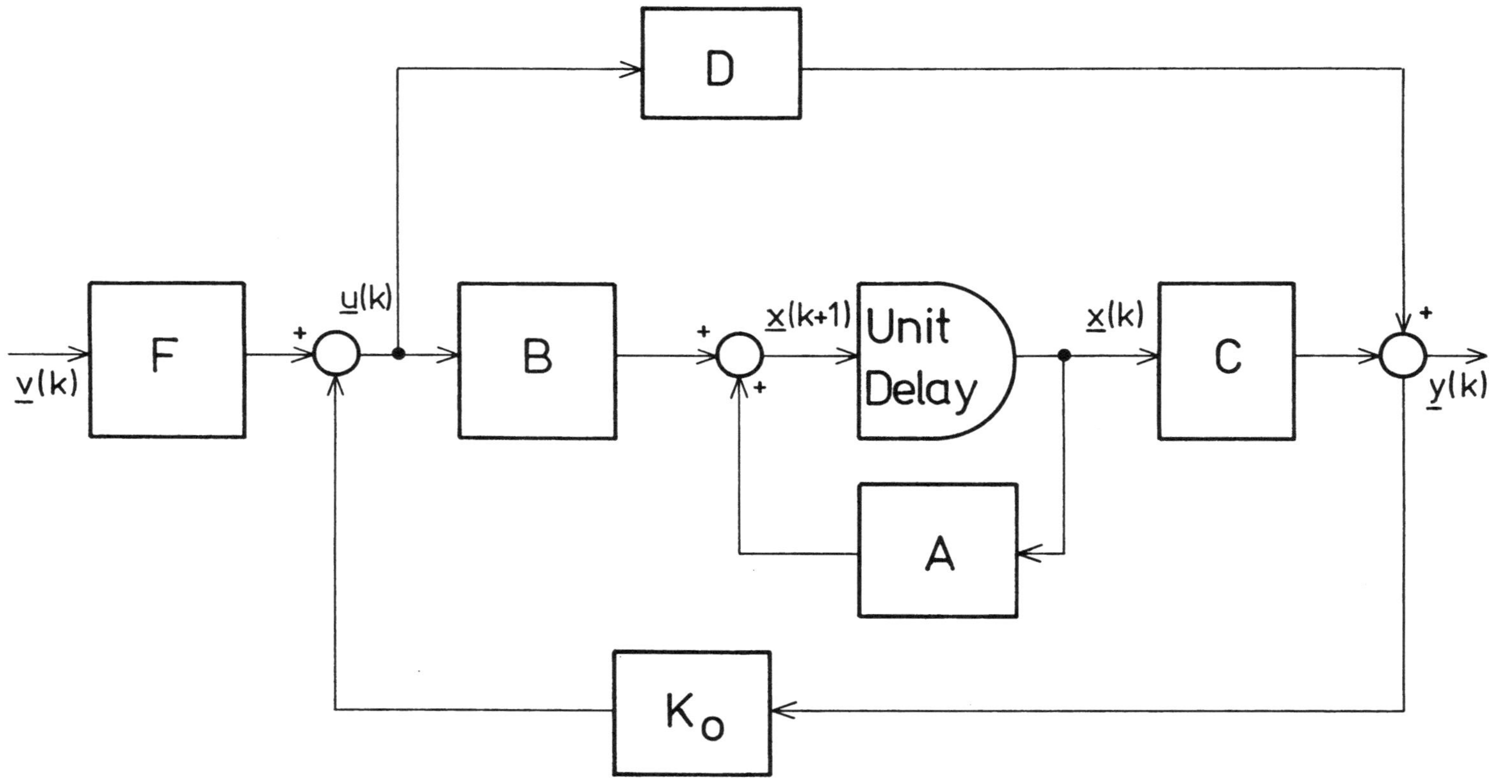

Fig. (4.6) Output feedback discrete control system

$$\underline{x}(k+1) = \{A - BK_O(I_p + DK_O)^{-1}C\}\ \underline{x}(k) + B(I_m + K_OD)^{-1}F_O\underline{s}(k) \tag{4.10}$$

We note that (4.10) and (4.8) are of the same form as (4.1a) and (4.1b). Their structural properties are determined by the triple $(\{A - BK_O(I_p + DK_O)^{-1}C\}, B(I_m + K_OD)^{-1}F_O, (I_p + DK_O)^{-1}C\}$. For simplicity, we assume that the matrix $(I_p + K_OD)$ is nonsingular and rank $(F_O) \doteq p$.

Recall that the reachability matrices $\tilde{W}_r$ and W_r have the same rank for any K, including the special case $K = BK_O(I_p + DK_O)^{-1}C$, therefore we can assert that:

"The reachability property is not altered when output feedback is applied".

In general, the open- and closed-loop systems have the same observability property. The output feedback observability (reconstructability) matrix is given by

$$\tilde{W}_O = \left(C^t(I_p+DK_O)^{-t} \vdots \{A-BK_O(I_p+DK_O)^{-1}C\}^t\ C^t(I_p+DK_O)^{-t} \vdots \right. \\ \left. \cdot\text{------} \vdots \{A-BK_O(I_p+DK_O)^{-1}C\}^{(n-1)^t} C^t(I_p-DK_O)^{-t}\right)$$

Due to the presence of the matrix C in the system matrix of (4.10) the columns of $C^t(I_p+DK_O)^{-t}K_O^tB^tC^t(I_p+DK_O)^{-t}$ can be shown to be linearly related to the columns C^t. A series of elementary column operations can therefore be used to reduce $\tilde{W}_O$ to W_O and to make rank $(\tilde{W}_O)$ = rank (W_O). Hence we assert that

"System reconstructability (observability) is preserved when output feedback is used".

We note here the difference between state and output feedback is affecting the observability property.

As for the state feedback, output feedback affects the system stability and can alter the open-loop eigenvalues. However, the degree of freedom one has in specifying closed-loop eigenvalue locations by choice of gain matrices depends on whether a state or output feedback is used. It is known that we can assign n eigenvalues to desired locations using state feedback [15], whereas we can move only a max (m,p) eigenvalues arbitrarily close to desired values using an output feedback [18].

We are now in a position to provide methods of determining the gain matrices K or K_O depending on whether we use state or output feedback.

4.2.3 *COMPUTATIONAL ALGORITHMS*

We have seen in the last subsections that if the discrete system is completely reachable, then the closed-loop eigenvalues can be placed at desired locations by the choice of a state feedback gain matrix. In the following, we first consider methods of computing the elements of the state feedback gain K. The problem is that given a desired closed-loop eigenvalue set $\{\gamma_i,\dots,\gamma_n\}$ determine the gain matrix K such that the eigenvalues of [A-BK] are $\{\gamma_i,\dots\gamma_n\}$. Let $\{\gamma_1,\dots,\gamma_n\}$, $[\underline{v}_1,\dots,\underline{v}_n]$ and $[\underline{w}_1^t,\dots,\underline{w}_n^t]$ be respectively the eigenvalues, eigenvectors and reciprocal eigenvectors of the matrix A. The first algorithm due to Young and Willems [19] is called the mapping approach which yields dyadic design. A summary of the steps involved is given below.

(i) Define K as the outer produce

$$K = \underline{f}\,\underline{h}^t \tag{4.11}$$

where $\underline{f}$ and $\underline{h}$ are m- and n-vectors.

(ii) Choose the vector $\underline{f}$ such that the resulting pseudo-single-input system [A,Bf] is completely reachable.

(iii) Calculate the open-loop and closed-loop system characteristic polynomials

$$\Delta_c(z) = \prod_{i=1} (z_i-\gamma_i) \qquad (4.12a)$$

$$= z^n+d_1 z^{n-1} +....+ d_n$$

$$\Delta_o(z) = \det[zI_n-A]$$

$$= z^n+a_1 z^{n-1} +....+ a_n \qquad (4.12b)$$

and obtain the difference polynomial

$$\Delta_d(z) = \Delta_c(z) - \Delta_o(z)$$

$$= (d_1-a_1)z^{n-1} +....+ (d_n-a_n) \qquad (4.13)$$

(iv) Define the matrices

$$W_r = [B\underline{f}, AB\underline{f}, ..., A^{n-1}B\underline{f}] \qquad (4.14a)$$

$$X = \begin{bmatrix} 1 & 0 & 0 & 0 \\ a_1 & 1 & \vdots & \vdots \\ \vdots & a_1 & \vdots & \vdots \\ a_{n-2} & a_{n-1} & 1 & \vdots \\ a_{n-1} & a_{n-2} \cdots & a_1 & 1 \end{bmatrix} \qquad (4.14b)$$

$$\underline{\Pi} = [(d_1-a_1), (d_2-a_2), ..., (D_n-a_n)]^t \qquad (4.14c)$$

then the required vector $\underline{h}^t$ is determined as

$$\underline{h} = [W_r^t]^{-1} X \underline{\Pi} \qquad (4.15)$$

We note that by definition the (nxn) matrix X is nonsingular and so is the matrix W_r. Also, these need only be calculated once for a variety of closed-loop system eigenvalue specification.

Another algorithm of finding a matrix K which yields any specified set of eigenvalues $\{\lambda_i\}$ is based on the fact that reachability is sufficient for arbitrary eigenvalue placement [20]. We know that the eigenvalues of the closed-loop state feedback system are roots of:

$$\Delta_c(\gamma) \triangleq \det[\gamma I_n - A + BK] = 0 \qquad (4.16)$$

An alternative way to write (4.16) is

$$\begin{aligned} \Delta_c(\gamma) &= \det[(\gamma I_n - A)\{I_n + (\gamma I_n - A)^{-1}BK\}] \\ &= \det[\gamma I_n - A]\det[I_n + (\gamma I_n - A)^{-1}BK] \\ &= \Delta_o(\gamma).\det[I_n + (\gamma I_n - A)^{-1}BK] \end{aligned} \qquad (4.17)$$

Using the determinant identity

$$\det[I_n + (\gamma I_n - A)^{-1}BK] =$$

$$\det[I_m + K(\gamma I_n - A)^{-1}B]$$

in (4.17) we have

$$\Delta_c(\gamma) = \Delta_o(\gamma)\det[I_m + K(\gamma I_n - A)^{-1}B] \qquad (4.18)$$

We emphasise that the gain matrix K must be selected such that $\Delta_c(\gamma_i) = 0$ for every γ_i. This will be accomplished by forcing the (mxn) determinant in (4.18) to vanish. If any desired γ_i is also a root of $\Delta_o(\gamma)$ the following procedure can be used [20].

Define

$$I_m = [\underline{e}_1, \ldots, \underline{e}_j, \ldots, \underline{e}_n]$$

$$\begin{aligned} \Psi(\gamma_i) &= (\gamma_i I_n - A)^{-1}B \\ &= [\Psi_1(\gamma_i), \ldots, \Psi_j(\gamma_i), \ldots, \Psi_n(\gamma_i)] \end{aligned}$$

Then γ_i is a root of $\Delta_c(\gamma)$ if K is selected to satisfy

$$\underline{e}_j + K\,\Psi_j(\gamma_i) = \underline{0}$$

since this forces column j to be zero. Thus

$$K\Psi_j(\gamma_i) = \underline{e}_j \qquad (4.19)$$

Since the pair (A,B) is reachable then rank $[\Psi(\gamma_i)] = m$ for each γ_i [20]. Two possible cases arise:

(1) All the desired eigenvalues are distinct.
In this case it is always possible to find n linearly independent columns $\underline{\Psi}_{j_1}(\gamma_1), \underline{\Psi}_{j_2}(\gamma_2), \ldots, \underline{\Psi}_{j_n}(\gamma_n)$ from the columns of the (nxm) matrix $[\Psi(\gamma_1), \Psi(\gamma_2), \ldots \Psi(\gamma_n)]$. Therefore

$$K = -EG^{-1} \qquad (4.20a)$$

where

$$E = [\underline{e}_{j_1} \underline{e}_{j_2} \ldots \underline{e}_{j_n}] \qquad (4.20b)$$

$$G = [\underline{\Psi}_{j_1}(\gamma_1)\underline{\Psi}_{j_2}(\gamma_2) \ldots \underline{\Psi}_{j_n}(\gamma_n)]^{-1} \qquad (4.20c)$$

(2) Repeated eigenvalues are desired.
If γ_i is desired to have algebraic multiplicity r_i, then

$$\left.\frac{d^t\Delta_c(\gamma)}{d\gamma^j}\right|_{\gamma=\gamma_i} = 0, \quad j=1,2,\ldots,r_i-1$$

But

$$\frac{d}{d\gamma}\Delta_c(\gamma) = \{\frac{d}{d\gamma}\Delta_0(\gamma)\}\det[I_m+K(\lambda I_m-A)^{-1}B]$$

$$+ \det\left[K\frac{d\underline{\Psi}_1}{d\gamma} \vdots\ \underline{e}_2 + K\underline{\Psi}_2 \ \vdots \ldots\ldots\ldots\right.$$

$$\left.\vdots \underline{e}_m + K\underline{\Psi}_m\right] \Delta_O(\gamma)$$

$$+ \det \left[\underline{e}_1 + K\underline{\Psi}_1 \vdots K \frac{d\underline{\Psi}_2}{d\gamma} \vdots \cdots\cdots\cdots \right.$$

$$\left.\vdots \underline{e}_m + K\underline{\Psi}_m\right] \Delta_O(\gamma) + \cdots$$

$$+ \det \left[\underline{e}_1 + K\underline{\Psi}_1 \vdots \underline{e}_2 + K\underline{\Psi}_2 \vdots \cdots\cdots \right.$$

$$\left.\vdots K \frac{d\underline{\Psi}_m}{d\gamma}\right] \Delta_O(\gamma) \qquad (4.21)$$

Examining (4.21) reveals that when $K\underline{\Psi}_j(\gamma_i) = -\underline{e}_j$ then every term in (4.21) has one zero column except the determinant containing $Kd\underline{\Psi}_j/d\gamma$. Thus an additional independent relation for forcing $\Delta_O(\lambda_i) = 0$ is

$$K \left.\frac{d\underline{\Psi}_j}{d\gamma}\right|_{\gamma=\gamma_i} = \underline{O} \qquad (4.22)$$

If necessary, higher-order derivatives can also be used such that a total of n linearly independent columns, either of $\underline{\Psi}_{j_i}(\gamma_i)$ or its derivatives, is selected.

The above two algorithms are useful for eigenvalue placement in discrete systems. For a wider coverage of pole assignment algorithms, the reader is referred to [21].

In case of eigenvalue assignment by output feedback, we will next present a version of the second algorithm of state feedback. The eigenvalue of the closed-loop output feedback system (4.10) are roots of:

$$\Delta_f(\gamma) = \det [\gamma I_n - A + BK_O(I_p + DK_O)^{-1}C] = 0 \qquad (4.23)$$

Manipulating (4.23) we arrive at

$$\Delta_f(\gamma) = \Delta_0(\gamma)\det[I_m + K_0(I_p+KD_0)^{-1}C(\gamma I_n-A)^{-1}B] \tag{4.24}$$

Define

$$\Omega(\gamma) = C(\gamma I_n-A)^{-1}B \tag{4.25}$$

Then it is well-known [22] that the number of linearly independent columns $\underline{\omega}_j(\gamma_i)$ (and their derivatives if γ_i is repeated) which can be found from the (p x mp) matrix

$$[\Omega(\gamma_1)\Omega(\gamma_2)\ldots\Omega(\gamma_n)]$$

will never exceed rank C. Provided the pair (A,B) is completely reachable and C has the full rank p ($p \leqq n$) we can assign p eigenvalues close to their desired positions. By similarity to (4.20) we have the output feedback gain

$$K_0 = -[I_m+E_0G_0^{-1}D]^{-1}E_0G_0^{-1} \tag{4.26a}$$

where

$$G_0 = [\underline{\omega}_{j_1}(\gamma_1)\underline{\omega}_{j_2}(\gamma_2)\ldots\ldots\underline{\omega}_{j_p}(\gamma_p)] \tag{4.26b}$$

$$E_0 = [\underline{e}_1\ldots\ldots\ldots\underline{e}_p] \tag{4.26c}$$

Here again, discussions of other pole assignment methods using output feedback are included in [21].

4.2.4 *EIGEN-STRUCTURE ASSIGNMENT*

It has been observed in the last section that specification of closed-loop eigenvalues for state feedback design of multivariable systems does not yield a unique closed-loop system. The nonuniqueness allows the designer to have additional freedom which can be exploited in different ways. Thus the eigenvalue assignment algorithms can be extended to define other features of the closed-loop systems. We shall describe hereafter some algorithms which assign the entire eigenstructure; that is the desired eigenvalues and their associated eigenvectors [46-50].

We start with an algorithm which is a generalisation of the second algorithm presented in the last sections and is due to Fahmy and O'Reilly [48]. Recall that the use of state feedback control $u(k) = -K\underline{x}(k)$ in the linear system (4.1a) yields the closed-loop system

$$\underline{x}(k+1) = (A-BK)\underline{x}(k)$$

whose characteristic polynomial is given by (4.18). Let the desired set of eigenvalues $\{\gamma_j\}$ be distinct and self-conjugate. Now, to determine the feedback gain matrix K we should look for a procedure to nullify the determinant in (4.18). It is shown [48] that one such procedure necessitates that the columns of the matrix

$$[I_m+K(\gamma_j I_n-A)^{-1}B] = [I_m+K\Psi(\gamma_j)] \tag{4.27a}$$

be linearly dependent. This, in turn, implies that for a non-zero set of vectors $\{\underline{r}_j\}$ we have

$$[I_m+K\Psi(\gamma_j)]\underline{r}_j = \underline{0}$$

or

$$K\Psi(\gamma_j)\underline{r}_j = -\underline{r}_j \tag{4.27b}$$

If the vectors $\{\underline{r}_j\}$ are arbitrarily selected such that the resulting set of vectors $\{\Psi(\gamma_j)\underline{r}_j \; ; \; j=1,\ldots,n\}$ are linearly independent, then from (4.27b) we get

$$K = -RG^{-1} \tag{4.28a}$$

where

$$R = [\underline{r}_1 \underline{r}_2 \cdots \underline{r}_n] \qquad (4.28b)$$

$$G = [\Psi(\gamma_1)\underline{r}_1 \quad \Psi(\gamma_2)\underline{r}_2 \cdots \Psi(\gamma_n)\underline{r}_n] \qquad (4.28c)$$

It is evident that the formula (4.28) is more general than (4.20) since $\underline{r}_j$ now replaces $\underline{e}_j$; the jth unit vector.

In order to examine the role of the set of vectors $\{\Psi(\gamma_j)\underline{r}_j\}$, we proceed as follows. Starting with the n-vector.

$$(A-BK)\Psi(\gamma_j)\underline{r}_j \qquad (4.29)$$

and utilising $\Psi(\gamma_j)$ from (4.27a), it becomes

$$(A-BK)\Psi(\gamma_j)\underline{r}_j =$$

$$(A-BK)(\gamma_j I_n - A)^{-1} B\underline{r}_j =$$

$$A(\gamma_j I_n - A)^{-1} B\underline{r}_j - BK(\gamma_j I_n - A)^{-1} B\underline{r}_j$$

This can be further simplified by substituting (4.27b) into

$$A(\gamma_j I_n - A)^{-1} B\underline{r}_j + B\underline{r}_j =$$

$$(A+\gamma_j I_n - A)(\gamma_j I_n - A)^{-1} B\underline{r}_j =$$

$$\gamma_j \Psi(\gamma_j)\underline{r}_j \qquad (4.30)$$

In view of (2.107), it is readily seen from (4.29) and (4.30) that the set of vectors $\{\Psi(\gamma_j)\underline{r}_j\}$ are characteristic vectors for the matrix (A-BK); the closed-loop system matrix, associated with the set of desired eigenvalues $\{\gamma_j\}$. This interesting result shows that

$$(A-BK)\underline{v}_j = \gamma_j \underline{v}_j \qquad j=1,\ldots,n \qquad (4.31)$$

which can be put in the form

$$[A-\gamma_j I_n \;\vdots\; B] \begin{pmatrix} \underline{v}_j \\ \\ \underline{f}_j \end{pmatrix} = \underline{0} \tag{4.32}$$

where $\underline{f}_j = -K\underline{v}_j$. It is well-known [20,39] that in order to satisfy (4.32), the (n+m)-vector $[\underline{v}_j^t \quad \underline{f}_j^t]$ must lie in the null space of the extended matrix

$$\Phi(\gamma_j) = [A-\gamma_j I_n \;\vdots\; B] \tag{4.33}$$

From the definition of $\underline{f}_j$, we can write

$$-[\underline{f}_1 \ldots\ldots \underline{f}_n] = [K\underline{v}_1 \ldots\ldots K\underline{v}_n] \tag{4.34}$$

Since K can be factored from (4.34) and the set of vectors $\{\underline{v}_j\}$ are linearly independent, we arrive at

$$K = -FV^{-1} \tag{4.35a}$$

where

$$F = [\underline{f}_1 \quad \underline{f}_2 \ldots\ldots \underline{f}_n] \tag{4.35b}$$

$$V = [\underline{v}_1 \quad \underline{v}_n \ldots\ldots \underline{v}_n] \tag{4.35c}$$

the freedom in selecting the vectors $\{\underline{r}_j\}$ is successfully exploited in assigning the entire eigenstructure. It should be noted that the closed-loop matrix (A-BK) will by now have a unique Jordan form given by

$$J = G^{-1}(A-BK)G$$

where G is defined by (4.28c).

The case of repeated eigenvalues can be treated in the same manner and can be found in [48].

It should be stressed that one of the basic ideas behind the development of the above algorithm is the fact that the set of vectors $\{\Psi(\gamma_j)\underline{r}_j\}$ belong to the range space of $(\gamma_j I_n - A)^{-1}B$. A slightly different idea has been pursued by Porter and D'Azzo [49-50] to develop an algorithm for entire eigenstructure assignment. To summarise their algorithm, we let $\{\gamma_j\}$ and $\{\underline{v}_j\}$ be the desired set of eigenvalues and associated eigenvectors for the closed-loop ststem; respectively. Thus we note that the main effort in implementing this algorithm lies in manipulating (4.32) to determine the vectors $\{\underline{v}_j^t \quad \underline{f}_j^t\}$ using efficient numerical routines. In both algorithms a real gain matrix is achieved whenever self-conjugate sets of eigenvalues and eigenvectors are specified a priori.

4.2.5 *REMARKS*

We now note certain aspects concerning the eigenstructure assignment algorithms. Initially, we must stress the fact that the eigenvectors, according to Fahmy and O'Reilly [48], arise quite naturally. More importantly, it can be easily shown that they satisfy the conditions posed in [47]. The method of Porter and D'Azzo [49,50] is based on the selection of an appropriate null space for an extended system matrix. By identifying a specific subspace in which the desired eigenvectors are located, the entire eigenstructure can be assigned. Looked at in this light, it is readily evident that the eigenvectors selected according to [48] constitute a basis for the null space considered in [49,50].

Therefore, it would seem that the first algorithm for eigenstructure assignment [48] requires less computational effort than the second algorithm [49,50]. In particular, the first algorithm is directly applicable to repeated eigenvalues, whereas the second algorithm needs some modifications. On the other hand, the procedure of the second algorithm allows the treatment of systems with unreachable modes. This is not the case for the first algorithm in which an additional effort is needed to isolate the unreachable modes (i.e. those which cannot be changed by state feedback) and assigning the associated eigenvectors.

4.2.6 *EXAMPLE*

The purpose of this example is to illustrate the use of state feedback eigenvalue assignment algorithms. Consider the discrete control system

$$\underline{x}(k+1) = \begin{bmatrix} .5 & .5 & 0 \\ 0 & 1 & 0 \\ .83 & 2.17 & -.33 \end{bmatrix} \underline{x}(k) + \begin{bmatrix} 3 & 1 \\ 2 & 0 \\ -1 & 1 \end{bmatrix} \underline{u}(k)$$

The characteristic polynomial is

$$\det[\gamma I_3 - A] = \det \begin{bmatrix} \gamma-.5 & -.5 & 0 \\ 0 & \gamma-1 & 0 \\ -.83 & -2.17 & \gamma+.33 \end{bmatrix}$$

$$= (\gamma-1)(\gamma-.5)(\gamma+.33)$$

and has roots at 1, .5 and .33. It is desired to relocate the eigenvalues at .167, .33 and .667. To apply the formula (4.27) we first compute

$$(\gamma I_3 - A)^{-1} =$$

$$\begin{bmatrix} (\gamma-1)(\gamma+.33) & .5(\gamma+.33) & 0 \\ 0 & (\gamma-.5)(\gamma+.33) & 0 \\ .83(\gamma-1) & .417-2.167(\gamma-.5) & (\gamma-1)(\gamma-5.) \end{bmatrix}$$

and then

$$\psi(\gamma) = (\gamma I_3 - A)^{-1} B$$

$$= \frac{1}{(\gamma-1)(\gamma-.5)(\gamma+.33)} \begin{pmatrix} 3\gamma(\gamma+.33) & (\gamma-1)(\gamma+.33) \\ 2(\gamma-.5)(\ +.33) & 0 \\ (\gamma-1)(3-\gamma)+.83 & (\gamma-1)(\gamma+.33) \\ -4.33(\gamma-.5) & \end{pmatrix}$$

At the desired eigenvalues, the above $\psi(\gamma)$ matrix are evaluated to yield

$$\psi(.167) = \begin{pmatrix} 1.8 & -0 \\ -2.4 & 0 \\ -.6 & -3 \end{pmatrix}$$

$$\psi(.33) = \begin{pmatrix} 9 & -12 \\ -3 & 0 \\ -10.5 & -12 \end{pmatrix}$$

$$\psi(.667) = \begin{pmatrix} -36 & 6 \\ -6 & 0 \\ 12 & 6 \end{pmatrix}$$

We construct the G matrix as:

$$G = [\psi_2(.167)\ \psi_1(.33)\ \psi_1(.667)]$$

$$= \begin{pmatrix} -3 & 9 & -36 \\ 0 & -3 & -6 \\ -3 & -10.5 & 12 \end{pmatrix}$$

The corresponding E matrix is given by

$$E = \begin{pmatrix} 0 & 1 & 1 \\ 1 & 0 & 0 \end{pmatrix}$$

Finally, we compute the gain matrix K from (4.27):

$$K = -E\,G^{-1}$$

$$= -\begin{bmatrix} 0 & 1 & 1 \\ 1 & 0 & 0 \end{bmatrix} \begin{bmatrix} -3 & 9 & -36 \\ 0 & -3 & -6 \\ -3 & -10.5 & -1 \end{bmatrix}^{-1}$$

$$= \begin{bmatrix} -.0115 & .2586 & -.0015 \\ -.1264 & .3448 & -.2069 \end{bmatrix}$$

To validate the obtained result, we compute the matrix

$$A + BK = \begin{bmatrix} .4081 & .069 & -.2534 \\ .023 & .4828 & -.023 \\ .6921 & 2.7734 & -.5254 \end{bmatrix}$$

whose eigenvalues are .67, .33 and .667 which is the desired result. Indeed, the gain matrix is not unique.

Next, we consider an important class of controllers, that is deadbeat controllers.

4.3 Deadbeat Controllers

4.3.1 *PRELIMINARIES*

The salient feature of deadbeat controllers [16,17] is that they can derive discrete ststems of the form (4.1) to the origin in a finite number of steps. To develop deadbeat controllers and examine their properties, we need the following definitions:

Definition 1 (Weiss, 1972)

The system (4.1) with rank (B) = m is completely θ-step reachable if there exists a control sequence

$$U_1 = \{\underline{u}(k-\theta+1), \underline{u}(k-\theta+2), \ldots . \underline{u}(k)\}$$

such that the state of (4.1) is transferred from the origin to $\underline{x}(k)$ in θ time steps under the action of U_1.

It has been shown in [23] that a necessary and sufficient condition for (4.1) to be completely θ-step reachable at time k is

$$\text{rank } [B, AB, \ldots . A^{\theta-1} B] = n \tag{4.36}$$

$$\text{for some } \theta \leq n$$

Definition 2:

The system (4.1) with rank (B) = m is completely θ-step controllable if there exists a control sequence

$$U_2 = \{\underline{u}(k), \underline{u}(k+\ ), \ldots , \underline{u}(k+\theta-1)\}$$

such that the state of (4.1) is transferred from $\underline{x}(k)$ to the origin in θ time steps under the action of U_2.

In general, criterion (4.36) is a sufficient condition for complete θ-step controllability. It is not however a necessary condition [23] unless A is invertible, and an alternative criterion is used. The controllability criterion is

$$\text{rank}[A^{-1}B, A^{-2}B, \ldots , A^{-\theta}B] = n$$

$$\text{for some } \theta \leq n \tag{4.37}$$

The scalar θ is known as the reachability index or controllability index (or both) of the system (4.1). In view of the above discussions, the deadbeat control problem is defined as [25]:

Find a control sequence $U_2 = \{\underline{u}(k), \underline{u}(k+1), \ldots, \underline{u}(k+\theta-1)\}$ *such that the state of (4.1) at any instant* k *is transferred from* $\underline{x}(k)$ *to the origin in at most* $\theta \leqq n$ *time steps.*

It should be noted that the formulation of the above problem is due to Kalman [16] who makes no a priori assumption about the nature of the control sequence U_2. A selection procedure for choosing n linearly independent columns of (4.37) to form a basis for the state space is provided by Mullis [24]:

Let $\{C_q\}$ be a sequence of nxm matrices.

Let $n_q = \text{rank}[C_1 \ldots C_q] - \text{rank}[C_1 \ldots C_{q-1}]$.

An ordered selection for $\{C_q\}$ is a sequence of matrices $\{D_q\}$ with $D_q(mxn_q)$ for which

$$R[C_1 D_1, \ldots, C_q D_q] = R[C_1, \ldots, C_q] \tag{4.38}$$

for each q, where $R[W]$ is the range space of W. The matrix $[C_1 D_1, \ldots, C_q D_q]$ has full rank and therefore every $\underline{x} \in R[C_1, \ldots, C_q]$ has a unique decomposition

$$\underline{x} = C_1 D_1 \underline{v}_1 + \ldots + C_q D_q \underline{v}_q \tag{4.39}$$

where $\underline{v}_j \in R^{nj}$. Next, we derive the general multi-input deadbeat controller.

4.3.2 *THE MULTI-INPUT DEADBEAT CONTROLLER*

The multi-input deadbeat controller, which is also known as the time-optimal controller, is a generalisation of the original work of Kalman [16]. It is based on the controllability criterion (4.37). Consider the linear discrete system (4.1) to be completely θ-step controllable. Expanding (4.1a) over $0 \leqq k \nleqq q$ given the controls $\underline{u}(0), \ldots, \underline{u}(q-1)$, one arrives at

$$\left.\begin{aligned} \underline{x}(1) &= A\underline{x}(0) + B\underline{u}(0) \\ \underline{x}(2) &= A\underline{x}(1) + B\underline{u}(1) \\ &\vdots \\ \underline{x}(q) &= A\underline{x}(q-1) + B\underline{u}(q-1) \end{aligned}\right\} \tag{4.40}$$

The set of initial states, represented by $\Gamma_q(A,B)$, that can be driven to the origin in q time steps is obtained from (4.40) by setting $\underline{x}(q) = \underline{O}$ as:

$$\Gamma_q(A,B) = \{\underline{x}(0) = - \sum_{j=1}^{q} A^{-j}B\underline{u}(j-1)\} \tag{4.41}$$

This implies that

$$\Gamma_q(A,B) = R[A^{-1}B,\ldots,A^{-q}B] \tag{4.42}$$

It can be readily seen from (4.42) that [24]:

$$\left.\begin{aligned} &\Gamma_0 = \{0\} \\ &\Gamma_{q+1} = A^{-1}(\Gamma_q + R[B]) \\ &\Gamma_0 \subset \Gamma_1 \subset \Gamma_2 \subset \ldots\ldots \subset R^n \end{aligned}\right\} \tag{4.43}$$

In the light of criterion (4.37) and condition (4.42), we have

$$\Gamma_q = \Gamma_\theta \qquad \text{for all } q \geq \theta \tag{4.44}$$

The interpretation of property (4.44) is that the initial state may be transferred to the origin in θ time steps and it then remains at the origin for $q \geq \theta$.

We now choose an ordered selection D_q for (4.42) in the manner of [24] such that

$$R[A^{-1}B,\ldots\ldots,A^{-\theta}B] = R[A^{-1}BD_1,\ldots\ldots A^{-\theta}BD_\theta] \tag{4.45}$$

By virtue of (4.41) and (4.45) we can write $\underline{x}(0)$ as

$$\underline{x}(0) = -S\underline{g} \tag{4.46}$$

where

$$S \triangleq [A^{-1}BD_1,\ldots.A^{-\theta}BD_\theta] \tag{4.47a}$$

$$\underline{g} = [\underline{v}^t(0),\ldots,\underline{v}^t(\theta-1)]^t \tag{4.47b}$$

$$\underline{u}(j) = D_{j+1}\underline{v}(j) \qquad j = 0,\ldots,\theta-1 \tag{4.47c}$$

The inversion of (4.46) yields

$$\underline{g} = \begin{bmatrix} \underline{v}(0) \\ \vdots \\ \underline{v}(\theta-1) \end{bmatrix} = S^{-1}\underline{x}(0)$$

$$\triangleq \begin{bmatrix} G_1 \\ \vdots \\ G_\theta \end{bmatrix} \underline{x}(0) \tag{4.48}$$

We observe that

$$- \begin{bmatrix} G_1 \\ \vdots \\ G_\theta \end{bmatrix} S = I_n \tag{4.49}$$

implies that the matrix G_1 is given by

$$G_1[A^{-1}BD_1,\ldots,A^{-\theta}BD_\theta] = -[I_{m_1},0_{m_2},\ldots,0_{m_\theta}] \tag{4.50}$$

Combining (4.47c), (4.48) and (4.450), we obtain the control input $\underline{u}(0)$ as [25]:

$$\underline{u}(0) = D_1\underline{v}(0)$$

$$= D_1G_1\underline{x}(0) \triangleq F\underline{x}(0) \qquad (4.51)$$

where the (mxn) gain matrix F satisfies

$$F[A^{-1}BD_1,\ldots\ldots,A^{-\theta}BD_\theta] = -[D_1,O_{m_2},\ldots\ldots,O_{m_\theta}] \qquad (4.52)$$

It is obvious that if the system (4.1) can be taken from the initial state $\underline{x}(0)$ to the origin in θ steps, it can be taken from the state $\underline{x}(1)$ to the origin in $(\theta-1)$ steps. In this case $\underline{v}(\theta) = \underline{0}$ and proceeding as above, we deduce that

$$\underline{x}(1) = -S\begin{bmatrix}\underline{v}(1)\\ \cdot\\ \cdot\\ \cdot\\ \cdot\\ \cdot\\ \underline{v}(\theta)\end{bmatrix} \qquad (4.53)$$

$$\begin{bmatrix}\underline{v}(1)\\ \cdot\\ \cdot\\ \cdot\\ \cdot\\ \underline{v}(\theta)\end{bmatrix} = -S^{-1}\underline{x}(1) = \begin{bmatrix}G_1\\ \cdot\\ \cdot\\ \cdot\\ \cdot\\ G_\theta\end{bmatrix}\underline{x}(1) \qquad (4.54)$$

and

$$\underline{u}(1) = F\underline{x}(1) \qquad (4.55)$$

where F satisfies (4.52). By the same token, it can be shown that

$$\underline{u}(k) = F\underline{x}(k), \qquad k = 0,1,\ldots\ldots,\theta-1 \qquad (4.56)$$

which establishes the general result:

If the linear system (4.1) is θ-step controllable in the sense of (4.37) and $\{D_j\}$*;* $j = 1,\ldots,\theta$ *is an ordered selection for (4.37) according to (4.45), the time-optimal (deadbeat) controller is of the linear form (4.56) where the feedback gain* F *satisfies (4.52).*

We emphasise that this general result hinges upon having a selection procedure for determining n linearly independent columns of the controllability matrix (4.37) to form a basis for the initial state vector in the n-dimensional space.

It is interesting to recall from (4.47a) and (4.52a) that

$$F = -[D_1, O_{r_2}, \ldots, O_{r_\theta}]S^{-1} \tag{4.57}$$

and the resulting closed-loop system

$$\underline{x}(k+1) = (A+BF)\underline{x}(k) \qquad k=0,1,\ldots\theta-1 \tag{4.58}$$

is driven to the origin in at most θ time steps. Since Γ_q is the set of states that can be transferred to the origin in q steps, this implies that $(A+BF)\Gamma_q$ can be transferred to the origin in (q-1) steps, that is

$$(A+BF)\ _q \subset \Gamma_{q-1} \tag{4.59}$$

More importantly, the time-optimal control sequence (4.56) is non-unique due to the fact that the ordered selection for (4.37) can be made in different ways [26]. Before discussing other related approaches, we examine the properties of the time-optimal controlled system (4.58).

4.3.3 BASIC PROPERTIES

We know that the discrete system (4.58) can be driven to the origin in at most θ time steps. By repeatedly applying (4.58), one gets:

$$\left.\begin{aligned} \underline{x}(1) &= (A+BF)\,\underline{x}(0) \\ \underline{x}(2) &= (A+BF)^2\underline{x}(0) \\ &\vdots \\ \underline{x}(\theta) &= (A+BF)^\theta\underline{x}(0) \end{aligned}\right\} \qquad (4.60)$$

Since (4.58) is θ-step controllable, then it follows from (4.60) that

$$(A+BF)^\theta\underline{x}(0) = \underline{0} \qquad (4.61)$$

for all initial state $\underline{x}(0)$. The interpretation of (4.61) is that the matrix $(A+BF)$ is nilpotent with index θ [27]. A direct consequence of (4.61) is that [25] all the n eigenvalues of $(A+BF)$ are zero and the eigenvectors span $N(A+BF)$, where $N[H]$ is the null space of H.

The eigenvalues of the discrete system (4.58) are given by

$$(A+BF)\underline{s} = \lambda\underline{s} \qquad (4.62)$$

where λ is the eigenvalue with corresponding eigenvector $\underline{s}$. Premultiplying (4.62) by $(A+BF)^{n-1}$ gives

$$(A+BF)^{n-1}(A+BF)\,\underline{s} = \lambda^n\underline{\theta} \qquad (4.63)$$

However, since $(A+BF)$ is nilpotent with index θ, the left-hand side of (4.63) is the zero vector which implies that all the eigenvalues of the closed-loop system (4.58) can be assigned to the origin with deadbeat feedback control. A direct consequence of this result is that the closed-loop matrix has zero trace.

By applying a similarity transformation to the matrix $(A+BF)$, it can be shown [28] that the corresponding Jordan matrix is composed of r nilpotent blocks J_{m_j} of order m_j, j=1,....,m, that is:

$$J = \begin{bmatrix} J_{m_1} & & 0 \\ & J_{m_2} & \\ 0 & & J_{m_o} \end{bmatrix} \qquad (4.64)$$

where J_{m_j} is an $(m_j x m_j)$ matrix with ones on the first super-diagonal and zeros elsewhere given by

$$J_{m_j} = \begin{bmatrix} 0 & 1 & & \\ & 0 & 1 & \\ & & & 1 \\ & & & 0 \end{bmatrix} \qquad (4.65)$$

The above discussion has illustrated some of the important properties of the deadbeat controlled system matrix. Being a nilpotent matrix, it represents a special class of linear transformations [28].

4.3.4 *OTHER APPROACHES*

In subsection 4.3.2 we presented a general procedure for constructing the multivariable deadbeat controller. There are other approaches to obtaining the deadbeat control law by using different ordered selection procedures. One such procedure [26] is based on examining the column vectors of (4.37), that is

$$A^{-1}b_1, A^{-1}b_2, \ldots, A^{-1}b_m, A^{-2}b_1, A^{-2}b_2, \ldots, A^{-\theta}b_m \qquad (4.66)$$

where b_j is the j-th column of B, $1 \leq j \leq m$. The vectors in

(4.66) are successively tested for linear independence from left to right. When n linearly independent vectors have been chosen, the selection procedure is terminated to form the (nxn) matrix S similar to (4.47a):

$$S = [A^{-1}B^{(1)}, A^{-2}B^{(2)}, \ldots, A^{-\theta}B^{(\theta)}] \tag{4.67}$$

where $A^{-q}B^{(q)}$ represents the m_q linearly independent columns of $A^{-q}B$ of (4.37), $m_1 = m$ and

$$\sum_{q=1}^{\theta} m_q = n \tag{4.68}$$

It has been shown [26] that the feedback gain matrix F satisfied

$$FS = [-I_{m_1}, O_{m_2}, \ldots, O_{m_\theta}] \tag{4.69}$$

which is obviously a special case of (4.52).

Another approach utilises the fact that deadbeat control can be achieved simply by choosing the gain matrix F in (4.58) so that all the eigenvalues of (A+BF) are assigned to the origin [29]. It follows from modal control theory [30] that the gain matrix F needed to do this job has the explicit form:

$$F = \underline{d} \sum_{j=1}^{n} \left\{ \frac{(-1)^n \lambda_j^n \underline{v}_j^t}{\beta_j \prod_{\substack{k=1 \\ j \neq k}} (\lambda_k - \lambda_j)} \right\} \tag{4.70}$$

where β_j, j=1,....,n are the elements of the (nx1) mode-controllability vector [30].

$$\underline{\beta} = V^t B \underline{d} \tag{4.71}$$

The vector $\underline{d}$ is any (mx1) vector for which $\underline{\beta}$ has no zero elements, $\underline{v}_j^t (j=1,\ldots,n)$ are the eigenvectors of A^t corresponding to the eigenvalues $\lambda_j (j=1,\ldots,n)$ of the nonsingular matrix A, and V is the matrix whose j-th column is $\underline{v}_j (j=1,\ldots,n)$. We observe that this deadbeat feedback control utilises the concept of dyadic design in eigenvalue assignment [21].

Still another approach to design deadbeat controllers has been developed [25,32,33] based on the canonical form of Luenberger [31]. By means of the state transformation $\underline{z}(k) = Q\underline{x}(k)$ the system (4.1) may be converted into the controllable canonical form [31]:

$$\underline{z}(k+1) = H\underline{z}(k) + L\underline{u}(k) \tag{4.72}$$

where $H \triangleq QAQ^{-1}$ and $L \triangleq QB$ are given by

$$H = \left[\begin{array}{cccccc|cccc}
0 & 1 & 0 & \cdots & 0 & & & & & \\
0 & 0 & 1 & \cdots & 0 & & & & & \\
\vdots & \vdots & & & & & & & & \\
0 & & & \cdots & 1 & & & & & \\
a_{m_1 2} & a_{m_1 2} & \cdots & a_{m_1 m_1} & & a_{m_1 m_1+1} & \cdots & & a_{m_1 n} & \\
\hline
 & & \vdots & \vdots & & & \vdots & & & \\
a_{m_{m-1}} & a_{m_{m-1}} & & & & & \cdots & & a_{m_{m-1} n} & \\
\hline
 & & & & & & 0 & 1 & 0 \cdots\cdots 0 & \\
 & & & & & & 0 & 0 & 1 \cdots\cdots 0 & \\
 & & & & & & 0 & \cdots & \cdots 1 & \\
a_{m_m 1} & a_{m_m 2} & & & & & & \cdots\cdots & &
\end{array}\right]
\begin{array}{l}
\left.\right\} m_1 \\
\leftarrow \delta_1 \\
\vdots \\
\leftarrow r-1 \\
\left.\right\} m_m \\
\leftarrow \delta_m
\end{array}$$

$$L = \left[\begin{array}{ccc|l}
0 & 0 & 0 & \\
\cdot & \cdot & \cdot & \\
\cdot & \cdot & \cdot & \\
\cdot & \cdot & \cdot & \} m_1 \\
\cdot & \cdot & \cdot & \\
0 & 0 & 0 & \\
1 & b_{m_1 2} & b_{m_1 m} & \leftarrow \delta_1 \\
\hline
0 & 0 & 0 & \\
\cdot & \cdot & \cdot & \\
\cdot & \cdot & \cdot & \\
\cdot & \cdot & \cdot & \} m_2 \\
0 & 0 & & \\
0 & 1 & b_{m_2 m} & \leftarrow \delta_2 \\
\hline
\cdot & \cdot & \cdot & \\
\cdot & \cdot & \cdot & \\
\cdot & \cdot & \cdot & \\
\hline
0 & 0 & \ldots 0 & \\
\cdot & \cdot & \cdot & \\
\cdot & \cdot & \cdot & \} m_r \\
\cdot & \cdot & \cdot & \\
0 & 0 & 0 & \\
0 & 0 & 1 & \leftarrow \delta_m
\end{array}\right]$$

In particular, $\sum_{j=1}^{m} m_j = n$, $\max m_j = \theta$; the reachability index. Also δ_j; $j=1,\ldots,m$ denotes the position of the non-trivial rows of H and L. The construction of the transformation matrix Q is quite similar to the selection procedure (4.66), but applied to the reachability matrix (4.36). From the non-trivial rows of H and L, we form the (mxn) matrix H_m and the upper triangular (mxm) matrix L_m, respectively [25]. Now, the gain matrix F_m can be chosen such that $(H_m + LF_m)$ is equal to the Jordan form (4.64) which, in turn, corresponds to zero value eigenvalue assignment. This means that [25]:

$$H_m + L_m F_m = 0$$

or

$$F_m = -L_m^{-1} H_m \tag{4.73}$$

We note that $(H_m + L_m F_m)^{\theta} = 0$. Transforming back to the original coordinate system, the time-optimal controller takes the form:

$$F = -\, L_m^{-1} H_m Q \tag{4.74}$$

which derives the system (4.1) to zero in at most θ time steps. We emphasise that all the deadbeat controllers derived in this subsection possess the basic properties of subsection 4.3.3. Identical results have been developed in [32-34] using different procedures.

We will conclude this section by recalling that the state and output feedback control laws have been developed to deal with the full-order discrete models. In the next sections we consider the analysis and control design based on a simplified model of the discrete system.

4.3.5 *EXAMPLES*

Example 1

A second-order model representing a digital positioning system is given by

$$\underline{x}(k+1) = \begin{bmatrix} 1 & .080 \\ 0 & .631 \end{bmatrix} \underline{x}(k) + \begin{bmatrix} .003 \\ .063 \end{bmatrix} \underline{u}(k)$$

The eigenvalues of the model are 1 and .631.

Using the similarity transformation

$$\underline{x}(k) = \begin{pmatrix} .003 & .003 \\ -.063 & .063 \end{pmatrix} \underline{z}(k)$$

the second order model is converted into the normal form

$$\underline{z}(k+1) = \begin{pmatrix} 0 & 1 \\ -.631 & 1.631 \end{pmatrix} \underline{z}(k) + \begin{pmatrix} 0 \\ 1 \end{pmatrix} \underline{u}(k)$$

In terms of the deadbeat gain $\underline{f} = (f_1 f_2)^t$, the closed-loop system matrix is given by:

$$A_c = \begin{pmatrix} 0 & 1 \\ -.631+f_1 & 1.631+f_2 \end{pmatrix}$$

It is readily seen that the deadbeat gain can be selected as

$$\underline{f} = (.631 \quad -1.631)^t$$

since this will yield zero eigenvalues for the system matrix A_c. The state deadbeat control law of the transformed system can be expressed as:

$$u(k) = (.631 \quad -1.631)\underline{z}(k)$$

Using $\underline{z}(k) = T^{-1}\underline{x}(k)$, the corresponding control law for the original system takes the form:

$$\underline{u}(k) = (-158.5 \quad -17.33)\underline{x}(k)$$

This feedback control will bring the digital positioning system to the origin in two steps.

Example 2

A linearised model describing the motion of an earth satellite in a circular, equatorial orbit is discretised using a sampling rate of twice per orbit. The discrete model matrices are given by:

$$A = \begin{bmatrix} 7 & 0 & 0 & .6363 \\ 0 & -1 & 0 & 0 \\ -18.86 & -.6364 & 1 & -1.5 \\ -75.43 & 0 & 0 & -7 \end{bmatrix}$$

$$B = \begin{bmatrix} 0 & .318 \\ -.5 & 0 \\ -.318 & -.75 \\ 0 & -3.5 \end{bmatrix}$$

This fourth-order model has the eigenvalues $\pm 1.$, $\pm .9998$. It is required to derive a state feedback control that makes any nonzero initial state return to zero in at most three steps. This can be achieved by computing the deadbeat gain of the form (4.74). In terms of the feedback gain matrix

$$F = \begin{bmatrix} f_1 & f_2 & f_3 & f_4 \\ f_5 & f_6 & f_7 & f_8 \end{bmatrix}$$

the closed-loop system matrix becomes

$$(A+BF) = \begin{bmatrix} 7+.318f_5 & .318f_6 & .318f_7 & .636+.318f_8 \\ -.5f_1 & -1-.5f_2 & -.5f_3 & -.5f_4 \\ -18.857-.318f_1 & -.636-318f_2 & 1-.318f_3 & -1.5-.318f_4 \\ -.75f_5 & -.75f_6 & -.75f_7 & -.75f_8 \\ -75.429-3.5f_5 & -3.5f_6 & -3.5f_7 & -7-3.5f_8 \end{bmatrix}$$

Using the state feedback eigenvalue assignment algorithm of section 4.2 with repeated eigenvalues, the gain matrix is calculated as

$$F = \begin{bmatrix} 0 & -2 & 3.143 & 0 \\ -22 & 0 & 0 & -2 \end{bmatrix}$$

Note that the deadbeat controlled system matrix is given by

$$(A+BF) = \begin{bmatrix} 0 & 0 & 0 & 0 \\ 0 & 0 & -1.571 & 0 \\ -2.357 & 0 & 0 & 0 \\ 1.571 & 0 & 0 & 0 \end{bmatrix}$$

It is easy to check that $\lambda(A+BF) = \{0,0,0,0\}$ as required. To verify the deadbeat property, we may compute

$$(A+BF)^2 = \begin{bmatrix} 0 & 0 & 0 & 0 \\ 3.704 & 0 & 0 & 0 \\ 0 & 0 & 0 & 0 \\ 0 & 0 & 0 & 0 \end{bmatrix}$$

$$(A+BF)^3 = [0]$$

which implies that the maximum required number of steps to achieve deadbeat is three or 1.5 orbits.

4.4 Development of Reduced-Order Models

Higher-order models emerge quite naturally when dealing with many physical and engineering problems. Direct application of the control theories and techniques requires a lot of computational effort and complicates the information processing in the implementation phase. It is therefore desirable and sometimes necessary in the analysis and design of control systems to develop simplified or lower-order models. Approaches to reduced-order modelling and control design have been known for quite some time; an overview of these has been given in a recent text [35]. In this section, we present the analytical development of reduced-order models of linear discrete control problems.

4.4.1 *ANALYSIS*

Let us consider a discrete-time model of an asymptotically stable control system of the form:

$$\begin{bmatrix} \underline{x}_1(k+1) \\ \underline{x}_2(k+1) \end{bmatrix} = \begin{bmatrix} A_1 & A_2 \\ A_3 & A_4 \end{bmatrix} \begin{bmatrix} \underline{x}_1(k) \\ \underline{x}_2(k) \end{bmatrix} + \begin{bmatrix} B_1 \\ B_2 \end{bmatrix} \underline{u}(k) \tag{4.75}$$

where $\underline{x}_1(k) \in R^r$, $\underline{x}_2(k) \in R^{n-r}$ are the state subvectors and $\underline{u}(k) \in R^n$ is the control vector. It is assumed that model (4.75) is obtained from (4.1) by reordering and/or rescaling of states such that $\underline{x}_1(k)$ contains the r states to be retained in the reduced-order model and $\underline{x}_2(k)$ comprises the $(n-r)$ states to be eliminated. This is equivalent to the situation where the original model (4.1) is of higher-order and has r-dominant modes and $(n-r)$-nondominant modes. The order r is either specified a priori based on physical grounds or determined from the eigenspectrum of the system. For the time being, we assume that r is given. We seek an r-th reduced-order model of the form

$$\underline{x}_r(k+1) = A_r \underline{x}_r(k) + B_r \underline{u}(k) \tag{4.76}$$

that approximates the long-term behaviour of the original model (4.75). For the purpose of the mathematical analysis to follow, we define

$$M = \begin{bmatrix} M_1 & M_2 \\ M_3 & M_4 \end{bmatrix} \tag{4.77a}$$

as the model matrix [30] whose columns are the corresponding right eigenvectors of A, and

$$N = \begin{bmatrix} N_1 & N_2 \\ N_3 & N_4 \end{bmatrix} \tag{4.77b}$$

as the matrix of reciprocal basis vectors whose rows are the left eigenvectors of A. It is well-known that $MN = NM = I_n$, from which using (4.77) we obtain

$$\left.\begin{aligned} M_1N_1 + M_2N_3 &= I_r \\ M_1N_2 + M_2N_4 &= O \\ M_3N_1 + M_4N_3 &= O \\ M_3N_2 + M_4N_4 &= I_{n-r} \end{aligned}\right\} \tag{4.78a}$$

$$\left.\begin{aligned} N_1M_1 + N_2M_3 &+ I_r \\ N_1M_2 + N_2M_4 &= O \\ N_3M_1 + N_4M_3 &= O \\ N_3M_2 + N_4M_4 &= I_{n-r} \end{aligned}\right\} \tag{4.78b}$$

We consider that the matrix M is arranged such that its columns, from left to right, are in order of decreasing significance of the corresponding eigenvalues of A. Since system (4.75) is assumed to be asymptotically stable, all the eigenvalues lie within the unit circle. Thus the columns of M on the left correspond to the eigenvalues of A which are nearest to unity in absolute value and those on the right correspond to eigenvalues of A nearest to zero. In general, the model matrix M consists of generalised eigenvectors.

By means of the similarity transformation

$$\begin{bmatrix} \underline{x}_1(k) \\ \underline{x}_2(k) \end{bmatrix} = \begin{bmatrix} M_1 & M_2 \\ M_3 & M_4 \end{bmatrix} \begin{bmatrix} \underline{z}_1(k) \\ \underline{z}_2(k) \end{bmatrix} \tag{4.79}$$

the system (4.75) is converted into the Jordan canonical form:

$$\begin{bmatrix} \underline{z}_1(k+1) \\ \underline{z}_2(k+1) \end{bmatrix} = \begin{bmatrix} J_1 & 0 \\ 0 & J_2 \end{bmatrix} \begin{bmatrix} \underline{z}_1(k) \\ \underline{z}_2(k) \end{bmatrix} + \begin{bmatrix} F_1 \\ F_2 \end{bmatrix} \underline{u}(k) \tag{4.80}$$

where $\underline{z}_1(k) \in R^r$, $\underline{z}_2(k) \in R^{n-r}$ and

$$J_1 = N_1(A_1M_1 + A_2M_3) + N_2(A_3M_1 + A_4M_3) \tag{4.81a}$$

$$J_2 = N_3(A_1M_2 + A_2M_4) + N_4(A_3M_2 + A_4M_4) \tag{4.81b}$$

$$P_1 = N_1B_1 + N_2B_2 \tag{4.81c}$$

$$P_2 = N_3B_1 + N_4B_2 \tag{4.81d}$$

It should be noted that the (rxr) Jordan block J_1 has the eigenvalues nearest to unity whereas the (n-r x n-r) Jordan block J_2 has the eigenvalues closest to zero. Therefore we say that $\underline{z}_1(k)$ comprises the slow modes while $\underline{z}_2(k)$ represents the fast modes, and according to (4.80) both modes are non-interacting.

4.4.2 *TWO SIMPLIFICATION SCHEMES*

The purpose of this subsection is to derive the simplified model (4.76) using two different schemes [36,37]. In the first scheme, the dynamics of the fast modes are neglected and the significant system modes are retained.

From (4.75), we have

$$\underline{x}_1(k+1) = A_1\underline{x}_1(k) + A_1\underline{x}_2(k) + B_1\underline{u}(k) \tag{4.82}$$

An expression for $\underline{x}_2(k)$ can be obtained by combining (4.77b) and (4.79) to yield:

$$\underline{x}_2(k) = N_4^{-1}\ \underline{z}_2(k) - N_3\underline{x}_1(k) \tag{4.83}$$

Since $\underline{z}_2(k)$ responds much faster than $\underline{z}_1(k)$ because the eigenvalues in J_2 are smaller in magnitude than those of J_1, we consider that this response takes place almost instantaneously. Therefore, from (4.80) we obtain

$$\begin{aligned}\underline{z}_2(k+1) &= \underline{z}_2(k) \\ &= (I_{n-r} - J_2)^{-1}P_2\underline{u}(k) \qquad (4.84)\end{aligned}$$

The substitution of (4.83) and (4.84) into (4.82) results in the desired simplified model (4.76) with

$$\begin{aligned}A_r &= A_1 - A_2N_4^{-1}N_3 \\ &= A_1 + A_2M_3M_1^{-1} \qquad (4.85)\end{aligned}$$

$$B_r = B_1 + A_2N_4^{-1}(I_{n-r} - J_2)^{-1}P_2 \tag{4.86}$$

and $\underline{x}_r(k)$ replaces $\underline{x}_1(k)$. We note that this scheme is simple and provides the same steady-state values of the dominant modes for constant inputs.

The second scheme is based on the assumption that the contribution of the insignificant eigenvalues of the original system to the system response is small and may be neglected. A reduced-model is then obtained in which the dominant modes are exited in the same proportion as in the original system. We know that the forced solution of (4.75), assuming that $\underline{x}(0) = 0$, is given by:

$$\begin{bmatrix} \underline{x}_1(k+1) \\ \underline{x}_2(k+1) \end{bmatrix} = \sum_{m=0}^{k-1} \begin{bmatrix} A_1 & A_2 \\ A_3 & A_4 \end{bmatrix}^m \begin{bmatrix} B_1 \\ B_2 \end{bmatrix} \underline{u}(m) \tag{4.87}$$

From the properties of similarity transformations [20], we can write (4.87) as:

$$\begin{bmatrix} \underline{x}_1(k+1) \\ \underline{x}_2(k+1) \end{bmatrix} = M \sum_{m=0}^{k-1} \begin{bmatrix} J_1^m & 0 \\ 0 & J_2^m \end{bmatrix} N \begin{bmatrix} B_1 \\ B_2 \end{bmatrix} \underline{u}(m) \tag{4.88}$$

To derive the simplified model (4.76), make the assumption that the fast modes have negligible effects on the system response, that is, we use $J_2 = 0$ to obtain:

$$\begin{bmatrix} \underline{x}_1(k+1) \\ \underline{x}_2(k+1) \end{bmatrix} = \begin{bmatrix} M_1 \\ M_3 \end{bmatrix} \left[\sum_{m=0}^{k-1} J_1^m [N_1 N_2] \begin{bmatrix} B_1 \\ B_2 \end{bmatrix} \underline{u}(m) \right] \tag{4.89}$$

Define an (rxl) vector $\underline{\alpha}$ as

$$\underline{\alpha} = \sum_{m=0}^{k-1} \{J_1^m [N_1 B_1 + N_2 B_2] \underline{u}(m)\} \tag{4.90}$$

then (4.89) reduces to

$$\begin{bmatrix} \underline{x}_1(k+1) \\ \underline{x}_2(k+1) \end{bmatrix} = \begin{bmatrix} M_1 \\ M_3 \end{bmatrix} \underline{\alpha} \tag{4.91}$$

or upon expansion, we get

$$\left.\begin{aligned} \underline{\alpha} &= M_1^{-1} \underline{x}_1(k+1) \\ \underline{x}_2(k+1) &= M_3 M_1^{-1} \underline{x}_1(k+1) \end{aligned}\right\} \tag{4.92}$$

To determine A_r and B_r of (4.76), we first express the control signal, that is, we use $\underline{u}(k) = \underline{0}$ in (4.82) to yield

$$\underline{x}_1(k+1) = [A_1 + A_2 M_3 M_1^{-1}] \underline{x}_1(k)$$

and consequently

$$A_r = A_1 + A_2 M_3 M_1^{-1} \tag{4.93}$$

which is identical to (4.85) of the first scheme. We then compare the solutions of the original model (4.75) and the reduced-order model (4.76) with $\underline{x}(0) = \underline{0}$ and $\underline{x}_r(k)$ replacing $\underline{x}_1(k)$. From (4.89), the response of $\underline{x}_1(k)$ is given by

$$\underline{x}_1(k+1) = M_1 \left\{ \sum_{m=0}^{k-1} \{J_1^m (N_1 B_1 + N_2 B_2) \underline{u}(m)\} \right\} \tag{4.94}$$

The solution of (4.76) is

$$\underline{x}_1(k+1) = \sum_{m=0}^{k-1} \{A_r^m B_r \underline{u}(m)\} \tag{4.95}$$

Using (4.77), (4.78), (4.81) in (4.95) and manipulating, we arrive at

$$A_r = M_1 J_1 M_1^{-1} \tag{4.96}$$

and

$$\underline{x}_1(k+1) = M_1 \left(\sum_{m=0}^{k-1} \{J_1^m M_1^{-1} B_r \underline{u}(m)\} \right) \tag{4.97}$$

Computing (4.94) and (4.97) we have

$$\begin{aligned} B_r &= M_1 (N_1 B_1 + N_2 B_2) \\ &= M_1 F_1 \end{aligned} \tag{4.98}$$

which is different from (4.86) of the first scheme.

Although the two simplification schemes are conceptionally

different they yield identical reduced-order system matrices. Since the reduced input matrices are not the same, we expect that the reduced-order model responses be different. This will be demonstrated later when some typical control problems are worked out.

4.4.3 *OUTPUT MODELLING APPROACH*

We have shown in the previous section that reduced-order models can be developed to approximate the behaviour of higher-order discrete models by neglecting the effects of non-dominant modes. In what follows, we present an alternative approach [38] to model simplification which is based on modelling a transformation of the state variables. The linear dynamic system to be considered here is described by the difference equations:

$$\underline{x}(k+1) = A\underline{x}(k) + B\underline{u}(k) \qquad (4.99)$$

$$\underline{y}(k) = C\underline{x}(k) \qquad (4.100)$$

where $\underline{x}(k) \in R^n$ is the state vector, $\underline{u}(k) \in R^m$ is the control vector and $\underline{y}(k) \in R^p$ is the output vector. The matrices A, B and C are of appropriate dimensions. We assume without loss of generality that rank (C) = p, which implies that there are no redundant measurements in the output vector. In general, $p < n$ which means that the output is a low-order linear transformation of the states. Since the output variables are directly measurable, it is desirable to seek a dynamic system which reproduces the output behaviour of the system (4.99) and (4.100). The proposed reduced-order dynamic system is described by:

$$\underline{y}(k+1) = F\underline{y}(k) + G\underline{u}(k) \qquad (4.101)$$

where $\underline{y}(0) = C\underline{x}(0)$ is known. The equivalence between the systems (4.99), (4.100) and (4.101) implies that the conditions

$$FC = CA \tag{4.102}$$

$$G = CB \tag{4.103}$$

are satisfied, whilst (4.103) poses no problem. (4.102) may not always be solvable for the (pxp) matrix F. To examine the conditions under which a solution of (4.102) exists, we define [38]:

$$(CA)^t \triangleq [\underline{\alpha}_1 \quad \underline{\alpha}_2 \cdots\cdots \quad \underline{\alpha}_p], \quad \underline{\alpha}_i \in R^n \tag{4.104a}$$

$$F^t \triangleq [\underline{f}_1 \quad \underline{f}_2 \cdots\cdots \underline{f}_p], \quad \underline{f}_i \in R^p \tag{4.104b}$$

By transposing (4.102) and using (4.104), we get

$$C^t \underline{f}_i = \underline{\alpha}_i \qquad ; \quad i=1,\ldots\ldots,p \tag{4.105}$$

It is well-known [39] that (4.105) will have a unique solution for $\underline{f}_i$ if and only if

$$\underline{\alpha}_i \in R[C^t] \tag{4.106a}$$

$$\text{rank } (C^t) = p \tag{4.106b}$$

Now, since the output matrix is assumed to be of full rank (4.102) will have a unique solution for F if and only if

$$R(A^t C^t) \subseteq R(C^t) \tag{4.107}$$

Suppose further that rank $(A^t C^t) = p$, then it spans a p-dimensional subspace and every row of C can be expressed as a linear combination of the rows of (CA), hence

$$(C^t) \subseteq (A^t C^t) \tag{4.108}$$

In view of (4.107) and (4.108), the solvability condition of

(4.102) is

$$R(C^t) = R(A^t C^t) \tag{4.109a}$$

or equivalently

$$N(C) = N(CA) \tag{4.109b}$$

Provided that rank (C) = rank (CA) = p, we can use (4.104) to solve (4.102) for $\underline{f}_i$ by minimising

$$||C^t \underline{f}_i - \alpha_i||^2$$

The result is known to be [38]

$$F = C\ A\ C^t (CC^t)^{-1} \tag{4.110}$$

Given the matrices A, B, C the reduced-order (4.101) can be completely determined by using (4.103) and (4.110). An interesting result that can be deduced from the above analysis is that in the light of (4.109b) the pair (A,C) is not observable [38]. This is to be expected since the model (4.110) just describes the observable (directly measurable) part of the system (4.99) and (4.100). It has been shown [35] that this approach is a special case of aggregation theory [40] in which the output matrix C plays the role of an aggregation matrix which relates the state vectors of the original and reduced-order models. We note that the output modelling approach relies upon the validity of the rank conditions. The simplification schemes described in section 4.4.2 require the calculation of the eigenstructure. Also, the output modelling provides an output behaviour which is close to the system output response, whereas the two simplifications described earlier give a good approximation to the dominant modes. Looked at in this light, it is clear that the model simplification schemes neglect the non-dominant modes, which, in effect, aim at reproducing the long-term behaviour of the original system. The output modelling approach represents the original system by a 'coarser' state-space description.

It has been shown [35] that the two simplification schemes are special cases of aggregation theory [40] with an aggregation matrix of the form

$$C_a = [I_r \vdots O_{rxn-r}]N \tag{4.111}$$

Consequently, the reduced-model matrix (4.85) inherits the r-dominant eigenvalues of the original system. However, it is shown in [38] that the matrix F has the same eigenvalues as the ones associated with the eigenvectors that span the row sub-space of the output matrix C.

4.4.4 *CONTROL DESIGN*

It is of interest to study the effect of applying a feedback control to the original discrete system which is designed for the reduced-order model. Consider the output modelling approach for which the reduced-order model is described by (4.101). Let a feedback control

$$\underline{u}(k) = K_O \underline{y}(k) \tag{4.112}$$

be designed to place p eigenvalues of the pair (F,G) at desired positions. This can be achieved by using any state-feedback eigenvalue assignment algorithm [21]. By implementing the control (4.112) on the system (4.99) using (4.100), we arrive at:

$$\underline{x}(k+1) = (A+BK_O C)\underline{x}(k) \tag{4.113}$$

This shows that the control law (4.112) is in fact an output feedback control for (4.99) as expected. A direct consequence of this result is that the remaining (n-p) eigenvalues may be located anywhere in the complex plane.

Now consider the simplified model (4.76) whose matrices are specified by (4.85) and (4.86). Let a state feedback control

$$\underline{u}(k) = K\underline{x}_r(k) \tag{4.114}$$

be designed to place r eigenvalues of the pair (A_r+B_rK) at desired locations. In view of (4.102) and (4.111), it can be shown [35] that

$$\underline{x}_r(k) = C_a\underline{x}(k) \tag{4.115}$$

Using (4.114) and (4.115) in (4.80), we have

$$\begin{bmatrix} \underline{z}_1(k+1) \\ \underline{z}_2(k+1) \end{bmatrix} = \begin{bmatrix} J_1 & O \\ O & J_2 \end{bmatrix} \begin{bmatrix} \underline{z}_1(k) \\ \underline{z}_2(k) \end{bmatrix} + \begin{bmatrix} P_1 \\ P_2 \end{bmatrix} KC_a \begin{bmatrix} \underline{x}_1(k) \\ \underline{x}_2(k) \end{bmatrix} \tag{4.116}$$

The substitution of (4.79) and (4.111) into (4.116) results in the closed-loop system:

$$\begin{bmatrix} \underline{z}_1(k+1) \\ \underline{z}_2(k+1) \end{bmatrix} = \begin{bmatrix} J_1+S_1 & O \\ S_2 & J_2 \end{bmatrix} \begin{bmatrix} \underline{z}_1(k) \\ \underline{z}_2(k) \end{bmatrix} \tag{4.117}$$

where S_1 and S_2 are given by

$$S_1 = (N_1B_1 + N_2B_2)K \tag{4.118a}$$

$$S_2 = (N_3B_1 + N_4B_2)K \tag{4.118b}$$

The eigenspectrum of (4.117) is

$$\begin{aligned} \sigma \triangleq \begin{bmatrix} J_1+S_1 & O \\ S_2 & J_2 \end{bmatrix} &= \det[\lambda I_r - (J_1+S_1)]\det[\lambda I_{n-r} - J_2] \\ &= \det[\lambda I_r - (J_1+S_1)] \prod_{j=r+1}^{n} (\lambda-\lambda_j) \end{aligned} \tag{4.119}$$

which shows that the eigenvalues $\{\lambda_{r+1},\ldots,\lambda_n\}$ does not change with the use of the feedback gain matrix K. It has been shown [35] that the reduced-order scheme (4.76) inherits the r-dominant eigenvalues and in the light of (4.102) it immediately follows that $A_r = J_1$ and $B_r = C_a B$. Hence,

$$\det[\, I_r - (J_1+S_1)] = \det[\, I_r - (A_r+B_r K)] \quad (4.120)$$

An examination of (4.119) and (4.120) reveals that the eigen-spectrum of the closed-loop system (4.75) consists of the eigen-spectrum of the reduced-order closed-loop system $\underline{x}_r(k+1) = (A_r+B_r K)\underline{x}_r(k)$ and the set of the eigenvalues remaining undistubed. In comparison, the reduced-order feedback control law (4.115) guarantees the stabilisation of the original system (4.75); however, this is not the case with the control law (4.112) for which the stabilisation is not always guaranteed.

4.4.5 *EXAMPLES*

We now present two examples to illustrate the development of reduced-order models.

Example 1

Consider the second-order model

$$\underline{x}(k+1) = \begin{bmatrix} .83 & .156 \\ -1.68 & -.27 \end{bmatrix} \underline{x}(k) + \begin{bmatrix} .097 \\ .14 \end{bmatrix} u(k)$$

The purpose is to demonstrate the application of the simplification techniques. The eigenvalues of the control model are .09 and .47, and the associated modal matrices are

$$M = \begin{bmatrix} -.32 & -.42 \\ 1.68 & 1.36 \end{bmatrix}; \quad N = \begin{bmatrix} 5.03 & 1.55 \\ -6.21 & -1.18 \end{bmatrix}$$

It is required to develop first-order model that approximates the steady-state response of the second-order model. Simple calculations of (4.85), (4.86) and (4.98) give

$$A_r = .47$$

$$B_r \text{ (first scheme)} = .133$$

$$B_r \text{ (second scheme)} = .177$$

which shows that the simplification schemes differ only in the input excitation matrices.

Example 2

A tenth-order evaporator model [37] of the form (4.75) is represented by the matrices:

$$A_1 = \begin{bmatrix} .9998 & -.4962\times10^{-7} & .6722\times10^{-3} & -.5016\times10^{-3} & .3296\times10^{-1} \\ .5755\times10^{-8} & .9988 & -.1168\times10^{-2} & -.2180\times10^{-3} & -.5731\times10^{-1} \\ -.3787\times10^{-6} & -.3054\times10^{-4} & .9610 & .3846\times10^{-1} & .5503\times10^{-1} \\ -.1835\times10^{-4} & .4812\times10^{-7} & -.6369\times10^{-3} & .9217 & -.3123\times10^{-1} \\ -.1113\times10^{-6} & -.5159\times10^{-6} & -.1436\times10^{-1} & .1806\times10^{-2} & .2560 \end{bmatrix}$$

$$A_2 = \begin{bmatrix} -.5753\times10^{-1} & -.1978\times10^{-1} & .5370\times10^{-2} & .2933\times10^{-2} & -.6923\times10^{-3} \\ -.2633\times10^{-1} & -.7329\times10^{-2} & -.6145\times10^{-2} & .7482\times10^{-2} & -.2556\times10^{-3} \\ ..2721\times10^{-1} & .7704\times10^{-2} & .5823\times10^{-2} & -.7216\times10^{-2} & .2687\times10^{-3} \\ .5487\times10^{-1} & .1878\times10^{-1} & -.5013\times10^{-2} & -.2799\times10^{-2} & .6573\times10^{-3} \\ .3140 & .9017\times10^{-1} & .3064\times10^{-1} & .2725\times10^{-1} & .3158\times10^{-2} \end{bmatrix}$$

$$A_3 = \begin{bmatrix} -.1851\times10^{-6} & -.1934\times10^{-6} & .2518\times10^{-2} & -.38\times10^{-2} & .1234 \\ -.2814\times10^{-6} & -.1553\times10^{-6} & .2090\times10^{-2} & .35\ 5\times10^{-2} & 1024 \\ -.1406\times10^{-5} & -.7147\times10^{-6} & .3745\times10^{-2} & .2438\times10^{-2} & .1837 \\ -.2511\times10^{-7} & -.2671\times10^{-6} & .1498\times10^{-2} & .6338\times10^{-3} & .7357\times10^{-1} \\ -.2568\times10^{-6} & -.414\times10^{-6} & .1911\times10^{-2} & .3219\times10^{-2} & .9376\times10^{-1} \end{bmatrix}$$

$$A_4 = \begin{bmatrix} .4860 & .1413 & .1967\times10^{-1} & .111\times10^{-1} & .4961\times10^{-2} \\ .4102 & .1193 & .1641\times10^{-1} & .9192\times10^{-2} & .4188\times10^{-2} \\ .2919 & .8418\times10^{-1} & .2297\times10^{-1} & .19\times10^{-1} & .295\times10^{-2} \\ .7682\times10^{-1} & .2197\times10^{-1} & .8611\times10^{-2} & .7936\times10^{-2} & .7696\times10^{-3} \\ .3757 & .1082 & .1501\times10^{-1} & .84\times10^{-2} & .3836\times10^{-2} \end{bmatrix}$$

$$B_1 = \begin{bmatrix} -.9644\times10^{-2} & -.809\times10^{-1} & .7565\times10^{-9} \\ -.2161\times10^{-2} & .8832\times10^{-1} & -.4063\times10^{-1} \\ .2283\times10^{-2} & -.4662\times10^{-1} & .6155\times10^{-6} \\ .9347\times10^{-2} & -.748\times10^{-3} & -.7402\times10^{-9} \\ .4901\times10^{-1} & .1867\times10^{-1} & .1546\times10^{-7} \end{bmatrix}$$

$$B_2 = \begin{bmatrix} .1338 & .3009\times10^{-2} & .3003\times10^{-8} \\ .128 & .2417\times10^{-2} & .2346\times10^{-8} \\ .5457\times10^{-1} & .1116\times10^{-1} & .1688\times10^{-7} \\ .9651\times10^{-2} & .4158\times10^{-2} & .5803\times10^{-8} \\ .1857 & .22\times10^{-2} & .2128\times10^{-8} \end{bmatrix}$$

The eigenvalues of the open-loop system arranged in descending order are $\{.9998, .9988, .9599, .9212, .7353, .1623, -1271\times10^{-3}, .1836\times10^{-4}, .5468\times10^{-5}, -.1542\times10^{-5}\}$. By inspection, it is clear that a fifth-order model preserving the first five dominant eigenvalues would seem a reasonable approximate model.

The modal matrix (4.77a) is given by the submatrices:

$$M_1 = \begin{bmatrix} .128\times10^{3} & -.3916\times10^{-2} & .8399\times10^{-3} & .2489\times10^{-2} & -.7499\times10^{-1} \\ -.1923\times10^{-1} & -.6407\times10^{2} & -.5607\times10^{-3} & -.2052\times10^{-2} & -.1568 \\ -.3017\times10^{-1} & .5054\times10^{-1} & -.2295\times10^{1} & -.2022\times10^{1} & .1647 \\ -.2997\times10^{-1} & -.195\times10^{-4} & -.5076\times10^{-4} & .2047\times10^{1} & .1012 \\ .5183\times10^{-3} & -.8817\times10^{-3} & .4682\times10^{-1} & .4047\times10^{-1} & -.4271 \end{bmatrix}$$

$$M_2 = \begin{pmatrix} .6147\times10^{-1} & .1235\times10^{-2} & .1946\times10^{-2} & -.2826\times10^{-2} & .3933\times10^{-3} \\ -.4965\times10^{-1} & -.1484\times10^{-1} & .1346\times10^{-2} & .1139\times10^{-2} & -.2089\times10^{-3} \\ .5221\times10^{-1} & .1493\times10^{-1} & -.1337\times10^{-2} & -.1164\times10^{-2} & .1895\times10^{-3} \\ -.6431\times10^{-1} & -.1249\times10^{-2} & -.1949\times10^{-2} & .2832\times10^{-2} & -.3927\times10^{-3} \\ -.8607 & -.1284 & .1319 & -.5501\times10^{-1} & -.9582\times10^{-2} \end{pmatrix}$$

$$M_3 = \begin{pmatrix} .1139\times10^{-3} & .8565\times10^{-4} & -.2769\times10^{-5} & -.183\times10^{-1} & -.4807 \\ -.12\times10^{-3} & .7211\times10^{-4} & -.4001\times10^{-5} & -.5090\times10^{-3} & -.4037 \\ -.8842\times10^{-4} & .1079\times10^{-3} & .5437\times10^{-5} & -.5597\times10^{-3} & -.3581 \\ -.2422\times10^{-4} & .3734\times10^{-4} & .6836\times10^{-5} & -.1937\times10^{-3} & -.1104 \\ -.1088\times10^{-3} & .6564\times10^{-4} & .9798\times10^{-6} & -.4714\times10^{-3} & -.3697 \end{pmatrix}$$

$$M_4 = \begin{pmatrix} .2623 & -.2302\times10^{-1} & .1714 & .9276\times10^{-1} & -.7321\times10^{-1} \\ .2289 & .9963\times10^{-1} & -.6306 & -.4911 & .8243\times10^{-1} \\ -.4795 & .7434\times10^{-1} & -.1228\times10^{1} & -.6148 & ..9694\times10^{-1} \\ -.2727 & .1078\times10^{1} & -.9025\times10^{-1} & -.8562\times10^{-1} & .1333\times10^{-1} \\ .2098 & -.1036 & .296\times10^{1} & .4026\times10^{1} & -4648\times10^{1} \end{pmatrix}$$

The substitution in (4.93) and (4.98) gives the matrices of the fifth-order simplified model

$$A_r = \begin{pmatrix} .9998 & -.1457\times10^{-6} & -.9457\times10^{-3} & -.1458\times10^{-4} & -.4634\times10^{-1} \\ .1386\times10^{-7} & .9988 & -.1990\times10^{-2} & .4827\times10^{-5} & -.9758\times10^{-1} \\ -.387k\times10^{-6} & -.3049\times10^{-4} & .9618 & .3823\times10^{-1} & .9646\times10^{-1} \\ -.1837\times10^{-4} & .1396\times10^{-6} & .9064\times10^{-3} & .9212 & .4441\times10^{-1} \\ -.2057\times10^{-6} & .6103\times10^{-8} & -.4623\times10^{-2} & -.8525\times10^{-3} & .7333 \end{pmatrix}$$

$$B_r = \begin{pmatrix} -.1396\times10^{-1} & -.8006\times10^{-1} & -.4499\times10^{-9} \\ .1309\times10^{-2} & .8754\times10^{-1} & -.4063\times10^{-1} \\ -.1365\times10^{-2} & -.4589\times10^{-1} & .6145\times10^{-6} \\ .1387\times10^{-1} & -.1627\times10^{-2} & .5220\times10^{-9} \\ .1058 & .6750\times10^{-2} & .3238\times10^{-7} \end{pmatrix}$$

This model inherits the dominant eigenvalues {.9988, .9988, .9599, .9212, .7353}. A low-order state feedback control can be designed based on the pair (A_r, B_r) such that the dominant eigenvalues are positioned at {.99, .96, .94, .9, .8}. The resulting feedback gain is given by

$$\begin{bmatrix} 4.897 & 71.05 & -.5315 & -.029 & -26.79 \\ 4.897 & 71.05 & -.5315 & -.029 & -26.79 \\ 4.897 & 71.05 & -.5315 & -.029 & -26.79 \end{bmatrix}$$

It is interesting to observe that the application of the lower-order feedback gain matrix K to the original model gives the closed-loop eigenvalues {.99, .942, .9586, .9027, .7967, .1623, $.1271\times10^{-3}$, $.1836\times10^{-4}$, $.5468\times10^{-5}$, $-.1373\times10^{-5}$}. This confirms the analysis of section 4.4 on control design using simplified models.

4.5 Control Systems with Slow and Fast Modes

Realistic models of large scale systems involve interacting dynamic phenomena of widely different speeds. Typical examples are found in the field of power engineering and electromechanical systems. The existence of multi-modes and multi-time-scales manifests itself in a broad class of physical models. From a practical point of view, it is considered beneficial to base the control design on separate regions of operation. The purpose of this section is to analyse the behaviour of discrete systems whose eigen-modes are separated by an order of magnitude. Then we provide a feedback control design using a two-stage procedure.

4.5.1 *TIME-SEPARATION PROPERTY*

Let the eigenspectrum of the asymptotically stable system (4.75) be arranged such that

$$1 > |\lambda \ | > \ldots . |\lambda_r| > |\lambda_{r+1}| > \ldots . |\lambda_n| \quad (4.121)$$

If the eigenvalue inequality

$$\mu = |\lambda_{r+1}| \ / |\lambda_r| \qquad << 1 \quad (4.122)$$

is satisfied, then the model (4.75) represents a two-time scale system with μ being a measure of the time-separation property. This class of systems consists of weakly-coupled subsystems whose modes are widely spaced.

In this case we can express the eigenspectrum $\delta(A)$ as

$$\sigma(A) = \sigma(A_s) \cup \sigma(A_f) \quad (4.123a)$$

where

$$\sigma(A_s) = \{\lambda_1, \ldots, \lambda_r\} \quad (4.123b)$$

are the eigenvalues of the slow (dominant) parts of system (4.75) and,

$$\sigma(A_f) = \{\lambda_{r+1}, \ldots, \lambda_n\} \quad (4.123c)$$

are the eigenvalues of the fast non-dominant parts of system (4.75). In the light of (4.121), (4.122), the interpretation of (4.123) is that $[A_s]^k$ tends to zero much more slowly than $[A_f]^k$, or equivalently the fast parts are only important during a short initial transients after which they die quickly.

4.5.2 *FAST AND SLOW SUBSYSTEMS*

In order to derive the fast and slow subsystems, we can use either of two approaches: explicitly invertible linear transformations [41,45] or the discrete quasi-steady state assumption [42-44].

The explicitly invertible linear transformation is a two-stage block-diagonalisation procedure. In the first stage we use the change of variables

$$\begin{bmatrix} \underline{x}_1(k) \\ \underline{x}_\ell(k) \end{bmatrix} = \begin{bmatrix} I_r & 0 \\ L & I_{n-r} \end{bmatrix} \begin{bmatrix} \underline{x}_1(k) \\ \underline{x}_2(k) \end{bmatrix} \quad (4.124)$$

to transform (4.75) into

$$\begin{bmatrix} \underline{x}_1(k+1) \\ \underline{x}_\ell(k+1) \end{bmatrix} = \begin{bmatrix} A_s & A_2 \\ 0 & A_f \end{bmatrix} \begin{bmatrix} \underline{x}_1(k) \\ \underline{x}_\ell(k) \end{bmatrix} + \begin{bmatrix} B_1 \\ LB_1+B_2 \end{bmatrix} \underline{u}(k) \quad (4.125)$$

where I_j is the $(n_j x n_j)$ identity matrix and

$$A_s = A_1 - A_2 L \quad (4.126a)$$

$$A_f = A_4 + LA_2 \quad (4.126b)$$

The (n-rxr) matrix L is a real root of:

$$A_4 L - LA_1 + LA_2 L - A_3 = 0 \quad (4.127)$$

The second stage decouples the block A_2 by applying the change of variables

$$\begin{bmatrix} \underline{x}_r(k) \\ \underline{x}_\ell(k) \end{bmatrix} = \begin{bmatrix} I_r & K \\ 0 & I_{n-r} \end{bmatrix} \begin{bmatrix} \underline{x}_1(k) \\ \underline{x}_\ell(k) \end{bmatrix} \quad (4.128)$$

to the system (4.125) to yield

$$\begin{bmatrix} \underline{x}_r(k+1) \\ \underline{x}_\ell(k+1) \end{bmatrix} = \begin{bmatrix} A_s & O \\ O & A_f \end{bmatrix} \begin{bmatrix} \underline{x}_r(k) \\ \underline{x}_\ell(k) \end{bmatrix} + \begin{bmatrix} G_1 \\ G_2 \end{bmatrix} \underline{u}(k) \tag{4.129}$$

where

$$G_1 = (I_r + KL)B_1 + KB_2 \tag{4.130a}$$

$$G_2 = LB_1 + B_2 \tag{4.130b}$$

and the $r \times (n-r)$ matrix K is a real root of

$$KA_f - A_sK + A_2 = O \tag{4.131}$$

We note that if the matrices L and K exist, then $\sigma(A_s)$ and $\sigma(A_f)$ correspond to the slow and fast eigenvalues, respectively. The combined transformations (4.124) and (4.127) give

$$\begin{bmatrix} \underline{x}_r(k) \\ \underline{x}_\ell(k) \end{bmatrix} = \begin{bmatrix} I_r+KL & K \\ L & I_{n-r} \end{bmatrix} \begin{bmatrix} \underline{x}_1(k) \\ \underline{x}_2(k) \end{bmatrix} \tag{4.132}$$

whose inverse is

$$\begin{bmatrix} \underline{x}_1(k) \\ \underline{x}_2(k) \end{bmatrix} = \begin{bmatrix} I_r & -K \\ -L & I_{n-r}+LK \end{bmatrix} \begin{bmatrix} \underline{x}_r(k) \\ \underline{x}_\ell(k) \end{bmatrix} \tag{4.133}$$

which requires no matrix inversion. It should be noted that $\underline{x}_r(k)$ represents the exact dominant (slow) component of the $\underline{x}_1(k)$ variable, so results obtained in the decoupled form (4.127) are easily interpreted.

Define

$$\left.\begin{aligned} L_o &= -(I_{n-r} - A_4)^{-1}A_3 \\ A_o &= A_1 - A_1L_o \\ K_o &= A_o^{-1}A_2 \end{aligned}\right\} \qquad (4.134)$$

The numerical solution of the transformation matrix L is usually sought to be of the form [41]:

$$L \doteq L_o + D \qquad (4.135a)$$

where D is a real root of

$$DA_o - (A_4+L_oA_2)D - DA_2D + L_oA_o - L_o = 0 \qquad (4.135b)$$

which can be iteratively solved using the updating scheme [41]:

$$D_{j+1} = \{L_o + (A_4+L_oA_2)D_j - L_oA_o + D_jA_2D_j\}A_o^{-1} \qquad (4.136)$$

For the K matrix, the successive approximation scheme

$$K_{j+1} = A_o^{-1}\{K_j(A_4-L_oA_2) + K_jDA_2 - A_2DK_j + A_2\} \qquad (4.137)$$

is used where the subscript j in (4.136) and (4.137) represents the iteration cycle. Analysis of the successive approximation schemes (4.136) and (4.137) using arguments of a fixed point theorem have shown [41] that if the norm inequalities

$$(||A_4||+||L_o||\,||A_2||)^2 \leqq 4(||L_o||\,||A_2||\,||I_r-A_o||) \qquad (4.138a)$$

$$||A_o^{-1}|| < \frac{1}{3(||A_4||+||L_o||\,||A_2||)} \qquad (4.138b)$$

are satisfied then the iterative schemes (4.136) and (4.137) possess unique real roots.

First order approximations of L and K are given by:

$$L = L_o + O(\mu)^{\dagger} \tag{4.139a}$$

$$K = K_o + O(\mu) \tag{4.139b}$$

By virtue of (4.134) and (4.139), it can be readily seen that

$$A_s = A_o\{1 + O(\mu)\} \tag{4.140a}$$

$$A_f = A_4 + O(\mu) \tag{4.140b}$$

which means that the eigenvalues of the matrices A_o and A_4 are only perturbations of the eigenvalues of the decoupled system (4.129) which, in turn, are equivalent to the eigenvalues of the discrete system (4.75). This is a very desirable result since the control analysis and feedback design of the discrete system (4.75) can be reasonably approximated to first order by two lower-order subsystems: a slow subsystem described by matrix A_o of order r and a fast subsystem characterised by the matrix A_4 of order (n-r).

An alternative approach to deriving the slow and fast subsystems is based on the concept of quasi-steady state [42-44]. As mentioned earlier, for asymptotically stable linear discrete systems having the time-separation property (4.122), the fast modes corresponding to the eigenvalues centred around the origin are important only during the first few discrete instants (transient period). After that period, they are negligible and the slow modes dominate the behaviour of the discrete system.

Neglecting the effect of the fast modes is expressed by formally letting $\underline{x}_2(k+1(= \underline{x}_2(k)$ in (4.75). Without these modes, the system (4.75) reduces to:

---

† A vector or matrix function $\Pi(\mu)$ of a positive scalar μ is said to be $O(\mu^m)$ if there exist positive constants d and μ^* such that $|\Pi(\mu)| \leq c\,\mu^m$ for all $\mu \leq \mu^*$

$$\underline{\overline{x_1}}(k+1) = A_1\underline{\overline{x_1}}(k) + A_2\underline{\overline{x_2}}(k) + B_1\underline{\overline{u}}(k);$$

$$\underline{\overline{x_1}}(0) = \underline{x}_{10} \qquad (4.141a)$$

$$\underline{\overline{x_2}}(k) = A_3\underline{\overline{x_1}}(k) + A_4\underline{\overline{x_2}}(k) + B_2\underline{\overline{u}}(k) \qquad (4.141b)$$

where a bar over a variable indicates a discrete quasi-steady state [43]. Assuming that $(I_{n-r} - A_4)^{-1}$ exists, we express $\underline{\overline{x_2}}(k)$ as:

$$\underline{\overline{x_2}}(k) \doteq (I_{n-r} - A_4)^{-1}\{A_3\underline{\overline{x_1}}(k) + B_2\underline{\overline{u}}(k)\} \qquad (4.142)$$

and substituting it into (4.141a), the slow subsystem of (4.75) is defined by:

$$\underline{x}_s(k+1) = A_0\underline{x}_s(k) + B_0\underline{u}_s(k) \qquad (4.143)$$

where

$$A_0 = A_1 + A_2(I_{n-r} - A_4)^{-1}A_3 \qquad (4.144a)$$

$$B_0 = B_1 + A_2(I_{n-r} - A_4)^{-1}B_2 \qquad (4.144b)$$

Hence $\underline{\overline{x_1}}(k) = \underline{x}_0(k)$, $\underline{x}_2(k)$ and $\underline{\overline{u}}(k) = \underline{u}_s(k)$ are the slow components of the corresponding variables in system (4.75).

The fast subsystem is derived by making the assumptions that $\underline{\overline{x_1}}(k) = \underline{x}_s(k)$ = constant and $\underline{\overline{x_2}}(k+1) = \underline{\overline{x_2}}(k)$. From (4.75) and (4.133) we get:

$$\underline{x}_2(k+1) - \underline{\overline{x_2}}(k+1) = A_4\{\underline{x}_2(k) - \underline{\overline{x_2}}(k)\} + B_2\{\underline{u}(k) - \underline{u}_s(k)\} \qquad (4.145)$$

Defining $\underline{x}_f(k) = \underline{x}_2(k) - \underline{\overline{x_2}}(k)$ and $\underline{u}_f(k) = \underline{u}(k) - \underline{u}_s(k)$, the fast subsystem of (4.75) can be expressed as

$$\underline{x}_f(k+1) = A_4\underline{x}_f(k) + B_2\underline{u}_f(k);$$

$$\underline{u}_f(0) = \underline{x}_{20} - \underline{\overline{x_2}}(0) \qquad (4.146)$$

It should be remarked that

(1) The assumptions used in deriving the fast subsystem are justified by noting that the slow modes of system (4.75) have magnitudes which are close to unity and during the short period of fast transients, these modes are changing very slowly with respect to the fast modes.

(2) Applying the linear transformations (4.130) with

$$L = -(I_{n-r} - A_4)^{-1} A_3, \quad K = A_o^{-1} A_2 \qquad (4.147)$$

to system (4.75) yields:

$$\underline{z}(k+1) = \begin{bmatrix} D_1 & D_2 \\ D_3 & D_4 \end{bmatrix} \underline{z}(k) + \begin{bmatrix} H_1 \\ H_2 \end{bmatrix} \underline{u}(k) \qquad (4.148)$$

where

$$D_1 = A_o + A_o^{-1} A_2 (I_{n-r} - A_4)^{-1} A_3 (I_r - A_o) \qquad (4.149a)$$

$$D_2 = A_o^{-1} A_2 (A_4 - (I_{n-r} - A_4)^{-1} A_3 A_o^{-1} A_2) \qquad (4.149b)$$

$$D_3 = (I_{n-r} - A_4)^{-1} A_3 (I_r - A_o) \qquad (4.149c)$$

$$D_4 = A_4 - (I_{n-r} - A_4)^{-1} A_3 A_o^{-1} A_2 \qquad (4.149d)$$

$$H_1 = B_o + [A_o^{-1} A_2 - A_2 (I_{n-r} - A_4)^{-1}] B_2 - A_o^{-1} A_2 (I_{n-r} - A_4)^{-1} A_3 B_1 \qquad (4.149e)$$

$$H_2 = B_2 - (I_{n-r} - A_4)^{-1} A_3 B_1 \qquad (4.149f)$$

The use of (4.147) in (4.127), (4.131) and (4.138) indicates that

$$D_1 = A_o + O(\mu) \qquad (4.150a)$$

$$D_2 = O(\mu) \quad , \qquad D_3 = O(\mu) \qquad (4.150b)$$

$$D_4 = A_4 + O(\mu) \qquad (4.150c)$$

$$H_1 = B_0 + O(\mu) \tag{4.150d}$$

$$H_2 = B_2 + O(\mu) \tag{4.150e}$$

This clearly shows that the two approaches to decoupling linear discrete systems are identical to first order perturbations. More importantly, it clarifies that the iterative scheme (4.136) and (4.137) are consistent with the results derived using the quasi-steady state concept [43, 44].

(3) In view of the above analysis and (4.140), it can be readily seen that (4.133) and (4.136) are first order perturbations to (4.129) in the sense that

$$\underline{x}_r(k) = \underline{x}_s(k) + O(\mu) \tag{4.151a}$$

$$\underline{x}\ (k) = \underline{x}_f(k) + O(\mu) \tag{4.151b}$$

To relate this result to the original discrete model (4.75) we use the linear transformation (4.133) to obtain:

$$\underline{x}_1(k) = \underline{x}_s(k) - A_0^{-1}A_2\underline{x}_f(k) + O(\mu) \tag{4.152a}$$

$$\underline{x}_2(k) = (I_{n-r} - A_4)^{-1}A_3\underline{x}_s(k) + \underline{x}_f(k) + O(\mu) \tag{4.152b}$$

which emphasises the closeness of the approximate trajectories $\{\underline{x}_s(k), \underline{x}_f(k)\}$ to $\{\underline{x}_1(k), \underline{x}_2(k)\}$; the original trajectories.

(4) By examining the eigenstructure of the discrete system (4.75) through modal analysis [30], it has been shown [42] that the matrices of the explicitly invertible linear transformations are given by:

$$\begin{aligned} L &= -M_3M_1^{-1} \\ &= N_4^{-1}N_3 \end{aligned} \tag{4.153a}$$

$$\begin{aligned} K &= M_1N_2 \\ &= M_2N_4 \end{aligned} \tag{4.153b}$$

The comparison of (4.85) with (4.126a) using (4.153a) reveals that $A_r = A_s$ which confirms the observation that the simplification schemes of section 4.4.2 preverses the dominant modes whereas the time-scale approach conserves the whole eigenspectrum. This motivates us to view the linear transformation (4.124) as an *extended* aggregation matrix [42].

4.5.3 *A FREQUENCY DOMAIN INTERPRETATION*

We now provide a frequency domain interpretation of the fast (nondominant) and slow (dominant) subsystems. To do this, we consider the control system (4.75) with the output

$$\underline{y}(k) = C\underline{x}(k) = [C_1 \quad C_2] \begin{bmatrix} \underline{x}_1(k) \\ \underline{x}_2(k) \end{bmatrix} \tag{4.154}$$

Note that (4.154) the output vector is split into components added together. By applying the z-transform theory to (4.75) and (4.154), we obtain the discrete transfer function matrix H(z) which can be put in the form:

$$H(z) = [C_1 \quad C_2] \begin{bmatrix} zI_r - A_1 & -A_2 \\ -A_3 & zI_{n-r} - A_4 \end{bmatrix}^{-1} \begin{bmatrix} B_1 \\ B_2 \end{bmatrix} \tag{4.155}$$

where I_j is the identity matrix of order j. One way to deal with H(z) is to expand it using the K-partitioning lemma [52] a well-known method for inverse of partitioned matrices. This yields:

$$H(z) = H_r(z) + H_n(z) \tag{4.156}$$

where

$$H_r(z) = C_0(z) \quad zI_r - A_0(z) \quad ^{-1}B_0(z) \tag{4.157a}$$

$$\left.\begin{aligned} A_0(z) &= A_1 + A_2(zI_{n-r} - A_4)^{-1}A_3 \\ B_0(z) &= B_1 + A_2(zI_{n-r} - A_4)^{-1}B_2 \\ C_0(z) &= C_1 + C_2(zI_{n-r} - A_4)^{-1}A_3 \end{aligned}\right\} \tag{4.157b}$$

$$H_n(z) = C_2(zI_{n-r} - A_4)^{-1}B_2 \tag{4.158}$$

It is interesting to observe from (4.158) that $H_n(z)$ is a function of the matrices $\{A_4, B_2, C_2\}$ describing the nondominant modes only. In view of the asymptotically stability of the original system (4.75), the nondominant transfer function $H_n(z)$ is stable. This is not generally true of all complex frequencies which correspond to different time-scales. Recall that the neighbourhood around $z=1$ represents a slow-time scale behaviour, whereas the small region around the origin provides a characterisation of the fast transient effects. By virtue of the fact that most reduced-order models are mainly based on the retention of the dominant modes, it is quite reasonable to consider the region around $z=1$ and consequently neglect $H_n(z)$. This means that the approximate model, in the frequency domain, will be described by $H_r(z=1)$. Substituting $z=1$ in (4.157) we obtain $A_0(z=1)$, $B_0(z=1)$ and $C_0(z=1)$, which would represent the approximate model that preserves the dominant models. It is significant to observe that this model corresponds exactly to (4.143) and (4.144). This confirms our analysis in the time-domain of deriving reduced-order models. The claim that the subsystems (4.143) and (4.145), corresponding to transfer functions (4.157) and (4.158), provide a slow-time scale approximation to the system (4.75) is now clearly justified.

We now consider the feedback control design for discrete two-time-scale systems.

4.5.4 *TWO-STAGE CONTROL DESIGN*

In view of the slow-fast separation, it would seem desirable in control design and implementation to derive the feedback signals using independent gain matrices. The design procedure is in two-stages where a set of desired fast eigenvalues is placed in the first stage to compute the fast feedback control. Then, in the second stage, the slow feedback control is derived by assigning a set of slow eigenvalues to desired locations. It has been shown [43] that the complete controllability (reachability) condition of (4.75) guarantees that the slow and fast subsystems are completely controllable (reachable). We seek a linear state feedback $\underline{u}(k) = K\underline{x}(k)$ that can be implemented in two-stages.

The following design procedure is proposed. In the first stage, we apply the transformation of (4.132) to system (4.75) to yield (4.129) and focus attention on the slow subsystem. An (nxr) feedback matrix K_1 is designed to place the eigenvalues of $(A_s + G_1K_1)$ at r desired locations. This is attained using any state-feedback eigenvalue algorithm [21]. The substitution of

$$\begin{aligned} \underline{u}(k) &= u_1(k) + \underline{u}_2(k) \\ &= [K_1 \quad 0] \begin{bmatrix} \underline{x}_r(k) \\ \underline{x}_\ell(k) \end{bmatrix} + u_2(k) \end{aligned} \tag{4.159}$$

into (4.120) results in

$$\begin{bmatrix} \underline{x}_r(k+1) \\ \underline{x}_\ell(k+1) \end{bmatrix} = \begin{bmatrix} A_s+G_1K_1 & 0 \\ G_2K_1 & A_f \end{bmatrix} \begin{bmatrix} \underline{x}_r(k) \\ \underline{x}_\ell(k) \end{bmatrix} + \begin{bmatrix} G_1 \\ G_2 \end{bmatrix} \underline{u}_2(k) \tag{4.160}$$

Since system (4.160) is a lower-triangular system, then to block-diagonalise it we can use a transformation of the type (4.124), that is

$$\begin{bmatrix} \underline{x}_r(k) \\ \underline{x}_g(k) \end{bmatrix} = \begin{bmatrix} I_r & 0 \\ L_2 & I_{n-r} \end{bmatrix} \begin{bmatrix} \underline{x}_r(k) \\ \underline{x}_\ell(k) \end{bmatrix} \tag{4.161}$$

where L_2 is a real root of

$$L_2(A_s + G_1K_1) - A_fL_2 + G_2K_1 = 0 \tag{4.162}$$

Applying the transformation (4.161) to (4.160) converts it to:

$$\begin{bmatrix} \underline{x}_r(k+1) \\ \underline{x}_g(k+1) \end{bmatrix} = \begin{bmatrix} A_s+G_1K_1 & 0 \\ 0 & A_f \end{bmatrix} \begin{bmatrix} \underline{x}_r(k) \\ x_g(k) \end{bmatrix} + \begin{bmatrix} G_1 \\ G_2+L_2G_1 \end{bmatrix} \underline{u}_2(k) \tag{4.163}$$

The pair $(A_f, G_2 + L_2G_1)$ is completely controllable (reachable) since (A_f, G_2) is completely controllable (reachable) [43]. In the second stage, an (mxn-r) feedback matrix K_2 is chosen such that the eigenvalues of $\{A_f + (G_2 + L_2G_1)K_2\}$ are positioned at (n-r) desired locations in the fast region. Substituting the feedback control

$$\underline{u}_2(k) = [0 \quad K_2] \begin{bmatrix} \underline{x}_r(k) \\ \underline{x}_g(k) \end{bmatrix} \tag{4.164}$$

into (4.163) gives

$$\begin{bmatrix} \underline{x}_r(k+1) \\ \underline{x}_g(k+1) \end{bmatrix} = \begin{bmatrix} A_s+G_1K_1 & G_1K_2 \\ 0 & A_f+(G_2+L_2G_1)K_2 \end{bmatrix} \begin{bmatrix} \underline{x}_r(k) \\ \underline{x}_g(k) \end{bmatrix} \tag{4.165}$$

which has its eigenvalues at the assigned locations. Combining (4.159), (4.161), (4.162) and making use of (4.132) we arrive at

$$\underline{u}(k) = [K_2L + (K_1 + K_2L_2)(I_r + KL) \;\vdots\; K_2 + (K_2L_2 + K_1)K] \; \underline{x}(k) \tag{4.166}$$

which is the desired result. This offers a general and systematic way of computing state feedback control of discrete systems with slow and fast modes.

Using the concept of quasi-steady state, a composite control of the form (4.166) can be derived [45] as:

$$\underline{u}(k) = \{[I_{n-r} - E_2(I_{n-r} - A_4)^{-1}B_2]E_1 + E_2(I_{n-r} - A_4)^{-1}A_3\}\underline{x}_1(k) + E_2\underline{x}_2(k) \tag{4.167}$$

where E_1 is designed to assign r slow eigenvalues of $(A_0 + B_0E_1)$ and E_2 is designed to assign fast eigenvalues of $(A_4 + B_2E_2)$ at (n-r) desired positions. It has been shown [42] that the use of (4.147) throughout (4.136 - 4.160) produces the control law (4.166). This once again confirms the similarity between the block-diagonalisation and quasi-steady state approaches in the analysis and control design of linear discrete systems with slow-fast separation.

Investigation of the structural properties of discrete two-time-scale systems [43] has emphasised that provided the fast subsystem is asymptotically stable, then the controllability (reachability) of the slow subsystem is invariant for a class of fast controls. Based on this, a reduced-order control can be designed from feedback of the slow states which guarantees the stabilisation of the original system.

4.5.5 *EXAMPLES*

We now present two control problems to illustrate the analysis and design of discrete two-time-scale systems.

Example 1

The discrete model of an eighth-order power system [41] has the state transition matrix

$$\begin{bmatrix} .835 & 0 & 0 & 0 & 0 & 0 & 0 & 0 \\ .96 & .861 & 0 & 0 & 0 & 0 & 0 & .029 \\ -.002 & -.005 & .882 & -.253 & .041 & -.003 & -.025 & -.001 \\ .007 & .014 & -.029 & .928 & 0 & .006 & .059 & .002 \\ -.03 & .061 & 2.028 & 2.303 & .088 & -.021 & -.224 & -.008 \\ .048 & .758 & 0 & 0 & 0 & .165 & 0 & .023 \\ -.012 & .027 & 1.209 & -1.4 & .161 & -.013 & .156 & .006 \\ .815 & 0 & 0 & 0 & 0 & 0 & 0 & .011 \end{bmatrix}$$

To put this matrix in two-time-scale form, we use the perturbation matrix

$$P = \{e_4, e_3, e_2, e_1, e_7, e_5, e_6, e_8\}$$

where e_i is the elementary column vector whose i-th entry is 1 and the scaling matrix

$$S = \text{diag}\ \{1, 1, .5, .33, .1, .055, .075, .05\}$$ to

obtain

$$A = \begin{bmatrix} .928 & -.029 & .028 & .0212 & .06 & 1.0727 & 0 & .04 \\ -.253 & .882 & -.01 & -.0061 & -.03 & 2.4545 & .5464 & -.02 \\ 0 & 0 & .861 & .1454 & 0 & 0 & 0 & .29 \\ 0 & 0 & 0 & .835 & 0 & 0 & 0 & 0 \\ 0 & 0 & .1516 & .0145 & .165 & 0 & 0 & .046 \\ -.077 & .0665 & -.003 & -.002 & -.0072 & .165 & .1181 & .0066 \\ -.1727 & .1521 & -.0092 & -.0068 & -.0158 & -.3055 & .088 & -.012 \\ 0 & 0 & 0 & .1235 & 0 & 0 & 0 & .011 \end{bmatrix}$$

The eigenspectrum of A {.8745±j.1696,.861,.845,.2866,.165,.0181,.011} suggests that the eighth-order has four slow and four fast variables. The static separation ratio of (4.113) has the value

$$\mu = |\lambda_5|/|\lambda_4| = .343$$

To derive the slow and fast subsystems, we solve (4.118) and (4.122) in the manner of (4.125) - (4.128) to obtain after four iterations the transformation matrix L as:

$$L = \begin{bmatrix} 0 & 0 & -.2178 & .0293 \\ .0843 & -.1238 & -.0012 & -.0037 \\ .1358 & -.1561 & .0053 & .0033 \\ 0 & 0 & 0 & -.1499 \end{bmatrix}$$

and after six iterations the transformation matrix K as:

$$K = \begin{bmatrix} .0947 & 1.9788 & .0421 & .0573 \\ -.0058 & .1009 & .6207 & .0039 \\ 0 & 0 & 0 & .3412 \\ 0 & 0 & 0 & 0 \end{bmatrix}$$

From (4.117) the slow and fast matrices are given by

$$A_s = \begin{bmatrix} .8376 & .1038 & .0424 & .0294 \\ -.2889 & .911 & -.0249 & -.0017 \\ 0 & 0 & .861 & .1889 \\ 0 & 0 & 0 & .835 \end{bmatrix}$$

with $\sigma(A_s) = \{.8745 \pm j.1696, .861, .835\}$, and

$$A_f = \begin{bmatrix} .165 & 0 & 0 & -.0172 \\ .0015 & .3027 & .0505 & .0121 \\ -.003 & -.0889 & .003 & .002 \\ 0 & 0 & 0 & .011 \end{bmatrix}$$

with $\sigma(A_f) = \{.2866, .1650, .0181, .011\}$.

It is obvious that (4.114a) is satisfied.

The spectral norms of the different matrices are:

$$\begin{aligned} ||A_1|| &= 1.057 \quad , \quad & ||A_2|| &= 1.1914 \\ ||A_3|| &= .2521 \quad , \quad & ||A_4|| &= .3446 \\ ||A_0|| &= 1.0011 \quad , \quad & ||A_0^{-1}|| &= 1.3167 \\ ||L_0|| &= .2523 \quad , \quad & ||I_4 - A_0|| &= .3455 \end{aligned}$$

Simple calculations reveal that inequality (4.129a) is satisfied whereas inequality (4.129b) is not satisfied with the factor 3. This implies that conditions (4.129) are conservative.

Example 2

A fifth-order power system model [42] of the type (4.75) is given by:

$$A = \begin{bmatrix} .915 & .051 & .038 & .015 & .038 \\ -.03 & .889 & -.0005 & .046 & .111 \\ -.006 & .468 & .247 & .014 & .048 \\ -.715 & -.022 & -.021 & .240 & -.024 \\ -.148 & -.003 & -.004 & .090 & .026 \end{bmatrix}$$

$$B = [.0098 \quad .122 \quad .036 \quad .562 \quad .115]^t$$

Evaluation of the system eigenspectrum reveals that this system has two slow states (r=2) and three fast states (n-r = 3), with a static separation ratio μ = .2646. The numerical solutions of (4.118) and (4.13) yield the transformation matrices as:

$$L = \begin{bmatrix} -.0685 & -.7161 \\ 1.0787 & -.074 \\ .2812 & -.0267 \end{bmatrix}$$

$$K = \begin{bmatrix} .0573 & .0146 & .0305 \\ .0132 & .0924 & .1329 \end{bmatrix}$$

The slow and fast models are described by

$$A_s = \begin{bmatrix} .8904 & .0796 \\ -.1105 & .8952 \end{bmatrix}, \quad G_1 = \begin{bmatrix} .0185 \\ .1889 \end{bmatrix}$$

$$A_f = \begin{bmatrix} .2443 & -.0198 & -.0339 \\ .0198 & .2527 & .0097 \\ .007 & .0934 & .0338 \end{bmatrix}, \quad G_2 = \begin{bmatrix} -.0522 \\ .5634 \\ .1148 \end{bmatrix}$$

Application of the approximate analysis gives the slow and fast subsystems to first-order perturbation:

$$A_O = \begin{bmatrix} .8901 & -.0727 \\ -.099 & .8858 \end{bmatrix}, \quad B_O = \begin{bmatrix} .0306 \\ .1761 \end{bmatrix}$$

$$A_4 = \begin{bmatrix} .2465 & .0139 & .048 \\ -.0207 & .2399 & -.0236 \\ -.0035 & .0904 & .0259 \end{bmatrix}, \quad B_2 = \begin{bmatrix} .0359 \\ .5619 \\ .1153 \end{bmatrix}$$

It is easy to check that

$$\sigma(A_s) \bigcup \sigma(A_f) = \{.8928 \pm j.0.0937, .2506 \pm j.0.0252, .0295\}$$

whereas $\sigma(A_0) \bigcup \sigma(A_4) = \{.8879 \pm j.0.0848, .2387 \pm j.0.0258, .0350\}$. Simple comparison confirms the validity of the approximation.

To develop a state feedback control, we assign two slow eigenvalues at .93, .85 and three fast eigenvalues at .22, .2 and .08. Application of the two-stage feedback control law (4.151) gives:

$$u(k) = [.6503 \;\; -.1419 \;\; .0565 \;\; .0065 \;\; -.2696]\underline{x}(k)$$

Using this control law in system (4.75) gives the closed-loop eigenvalues {.93,.85,.2201,.1998,.0801} which are very close to the desired ones. For comparison, the approximate feedback control law (4.152) takes the form:

$$u(k) = [.6453 \;\; -.0854 \;\; .0134 \;\; .0324 \;\; -.2345]\underline{x}(k)$$

The corresponding closed-loop eigenvalues are {.9264,.8572, .2411,.211,.0682} which are close to the desired ones. This, once again, supports the validity of the first-order approximation in control analysis and design of discrete two-time-scale systems.

4.6 Concluding Remarks

In this chapter, we have presented different methodologies for designing linear, deterministic, discrete control systems using state or output feedback. Three main subjects are considered, namely deadbeat controllers, model simplification and two-time-scale systems. Although, we have put emphasis on the mathematical treatment of the control problems however, several illustrative examples are worked out in detail to demonstrate the application of the different techniques.

4.7 Problems

1. A third-order control system is described by

$$A = \begin{pmatrix} -2 & 1 & 0 \\ 0 & -2 & 0 \\ 0 & 0 & 4 \end{pmatrix} \quad ; \quad B = \begin{pmatrix} 0 & 0 \\ 0 & 1 \\ 1 & 0 \end{pmatrix}$$

Design a constant state feedback matrix K which yields closed-loop eigenvalues $\sigma_d = \{-.1,-.2,.3\}$

2. Repeat problem 1 if $\sigma_d = \{-.1,-.1,.8\}$

3. Repeat problem 1 if $\sigma_d = \{.2,.2,.2\}$

(Hint: in problems 1-3 consider the assignment of eigenvalues only)

4. A discretised model of an engine/dynamometer test rig has the following matrices [51]

$$A = \begin{pmatrix} .7189 & .0866 & .0733 \\ .4312 & .4704 & -.4206 \\ -.3262 & .1731 & .2027 \end{pmatrix}$$

$$B = \begin{pmatrix} .1637 & -.2056 \\ .2010 & -.2155 \\ .0169 & .0152 \end{pmatrix}$$

$$C = \begin{pmatrix} 1 & 0 & 1 \\ 0 & 0 & 1 \end{pmatrix}$$

where the state variables are the dynamometer rotor speed, engine speed and shaft torque. The control variables are the throttle servo voltage and dynamometer source current. Design an output feedback scheme to place the closed-loop eigenvalues at $.13 \pm j.3$. What will be the third eigenvalue?

5. A model for milk supply by three different dairies is given by [39]

$$A = \begin{pmatrix} .8 & .2 & .1 \\ .1 & .7 & .3 \\ .1 & .1 & .6 \end{pmatrix} \quad ; \quad B = \begin{pmatrix} 1 & 0 \\ 0 & 1 \\ 1 & 1 \end{pmatrix}$$

a) Calculate the eigenvalues of the system and show that it is marginally stable.

b) Design a stabilising feedback control policy to locate the closed-loop eigenvalue at $.4, \pm.7$.

c) Compute the gain matrix which yield a closed-loop system matrix of the form

$$J = \begin{pmatrix} .3 & 1 & 0 \\ 0 & .3 & 0 \\ 0 & 0 & .8 \end{pmatrix}$$

6. Design a deadbeat controller for the systems described by:

(a)

$$A = \begin{pmatrix} 0 & 1 & 0 \\ 0 & 0 & 1 \\ 4 & 4 & -1 \end{pmatrix} \quad ; \quad B = \begin{pmatrix} 1 & 0 \\ 0 & 0 \\ 0 & 1 \end{pmatrix}$$

(b)

$$B = \begin{pmatrix} 1 & 1 & 0 \\ 0 & 1 & 0 \\ 0 & 0 & 1 \end{pmatrix} \quad ; \quad B = \begin{pmatrix} 0 & 1 \\ 1 & 0 \\ 1 & 0 \end{pmatrix}$$

7. In the system model considered in problem 4, by including the dynamics of dynamometer field-current amplifier, we obtain a fifth-order model described by [51]:

$$A = \begin{bmatrix} .8070 & 0 & 0 & .0092 & 0 \\ -.0267 & .5527 & .0171 & -.0002 & .0012 \\ -.1998 & 5.956 & .1599 & -.0018 & -.2576 \\ -5.0795 & 0 & 0 & -.0381 & 0 \\ .0243 & -6.8493 & .2311 & .0003 & -.3805 \end{bmatrix}$$

$$B = \begin{bmatrix} 0 & .8511 \\ .0766 & -.0106 \\ .7019 & -.0832 \\ 0 & 22.3995 \\ .1418 & .0257 \end{bmatrix}$$

$$C = \begin{bmatrix} 0 & 1 & 0 & 0 & 0 & 0 \\ 0 & 0 & 0 & 0 & 0 & 1 \end{bmatrix}$$

a) Calculate the eigenvalues of the open-loop system and identify the number of dominant and nondominant modes.

b) Develop the reduced-order models to retain the dominant eigenvalues. Make a comparison between the two models.

c) Develop a second-order approximate model based on the output-modelling approach.

d) By evaluating the step response of the above three reduced models, draw conclusions about the main differences between them.

8. Consider the fifth-order model presented in problem 7. Put the model in block-diagonal form by separating the fast and slow modes. Design a two-stage feedback control which yields closed-loop eigenvalues at .8, .7, .1, -.2026, -.2173 to eliminate system oscillation.

4.8 References

[1] Bishop, A.B.
"Introduction fo Discrete Linear Controls", Academic Press, New York, 1975.

[2] Cadzow, J.A. and H.R. Martens,
"Discrete-Time and Computer Control Systems", Prentice-Hall, Inc., New Jersey, 1970.

[3] Cadzow, J.A.
"Discrete-Time Systems", Prentice-Hall, Inc., New Jersey, 1973.

[4] Gill, A.
"Linear Sequential Circuits: Analysis, Synthesis and Applications", McGraw-Hill, New York, 1967.

[5] Ogata, K.
"Modern Control Engineering", Prentice-Hall, Inc., New Jersey, 19

[6] Power, H.M. and R.J. Simpson,
"Introduction to Dynamics and Control", McGraw-Hill, London, 1978.

[7] Truxal, J.G.
"Introductory System Engineering", McGraw-Hill, New York, 1972.

[8] Wolovich, W.A.
"Linear Multivariable Systems", Springer-Verlag, New York, 1974.

[9] Young, P. and B. Beck,
"The Modelling and Control of Water Quality in a River System", Automatica, Vol. 10, 1974, 455-468.

[10] Singh, M.G. and A. Titli,
"Systems: Decomposition, Optimisation and Control", Pergamon Press, Oxford, 1978.

[11] Runyan, H.M.
"Cybernetics of Economic Systems", IEEE Trans. Systems, Man and Cybernetics, Vol. SMC-1, 1971, 8-18.

[12] Perkins, W.R., J.B. Cruz, Jr. and N. Sundarajan, "Feedback Control of a Macroeconomic System using an Observer", IEEE Trans. Systems, Man and Cybernetics, Vol. SMC-2, 1972, 275-278.

[13] Allen, R.G.D.
"Macro-Economic Theory - A Mathematical Treatment", St. Martin's, New York, 1968.

[14] Singh, M.G. and H. Tamura,
"Modelling and Hierarchical Optimisation for Oversaturated Urban Road Traffic Networks", Int. J. Control, Vol. 20, 1974, 913-934.

[15] Mitter, S.K. and R. Foulkes,
"Controllability and Pole Assignment for Discrete Time Linear Systems Defined over Arbitrary Fields", SIAM J. Control, Vol. 9, 1971, 1-7.

[16] Kalman, R.E.
"On the General Theory of Control Systems", Preprint First IFAC Congr. Automat. Contr. Moscow, 4, 1960, 2020-2030.

[17] Kalman, R.E., P.L. Falb and M. Arbib,
"Topics in Mathematical System Theory", McGraw-Hill, New York, 1969.

[18] Davison, E.J.
"On Pole Assignment in Linear Systems with Incomplete State-Feedback", IEEE Trans. Automat. Contr., Vol. AC-15, 1970, 348-351.

[19] Young, P.C. and J.C. Willems,
"An Approach to the Linear Multivariable Servomechanism Problem", Int. J. Control, Vol. 15, 1972, 961-979.

[20] Brogan, W.L.
"Modern Control Theory", Quantum Publishers, Inc., New York, 1974.

[21] Munro, N.
"Pole Assignment", Proc. IEE, Vol. 126, 1979, 549-554.

[22] Chen, C.T.
"Introduction to Linear System Theory", Holt, Rinehart and Winston, New York, 1970.

[23] Weiss, L.
"Controllability, Realisation and Stability of Discrete-Time Systems", SIAM J. Control, Vol. 10, 1972, 230-251.

[24] Mullis, C.T.
"Time-Optimal Discrete Regulator Gains", IEEE Trans. Automat. Contr., Vol. AC-17, 1972, 265-266.

[25] O'Reilly, J.
"The Discrete Linear Time-Invariant Time-Optimal Control Problem - An Overview", Automatica, Vol. 17, 1981, 363-370.

[26] Leden, B.
"Multivariable Deadbeat Control", Automatica, Vol. 13, 1977, 185-188.

[27] Cadzow, J.A.
"Nilpotency Property of the Discrete Regulator", IEEE Trans. Automat. Contr., Vol. AC-13, 1968, 734-735.

[28] Halmos, P.R.
"Finite-Dimensional Vector Spaces", Van Nostrand, New Jersey, 1958.

[29] Porter, B.
"Deadbeat Control of Linear Multivariable Discrete-Time Systems", Electr. Lett., Vol. 9, 1973, 83-84.

[30] Porter, B. and T.R. Crossley,
"Modal Control", Taylor & Francis, London, 1972.

[31] Luenberger, D.G.
"Observers for Multivariable Systems", IEEE Trans. Automat. Contr., Vol. AC-11, 1966, 190-197.

[32] Fahmy, M.M., A.A.R. Hanafy and M.F. Sakr,
"On the Discrete Time-Optimal Regulator Control Problem", Information and Control, Vol. 44, 1980, 223-235.

[33] O'Reilly, J.
"The Deadbeat Control of Linear Multivariable Systems with Inaccessible States", Int. J. Control, Vol. 31, 1980, 645-654.

[34] Ichikawa, K.
"Discrete-Time Fast Regulator with Fast Observer", Int. J. Control, Vol. 28, 1978, 733-742.

[35] Mahmoud, M.S. and M.G. Singh,
"Large Scale Systems Modelling", Pergamon Press, Oxford, 1981.

[36] Wilson, R.G., D.G. Fisher and D.E. Seborg,
"Model Reduction for Discrete-Time Dynamic Systems", Int. J. Control, Vol. 16, 1972, 549-558.

[37] Wilson, R.G., D.G. Fisher and D.E. Seborg, "Model Reduction and the Design of Reduced-Order Control Laws", AIChE Journal, Vol. 20, 1974.

[38] Arbel, A. and E. Tse, "Reduced-Order Models, Canonical Forms and Observers", Int. J. Control, Vol. 30, 1979, 513-531.

[39] Noble, B. "Applied Linear Algebra", Prentice-Hall, New Jersey, 1969.

[40] Aoki, M. "Control of Large Scale Dynamic Systems by Aggregation", IEEE Trans. Automat. Contr., Vol. AC-13, 1968, 246-253.

[41] Mahmoud, M.S., Y. Chen and M.G. Singh, "Discrete Two-Time-Scale Systems", UMIST Control Systems Centre Report No. 497, December, 1980.

[42] Mahmoud, M.S. "Order Reduction and Control of Discrete Systems", Proc. IEE, Vol. 129, 1982, 129-135.

[43] Mahmoud, M.S. "Structural Properties of Discrete Systems with Slow and Fast Modes", Large Scale Systems, Vol. 3, 1982, 227-236.

[44] Mahmoud, M.S., Y. Chen and M.G. Singh, "A Two-Stage Output Feedback Design", UMIST Control Systems Centre Report No. 516, May, 1981.

[45] Mahmoud, M.S., Y. Chen and M.G. Singh, "On the Eigenvalue Assignment in Discrete Systems with Fast and Slow Modes", UMIST Control Systems Centre Report No. 499, February, 1981.

[46] Moore, B.C. "On the Flexibility Offered by State Feedback in Multivariable Systems Beyond Closed Loop Eigenvalue Assignment" IEEE Trans. Automat. Contr., Vol. AC-21, 1976, 689-692.

[47] Klein, G. and B.C. Moore, "Eigenvalue-Generalised Eigenvector Assignment with State Feedback", IEEE Trans. Automat. Contr., Vol. AC-22, 1977, 140-141.

[48] Fahmy, M.M. and J. O'Reilly, "On Eigenstructure Assignment in Linear Multivariable Systems", IEEE Trans. Automat. Contr., Vol. AC-27, 1982, 690-693.

[49] Porter, B. and J.J. D'Azzo "Algorithm for the Synthesis of State-Feedback Regulators by Entire Eigenstructure Assignment", Electron. Lett., Vol. 13, 1977, 230-231.

[50] Porter, B. and J.J. D'Azzo,
"Closed-Loop Eigenstructure Assignment by State Feedback in Multivariable Linear Systems", Int. J. Control, Vol. 27, 1978, 487-492.

[51] Mahmoud, M.S. and Y. Chen,
"Design of Feedback Controllers by Two-Stage Methods", Applied Mathematical Modelling, Vol. 7, 1983, pp. 163-=68.

[52] Nicholson, H.
"Structure of Interconnected Systems", Peregrinus, London, 1978.

Chapter 5

Control of Systems with Inaccesible States

5.1 Introduction

In Chapter 4, we considered the feedback design problem of discrete-time systems, and developed linear control laws which required access to the system state vector for implementation. It has been shown that the feedback gain matrices can be determined by assigning the closed-loop eigenvalues to desired locations. In cases where only a small number of system variables (outputs) are directly available for forming feedback signals, some of the closed-loop eigenvalues can be positioned close to desired locations. It was shown that the remaining eigenvalues may take arbitrary values and this may lead to undesirable performance for the overall system.

The aim of this chapter is to present methods of reconstructing the state vector, or finding approximations to the state vector, from the input and output records. The reconstructed state vector can then be used in generating linear feedback control laws.

The material covered in this chapter is organised into four basic sections. In Section 5.2 we examine state reconstruction schemes, using full-order and reduced-order deterministic observers. Two different measurement patterns are emphasised: instantaneous recording and one-step delay in recording. The various properties of interest of state reconstructors and the relationships between them are pointed out.

The problem of designing observer-based controllers is discussed in Section 5.3, and the separation principle of design is developed. In both Sections 5.2 and 5.3, the important notion of basing the design at deadbeat response is emphasised.

Sections 5.4 and 5.5 are devoted to investigating the problems of state observation and observer-based controllers for higher order systems. In Section 5.4, the class of interconnected systems with linear coupling effects is studied. A two-level state reconstruction scheme is developed to take advantage of the subsystem dynamics. Discrete two-time-scale systems are considered in Section 5.5 . The slow-fast separation property is used to develop full-order and reduced-order observers and then observer-based controllers are constructed. Several examples are given in the chapter to illustrate the theoretical analysis.

5.2 State Reconstruction Schemes

The problem of designing state reconstructors (deterministic observers) for dynamical systems has received considerable attention over the past decade. The original work of Luenberger *[1-3]* established the basic foundations of observer theory. Since then, several other researchers have simplified the design procedure and provided further insights into the subject *[4-12]*. For discrete-time systems, the state reconstruction problem is that of determining an estimate of the state vector whose dynamic description is governed by:

$$\underline{x}(k+1) = A\underline{x}(k) + B\underline{u}(k) \tag{5.1}$$

$$\underline{y}(k) = C\underline{x}(k) \tag{5.2}$$

where at the discrete instant k,

$\underline{x}(k)$ is the n-dimensional state vector
$\underline{u}(k)$ is the m-dimensional input vector
$\underline{y}(k)$ is the p-dimensional output vector
A is the (nxn) system matrix
B is the (nxm) input matrix
C is the (pxn) output matrix

Without loss of generality, it is assumed from now onwards that rank $[C] = p < n$ which implies that the output variables are independent. It is required to reconstruct the state vector $\underline{x}(k)$ from the available input and output records, $\{\underline{u}(k), y(k)\}$. The solution of the state reconstruction problem depends on the type of output records, and whether we wish to develop full-order or reduced-order deterministic observers. We first consider the design of full-order deterministic observers.

5.2.1 *FULL-ORDER STATE RECONSTRUCTORS*

An essential distinction should be made between the following two problem statements *[8]*:

(a) *the state* $\underline{x}(k)$ *is to be reconstructed from the observation records* $\{\underline{y}(k-1), \underline{y}(k-2), \ldots, \underline{y}(0)\}$

(b) *the state* $\underline{x}(k)$ *is to be reconstructed from the observation records* $\{\underline{y}(k), \underline{y}(k-1), \ldots, \underline{y}(0)\}$

We note that problem statement (a) entails a one-step delay in the measurement pattern and problem statement (b) considers the situation where the outputs are recorded instantaneously. The solution of problem (a) is considered first. By virtue of the fact that the state reconstructor must be a linear dynamical system with the vectors $\underline{u}(k)$ and $\underline{y}(k)$ being the external inputs, we can write it as *[3,5]*:

$$\hat{\underline{x}}(k+1) = \hat{A}\hat{\underline{x}}(k) + \hat{B}\underline{u}(k) + \hat{C}\underline{y}(k) \qquad (5.3)$$

where $\hat{\underline{x}}(k)$ is the $(n\times 1)$ vector approximation to $\underline{x}(k)$. The unknown matrices $\hat{A}$, $\hat{B}$, $\hat{C}$ are of dimensions $(n\times n)$, $(n\times m)$ and $(n\times p)$, respectively. It has been shown *[5]* that system (5.3) is a *full-order state reconstructor* of the system (5.1) and (5.2) if $\hat{\underline{x}}(0) = \underline{x}(0)$ implies $\hat{\underline{x}}(k) = \underline{x}(k)$ for all $\underline{u}(k)$, $k > 0$. This is satisfied if and only if the pair (A,C) is completely observable, that is:

$$\text{rank}[C^t, A^tC^t, \ldots, A^{(n-1)}C^t] = n \qquad (5.4)$$

and
$$\hat{A} = A - HC \quad (5.5a)$$
$$\hat{B} = B \quad (5.5b)$$
$$\hat{C} = C \quad (5.5c)$$

where the (n×p) matrix H contains the design parameters.

With (5.5), the state reconstructor (5.3) can be represented as follows:

$$\begin{aligned}\hat{\underline{x}}(k+1) &= A\hat{\underline{x}}(k) + B\underline{u}(k) + H[\underline{y}(k) - C\hat{\underline{x}}(k)] \\ &\equiv [A-HC]\hat{\underline{x}}(k) + B\underline{u}(k) + H\underline{y}(k) \quad (5.6)\end{aligned}$$

This shows that the state reconstructor consists of a model of the system, with as an extra driving variable an input which is proportional to the difference $[\underline{y}(k)-\hat{\underline{y}}(k)]$ of the recorded variable $\underline{y}(k)$ and its predicted value

$$\hat{\underline{y}}(k) = C\hat{\underline{x}}(k) \quad (5.7)$$

A block diagram of the full-order state reconstructor (5.6) is given in Fig. 5.1 .

We now discuss the stability of the state reconstructor and the behaviour of the reconstruction error

$$\hat{e}(k) = \underline{x}(k) - \hat{\underline{x}}(k) \quad (5.8)$$

Upon subtracting the state difference equations (5.1) and (5.6) and using (5.2) we obtain:

$$\hat{\underline{e}}(k+1) = [A - HC]\hat{\underline{e}}(k) \quad (5.9)$$

If the eigenvalues of $[A - HC]$ all have absolute values less than one, then the reconstruction error is asymptotically stable. This indicates that $\hat{\underline{e}}(k) \simeq \underline{0}$, or $\hat{\underline{x}}(k) \simeq \underline{x}(k)$ as $k \to \infty$. Since it is assumed that the system (5.1), (5.2) is completely observable, it is always possible to determine a

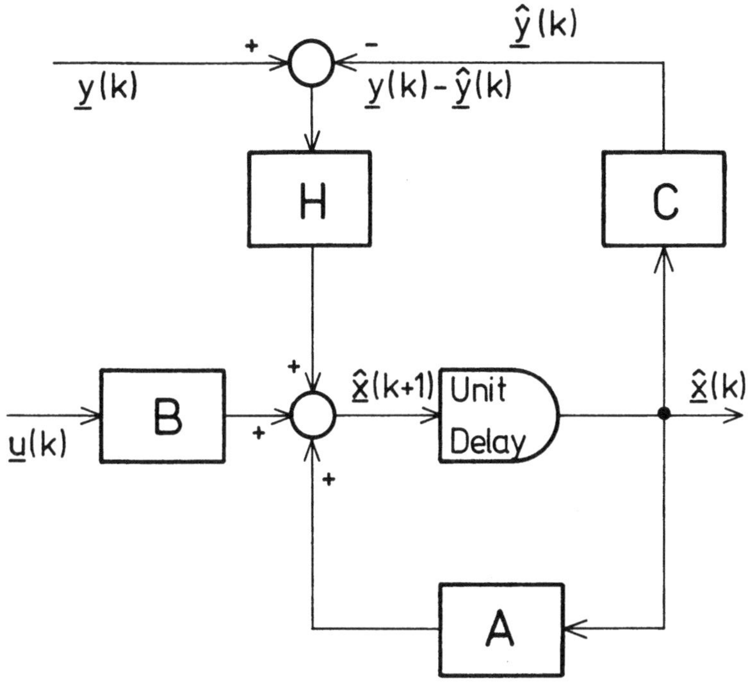

Fig. (5.1) A block-diagram of the full-order state reconstructor with one-step delay in the measurements

matrix H which will yield any set of desired eigenvalues for $[A-HC]$ within the restriction that complex eigenvalues occur in conjugate pairs. Thus it is possible to control the rate at which $\hat{\underline{x}}(k) \to \underline{x}(k)$. Recall that the characteristic polynomial of the system (5.6) is given by:

$$\begin{aligned} \Delta_c(\gamma) &= \det[\gamma I_n - A + HC] \\ &= \det[\gamma I_n - A]\det[I_n + (\gamma I_n - A)^{-1}HC] \\ &= \det[\gamma I_n - A]\det[I_p + C(\gamma I_n - A)^{-1}H] \end{aligned} \tag{5.10}$$

In view of (5.4), p linearly independent rows can be selected from the rows of $C(\gamma I_n - A)^{-1}$ (or its derivatives if necessary when repeated eigenvalues are desired). To determine H, we can therefore use any of the eigenvalue assignment algorithms discussed in Chapter 4.

We refer to the characteristic values of (5.10) as the observer eigenvalues. Intuitively, to obtain fast convergence of the reconstruction error to zero, H should be chosen such that the observer eigenvalues are close to the origin in the complex plane. This, however, generally must be achieved by making the gain matrix H large, which in turn makes the state reconstructor very sensitive to any external measurement output noise. On the other hand, observer eigenvalues with larger magnitudes tend to yield slow convergence of the reconstruction error. It is therefore up to the designer to conceive a compromise between the two situations.

Next, we consider the solution of problem (b) in which the present output records are available for the reconstruction of the present state. In this case, the driving term, proportional to $[\underline{y}(k) - C\hat{\underline{x}}(k)]$ in (5.6), should be replaced by the difference between the present output record $\underline{y}(k+1)$ and its estimate

$$C[A\tilde{\underline{x}}(k) + B\underline{u}(k)]$$

where $\tilde{\underline{x}}(k)$ is the (n×1) vector approximation to $\underline{x}(k)$ when the observation records are instantaneous. Similarly, the full-order state reconstructor is governed by:

$$\underline{\tilde{x}}(k+1) = A\underline{\tilde{x}}(k) + B\underline{u}(k) + P[\underline{y}(k+1) - CA\underline{\tilde{x}}(k) - CB\underline{u}(k)]$$

or

$$\underline{\tilde{x}}(k+1) = [A - PCA]\underline{\tilde{x}}(k) + [B - PCB]\underline{u}(k) + P\underline{y}(k+1) \qquad (5.11)$$

where P is an n×p matrix of design parameters. In terms of the reconstruction error

$$\underline{\tilde{e}}(k) = \underline{x}(k) - \underline{\tilde{x}}(k) \qquad (5.12)$$

we obtain from (5.1), (5.2) and (5.11) the dynamics of the reconstruction error as:

$$\underline{\tilde{e}}(k+1) = [A - PCA]\underline{\tilde{e}}(k) \qquad (5.13)$$

In Fig. 5.2, a block diagram of the state reconstructor (5.11) is presented.

In contrast to (5.9), the possibility of assigning the eigenvalues of $[A - PCA]$ at desired locations does not depend on the observability of the pair (A,C), but on the observability of the pair (A,CA). The implication of this situation is analysed in the following:

1. For the pair (A,CA) to be observable, it is required that

$$\text{rank}[A^tC^t, A^tA^tC^t, \ldots, A^tA^{(n-1)^t}C^t] = n \qquad (5.14)$$

   Provided the system A is nonsingular, it can be shown [7] that condition (5.4) is equivalent to (5.14).

   This shows that for sampled-data systems, in which the state transition matrix is always nonsingular, the con-

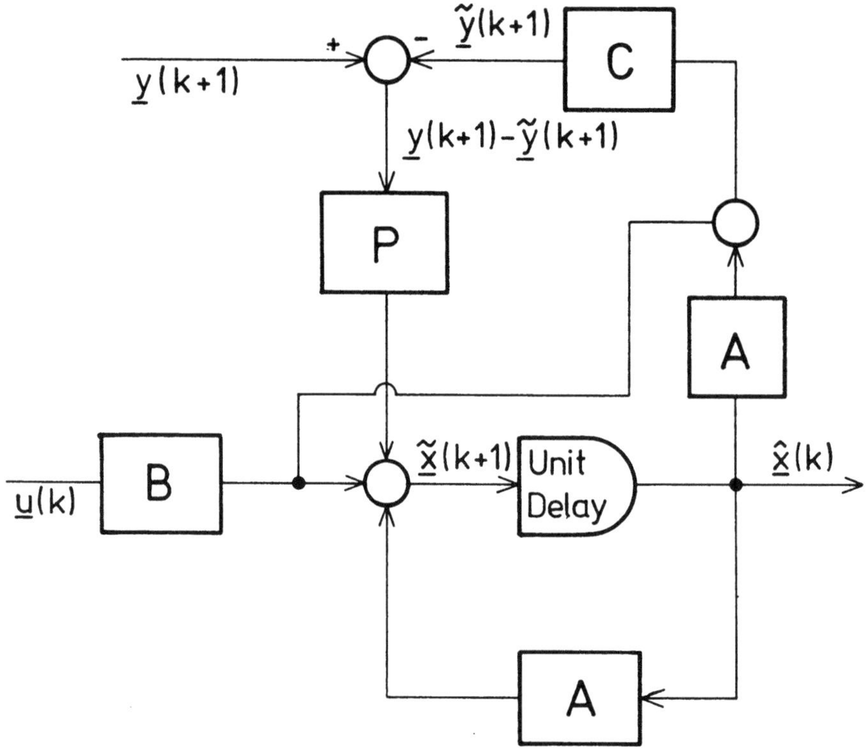

Fig. (5.2) A block-diagram of the full order state reconstructor with instantaneous measurements

dition for solving problem (a) is identical to that of problem (b).

2. In the case where A is singular and the pair (A,C) is observable in the sense of (5.4), the pair (A,CA) is not observable. However, it can be shown *[8]* that all unobservable modes of (A,CA) correspond to zero eigenvalues. This means that the pair (A,CA) is both detectable (all non-observable modes are asymptotically stable) and reconstructable (all non-observable modes are deadbeat).

   Geometrically, this implies that *[5]*

$$N[A] = N[(A,CA)]$$

   where $N[W]$ is the null space of the matrix W.

In the light of the above discussion, we can determine the gain matrix P using an eigenvalue algorithm similar to the one that can be used to determine the gain matrix H. since the difference between problems (a) and (b) is in the type of output records, an immediate question arises: what is the relationship between the two state reconstruction schemes? The answer to this question can be obtained by considering the following two-step state reconstruction procedure: *[8]*

Step 1:

Reconstruct the state $\underline{x}(k)$ using the scheme (5.11). The result will be $\tilde{\underline{x}}(k)$.

Step 2:

The reconstructed state $\hat{\underline{x}}(k)[1]$ is derived from the dynamic system (5.1), using $\tilde{\underline{x}}(k)$.

By virtue of Step 1, we get:

$$\tilde{\underline{x}}(k) = [A - PCA]\tilde{\underline{x}}(k-1) + [B - PCB]u(k-1) + P\underline{y}(k) \qquad (5.15)$$

From Step 2, we have:

$$\hat{\underline{x}}(k+1) = A\tilde{\underline{x}}(k) + B\underline{u}(k) \qquad (5.16)$$

The substitution of (5.16) into (5.15) yields:

$$\hat{\underline{x}}(k+1) = [A^2 - APCA]\tilde{\underline{x}}(k-1) + [AB - APCB]\underline{u}(k-1) + AP\underline{y}(k) + B\underline{u}(k) \quad (5.17)$$

But

$$\hat{\underline{x}}(k) = A\tilde{\underline{x}}(k-1) + B\underline{u}(k-1) \quad (5.18)$$

therefore, from (5.17) and (5.18) we arrive at:

$$\hat{\underline{x}}(k+1) = [A - APC]\hat{\underline{x}}(k) + B\underline{u}(k) + AP\underline{y}(k) \quad (5.19)$$

Comparing (5.6) and (5.19) we see that they are identical provided the condition

$$H = AP \quad (5.20)$$

holds. The interpretation of this result is that the eigenvalues of the matrix $[A - HC]$ cannot be arbitrarily assigned unless A is nonsingular. This also implies that in this case one is able to reconstruct the state $\underline{x}(k+1)$ from the output records $\{\underline{y}(k), y(k-1), \ldots, \underline{y}(0)\}$. In the case where A is singular, we know that the fixed modes are all zero and they disappear after one step *[8]*.

Next, we consider the design of reduced-order state reconstructors.

5.2.2 *REDUCED-ORDER STATE-RECONSTRUCTORS*

In this section we show that it is possible to find state reconstructors of dimension less than the dimension of the system to be observed. Such state reconstructors are called ***reduced-order state reconstructors***.

Consider the nth-dimensional discrete system (5.1), (5.2). Since the measurement equation $\underline{y}(k) = C\underline{x}(k)$ provides us with p linear equations in the unknown state $\underline{x}(k)$, it is necessary to reconstruct only (n-p) linear combinations of the components of the state.

We now discuss the measurement patterns in the light of problems (a) and (b) stated earlier. According to problem (a), the output $\underline{y}(k)$ is not available for the reconstruction of the state $\underline{x}(k)$ but only for the reconstruction of $\underline{x}(k+1)$. Therefore, all state variables are to be reconstructed and, in general, no state variables (or their linear functions) can be directly derived from the output records. The concept of reduced-order state reconstruction would not appear to be directly applicable in the case of problem (a).

In problem (b), the output $\underline{y}(k)$ is available for the reconstruction of the state $\underline{x}(k)$. Here, the concept of reduced-order state reconstruction is applicable since some state variables (or their linear functions) can be directly computed from the output $\underline{y}(k)$. Since C is of full rank (p), we assume without loss of generality that it can be expressed in the form

$$C = [C_1 \quad C_2]$$

where the $(p \times p)$ matrix C_1 is nonsingular. By introducing the coordinate transformation *[13]*

$$T = \begin{pmatrix} C_1 & C_2 \\ 0 & I_{n-p} \end{pmatrix} ; \quad T^{-1} = \begin{pmatrix} C_1^{-1} & -C_1^{-1}C_2 \\ 0 & I_{n-p} \end{pmatrix} \tag{5.21}$$

and using

$$\begin{pmatrix} \underline{v}(k) \\ \underline{w}(k) \end{pmatrix} = T\,\underline{x}(k) \tag{5.22}$$

the discrete system (5.1), (5.2) reduces to:

$$\begin{pmatrix} \underline{v}(k+1) \\ \underline{w}(k+1) \end{pmatrix} = \begin{pmatrix} A_1 & A_2 \\ A_3 & A_4 \end{pmatrix} \begin{pmatrix} \underline{v}(k) \\ \underline{w}(k) \end{pmatrix} + \begin{pmatrix} B_1 \\ B_2 \end{pmatrix} \tag{5.23}$$

$$\underline{y}(k) = \begin{bmatrix} I_p & \vdots & 0_{(n-p)\times p} \end{bmatrix} \begin{bmatrix} \underline{v}(k) \\ \underline{w}(k) \end{bmatrix} \qquad (5.24)$$

where A_1 is a p×p matrix
A_2 is a p×(n-p) matrix
A_3 is a (n-p)×p matrix
A_4 is a (n-p)×(n-p) matrix

and the pair (A_4, A_2) is completely observable *[9]*. We note that the vector $\underline{v}(k)$ need not be reconstructed since it is measured as the output. Hence, only the vector $\underline{w}(k)$ needs to be reconstructed, and this requires a state reconstructor of dimension (n-p). Upon examining (5.23), (5.24), it is readily seen that the output records $\underline{y}(k+1)$ and $\underline{y}(k)$ together with the knowledge of the input $\underline{u}(k)$ are sufficient to compute the vector $A_2\underline{w}(k)$. Therefore, we consider this vector as an output component of the dynamic system

$$\underline{w}(k+1) = A_4\underline{w}(k) + A_3\underline{v}(k) + B_2\underline{u}(k) \qquad (5.25)$$

Applying the state reconstruction scheme (5.6) with $\underline{w}(k)$ replacing $\underline{x}(k)$, $A_2\underline{w}(k)$ replacing $\underline{y}(k)$, we obtain the reconstruction algorithm

$$\begin{aligned} \tilde{\underline{v}}(k+1) &= \underline{v}(k+1) \\ \tilde{\underline{w}}(k+1) &= A_4\tilde{\underline{w}}(k) + A_3\underline{v}(k) + B_2\underline{u}(k) \\ &\quad + S[A_2\underline{w}(k) - A_2\tilde{\underline{w}}(k)] \end{aligned} \qquad (5.26)$$

where the (n-p)×p matrix S contains the design parameters. Using (5.23) in (5.26), this reduces to:

$$\begin{aligned} \tilde{\underline{w}}(k+1) &= A_4\tilde{\underline{w}}(k) + A_3\underline{v}(k) + B_2\underline{u}(k) + S[\underline{v}(k+1) \\ &\quad - A_1\underline{v}(k) - B_1\underline{u}(k) - A_2\tilde{\underline{w}}(k)] \end{aligned} \qquad (5.27)$$

Defining:

$$\underline{z}(k) = \tilde{\underline{w}}(k) - S\underline{v}(k) \qquad (5.28)$$

then the state reconstruction scheme (5.27) becomes

$$\underline{z}(k+1) = [A_4-SA_2]\underline{z}(k) + [B_2-SB_i]\underline{u}(k) + [A_3-SA_1+A_4S-SA_2S]\underline{y}(k) \tag{5.29}$$

with the output

$$\tilde{\underline{w}}(k) = \underline{z}(k) + S\underline{v}(k) \tag{5.30}$$

The initial condition $\underline{z}(0)$ should be taken as

$$\underline{z}(0) = \overline{\underline{w}}(0) - S\overline{\underline{v}}(0) \tag{5.31}$$

where $\overline{\underline{w}}(0)$ and $\overline{\underline{v}}(0)$ denote the *a priori* estimates of $\underline{w}(0)$ and $\underline{v}(0)$ when no output measurement is available [8]. To recover the estimate of the original state vector we use

$$\tilde{\underline{x}}(k) = T^{-1}\begin{pmatrix} \underline{y}(k) \\ \underline{z}(k) + S\underline{y}(k) \end{pmatrix} \tag{5.32}$$

Following the analysis of Section 5.2.1, it is easy to see that the state reconstruction error $\underline{e}(k) = \underline{w}(k) - \tilde{\underline{w}}(k)$ evolves according to:

$$\underline{e}(k+1) = [A_3 - SA_2]\underline{e}(k) \tag{5.33a}$$

with the initial condition

$$\underline{e}(0) = \underline{w}(0) - \overline{\underline{w}}(0) + S\underline{y}(0) - S\overline{\underline{v}}(0) \tag{5.33b}$$

It is obvious that the eigenvalues of the reduced-order state reconstruction error (5.33a) can be arbitrarily located since the observability of the pair (A,C) implies that [9] the pair (A_4,A_2) is observable.

We can now return to the problem (a) in which only the output records $\{\underline{y}(k-1),\underline{y}(k-2),\ldots,\underline{y}(0)\}$ are available to reconstruct the state at time k. It was pointed out earlier that in this case no state variables are directly recorded and as such a full-order state reconstructor would seem to be required to perform the reconstruction scheme. In order to get around this difficulty and derive a reduced-order state reconstructor, we adopt a two-step procedure *[8]* which is analogous to the one developed in Section 5.2.1. By noting that the available records at the (k+1)th discrete instant are $\{\underline{y}(k),\underline{y}(k-1),\ldots \ldots,\underline{y}(0)\}$, we have:

Step 1:

The state $\underline{x}(k)$ is estimated using the reduced-order reconstruction scheme (5.29) to yield $\tilde{\underline{w}}(k)$.

Step 2:

The state $\underline{x}(k+1)$ is reconstructed from the dynamic system (5.1) using $\tilde{\underline{w}}(k)$.

From (5.23), (5.24), Step 1 of the proposed reconstruction algorithm leads to:

$$\underline{z}(k) = [A_4 - LA_2]\underline{z}(k-1) + [B_2 - LB_1]\underline{u}(k-1) + [A_3 - LA_1 + A_4L - LA_2L]\underline{y}(k-1) \quad (5.34a)$$

$$\tilde{\underline{w}}(k) = \underline{z}(k) + L\underline{y}(k) \quad (5.34b)$$

$$\tilde{\underline{v}}(k) = \underline{y}(k) \quad (5.34c)$$

where L is the (n-p)xp matrix of design parameters. According to Step 2, we have

$$\hat{\underline{x}}(k+1) = A\tilde{\underline{x}}(k) + B\underline{u}(k) \quad (5.35)$$

Defining

$$\underline{f}(k) = \underline{z}(k-1) \quad (5.36)$$

Then, combining (5.34) - (5.36), we get:

$$\underline{f}(k+1) = A_4 - LA_3\underline{f}(k) + B_2 - LB_1\underline{u}(k-1)$$
$$+ A_3 - LA_1 + A_4L - LA_2L\underline{y}(k-1) \qquad (5.37)$$

with output

$$\hat{\underline{x}}(k) = AT^{-1}\begin{bmatrix} \underline{y}(k-1) \\ \underline{f}(k) + L\underline{y}(k-1) \end{bmatrix} + B\underline{u}(k-1) \qquad (5.38)$$

which yields $\hat{\underline{x}}(k)$ as the estimate of $\underline{x}(k)$ using the records $\{\underline{y}(k-1), \underline{y}(k-2), \ldots, \underline{y}(0)\}$ using a reduced-order state reconstructor. To initialise the algorithm (5.37) we use

$$\underline{f}(1) = \overline{\underline{w}}(0) - L\overline{\underline{v}}(0)$$

where $\overline{\underline{w}}(0)$ and $\overline{\underline{v}}(0)$ are the *a priori* estimates of $\underline{w}(0)$ and $\underline{v}(0)$ when no output records are available. Finally, the state reconstruction errors satisfy the dynamic equation

$$\begin{bmatrix} \underline{e}_1(k) \\ \underline{e}_2(k) \end{bmatrix} = \begin{bmatrix} -A_2L & A_2 \\ -A_4L & A_4 \end{bmatrix} \begin{bmatrix} \underline{e}_1(k) \\ \underline{e}_2(k) \end{bmatrix} \qquad (5.39)$$

where $\underline{e}_1(k) \triangleq \underline{v}(k) - \tilde{\underline{v}}(k)$ and $\underline{e}_2(k) \triangleq \underline{w}(k) - \tilde{\underline{w}}(k)$. It should be emphasized that the gain matrices S or L can be determined using standard eigenvalue algorithms like those discussed in Chapter 4.

Next, we examine the relationship between full-order and reduced-order state reconstruction schemes.

5.2.3 *DISCUSSION*

The purpose of this section is to show that the reduced-order state reconstruction schemes can be deduced from the corresponding full-order schemes. We first consider that the measurement records are $\{\underline{y}(k), \underline{y}(l-1), \ldots, \underline{y}(0)\}$ which corresponds to the problem statement (b). Let the discrete system (5.1), (5.2) be converted to (5.23), (5.24) using the coordinate transformation (5.21). For the transformed system we use $P = [P_1^t \quad P_2^t]^t$ where P_1 and P_2 have dimensions (pxp) and (n-p)xp, respectively. Hence, the full-order state reconstruction scheme (5.11) becomes:

$$\begin{bmatrix} \tilde{\underline{v}}(k+1) \\ \tilde{\underline{w}}(k+1) \end{bmatrix} = \begin{bmatrix} A_1 - P_1A_1 & A_2 - P_1A_2 \\ A_3 - P_2A_1 & A_4 - P_2A_2 \end{bmatrix} \begin{bmatrix} \tilde{\underline{v}}(k) \\ \tilde{\underline{w}}(k) \end{bmatrix} + \begin{bmatrix} B_1 - P_1B_1 \\ B_2 - P_2B_1 \end{bmatrix} \underline{y}(k) + \begin{bmatrix} P_1 \\ P_2 \end{bmatrix} \underline{y}(k+1) \quad (5.40)$$

We know that in reduced-order reconstruction schemes, the gain matrices are of the order (n-p)xp. Therefore, we choose P_1 as I_p, the identity matrix of order p. With this choice (5.40) reduces to

$$\begin{aligned} \tilde{\underline{v}}(k+1) &= \underline{y}(k+1) \\ \tilde{\underline{w}}(k+1) &= [A_4 - P_2A_2]\tilde{\underline{w}}(k) + [A_3 - P_2A_1]\underline{y}(k) \\ &\quad + [B_2 - P_2B_1]\underline{u}(k) + P_2\underline{y}(k+1) \end{aligned} \quad (5.41)$$

Comparison of (5.27) and (5.41) reveals that both reconstruction schemes are identical if $P_2 = S$. This means that we can deduce the required scheme by an appropriate choice of the gain matrices of the state reconstruction algorithm in which the output measurements are recorded instantaneously.

We now direct our attention to the problem statement (a) in which there is a one-step delay in recording the output measurements. Analogously to the above analysis, we consider the transformed system (5.23), (5.24) and apply the state reconstruction scheme (5.6) with $H = [H_1^t \quad H_2^t]^t$ where H_1 and H_2 are block-matrices of dimensions (pxp), (n-p)xp, respectively. The scheme (5.6) becomes

$$\begin{bmatrix} \hat{\underline{v}}(k+1) \\ \hat{\underline{w}}(k+1) \end{bmatrix} = \begin{bmatrix} A_1-H_1 & A_2 \\ A_3-H_2 & A_4 \end{bmatrix} \begin{bmatrix} \hat{\underline{v}}(k) \\ \hat{\underline{w}}(k) \end{bmatrix} + \begin{bmatrix} B_1 \\ B_2 \end{bmatrix} \underline{u}(k) + \begin{bmatrix} H_1 \\ H_2 \end{bmatrix} \underline{y}(k) \quad (5.42)$$

Willems [8] has proposed selecting H_1 and H_2 according to:

$$\begin{aligned} H_1 &= A_1 + A_2 L \\ H_2 &= A_3 + A_4 L \end{aligned} \tag{5.43}$$

In which case (5.42) becomes

$$\begin{bmatrix} \hat{\underline{v}}(k+1) \\ \hat{\underline{w}}(k+1) \end{bmatrix} = \begin{bmatrix} -A_2 L & A_2 \\ -A_4 L & A_4 \end{bmatrix} \begin{bmatrix} \hat{\underline{v}}(k) \\ \hat{\underline{w}}(k) \end{bmatrix} + \begin{bmatrix} B_1 \\ B_2 \end{bmatrix} \underline{u}(k) + \begin{bmatrix} A_1 + A_2 L \\ A_3 + A_4 L \end{bmatrix} \underline{y}(k) \tag{5.44}$$

Introducing the (n-p) vector $\underline{g}(k)$

$$\underline{g}(k) = \hat{\underline{w}}(k) - L\hat{\underline{v}}(k) \tag{5.45}$$

then the scheme (5.44) reduces to

$$\begin{aligned} \underline{g}(k) = {} & [A_4 - LA_2]\underline{g}(k) + [B_2 - LB_1]\underline{u}(k) \\ & + [A_3 - LA_1 + A_4 L - LA_2 L]\underline{y}(k-1) \end{aligned} \tag{5.46}$$

which is exactly the reduced-order scheme (5.38). The reconstructed state $\underline{x}(k)$ is given by:

$$\hat{\underline{x}}(k) = \begin{bmatrix} A_2 \\ A_4 \end{bmatrix} \underline{g}(k) + \begin{bmatrix} A_1 + A_2 L \\ A_3 + A_4 L \end{bmatrix} \underline{y}(k-1) + \begin{bmatrix} B_1 \\ B_2 \end{bmatrix} \underline{u}(k-1) \tag{5.47}$$

Simple manipulation shows that (5.47) is equivalent to (5.38).

In the light of the above analysis, it is interesting to observe that the gain matrices P and H chosen to deduce the reduced-order state reconstructors from the corresponding full-order ones need to satisfy (5.20).

We should emphasize that the relationship between full-order and reduced-order deterministic observers is a unique feature of discrete systems [8] and does not have a counterpart in continuous systems.

5.2.4 *DEADBEAT STATE RECONSTRUCTORS*

The deadbeat control solutions derived in Section 4.3 presume that the system state vector is completely available for feedback control. As pointed out in the previous sections, only certain linear combinations of the states of (5.1) are usually available through the system outputs (5.2). This gives rise to the deadbeat control problem with inaccessible states, an important part of which is the deadbeat state reconstruction scheme. In such a scheme, the objective is to design either a full-order or reduced-order deadbeat state reconstructor to estimate the system state $\underline{x}(k)$ in a minimum number of discrete steps. For simplicity of exposition, we will consider only problem (a) of Section 5.2.1 in which there is a one-step delay in the output records. Recall that the full-order state reconstructor which will estimate the state $\underline{x}(k)$ is described by:

$$\begin{aligned}\hat{\underline{x}}_d(k+1) &= A\hat{\underline{x}}_d(k) + B\underline{u}(k) + H_d[\underline{y}(k) - C\hat{\underline{x}}_d(k)] \\ &= [A - H_dC]\hat{\underline{x}}_d(k) + B\underline{u}(k) + H_d\underline{y}(k) \qquad (5.48)\end{aligned}$$

where the $(n\times p)$ matrix H_d contains the design parameters and $\hat{\underline{x}}_d(k)$ is the $(n\times 1)$ vector which approximates $\underline{x}(k)$ at deadbeat response. By (5.1), (5.2) and (5.48), the state reconstruction error $\underline{e}(k) \triangleq \underline{x}(k) - \hat{\underline{x}}_d(k)$ propogates according to

$$\underline{e}(k+1) = [A - H_dC]\underline{e}(k) \qquad (5.49)$$

so that at any instant $\Theta \geq 0$

$$\underline{e}(\Theta) = [A - H_dC]^{\Theta}\underline{e}(0) \qquad (5.50)$$

The deadbeat state reconstruction problem [14-18] is one of choosing the gain matrix H_d such that $\underline{e}(\mu_0) = \underline{0}$ where μ_0 is known as the observability index of the system (5.1).

It is interesting to note that the state reconstruction problem addressed above is the dual of the deadbeat control problem [5], namely, find the gain matrix H_d^t such that $[A^t - C^t H_d^t]$ is a nilpotent matrix with an index of nilpotency μ_0. We see this by noting that we wish to choose H_d such that all the eigenvalues of $[A-H_dC]$ are assigned to the origin, that is

$$\det[\gamma I_n - A + H_d C] = \gamma^n = 0 \tag{5.51}$$

In view of the Cayley-Hamilton theorem [7], this means that

$$[A - H_d C]^{\mu_0} = 0 \tag{5.52}$$

It should be remarked although the characteristic polynomial (5.51) is γ^n due to the observability requirement its minimal polynomial is γ^{μ_0}.

To compute the deadbeat gain matrix, one can use the dual result of the deadbeat controller design [14] to arrive at:

$$H_d^t = \underline{g} \sum_{j=1}^{n} \left\{ \frac{(-1)^n \gamma_j^n \underline{u}_j^t}{\alpha_j \prod_{m \neq j} (\gamma_m - \gamma_j)} \right\} \tag{5.53}$$

where α_j $(j = 1,\ldots,n)$ are the elements of the $n\times 1$ vector

$$\underline{\alpha} = U^t C^t \underline{g} \tag{5.54}$$

$\underline{g}$ is any $(p\times 1)$ vector for which $\underline{\sigma}$ has no zero elements, $\underline{u}_j$ $(j = 1,2,\ldots,n)$ are the eigenvectors of A corresponding to the distinct eigenvalues λ_j $(j = 1,2,\ldots,n)$ of the nonsingular matrix A, and U is the matrix whose jth column is $\underline{u}_j$ $(j = 1,2,\ldots,n)$. It is clear [14] that the $(n\times p)$ deadbeat state reconstructor H_d given by (5.53) has a dyadic structure.

This algorithm is restricted to system matrices with distinct eigenvalues.

Another algorithm developed by O'Reilly *[15]* and Ichikawa *[16]* is based on transforming the system matrices into observable companion forms *[19]*. Let Q^t denote the matrix which transforms the system (5.1), (5.2) into the observable canonical form *[19]* with

$$A_c = \left(\begin{array}{cccccccc}
0 & 0 \dots 0 & x & & & x & & x \\
1 & 0 \dots 0 & x & & & x & & x \\
\vdots & & \vdots & & & \vdots & & \vdots \\
0 & 0 \dots 1 & x & & & & & \\
 & & x & 0 & 0 \dots & x & & \\
 & & x & 1 & 0 \dots & x & & \\
 & & \vdots & \vdots & & & & \\
 & & \vdots & 0 & \dots 1 & x & & \\
 & & \vdots & & & x & & \\
 & & \vdots & & & x & & \\
 & & \vdots & & & \vdots & & \\
 & & x & & & & & \\
 & & x & & & & 0 \dots & x \\
 & & \vdots & & & & 1 \dots & x \\
 & & \vdots & & & & & \vdots \\
 & & x & & & & 0 \quad 1 & x
\end{array}\right)$$

$$\qquad\qquad \underbrace{\qquad\qquad\qquad}_{\mu_1 \ \uparrow\ \bar{\sigma}_p} \qquad\qquad \underbrace{\qquad\qquad}_{\mu_1 \ \uparrow\ \bar{\sigma}_p}$$

$$C_c = \left(\begin{array}{cccccccc}
0 \dots & 0 & 1 & 0 \dots & 0 & & 0 \dots & 0 \\
0 & 0 & x & 0 \dots & 1 & \dots & 0 \dots & 0 \\
\vdots & \vdots & \vdots & \vdots & \vdots & & & \vdots \\
0 & 0 & x & 0 & x & & 0 & 1
\end{array}\right)$$

In A_c and C_c above, $\sum_{i}^{p} \bar{\mu}_j = n$, $\max_{1 \le j \le p} \bar{\mu}_j$ is the observability index μ_0, $\bar{\sigma}_i = \sum_{j=1}^{i} \mu_j$, $i = 1,2,\dots,m$ denotes the position of the non-trivial columns of A_c and C_c. These columns constitute the $n \times p$ matrix A_p and the $p \times p$ matrix C_p, where C_p is clearly nonsingular. In the transformed situation, one way to determine H_p is by enforcing the closed-loop system matrix $[A_p - H_pC_p]$ to have zero columns in all non-trivial columns [16]. This yields

$$H_p = A_pC_p^{-1} \qquad (5.55)$$

By transforming back to the original system, the deadbeat gain matrix takes the form

$$H_c = - Q^tA_pC_p^{-1} \qquad (5.56)$$

which ensures that

$$[A_p - H_cC_p]^{\mu_0} = 0 \qquad (5.57)$$

Obviously (5.57) is equivalent to (5.52). We note that the derivation of (5.56) is based on using a suitable procedure for selection from amongst the comparison matrices. Another way of doing the same [15] is to utilise the observability matrix to choose a set of n linearly independent columns as a basis to force the reconstruction error (5.50) to zero in a finite number of steps. Let the n linearly independent columns of the observability matrix form the (nxn) nonsingular matrix R:

$$R \triangleq \left[\underline{c}_1^t \ldots \underline{c}_p^t \,|\, A^t\underline{c}_1 \ldots A^t\underline{c}_{p_1} \,|\, \ldots \,|\, A^{(\mu_0-1)^t}\underline{c}_1^t \ldots A^{(\mu_0-1)^t}\underline{c}_{p_{\mu_0-1}} \,|\right] \qquad (5.58)$$

where $\underline{c}_j$ is the jth row of C, $p = p_0 \geq p_1 \geq \ldots \geq p_{\mu_0-1}$ and $\sum_{j=0}^{n} p_j = n$. The column vectors of the inverse matrix $S \triangleq R^{-t}$ are denoted by:

$$S = [\underline{s}_{10} \ldots \underline{s}_{p_0} \,|\, \underline{s}_{11} \ldots \underline{s}_{p_1 1} \,|\, \ldots \,|\, \underline{s}_{1_{\mu_0-1}} \ldots \underline{s}_{p_{\mu_0-1}} \,|] \qquad (5.59)$$

By replacing $\underline{s}_{ij}$ in S by $A^{v_i - j}\underline{s}_{i,v_i}$ $(i = 1,\ldots,p_j\,;\; j = 0,1,\ldots,v_1)$ we obtain

$$W \triangleq [A^{v_1}\underline{s}_{1,v_1}\ldots A^{v_p}\underline{s}_{p,v_p} \,|\, A^{v_1-1}\underline{s}_{1,p_1}\ldots A^{v_{p_1}-1}\underline{s}_{p_1,v_{p_1}}]$$

$$\triangleq [W_0, W_1, \ldots W_j, \ldots W_{v_1}] \tag{5.60}$$

where $\mu_0 - 1 = v_1 \geq v_2 \geq \ldots \geq v_p > 0$ and $p + \sum_{j=1}^{p} v_j = n$.
O'Reilly *[15]* has shown that the deadbeat gain matrix is given by:

$$H_d = AW_0(CW_0)^{-1} \tag{5.61}$$

which ensures that the state vector $\underline{x}(k)$ is reconstructed exactly in μ_0 discrete steps. It is interesting to note that for single-input single-output systems in which the companion matrices are:

$$A_c = \begin{pmatrix} 0 & 0 \ldots & 0 & -a_0 \\ 1 & 0 \ldots & 0 & -a_1 \\ \vdots & \vdots & \vdots & \vdots \\ 0 & 0 & 0 & -a_{n-1} \end{pmatrix} \tag{5.62a}$$

$$\underline{c}_c^t = [\,0 \;\ldots.\; 0 \quad 1\,] \tag{5.62b}$$

the deadbeat gain matrix reduces to the (nxl) vector $\underline{h}_d$ given by

$$\underline{h}_d = Q^t\underline{a} \tag{5.63}$$

where $\underline{a} = [a_0\; a_1 \ldots a_{n-1}]^t$ and Q^t is the transformation matrix which converts (5.1), (5.2) into the observable companion form (5.62). It should be emphasized that the case in which the measurement pattern is instantaneous, corresponding to problem (b), can be treated in an analogous way using the results of Section 5.2.1.

The development of reduced-order deadbeat state reconstructors proceeds in a similar manner to that in Section 5.2.4. In the case of one-step delay measurement patterns, problem (a), the reconstruction scheme is (5.37), whereas in the case of instantaneous measurement patterns problem (b), the reconstruction scheme is (5.29). Recall that the pair (A_4, A_2) is required to be observable. It is known *[9]* that if the observability index of the pair (A,C) is μ_0 then the pair (A_4, A_2) is also an observable pair of observability index $\mu_0 - 1$. This implies that the state vector $\underline{x}(k)$ may be reconstructed by a reduced-order state reconstructor (5.29) or (5.37) in a minimum number of steps equal to $\mu_0 - 1$. The reduced order deadbeat state reconstruction error is required to satisfy:

$$\underline{e}(\mu_0 - 1) = (A_4 - S_d A_2)^{\mu_0 - 1} = 0 \tag{5.64}$$

by an appropriate selection of the gain matrix S_d. By analogy to (5.61) the deadbeat gain matrix S is given by *[15]*:

$$S_d = A_4 W_0 (A_2 W_0)^{-1} \tag{5.65}$$

where W can be obtained by analogy to (5.58)-(5.60).

5.2.5 *EXAMPLES*

The deadbeat full-order and reduced-order state reconstruction designs are illustrated respectively by means of the following two examples *[15]*.

Example 1:

It is required to obtain a full-order state reconstructor for the fifth-order system of the type given by (5.1), (5.2).

$$\underline{x}(k+1) = \begin{bmatrix} 1 & 0 & 0 & 1 & 1 \\ 0 & 0 & 0 & 0 & 0 \\ 0 & 0 & 0 & 1 & 0 \\ 1 & 0 & 0 & 0 & 1 \\ 1 & 1 & 0 & 1 & 0 \end{bmatrix} \underline{x}(k)$$

$$\underline{y}(k) = \begin{bmatrix} 1 & 0 & 1 & 0 & 0 \\ 0 & 1 & 0 & 0 & 0 \end{bmatrix} \underline{x}(k)$$

There is a one-step delay in recording the outputs. Examining the observability matrix, one finds that the observability index $\mu_0 = 4$. From (5.58) the R matrix is

$$R = [\underline{c}_1^t \; \underline{c}_2^t \;\vdots\; A^t \; \underline{c}_1^t \;\vdots\; A^{2^t} \; \underline{c}_1^t \;\vdots\; A^{3^t} \; \underline{c}_1^t]$$

$$\begin{bmatrix} 1 & 0 & 1 & 4 & 9 \\ 0 & 1 & 0 & 1 & 3 \\ 1 & 0 & 0 & 0 & 0 \\ 0 & 0 & 2 & 2 & 7 \\ 0 & 0 & 1 & 3 & 6 \end{bmatrix}$$

whose transposed inverse is given by

$$S = R^{-t} = \frac{1}{7} \begin{bmatrix} 0 & -7 & -9 & -5 & 4 \\ 0 & 7 & 0 & 0 & 0 \\ 7 & 7 & 9 & 5 & -4 \\ 0 & 0 & 3 & -3 & 1 \\ 0 & 7 & 10 & 11 & -6 \end{bmatrix}$$

$$[\underline{s}_{1,0} \quad \underline{s}_{20} \quad \underline{s}_{1,1} \quad \underline{s}_{1,2} \quad \underline{s}_{1,3}]$$

Thus

$$W_0 = [A^3 s_{1,0} \quad \underline{s}_{2,0}]$$

$$\begin{bmatrix} 3 & -7 \\ 0 & 7 \\ 4 & 7 \\ -1 & 0 \\ 6 & 7 \end{bmatrix}$$

Applying (5.61) we obtain the deadbeat gain matrix

$$H_d = \frac{1}{7} \begin{bmatrix} 8 & 0 \\ 0 & 0 \\ -1 & 0 \\ 9 & 0 \\ 2 & 0 \end{bmatrix}$$

The state reconstructor matrix takes the form

$$[A - H_d C] = \begin{bmatrix} -1 & 0 & -8 & 7 & 7 \\ 0 & 0 & 0 & 0 & 0 \\ 1 & 0 & 1 & 7 & 0 \\ -2 & 0 & 9 & 0 & 7 \\ 5 & 7 & -2 & 7 & 0 \end{bmatrix}$$

Evaluating this matrix, we find that $[A - H_d C]^4 = [0]$ which is the nilpotency index $= 4 = \mu_0$ as required.

Example 2:

It is required to design a reduced-order state reconstructor for the seventh-order system given by (5.23), (5.25) in which the measurements are recorded with a one-step delay. The sub-matrices are given by:

$$A_1 = \begin{bmatrix} 0 & 0 & 0 \\ 0 & 0 & 0 \\ 0 & 0 & 0 \end{bmatrix} \; ; \; A_2 = \begin{bmatrix} -2 & 1 & 1 & 1 \\ 0 & -1 & 0 & 0 \\ 1 & -1 & 1 & 0 \end{bmatrix}$$

$$A_3 = \begin{bmatrix} 0 & 0 & 0 \\ 0 & 0 & 0 \\ 0 & 0 & 1 \\ 0 & 1 & 0 \end{bmatrix} \; ; \; A_4 = \begin{bmatrix} 0 & 1 & 0 & 0 \\ 1 & 0 & 1 & 0 \\ 0 & -1 & 0 & 1 \\ 0 & 0 & 0 & 0 \end{bmatrix}$$

$$C = [I_3 \;\vdots\; 0]$$

It is easy to check that the seventh-order system is completely observable with an observability index $\mu_0 = 3$. The nonsingular matrix W_0 in (5.56) is computed as follows:

$$R = [\underline{a}_1^t \quad \underline{a}_2^t \mid \underline{a}_3^t \; A_4^t \underline{s}_1^t]$$

$$= \left[\begin{array}{ccc|c} 2 & 0 & 1 & 0 \\ 0 & -1 & -1 & 1 \\ 1 & 0 & 1 & 0 \\ 1 & 0 & 0 & 1 \end{array}\right]$$

where $\underline{a}_j$ is the jth row of A_2.

$$S = R^{-t} = \begin{bmatrix} -1 & 0 & 1 & 1 \\ 0 & 1 & 0 & 0 \\ 1 & 1 & -2 & -1 \\ 0 & -1 & 0 & -1 \end{bmatrix}$$

$$[\underline{s}_{1,0} \quad \underline{s}_{2,0} \quad \underline{s}_{3,0} \quad \underline{s}_{1,1}]$$

$$W_0 = [A\underline{s}_{1,1} \quad \underline{s}_{2,0} \quad \underline{s}_{3,0}]$$

$$= \begin{bmatrix} 0 & 0 & 1 \\ 0 & 1 & 0 \\ -1 & 1 & -2 \\ 0 & -1 & 0 \end{bmatrix}$$

Therefore the gain matrix S_d of (5.65) has the value

$$S_d = \begin{bmatrix} 0 & -1 & 0 \\ 0 & -1 & 1 \\ 0 & 2 & 0 \\ 0 & 0 & 0 \end{bmatrix}$$

from which the state reconstructor matrix is

$$(A_4 - S_d A_2) = \begin{bmatrix} 0 & 0 & 0 & 0 \\ 0 & 0 & 0 & 0 \\ 0 & 1 & 0 & 1 \\ 0 & 0 & 0 & 0 \end{bmatrix}$$

A straightforward calculation shows that the matrix $[A_4 - S_d A_2]^2$ is the null matrix which confirms that the degree of nilpotency is equal to $2 = \mu_0 = 1$ as expected.

5.3 Observer-Based Controllers

So far, we have been concerned with the development of state reconstruction (observation) schemes. The previous section has been devoted to the design of full-order and reduced-order deterministic observers under different measurement patterns. We have shown that a deterministic observer produces an estimate of the state vector of the linear discrete system. In this section, we will examine the use of this approximate state vector as an input to a feedback controller to yield a closed-loop discrete system. Then, we will study the properties of such a system.

5.3.1 *STRUCTURE OF CLOSED-LOOP SYSTEMS*

The basic idea of designing observer-based controllers is to use the output of the deterministic observer, which is the asymptotic estimate of the state instead of the actual state, in forming feedback signals. This is performed through a constant state feedback matrix. The composite closed-loop system is of order 2n and is shown in Fig. 5.3, for the case of the full-order state reconstructor. We note that this composite structure contains two gain matrices to be determined according to the design specifications. Next, we analyse the closed-loop dynamics.

5.3.2 *THE SEPARATION PRINCIPLE*

With reference to Fig. 5.3, let the discrete system and the full-order state reconstructor be described by:

$$\underline{x}(k+1) = A\underline{x}(k) + B\underline{u}(k) \tag{5.66a}$$

$$\underline{y}(k) = C\underline{x}(k) \tag{5.66b}$$

$$\underline{x}_b(k+1) = A_b\underline{x}_b(k) + B_b\underline{u}(k) + K_b\underline{y}(k) \tag{5.67}$$

where, in the case of a one-step delay in the measurement pattern,

$$\left.\begin{aligned} \underline{x}_b(k) &= \hat{\underline{x}}(k) \\ A_b &= A - HC \\ B_b &= B \\ K_b &= H \end{aligned}\right\} \tag{5.68}$$

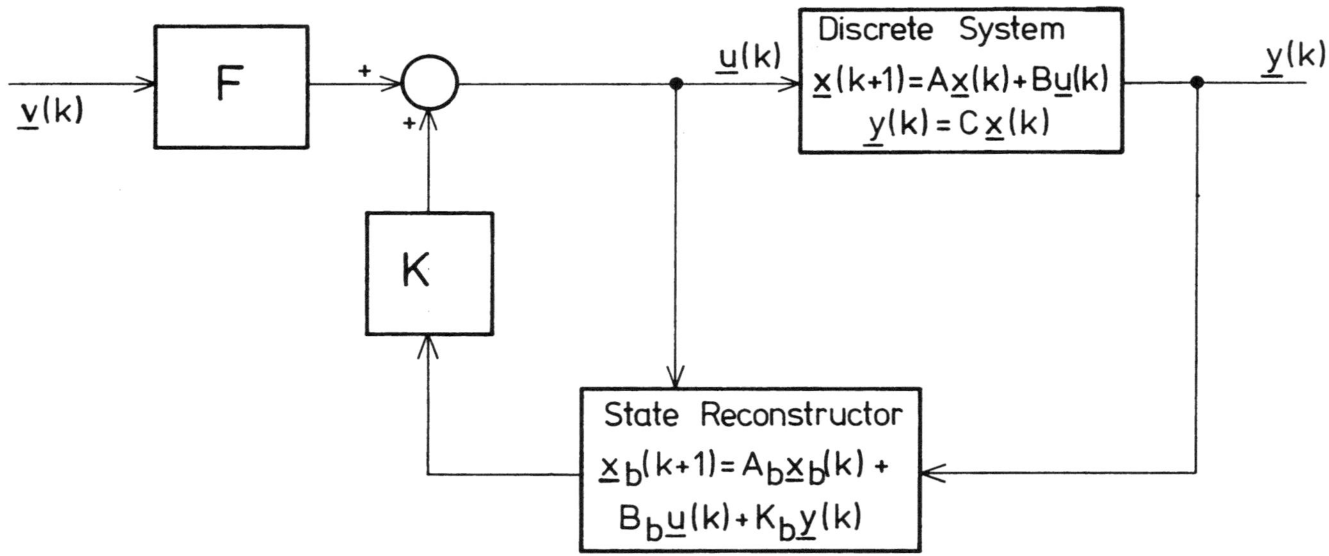

Fig. (5.3) Observer-based controller structure

and in the case of instantaneous measurements

$$\left.\begin{aligned} \underline{x}_b(k) &= \tilde{\underline{x}}(k) \\ A_b &= A - PCA \\ B_b &= B - PCB \\ K_b\underline{y}(k) &= P\underline{y}(k+1) \end{aligned}\right\} \qquad (5.69)$$

We consider that the closed-loop control is given by the expression:

$$\underline{u}(k) = F\underline{v}(k) - K\underline{x}_b(k) \qquad (5.70)$$

where $\underline{v}(k)$ is the (nxl) vector of the new input and F is a constant (mxm) feedforward matrix. By combining (5.66),(5.67), (5.68) and (5.70) for the case of one-step delays in the measurements, we get the composite system:

$$\begin{pmatrix} \underline{x}(k+1) \\ \hat{\underline{x}}(k+1) \end{pmatrix} = \begin{pmatrix} A & -BC \\ HC & A-HC-BK \end{pmatrix} \begin{pmatrix} \underline{x}(k) \\ \hat{\underline{x}}(k) \end{pmatrix} + \begin{pmatrix} BF \\ BF \end{pmatrix} \underline{v}(k) \qquad (5.71)$$

which is of order 2n, and has two unknown gain matrices, namely K and H. The 2n closed-loop eigenvalues are roots of:

$$\Delta_T(\gamma) = \det \begin{pmatrix} \gamma I_n - A & BK \\ -HC & \gamma I_n - A + HC + BK \end{pmatrix} = 0 \qquad (5.72)$$

Since any similarity transformation preserves the eigenvalue distribution *[7]*, the roots of (5.72) are the same under the eigenvalue transformation:

$$\Delta_T(\gamma) = \det\left\{ \begin{pmatrix} I_n & 0 \\ I_n & -I_n \end{pmatrix} \begin{pmatrix} \gamma I_n - A & BK \\ -HC & \gamma I_n - A + HC + BK \end{pmatrix} \begin{pmatrix} I_n & 0 \\ I_n & -I_n \end{pmatrix} \right\} = 0$$

$$= \det \begin{pmatrix} \gamma I_n - A + B & -BK \\ 0 & \gamma I_n - A + HC \end{pmatrix} = 0$$

$$= \det[\gamma I_n - A + BK] \det[\gamma I_n - A + HC] = 0 \qquad (5.73)$$

which shows that by proper selection of K, n of the closed-loop eigenvalues can be specified and the remaining n eigenvalues can be chosen independently through an appropriate selection of H.

We now consider combining (5.66), (5.67), (5.69) and (5.70) for the case of instantaneous measurements to obtain the composite system:

$$\begin{pmatrix} \underline{x}(k+1) \\ \tilde{\underline{x}}(k+1) \end{pmatrix} = \begin{pmatrix} A & -BK \\ PCA & A-PCA-BK \end{pmatrix} \begin{pmatrix} \underline{x}(k) \\ \tilde{\underline{x}}(k) \end{pmatrix} + \begin{pmatrix} BF \\ BF \end{pmatrix} \underline{v}(k) \tag{5.74}$$

which is also of order 2n, and has two unknown gain matrices K and P. The closed-loop eigenvalues are determined by:

$$\bar{\Delta}_T(\gamma) = \det \begin{pmatrix} \gamma I_n - A & BK \\ -PCA & \gamma I_n - A+PCA+BK \end{pmatrix} = 0 \tag{5.75}$$

which, under the equivalence transformation as in (5.73), reduces to:

$$\bar{\Delta}_T(\gamma) = \det \begin{pmatrix} \gamma I_n - A+BK & -BK \\ 0 & \gamma I_n - A+PCA \end{pmatrix}$$

$$= \det[\gamma I_n - A + BK] \det[\gamma I_n - A + PCA] = 0 \tag{5.76}$$

Here again, by an appropriate selection of the gain matrices K and P, two sets of n eigenvalues can be specified separately. In the light of (5.73) and (5.76), we can state the *separation principle* which provides the basis for designing observer-based controllers:

> *Provided the linear discrete system is completely reachable and completely reconstructible (observable), the 2n closed-loop eigenvalues of the observer-based controller system can be arbitrarily specified as two separate sets, one by an appropriate selection of the observer gains and the other by an appropriate selection*

of the state feedback gains.

In general, there is no restriction on the relative magnitudes of the observer eigenvalues and the state feedback eigenvalues. However, experience indicates that a good design results if the observer eigenvalues are selected to be a little closer to the origin than the closed-loop state feedback eigenvalues. The reason for this is that the observation scheme will converge faster for the case where we have selected the eigenvalues further from the origin. We emphasize that, by pursuing similar arguments, an observer-based controller structure can be designed *[16]* using a reduced-order state reconstructor of the type (5.29) or (5.37). The difference lies in the definition of the matrices A_b, B_b and K_b which characterise the state reconstructor. A straightforward analysis shows that the observer-based controlled system, using a reduced-order state reconstructor of the form (5.29) with instantaneous measurements, has the characteristic polynomial

$$\Delta_R(\gamma) = \det \begin{bmatrix} \gamma I_n - A + BK & -BKF_R \\ 0 & \gamma I_n - A_4 + S_d A_2 \end{bmatrix} = 0 \tag{5.77}$$

where the state vector of the composite vector is given by:

$$\underline{x}_R(k) = [\underline{x}^t(k) \quad \underline{e}^t(k)]^t \tag{5.78a}$$

and F_R has the form

$$F_R = \begin{bmatrix} -C_1^{-1} C_2 \\ I_{n-p} \end{bmatrix} \tag{5.78b}$$

In this case the dimension of the composite system is 2n-m. This verifies that the separation principle is applicable here as well. In general, the design of the observer-based controller rests upon the determination of the unknown gains K and H (P or S) which can be achieved using any of the standard eigenvalue algorithms previously discussed in Chapter 4.

5.3.3 *DEADBEAT TYPE CONTROLLERS*

We now examine the problem of designing deadbeat observer-based controllers. The interest in this problem stems from the fact that the closed-loop eigenvalues of the composite system have to be placed at the origin. It is clear from (5.73) that the free part of the composite system can be written as:

$$\begin{pmatrix} \underline{x}(k+1) \\ \hat{\underline{e}}(k+1) \end{pmatrix} = \begin{pmatrix} A-BK & BK \\ 0 & A-HC \end{pmatrix} \begin{pmatrix} \underline{x}(k) \\ \hat{\underline{e}}(k) \end{pmatrix} \tag{5.79}$$

The solution of (5.79) takes the form:

$$\begin{pmatrix} \underline{x}(k) \\ \hat{\underline{e}}(k) \end{pmatrix} = \begin{pmatrix} (A-BK)^k & \sum_{j=1}^{k} (A-BK)^{j-1} BK (A-HC)^{k-j} \\ 0 & (A-HC)^k \end{pmatrix} \begin{pmatrix} \underline{x}(0) \\ \hat{\underline{e}}(0) \end{pmatrix} \tag{5.80}$$

Now, if K is designed such that the matrix $(A-BK)$ is nilpotent with index π_0 and H is designed such that the matrix $(A-HC)$ is nilpotent with index ψ_0, it can be seen from (5.80) that *[20]*:

$$\hat{\underline{e}}(\psi_0) = \underline{0} \tag{5.81a}$$

$$\underline{x}(\pi_0 + \psi_0) = \underline{0} \tag{5.81b}$$

for any initial states $\underline{x}(0)$ and $\hat{\underline{e}}(0)$. Porter and Bradshaw *[20]* have shown that

$$\pi_0 \geq \mu_c \tag{5.82a}$$

$$\psi_0 \geq \mu_0 \tag{5.82b}$$

where μ_c is the reachability index of the pair (A,B) and μ_0 is the observability index of the pair (A,C). By determining the feedback matrix K and the observer matrix H such that $\pi_0 = \mu_c$, $\psi_0 = \mu_0$, the composite system (5.80) will be driven from any initial state to the origin in precisely $(\mu_0 + \mu_c)$ steps. Numerical algorithms to compute the unknown gains can be constructed by dualising and generalising those developed for the deadbeat controllers based on the canonical forms *[20]* or dyadic pole placement *[21]*.

We emphasize that the above analysis is equally applicable to the composite system (5.76) when the outputs are recorded instantaneously and to the composite system whose characteristic polynomial is (5.77) and in which a reduced-order state reconstructor is used. In the latter case, the state $\underline{x}(k)$ is transferred to the origin in no more than $\mu_0 + (\nu_0 - 1)$ steps, where $(\nu_0 - 1)$ is the observability index of the pair (A_4, A_2). This implies that after at most $(\nu_0 - 1)$ steps perfect observer state reconstruction ensues, the control law (5.70) provides the feedback action and the linear regulator transfers $\underline{x}(\nu_0 - 1)$ to the origin in at most μ_0 further steps.

5.3.4 *EXAMPLE*

The design of a deadbeat observer-based control structure is now illustrated by a third order system described by:

$$\underline{x}(k+1) = \begin{pmatrix} 0 & 1 & 2 \\ -2 & 3 & 0 \\ -2 & -1 & 0 \end{pmatrix} \underline{x}(k) + \begin{pmatrix} 1 & 2 \\ 1 & 0 \\ 0 & 0 \end{pmatrix} \underline{u}(k)$$

$$\underline{y}(k) = \begin{pmatrix} 0 & -1 & 1 \\ -2 & 0 & 0 \end{pmatrix} \underline{x}(k)$$

For this system, the reachability index $\mu_c = 2$ and the observability index $\mu_0 = 2$. To place the eigenvalues of the matrix $(A-BK)$ at the origin, the feedback gain is given by:

$$K = \begin{pmatrix} -2 & 3 & 0 \\ 1 & -1 & 1 \end{pmatrix}$$

and it is readily seen that

$$(A-BK) = \begin{pmatrix} 0 & 0 & 0 \\ 0 & 0 & 0 \\ -2 & -1 & 0 \end{pmatrix}$$

and $(A-BK)^2$ is the null matrix. Assuming a one-step delay in the measurements, the deadbeat gain matrix H is computed as

$$H = \begin{pmatrix} -.25 & 0 \\ -2.25 & 1 \\ .75 & 1 \end{pmatrix}$$

and gives the state reconstruction matrix

$$(A-HC) = \begin{pmatrix} 0 & .75 & 2.25 \\ 0 & .75 & 2.25 \\ 0 & -.25 & -.75 \end{pmatrix}$$

We note that $(A-HC)^2 = (0)$. The composite system (5.79) becomes:

$$\begin{pmatrix} \underline{x}(k+1) \\ \hat{\underline{e}}(k+1) \end{pmatrix} = \left(\begin{array}{ccc:ccc} 0 & 0 & 0 & 0 & 1 & 2 \\ 0 & 0 & 0 & -2 & 3 & 0 \\ -2 & -1 & 0 & 0 & 0 & 0 \\ \hdashline 0 & 0 & 0 & 0 & .75 & 2.25 \\ 0 & 0 & 0 & 0 & .75 & 2.25 \\ 0 & 0 & 0 & 0 & -.25 & -.75 \end{array}\right) \begin{pmatrix} \underline{x}(k) \\ \hat{\underline{e}}(k) \end{pmatrix}$$

and it can be readily verified that

$$\begin{pmatrix} 0 & 0 & 0 & 0 & 1 & 2 \\ 0 & 0 & 0 & -2 & 3 & 0 \\ -2 & -1 & 0 & 0 & 0 & 0 \\ 0 & 0 & 0 & 0 & .75 & 2.25 \\ 0 & 0 & 0 & 0 & .75 & 2.25 \\ 0 & 0 & 0 & 0 & -.25 & -.75 \end{pmatrix}^k = [0]$$

for all $k \geq 4$ as expected. The interpretation of the results is that the state-reconstruction error is eliminated after two steps and the state of the discrete system is driven to the origin after a further two steps.

Next, we investigate the design of state-reconstruction schemes and observer-based controllers of high-order discrete time systems.

5.4 Two-Level Observation Structures

The purpose of this section is to develop state reconstruction structures that produce estimates of the state vector of linear interconnected discrete systems. We will consider only the case where the interconnection pattern is a linear function of the state vector components. A state-space model of linear systems composed of N interconnected discrete subsystems is given by:

$$\underline{x}_i(k+1) = A_i\underline{x}_i(k) + B_i\underline{u}(k) + \underline{h}_i(k,\underline{x}) \tag{5.83a}$$

$$\underline{y}_i(k) = C_i\underline{x}_i(k) \tag{5.83b}$$

$$h_i = \sum_{j=1}^{N} D_{ij}\underline{x}_j(k) \quad ; \quad i = 1,\ldots,N \tag{5.83c}$$

where for the ith subsystem

$\underline{x}_i(k)$ is the n_i-th dimensional state vector,
$\underline{u}_i(k)$ is the m_i-th dimensional input vector,
$\underline{y}_i(k)$ is the p_i-th dimensional output vector,
$\underline{h}_i(k,x)$ is the coupling vector from the other subsystems,

and

$$\sum_{i=1}^{N} n_i = n, \quad \sum_{i=1}^{N} m_i = m, \quad \sum_{i=1}^{N} p_i = p$$

We will assume from now onwards that the N-pairs (A_i,C_i) are completely observable in the sense of (5.4). Instead of building up an integrated state reconstructor, our objective is to consider the design problem of N-independent state reconstructors based on the subsystem dynamics. This obviously has the merit of distributing the computational efforts, and reducing the associated wirings in the implementation.

5.4.1 *FULL-ORDER LOCAL STATE RECONSTRUCTORS*

Our approach to designing full-order state reconstruction schemes for the system (5.83) proceeds in two basic steps:

Step 1

Consider the subsystems in (5.83) to be decoupled and build a full-order state reconstructor for each of them.

Step 2

Modify the local state-reconstructors such that they can perform satisfactorily for the overall system.

Without the coupling effects, the ith subsystem is described by:

$$\underline{x}_i(k+1) = A_i\underline{x}_i(k) + B_i\underline{u}_i(k) \tag{5.84a}$$

$$\underline{y}_i(k) = C_i\underline{x}_i(k) \tag{5.84b}$$

We note that (5.84) is similar to (5.1), (5.2), and thus all the results derived in Section 5.2.1 carry over to our local subsystem. In particular, the measurement patterns could be with one-step delay or they may be instantaneous.

Consider first that the output records are $\{\underline{y}_i(k-1), \underline{y}_i(k-2), \ldots, \underline{y}_i(0)\}$ which corresponds to problem (a) in Section 5.2.1. Therefore, by analogy with (5.6), the full-order local state reconstructor takes the form:

$$\hat{\underline{x}}_i(k+1) = A_i\hat{\underline{x}}_i(k) + B_i\underline{u}_i(k) + H_i[\underline{y}_i(k) - C\hat{\underline{x}}_i(k)] \tag{5.85}$$

where $\hat{\underline{x}}_i(k)$ is the $(n_i \times 1)$ vector approximation to $\underline{x}_i(k)$ and H_i is the $(n_i \times p_i)$ matrix of design parameters. The reconstruction error $\hat{\underline{e}}_i(k) = \underline{x}_i(k) - \hat{\underline{x}}_i(k)$ propagates according to:

$$\hat{\underline{e}}_i(k+1) = [A_i - H_iC_i]\hat{\underline{e}}_i(k) \tag{5.86}$$

Here again, $\hat{\underline{e}}_i(k) \to \underline{0}_i$ as $k \to \infty$ provided the eigenvalues of the matrix $[A_i - H_i C_i]$ are located within the unit circle in the complex plane. The design of H_i to achieve asymptotic decay of the error can be done using any of the standard eigenvalue assignment algorithms.

We now consider problem (b) stated in Section 5.2.1, in which the output records consist of $\{\underline{y}_i(k), \underline{y}_i(k-1), \ldots, \underline{y}_i(0)\}$. In the manner of (5.11), the full-order local state reconstructor with instantaneous measurement is given by:

$$\tilde{\underline{x}}_i(k+1) = [A_i - P_i C_i A_i]\tilde{\underline{x}}_i(k) + [B_i - P_i C_i B_i]\underline{u}_i(k) + P_i \underline{y}_i(k+1) \qquad (5.87)$$

where $\tilde{\underline{x}}_i(k)$ is the $(n_i \times 1)$ vector approximation to $\underline{x}_i(k)$ when the measurements are instantaneous and P_i is the $(n_i \times p_i)$ matrix of design parameters. Also, it can be easily seen that the reconstruction error $\tilde{\underline{e}}_i(k) = \underline{x}_i(k) - \tilde{\underline{x}}_i(k)$ satisfies the dynamic model:

$$\tilde{\underline{e}}_i(k+1) = [A_i - P_i C_i A_i]\tilde{\underline{e}}_i(k) \qquad (5.88)$$

By similar arguments, the error $\tilde{\underline{e}}_i(k)$ dies out as k increases indefinitely provided that the eigenvalues of the matrix $[A_i - P_i C_i A_i]$ are positioned within the unit circle of the complex plane. We emphasize that all the comments regarding the observability of the pair $(A_i, C_i A_i)$ and the singularity of A_i, which were discussed in Section 5.2.1, apply here as well.

This completes the solution of *Step 1* and we now direct our attention to *Step 2*.

5.4.2 *MODIFICATIONS TO ENSURE OVERALL ASYMPTOTIC RECONSTRUCTION*

In the following, we propose modifications to the full-order local state reconstructor (5.85) or (5.87) which will ensure that the overall state reconstruction structure converges in an asymptotic way. First, consider the scheme (5.85) which when summed up over the N subsystems results in:

$$\underline{\hat{x}}(k+1) = [A - HC]\underline{\hat{x}}(k) + B\underline{u}(k) + H\underline{y}(k) \qquad (5.89)$$

where $\underline{\hat{x}}(k) = [\underline{\hat{x}}^t(k)\ \underline{\hat{x}}_2^t(k)\ \ldots\ \underline{\hat{x}}_N^t(k)]^t$ and $H = diag[K_1\ K_2 \ldots \ldots K_N]$. Let the matrix $D = \{D_{ij}\}$ of coupling coefficients satisfy the rank condition [22]

$$rank \begin{bmatrix} C \\ D \end{bmatrix} = rank\ [C] = p \qquad (5.90)$$

It is well known [23] that (5.90) implies that

$$D = GC \qquad (5.91)$$

or equivalently $G = DC^t[CC^t]^{-1}$ and hence [7]:

$$rank \begin{bmatrix} C \\ CA \\ CA^2 \\ \vdots \\ CA^{n-1} \end{bmatrix} = rank \begin{bmatrix} C \\ C(A+D) \\ C(A+D)^2 \\ \vdots \\ C(A+D)^{n-1} \end{bmatrix} \qquad (5.92)$$

However, since each pair (A_i, C_i) is completely observable and

$$A = diag[A_i], \quad C = diag[C_i] \qquad (5.93)$$

the composite observability matrix in (5.92) is of full rank. Accordingly, the pair $[(A+D), C]$ is completely observable. In this way the state of the integrated system

$$\underline{x}(k+1) = [A + D]\underline{x}(k) + B\underline{u}(k) \qquad (5.94)$$

can be reconstructed by:

$$\underline{\hat{x}}(k+1) = [A + D - WC]\underline{x}(k) + B\underline{u}(k) + W\underline{y}(k) \qquad (5.95)$$

The choice of $W = H+G$ would eventually yield a similar eigenvalue distribution for $(A+D-WC)$ and $(A-HC)$, hence ensuring the prescribed convergence for the state reconstruction scheme (5.95). However, by the same choice, it is evident that (5.95) reduces to:

$$\hat{\underline{x}}(k+1) = [A-HC]\hat{\underline{x}}(k) + B\underline{u}(k) + H\underline{y}(k) + G\underline{y}(k) \quad (5.96)$$

which verifies that the required modification to each local state reconstructor is the term $\sum_{j=1}^{N} G_{ij}\underline{y}_j(k)$ which acts as an additional input. In Fig. 5.4 we display the wiring diagram of this observation scheme. It should be remarked that:

(1) All the computations are performed at the subsystem level which comprises lower-order design problems. Thus, this scheme yields a saving in the computational effort, and it also provides more flexibility in allocating the required eigenvalues.

(2) Only the observability condition of the decoupled subsystems is needed for the above analysis.

(3) The rank qualification (5.90) identifies a class of interconnection patterns that will guarantee the asymptotic behaviour of the state reconstruction scheme.

Now, to develop modifications to the scheme (5.87) similar to the above, we consider the composite state reconstructor

$$\tilde{\underline{x}}(k+1) = [A - PCA]\tilde{\underline{x}}(k) + [B - PCB]\underline{u}(k) + P\underline{y}(k+1) \quad (5.97)$$

where

$$\begin{aligned} \tilde{\underline{x}}(k) &= [\,\tilde{\underline{x}}_1^t(k)\ \tilde{\underline{x}}_2^t(k)\ \dots\ \tilde{\underline{x}}_N^t\,] \\ P &= \mathrm{diag}[P_1\ \ P_2\ \dots\ P_N] \\ \underline{u}(k) &= [\underline{u}_1^t(k)\ \underline{u}_2^t(k)\ \dots\ \underline{u}_N^t(k)] \\ \underline{y}(k) &= [\underline{y}_1^t(k)\ \underline{y}_2^t(k)\ \dots\ \underline{y}_N^t(k)] \end{aligned} \quad (5.98)$$

It has been shown [22] that if A_i is nonsingular and condition

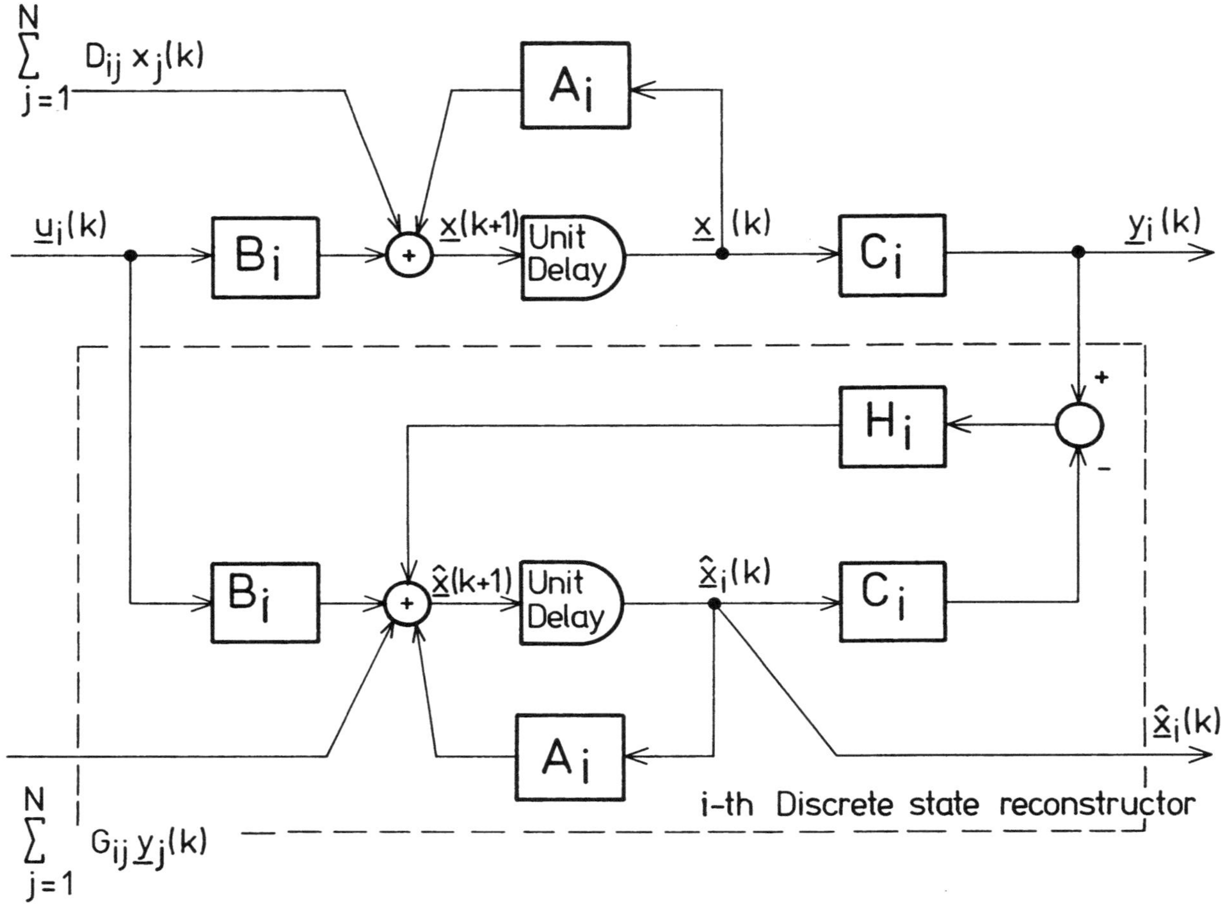

Fig. (5.4) Full-order local state reconstructor

(5.90) is satisfied, then the required modification is the same as in the previous case. That is, the addition of $\sum_{j=1}^{N} G_{ij}\underline{y}_j(k)$ as an external input will ensure the asymptotic convergence of the reconstructed state vector of the integrated system. This result is to be expected in the light of the correspondence between the statements of problems (a) and (b). *Step 2* is now completed. One can visualise the full-order state reconstruction structure as consisting of two levels. At the first level there is a set of local state reconstructors. Each one is designed to approximate a part of the state vector. A coordinating unit is employed by the second level which supplies the required modifications to ensure the overall asymptotic convergence of the reconstruction scheme.

Although we have considered only the design of full-order local state reconstructors and their modifications, the design of reduced-order local state reconstructors and the corresponding modifications can be obtained by a straightforward analysis *[22]*. It is interesting to point out that the required modifications are exactly the same.

5.4.3 *EXAMPLES*

In order to demonstrate the application of the design procedure for the two-level state reconstruction, two examples will be considered.

Example 1:

An interconnected system consisting of three subsystems is described by the matrices:

$$A_1 = \begin{bmatrix} .7 & .2 \\ .1 & .8 \end{bmatrix} \qquad B_1 = \begin{bmatrix} 0 \\ 1 \end{bmatrix}$$

$$C_1 = [1 \quad 1]$$

$$A_2 = \begin{pmatrix} .5 & .15 \\ 0 & .4 \end{pmatrix} \qquad B_2 = \begin{pmatrix} 1 \\ 0 \end{pmatrix}$$

$$C_2 = [2 \quad -1]$$

$$A_3 = \begin{pmatrix} .6 & 0 \\ .3 & .1 \end{pmatrix} \qquad B_3 = \begin{pmatrix} 1 \\ 1 \end{pmatrix}$$

$$C_3 = [-1 \quad 1]$$

and the interconnection matrix is given by:

$$D = \begin{pmatrix} 0 & 0 & -6 & 3 & 0 & 0 \\ 0 & 0 & 0 & 0 & -3 & 3 \\ 0 & 0 & 0 & 0 & 2 & -2 \\ -2 & -2 & 0 & 0 & 0 & 0 \\ 1 & 1 & 0 & 0 & .6 & 0 \\ 0 & 0 & 4 & -2 & .3 & .1 \end{pmatrix}$$

We note that the matrices A_1, A_2 and A_3 are nonsingular, rank $[C_1]$ = rank $[C_2]$ = rank $[C_3]$ = 1 and the pairs (A_1,C_1), (A_2,C_2) and (A_3,C_3) are observable. In case of one-step-delay in measurement records, we compute the gains H_1, H_2 and H_3 by assigning three eigenvalues at $\gamma_1 = .25$, $\gamma_2 = .45$ and $\gamma_3 = .65$ to yield:

$$H_1 = \begin{pmatrix} .506 \\ 0 \end{pmatrix}, \quad H_2 = \begin{pmatrix} .025 \\ 0 \end{pmatrix}, \quad H_3 = \begin{pmatrix} .11 \\ 0 \end{pmatrix}$$

Because the rank condition (5.90) is satisfied, the required modification term is computed as:

$$G = DC^t(CC^t)^{-1}$$

$$\begin{pmatrix} 0 & -3 & 0 \\ 0 & 0 & 3 \\ 0 & 0 & -2 \\ -2 & 0 & 0 \\ 1 & 0 & 0 \\ 0 & 2 & 0 \end{pmatrix}$$

If the measurements were instantaneous, and due to the nonsingularity of the subsystem matrices, the reconstruction scheme (5.87) can be used and the associated gain matrices are computed as

$$P_1 = \begin{bmatrix} .749 \\ -.094 \end{bmatrix}, \quad P_2 = \begin{bmatrix} .05 \\ 0 \end{bmatrix}, \quad P_3 = \begin{bmatrix} .183 \\ -.55 \end{bmatrix}$$

which will place the reconstructor's eigenvalues at $\gamma_1 = .25$, $\gamma_2 = .45$, and $\gamma_3 = .65$.

Example 2:

The sixth order system of Example 1 is considered again but with

$$C_1 = \begin{bmatrix} 1 & 0 \\ 0 & 1 \end{bmatrix}, \quad C_2 = \begin{bmatrix} 2 & 0 \\ 0 & -1 \end{bmatrix}, \quad C_3 = \begin{bmatrix} -1 & 0 \\ 0 & 1 \end{bmatrix}$$

and the interconnection matrix

$$D = \begin{bmatrix} 0 & 0 & 6 & 0 & 0 & 0 \\ 0 & 0 & 0 & 0 & 0 & 3 \\ 0 & 0 & 0 & 0 & 0 & 0 \\ 0 & -2 & 0 & 0 & 2 & 0 \\ 1 & 0 & 0 & 0 & 0 & 0 \\ 0 & 0 & 0 & -2 & 0 & 0 \end{bmatrix}$$

Here, rank $[C_1]$ = rank $[C_2]$ = rank $[C_3]$ = 2, the pairs (A_1,C_1), (A_2,C_2) and (A_3,C_3) are observable and

$$\text{rank} \begin{bmatrix} C \\ 0 \end{bmatrix} = 2$$

Consider first that the measurements are recorded with a one-step-delay. The desired eigenvalues are:

$\gamma_{11} = .1, \quad \gamma_{12} = -.7,$ for subsystem 1

$\gamma_{21} = .3, \quad \gamma_{22} = .4,$ for subsystem 2

$$\gamma_{31} = -.6, \quad \gamma_{32} = -.2, \qquad \text{for subsystem 3}$$

The corresponding gain matrices are:

$$H_1 = \begin{bmatrix} .583 & .083 \\ .217 & 1.517 \end{bmatrix}, \quad H_2 = \begin{bmatrix} .1 & .3 \\ 0 & 0 \end{bmatrix}, \quad H_3 = \begin{bmatrix} -.8 & 0 \\ -.7 & .7 \end{bmatrix}$$

and the modification matrix G is given by:

$$G = \begin{bmatrix} 0 & 0 & -3 & 0 & 0 & 0 \\ 0 & 0 & 0 & 0 & 0 & 3 \\ 0 & 0 & 0 & 0 & -2 & 0 \\ 0 & -2 & 0 & 0 & 0 & 0 \\ 1 & 0 & 0 & 0 & 0 & 0 \\ 0 & 0 & 0 & 2 & 0 & 0 \end{bmatrix}$$

If instantaneous measurement records were available, the gain matrices could be obtained to place the desired six eigenvalues as above:

$$P_1 = \begin{bmatrix} .784 & -.438 \\ .173 & 1.951 \end{bmatrix}, \quad P_2 = \begin{bmatrix} .2 & .6 \\ 0 & 0 \end{bmatrix}, \quad P_3 = \begin{bmatrix} -1.33 & 0 \\ -3 & 7 \end{bmatrix}$$

Next, we consider another class of higher-order discrete time systems.

5.5 Discrete Two-Time-Scale Systems

5.5.1 *INTRODUCTION*

In the previous section we have examined the state reconstruction problem for discrete systems which are composed of interconnected subsystems. The coupling pattern amongst subsystems is a measure of the physical links between the dynamical variables in the state-space description. Another type of coupling would be the interaction between the system modes in the time dimension. This situation results in physical systems which possess multi-time-scale phenomena [24]. A good example of such systems is the discrete time-scale systems which have been analysed in Chapter 4 for feedback control purposes with accessible states.

In this section, we consider the problem of designing observers and observer-based controllers for discrete two-time-scale systems with inaccessible states. Initially, we recall from Chapter 4 that the discrete-two-time-scale system of order $(n_1 + n_2)$,

$$\underline{x}_1(k+1) = A_1\underline{x}_1(k) + A_2\underline{x}_2(k) + B_1\underline{u}(k) \quad (5.99a)$$

$$\underline{x}_2(k+1) = A_3\underline{x}_1(k) + A_4\underline{x}_2(k) + B_2\underline{u}(k) \quad (5.99b)$$

$$\underline{y}(k) = C_1\underline{x}_1(k) + C_2\underline{x}_2(k) \quad (5.99c)$$

has a slow subsystem of order n_1 given by:

$$\underline{x}_s(k+1) = A_0\underline{x}_s(k) + B\underline{u}_s(k) \quad (5.100a)$$

$$\underline{y}_s(k) = C_0\underline{x}_s(k) + D_0\underline{u}_s(k) \quad (5.100b)$$

and a fast subsystem of order n_2 described by:

$$\underline{x}_f(k+1) = A_4\underline{x}_f(k) + B_2\underline{u}_f(k) \quad (5.101a)$$

$$\underline{y}_f(k) = C_2\underline{x}_f(k) \quad (5.101b)$$

where

$$A_0 = A_1 + A_2(I_2 - A_4)^{-1}A_3 \quad (5.102a)$$

$$B_0 = B_1 + A_2(I_2 - A_4)^{-1}B_2 \quad (5.102b)$$

$$C_0 = C_1 + C_2(I_2 - A_4)^{-1}A_3 \quad (5.102c)$$

$$D_0 = C_2(I_2 - A_4)^{-1}B_2 \quad (5.102d)$$

$$\underline{u}(k) = \underline{u}_s(k) + \underline{u}_f(k) \quad (5.102e)$$

It should be noted that the n_1 eigenvalues of the slow subsystem (5.100) are distributed near the unit circle and the n_2 eigenvalues of the fast subsystem (5.101) are centred around the origin, in the complex plane.

5.5.2 *TWO-STAGE OBSERVER DESIGN*

The problem of designing deterministic observers in order to estimate the slow and fast states is now considered. For simplicity in exposition, we assume that there is a one-step-delay between measuring and processing the information records. Thus, a full-order deterministic observer for the discrete system (5.99) can be constructed in the manner of (5.6) to yield:

$$\hat{\underline{x}}_1(k+1) = A_1\hat{\underline{x}}_1(k) + A_2\hat{\underline{x}}_2(k) + B_1\underline{u}(k) + H_1[\underline{y}(k) - C_1\hat{\underline{x}}_1(k) - C_2\hat{\underline{x}}_2(k)] \quad (5.103a)$$

$$\hat{\underline{x}}_2(k+1) = A_3\hat{\underline{x}}_1(k) + A_4\hat{\underline{x}}_2(k) + B_2\underline{u}(k) + H_2[\underline{y}(k) - C_1\hat{\underline{x}}_1(k) - C_2\hat{\underline{x}}_2(k)] \quad (5.103b)$$

where $\hat{\underline{x}}_1(k)$, $\hat{\underline{x}}_2(k)$ are the $(n_1\times1)$, $(n_2\times1)$ vectors approximating to $\underline{x}_1(k)$, $\underline{x}_2(k)$, respectively, and H_1, H_2 are the gain matrices of the design parameters. In terms of the state reconstruction error vectors, $\hat{\underline{e}}_1(k) = \underline{x}_1(k) - \hat{\underline{x}}_1(k)$ and $\hat{\underline{e}}_2(k) = \underline{x}_2(k) - \hat{\underline{x}}_2(k)$, it follows from (5.99) and (5.103) that

$$\hat{\underline{e}}_1(k+1) = [A_1-H_1C_1]\hat{\underline{e}}_1(k) + [A_2-H_1C_2]\hat{\underline{e}}_2(k) \quad (5.104a)$$

$$\hat{\underline{e}}_2(k+1) = [A_3-H_2C_1]\hat{\underline{e}}_1(k) + [A_4-H_2C_2]\hat{\underline{e}}_2(k) \quad (5.104b)$$

We know from the previous sections that system (5.104) will function as a deterministic observer for the discrete system (5.99) if the (n_1xp) matrix H_1 and the (n_2xp) matrix H_2 can be chosen such that system (5.104) is asymptotically stable.

Using the concept of quasi-steady-state *[24]*, the slow and fast subsystems (5.104) are given by:

$$\hat{\underline{e}}_s(k+1) = H_2\hat{\underline{e}}_s(k) \quad (5.105a)$$

$$\hat{\underline{e}}_f(k+1) = H_f\hat{\underline{e}}_f(k) \quad (5.105b)$$

where

$$H_s = (A_1-H_1C_1) + (A_2-H_1C_2)(I_2-A_4+H_2C_2)^{-1}(A_3-H_2C_1) \tag{5.106a}$$

$$H_f = (A_4 - H_2C_2) \tag{5.106b}$$

where I_2 is the $(n_2 x n_2)$ identity matrix. Now, if the pair (A_4, C_2) is completely observable then the gain matrix H_2 can be chosen such that the matrix $[A_4-H_2C_2]$ has a spectral norm of less than one. This means that the fast subsystem (5.105b), (5.106b) is asymptotically stable.

Consider the slow subsystem. Using the matrix identities

$$\begin{aligned}(R + QS)^{-1} &= R^{-1}(I + QSR)^{-1} \\ &= R^{-1}(I - Q(I + SR^{-1}Q)^{-1}SR^{-1})\end{aligned}$$

with $R = (I_2-A_4)$, $Q = H_2$, $S = C_2$ and manipulating (5.106a) along with (5.102a) and (5.102c), it follows that:

$$H_s = A_0 - H_0C_0 \tag{5.107a}$$

where

$$H_0 = H_1 + (A_2-H_1C_2)(I-A_4)^{-1}H_2[I+C_2(I_2-A_4)^{-1}H_2]^{-1} \tag{5.107b}$$

It is therefore evident that the $(n_1 xp)$ H_0 can be chosen such that $(A_0-H_0C_0)$ is asymptotically stable provided the pair (A_0, C_0) is completely observable. Since the asymptotic stability of the fast and slow subsystems guarantees the asymptotic stability of system (5.104) under the fast-slow separation property [25], the design of the state reconstructor (5.103) can be accomplished in the following two stages:

(i) Compute the gain H_2 to place the n_2 eigenvalues of $(A_4-H_2C_2)$ at desired locations,

(ii) Compute the gain H_0 to place the n_1 eigenvalues of $(A_0-H_0C_0)$ at desired locations. Then compute the gain H_1 using the formula:

$$H_0 = H_1 + (A_2 - H_1C_2)(I - A_4)^{-1}H_2[I + C_2(I_2 - A_4)^{-1}H_2]^{-1} \tag{5.108}$$

We emphasize that the two-stage observer design rests upon the observability of the pairs (A_0, C_0) and (A_4, C_2) and the invertibility of the matrix $(I_2 - A_4)$. The latter condition is always satisfied for discrete two-time-scale systems *[24]*.

In asymptotically stable discrete systems with a slow-fast separation property, the matrix A_4 is a stable matrix. If this is the case, then $H_2 = 0$ is an admissible choice for H_2 in (5.103). From (5.108), it follows that $H_1 = H_0$ and we obtain the reduced-order deterministic observer:

$$\begin{aligned}\hat{\underline{x}}_1(k+1) &= A_1\hat{\underline{x}}_1(k) + A_2\hat{\underline{x}}_2(k) + B_1\underline{u}(k) \\ &\quad + H_0[\underline{y}(k) - C_1\hat{\underline{x}}_1(k) - C_2\hat{\underline{x}}_2(k)]\end{aligned} \tag{5.109a}$$

$$\hat{\underline{x}}_2(k+1) = A_3\hat{\underline{x}}_1(k) + A_4\hat{\underline{x}}_2(k) + B_2\underline{u}(k) \tag{5.109b}$$

It is interesting to note that the two-stage procedure for the design of full-order observers is the dual of the procedure developed in Section 4.5 for the design of stabilising state-feedback controllers for discrete systems with slow and fast modes.

5.5.3 *DYNAMIC STATE FEEDBACK CONTROL*

When the fast subsystem (5.101) of the discrete two-time-scale system (5.99) is asymptotically stable, a reduced-order state reconstructor *[25]* can be used to estimate the slow states. In the light of the above analysis, this reduced-order observer is given by:

$$\begin{aligned}\hat{\underline{x}}_1(k+1) &= A_0\hat{\underline{x}}_1(k) + B_0\underline{u}(k) + H_0[\underline{y}_k - C_0\hat{\underline{x}}_1(k) \\ &\quad + D_0\underline{u}(k)]\end{aligned} \tag{5.110}$$

A dynamic state feedback control is described by:

$$\underline{u}(k) = C_0\hat{\underline{x}}_1(k) \tag{5.111}$$

where the $(n_1 x p)$ matrix H_0 and the $(m x n_1)$ matrix C_0 are unknown gains to be determined. Consider the composite system (5.99), (5.110) and (5.111):

$$\begin{bmatrix} \underline{x}_1(k+1) \\ \hat{\underline{x}}_1(k+1) \end{bmatrix} = \left[\begin{array}{c|c} A_1 & B_1G_0 \\ \hline H_0C_1 & A_0+B_0G_0-H_0C_0-H_0D_0C_0 \end{array}\right] \begin{bmatrix} \underline{x}_1(k) \\ \hat{\underline{x}}_1(k) \end{bmatrix}$$

$$+ \begin{bmatrix} A_2 \\ H_0C_2 \end{bmatrix} \underline{x}_2(k) \tag{5.112a}$$

$$\underline{x}_2(k+1) = [A_3 \quad B_2G_0] \begin{bmatrix} \underline{x}_1(k) \\ \hat{\underline{x}}_1(k) \end{bmatrix} + A_4\underline{x}_2(k) \tag{5.112b}$$

Using the quasi-steady state concept *[24]*, the $(2n_1+n_2)$th-order system (5.112) has a fast subsystem described by:

$$\underline{x}_f(k+1) = A_4\underline{x}_f(k) \tag{5.113a}$$

and a slow subsystem given by:

$$\underline{x}_s(k+1) = F_0\underline{x}_s(k) \tag{5.113b}$$

where

$$F_0 = \left[\begin{array}{c|c} A_1 & B_1G_0 \\ \hline H_0C_1 & A_0+B_0G_0-H_0C_0-H_0D_0C_0 \end{array}\right]$$

$$+ \begin{bmatrix} A_2 \\ H_0C_2 \end{bmatrix} (I_2 - A_4)^{-1} [A_3 \quad B_2G_0] \tag{5.113c}$$

Now, if the pairs (A_0,B_0) and (A_0,C_0) are completely controllable and observable, we can determine the unknown gains H_0 and G_0 as follows. In view of (5.102), it can be readily seen that

$$F_0 = \begin{bmatrix} A_0 & B_0G_0 \\ H_0C_0 & A_0+B_0G_0-H_0C_0 \end{bmatrix}$$

and under the equivalence transformation, it becomes:

$$\begin{bmatrix} I_n & 0 \\ I_n & -I_n \end{bmatrix} F_0 \begin{bmatrix} I_n & 0 \\ I_n & -I_n \end{bmatrix} = \begin{bmatrix} A_0+B_0G_0 & -B_0G_0 \\ 0 & A_0-H_0C_0 \end{bmatrix} \qquad (5.114)$$

So that the eigenvalues of F_0 are the eigenvalues of $(A_0+B_0G_0)$ together with those of $(A_0-H_0C_0)$. We note that (5.114) is a version of the *separation principle* discussed in Section 5.3.2. The design procedure can be implemented in two stages. In the first stage, the observer gain matrix K_0 is computed to place n_1 arbitrary eigenvalues, and in the second stage the controller gain G_0 is computed to place n_1 arbitrary closed-loop eigenvalues.

Next, we illustrate the application of this two-stage feedback control design on a ninth-order discrete model.

5.5.4 *EXAMPLE*

The state variables of a boiler system comprising a superheater and riser in series with each other *[26]* are: density and temperature of output steam flow, temperature of the superheater, riser outlet mixture quality, water flow in riser, pressure, riser tube-wall temperature, and temperature and level of water in boiler. The variables which can be manipulated are input fuel and input water flows, whereas the directly measurable output variables are temperature of output steam flow, riser outlet mixture quality, pressure and water level. By simulating the ninth-order linear continuous model *[26]* and its discretised version, it is found that a sampling period of 0.5 sec yields a discrete model whose response matches very closely that of the continuous model. Using the permutation matrix

$$P = (e_9, e_8, e_2, e_5, e_7, e_6, e_4, e_3, e_1)$$

and the scaling matrix

$$S = \text{diag}(.015,\ 0.15,\ .05,\ .1,\ 2,\ .5\text{x}10^{-4},\ .15,\ 5,\ .2\text{x}10^{4})$$

where e_i is the elementary column vector whose ith entry is 1, the transformed discrete system has the eigenvalues (1.0, $0.1452 \pm 0.0726j$, 0.2298, 0.98, 0.996, $0.9741 \pm 0.0905j$, 0.8461) and it is estimated to have six slow and three fast variables.

In terms of the model (5.99) the subsystem matrices are:

$$A_1 = \begin{pmatrix} 1.0 & -0.1489\text{x}10^{-3} & 0.1051\text{x}10^{-3} & 0.1051\text{x}10^{-3} & -0.2894\text{x}10^{-1} & 0.3127\text{x}10^{-3} \\ 0 & 0.9866 & -0.3555\text{x}10^{-3} & -0.2745\text{x}10^{-3} & 0.9544\text{x}10^{-5} & -0.1949\text{x}10^{-1} \\ 0 & -0.1389\text{x}10^{-2} & 0.9686 & 0.3165\text{x}10^{-3} & -0.3907\text{x}10^{-1} & 0.2572\text{x}10^{-1} \\ 0 & 0.8084\text{x}10^{-2} & 0.2865\text{x}10^{-2} & 0.9057 & -0.7275\text{x}10^{-4} & 0.1951 \\ 0 & -0.2065\text{x}10^{-2} & 0.3328\text{x}10^{-2} & 0.7091\text{x}10^{-3} & 0.8829 & 0.1479\text{x}10^{-1} \\ 0 & 0.7152\text{x}10^{-2} & 0.2589\text{x}10^{-1} & 0.1980\text{x}10^{-1} & -0.8358\text{x}10^{-3} & 0.8705 \end{pmatrix}$$

$$A_2 = \begin{pmatrix} -0.2667\text{x}10^{-5} & -0.5914\text{x}10^{-6} & -0.3823\text{x}10^{-5} \\ -0.1585\text{x}10^{-7} & 0.4712\text{x}10^{-2} & 0.5030\text{x}10^{-4} \\ 0.8717\text{x}10^{-4} & 0.9676\text{x}10^{-5} & -0.1144\text{x}10^{-5} \\ 0.1169\text{x}10^{-6} & 0.3265\text{x}10^{-5} & 0.1673\text{x}10^{-4} \\ -0.1071\text{x}10^{-4} & -0.9028\text{x}10^{-5} & 0.1334\text{x}10^{-4} \\ 0.1445\text{x}10^{-5} & 0.1345\text{x}10^{-4} & 0.1143\text{x}10^{-3} \end{pmatrix}$$

$$A_3 = \begin{pmatrix} 0 & -6.0165 & 0.3120\text{x}10^{2} & -0.1336\text{x}10^{1} & -0.231\text{x}10^{3} & -0.10006\text{x}10^{3} \\ 0 & 0.2490\text{x}10^{3} & -0.8749 & -0.6724 & 0.2564\text{x}10^{-1} & -0.2105\text{x}10^{2} \\ 0 & -0.5153\text{x}10^{2} & 6.2408 & 0.4815\text{x}10^{1} & -0.1692 & 0.3291\text{x}10^{3} \end{pmatrix}$$

$$A_4 = \begin{pmatrix} 0.2375 & 0.6427\times10^{-1} & -0.2622\times10^{-1} \\ -0.4447\times10^{-4} & 0.1998 & 0.8275\times10^{-1} \\ 0.2825\times10^{-3} & -0.1018 & 0.1490 \end{pmatrix}$$

$$B_1 = \begin{pmatrix} 0.1777\times10^{-4} & 0.4490\times10^{-5} \\ -0.319\times10^{-3} & 0.1159\times10^{-1} \\ 0.2177\times10^{-3} & 0.3889\times10^{-4} \\ -0.6494\times10^{-4} & 0.1109 \\ -0.1159\times10^{-3} & 0.2689\times10^{-4} \\ -0.7698\times10^{-3} & 0.1239\times10^{-2} \end{pmatrix}$$

$$B_2 = \begin{pmatrix} 2.308 & 0.1651 \\ -0.7292 & 1.8098 \\ -0.4393 & -0.5085\times10^{-1} \end{pmatrix}$$

$$C_1 = \begin{pmatrix} 1 & 0 & 0 & 0 & 0 & 0 \\ 0 & 0 & 0 & 0 & 1 & 0 \\ 0 & 0 & 0 & 0 & 0 & 1 \\ 0 & 0 & 0 & 0 & 0 & 0 \end{pmatrix}$$

$$C_2 = \begin{pmatrix} 0 & 0 & 0 \\ 0 & 0 & 0 \\ 0 & 0 & 0 \\ 0 & 1 & 0 \end{pmatrix}$$

The slow subsystem is described by the variables: water level, temperature of the superheater, temperature of water in the boiler, temperature of riser, riser outlet mixture quality and pressure, in order of dominance. On the other hand, the fast subsystem is represented by the variables: water flow in riser, temperature and density of output steam flow.

In terms of (5.101), the fast subsystem is described by the triple (A_4, B_2, C_2) and with reference to (5.100) the slow subsystem is characterised by:

$$A_0 = \begin{bmatrix} 1.0 & -0.1334\times10^{-4} & 0.5951\times10^{-3} & 0.8872\times10^{-4} & -0.2813\times10^{-1} & -0.7785\times10^{-3} \\ 0 & 0.9959 & -0.299\times10^{-6} & -0.2274\times10^{-6} & 0.1616\times10^{-7} & 0.6003\times10^{-5} \\ 0 & 0.345\times10^{-2} & 0.9721 & 0.1350\times10^{-3} & -0.6551\times10^{-1} & 0.1263\times10^{-1} \\ 0 & 0.7419\times10^{-2} & 0.2983\times10^{-2} & 0.9058 & -0.1132\times10^{-3} & 0.2015 \\ 0 & -0.6295\times10^{-2} & 0.2994\times10^{-2} & 0.8083\times10^{-3} & 0.8862 & 0.2149\times10^{-1} \\ 0 & 0.1967\times10^{-3} & 0.2679\times10^{-1} & 0.2045\times10^{-1} & -0.1308\times10^{-2} & 0.9145 \end{bmatrix}$$

$$B_0 = \begin{bmatrix} 0.2953\times10^{-3} & 0.3311\times10^{-5} \\ -0.6407\times10^{-3} & 0.1168\times10^{-1} \\ 0.4760\times10^{-3} & 0.9704\times10^{-4} \\ -0.1528\times10^{-3} & 0.1109 \\ -0.2016\times10^{-3} & -0.2062\times10^{-5} \\ -0.1355\times10^{-2} & 0.1232\times10^{-2} \end{bmatrix}$$

$$C_0 = \begin{bmatrix} 1.0 & 0 & 0 & 0 & 0 & 0 \\ 0 & 0 & 0 & 0 & 1.0 & 0 \\ 0 & 0 & 0 & 0 & 0 & 1.0 \\ 0 & 0.3012\times10^{3} & -0.3316 & -0.250 & 0.1771\times10^{-1} & 1.1668 \end{bmatrix}$$

It is easy to check that both slow and fast subsystems are completely controllable and observable. A full-order observer can be designed to reconstruct the slow and fast states. Assigning three eigenvalues at (0.15, 0.13, 0.11) yields:

$$K_2 = \begin{bmatrix} 0 & 0 & 0 & -61.6373 \\ 0 & 0 & 0 & -0.1962 \\ 0 & 0 & 0 & -0.9500 \end{bmatrix}$$

and positioning six eigenvalues at (0.99, 0.97, 0.95, 0.93, 0.91, 0.89) gives:

$$K_0 = \begin{bmatrix} -0.3992\times10^{-1} & 0.8283\times10^{-2} & -0.7771\times10^{-2} & 0.1325\times10^{-4} \\ -0.1301\times10^{-2} & -0.2002\times10^{-3} & 0.3398\times10^{-3} & -0.2193\times10^{-3} \\ 0.4738\times10^{-1} & 0.3095\times10^{-1} & -0.1646\times10^{-1} & -0.3697\times10^{-4} \\ 0.6404 & 0.1308 & -0.2290 & 0.1758\times10^{-3} \\ 0.8904\times10^{-1} & 0.1859\times10^{-1} & -0.2819\times10^{-1} & 0.3910\times10^{-4} \\ -0.3736 & -0.4547\times10^{-1} & 0.5312\times10^{-1} & -0.1472\times10^{-3} \end{bmatrix}$$

From (5.108), the gain matrix K_1 is computed as:

$$K_1 = \begin{bmatrix} -0.3992\text{x}10^{-1} & 0.8283\text{x}10^{-2} & -0.7771\text{x}10^{-2} & 0.1996\text{x}10^{-4} \\ -0.1301\text{x}10^{-2} & -0.2002\text{x}10^{-3} & 0.3398\text{x}10^{-3} & -0.2585\text{x}10^{-3} \\ 0.4738\text{x}10^{-1} & 0.3095\text{x}10^{-1} & -0.1646\text{x}10^{-1} & 0.7004\text{x}10^{-2} \\ 0.6404 & 0.1308 & -0.2290 & 0.2321\text{x}10^{-3} \\ 0.8904\text{x}10^{-1} & 0.1859\text{x}10^{-1} & -0.2819\text{x}10^{-1} & -0.8178\text{x}10^{-3} \\ -0.3736 & -0.4547\text{x}10^{-1} & 0.5312\text{x}10^{-1} & -0.5173\text{x}10^{-4} \end{bmatrix}$$

This completes the construction procedure of the full-order state observer.

Since A_4 is a stable matrix, (A_0, B_0) is a controllable pair and (A_0, C_0) is an observable pair, we can proceed to design a lower-order observer-based controller. Placing the observer eigenvalues at (0.83, 0.82, 0.81, 0.80, 0.79, 0.78) gives:

$$K_0 = \begin{bmatrix} -0.5879 & 0.2398 & -0.2538\text{x}10^{-1} & 0.2100\text{x}10^{-4} \\ -0.9063\text{x}10^{-1} & 0.5282\text{x}10^{-1} & -0.6228\text{x}10^{-2} & -0.6453\text{x}10^{-3} \\ -0.1249\text{x}10^{3} & 0.6905\text{x}10^{2} & -0.8445\text{x}10^{1} & 0.6829\text{x}10^{-2} \\ 0.3669\text{x}10^{2} & -0.1807\text{x}10^{2} & 0.1703\text{x}10^{1} & -0.1813\text{x}10^{-2} \\ -0.1932\text{x}10^{1} & 0.9968 & -0.1554 & 0.1285\text{x}10^{-3} \\ -0.1232\text{x}10^{2} & 0.7272\text{x}10^{1} & -0.1058\text{x}10^{1} & 0.7215\text{x}10^{-3} \end{bmatrix}$$

and selecting the desired closed-loop eigenvalues to be (0.99, 0.97, 0.95, 0.93, 0.91, 0.89) results in:

$$G_0 = \begin{bmatrix} -0.2035\text{x}10^{3} & -0.7295\text{x}10^{2} & 0.6145\text{x}10^{2} & 0.6797\text{x}10^{1} & 0.7904\text{x}10^{1} & 0.3152\text{x}10^{2} \\ -0.3760\text{x}10^{2} & -0.8199\text{x}10^{1} & 0.5847\text{x}10^{1} & 0.8749 & 0.5298\text{x}10^{1} & 0.6536\text{x}10^{1} \end{bmatrix}$$

The matrices K_0 and G_0 are then the required gains to implement the dynamic state feedback controller (5.111) to the discrete boiler system model.

5.6 Concluding Remarks

The purpose of this chapter has been the development of state reconstructors and observer-based controllers for both low- and high-order discrete systems. Two important aspects have been emphasized throughout. The first aspect is the effect of measurement patterns, i.e. instantaneous recording or a one-step delay in recording. The second aspect is the derivation of the deadbeat observers and deadbeat controllers. Several examples have been worked out to illustrate the theoretical developments. When dealing with high-order systems, attention was focused on the use of the fundamental results at the subsystem level.

5.7 Problems

1. A third order system is described by:

$$A = \begin{bmatrix} -2 & -2 & 0 \\ 0 & 0 & 1 \\ 0 & -3 & -4 \end{bmatrix} ; \quad B = \begin{bmatrix} 1 & 0 \\ 0 & 0 \\ 0 & 1 \end{bmatrix}$$

$$C = \begin{bmatrix} 1 & 0 & 1 \\ 0 & 1 & 0 \end{bmatrix}$$

   (a) Design a state-reconstructor whose output settles, within an accuracy of 10^{-5}, to the actual state vector after four discrete steps.

   (b) Design a deadbeat state-reconstructor.

2. Compute the gains of an observer-based controller for the system

$$A = \begin{bmatrix} 1 & 1 & 0 \\ 0 & 1 & 0 \\ 0 & 0 & 1 \end{bmatrix} ; \quad B = \begin{bmatrix} 0 & 1 \\ 1 & 0 \\ 1 & 0 \end{bmatrix} ; \quad C = \begin{bmatrix} 1 & 1 & -1 \\ 0 & 1 & 0 \end{bmatrix}$$

   such that the resulting closed-loop eigenvalues are:

{.1,.1,.1} for the observer part and {.6,.7,.8} for the controller part.

3. Design a deadbeat observer-based controller for the system of problem 2.

4. Consider the system represented by the matrices

$$A = \begin{pmatrix} 0 & 1 & 0 \\ 0 & 0 & 1 \\ 1 & 0 & 0 \end{pmatrix} \qquad B = \begin{pmatrix} 0 \\ 1 \\ 0 \end{pmatrix}$$

$$C = \begin{pmatrix} 1 & 0 & 0 \\ 1 & 1 & 0 \end{pmatrix}$$

Find the gain matrices of a full-order state reconstructor such that the desired eigenvalues at $\gamma_1 = \varepsilon$, $\gamma_2 = -\varepsilon$ where ε is arbitrarily close to zero. Compare the results with a deadbeat type state reconstructor.

5.8 References

[1] Luenberger, D.G.
"Observing the State of a Linear System",
IEEE Trans. Military Electr., vol. MIL-8, 1964, 74-80.

[2] Luenberger, D.G.
"Observers for Multivariable Systems", IEEE Trans. Automat. Contr., vol. AC-11, 1966, 190-197.

[3] Luenberger, D.G.
"An Introduction to Observers", IEEE Trans. Automat. Contr., vol. AC-16, 1971, 596-602.

[4] Leondes, C.T. and L.M. Novak
"Optimal Minimal-Order Observers for Discrete Systems - A Unified Theory", Automatica, vol. 8, 1972, 379-387

[5] Kwakernaak, H. and R. Sivan
"Linear Optimal Control Systems", Wiley Inter-Science, N.Y., 1972.

[6] Yuksel, Y.O. and J.J. Bongiorno
"Observers for Linear Multivariable Systems With Applications", IEEE Trans. Automat Contr., vol. AC-16, 1971, 603-613.

[7] Chen, C.T.
"Introduction to Linear System Theory", Holt, Rinehart and Winston, N.Y., 1970.

[8] Willems, J.L.
"Design of State Observers for Linear Discrete-Time Systems, Int. J. Syst. Sci., vol. 11, 1980, 139-147.

[9] Gopinath, B.
"On the Control of Linear Multiple Input-Output Systems", Bell Syst. Tech. J., vol. 50, 1971, 1063-1081.

[10] Ramakrishna, K. and V. Gourishankar
"Optimal Observers with Specified Eigenvalues", Int. J. Control, vol. 27, 1978, 239-244.

[11] Gourishanker, V. and P. Kudva,
"Optimal Observers for State Regulation of Linear Discrete-Time Plants", Int. J. Control, vol. 26, 1977, 359-368.

[12] Kudva, P. and V. Gourishankar
"Observers for Linear Discrete Multivariable Systems With Inaccessible Inputs", Int. J. Control, vol. 27, 1978, 539-546.

[13] Arbel, A. and E. Tse
"Observer Design for Large-Scale Linear Systems", IEEE Trans. Automat. Contr., vol. AC-24, 1979, 469-476.

[14] Porter, B.
"Deadbeat State Reconstruction of Linear Multivariable Discrete-Time Systems", Electr. Lett., vol. 9, 1973, 176-177.

[15] O'Reilly, J.
"Observer Design for the Minimum-Time State Reconstruction of Linear Discrete-Time Systems", J. Dynamic Systems, Measurement and Control, vol. 101, 1979, 350-354.

[16] Ichikawa, K.
"Discrete-Time Fast Regulator with Fast Observer", Int. J. Control, vol. 28, 1978, 733-742.

[17] Ichikawa, K.
"Synthesis of Optimal Feedback Control Systems With Model Feedback Observers", J. Dynamic Systems, Measurement and Control, vol. 96, 1974, 470-474.

[18] Ichikawa, K.
"Design of Discrete-Time Deadbeat Reduced-Order State Observer", Int. J. Control, vol. 29, 1979, 93-101.

[19] Luenberger, D.G.
"Canonical Forms for Linear Multivariable Systems", IEEE Trans. Automat. Contr., vol. AC-12, 1967, 290-293.

[20] Porter, B. and A. Bradshaw
"Design of Deadbeat Controllers and Full-Order Observers for Linear Multivariable Discrete-Time Plants", Int. J. Control, vol. 22, 1975, 149-155.

[21] Porter, B.
"Deadbeat Control of Discrete-Time Systems Incorporating Deadbeat Discrete-Time Observers", Electr. Lett., vol. 9, 1973, 547-548.

[22] Mahmoud, M.S. and M.G. Singh,
"Decentralised State Reconstruction of Interconnected Discrete Systems", Large Scale Systems, vol. 2, 1981, 151-158.

[23] Ben-Israel, A. and T.N.E. Greville
"Generalised Inverses: Theory and Applications", Wiley, N.Y., 1974.

[24] Mahmoud, M.S. and M.G. Singh
"Large Scale Systems Modelling", Pergamon Press, Oxford, 1981, Ch. 6

[25] Mahmoud, M.S.
"Structural Properties of Discrete Systems With Slow and Fast Modes", Large Scale Systems, vol. 3, 1982, 227-236.

[26] Wilson, D.A.
"Model Reduction for Multivariable Systems", Int. J. Control, vol. 20, 1974, 57-64.

Chapter 6
State and Parameter Estimation

6.1 Introduction

The purpose of this chapter is to study the behaviour of discrete-time dynamical systems under the influence of external effects which can be described in a statistical way. It can be argued that all real systems operate in a stochastic environment where they are subject to noise (unknown disturbances) and, in addition, the controller has to rely, in practice, on imperfect measurements. The noise may arise due to unpredictable changes at the input end of the system, and/or due to inaccurate measurements at the output end. In either case, exact information about the state of the system is not available, and we should therefore seek methods to *estimate* the state of the system on the basis of statistically related data. This leads to the *state estimation problem*. In other applications, the coefficients of the models need to be determined on the basis of the input and output records which are corrupted by noise components. This defines the *parameter estimation problem*. Both these problems are examined in this chapter and techniques for their solutions are developed.

6.2 Random Variables and Gauss-Markow Processes

Before we begin our study of state estimation, we will introduce some fundamental notions of probability theory, random variables and Gaussian processes. The material covered will be sufficient for our later development of estimation techniques and the reader, interested in more advanced topics in estimation theory is referred to *[1-5]*.

6.2.1 *BASIC CONCEPTS OF PROBABILITY THEORY*

We start by considering an experiment with a number of possible outcomes, examples of which are the throwing of a dice, the drawing of a card from a card deck or the picking of a coloured ball from a basket of balls. Three fundamental concepts are introduced. The first is that of *sample space* Ω, the set of possible outcomes of an experiment. We will call any particular member of this set ω. As an example, for the case of throwing of a dice $\Omega = \{1,2,3,4,5,6\}$.

The second fundamental concept is that of an *event*, which is defined as any subset of the sample space. For example, "obtain the number 5" or "obtain a red ball" are events.

The third fundamental concept is that of *probability*. A probability measure $P(\cdot)$ is a mapping from events into the real line satisfying the following axioms:

1. $P(A) \geq 0$
2. $P(\Omega) = 1$
3. For a countable set $\{A_j\}$ of mutually disjoint events, that is $A_j \cap A_m = \phi$ for all j,m, $P(\cup A_j) = \sum_j P(A_j)$. Here ϕ denotes the empty set.

Some important formulae which arise from these axioms are:

(i) $P(\phi) = 0$

(ii) $P(A) \leq 1$

(iii) $P(A^*) = 1 - P(A)$; A^* is the complement of A.
Since $A^* \cap A = \phi$ and $A^* \cup A = \Omega$.

(iv) $P(A \cap B^*) = P(A) - P(A \cap B)$.
This is true in view of the fact that the events $(A \cap B^*)$ and $(A \cap B)$ are mutually disjoint and their union is A.

If we write the event $(A \cup B)$ as the union of two mutually exclusive event, we get :

$$P(A \cup B) = P(A) + P(A^* \cap B)$$

In view of the previous result, (iv), we have

$$P(A^* \cap B) = P(B) - P(A \cap B)$$

Combining the above two relations yields:

(v) $$P(A \cup B) = P(A) + P(B) - P(A \cap B)$$

We note that result (v) reduces to the third axiom when A and B are mutually disjoint.

Suppose A and B are two events and an experiment is conducted with the result that event B occurs. The probability that event A has also occurred, *the conditional probability of A given B*, written as $P(A|B)$ is given by:

$$P(A|B) = P(AB)/P(B) \tag{6.1}$$

where $P(AB) = P(A \cap B)$ is the joint probability of A and B.

We point out that

(a) $P(B) \neq 0$ in (6.1), otherwise the definition of $P(A|B)$ would be meaningless.

(b) $P(A|B)$ for fixed B and variable A satisfies the probability measure axioms.

We now consider the notion of independence. Events $A_1, A_2, \ldots, A_n$ are *mutually independent* if and only if

$$P(A_{j_1} \cap A_{j_2} \cap \ldots \cap A_{j_m}) = P(A_{j_1})P(A_{j_2})\ldots(P(A_{j_m}) \tag{6.2}$$

for all integers $j_1, j_2, \ldots, j_m$ selected from the set of integers $[1,2,\ldots,n]$ where no two are the same. We caution the reader to distinguish between the notions of independence and of mutually disjoint events.

In the case of two independent events A and B, then (6.2) becomes:

$$P(AB) = P(A)P(B) \tag{6.3}$$

which, when used in (6.1), yields:

$$P(A|B) = P(A)$$

This result agrees with out intuitive idea of independence and conditional probability in that, since B and A are independent, we do not need to know B to arrive at the probability $P(A|B)$. Consider the situation of three events A, B, C such that each pair is mutually independent, that is

$$P(AB) = P(A)P(B)$$

$$P(BC) = P(B)P(C)$$

$$P(CA) = P(C)P(A)$$

It is easy to show that these conditions do not imply that A, B, C are mutually independent.

We say that two events A and B are *conditionally independent given an event C* when

$$P(AB|C) = P(A|C)P(B|C) \tag{6.4}$$

If A_j, $j = 1,2,\dots,n$ are mutually disjoint and $\cup A_j = \Omega$, then for arbitrary B we have

$$P(B) = \sum_j P(B|A_j)P(A_j) \tag{6.5}$$

An important consequence of (6.1) is *Bayes' Rule:*

$$P(A|B) = P(B|A)P(A)/P(B) \qquad (6.6)$$

provided that $P(B) \neq 0$. Again consider n mutually disjoint events A_j with $\bigcup_j A_j = \Omega$. By virtue of (6.5) and (6.6) we have:

$$P(A_j|B) = P(B|A_j)P(A_j)/\{\sum_j P(B|A_j)P(A_j)\} \qquad (6.7)$$

We now proceed to consider random variables and examine their mathematical properties.

6.2.2 *MATHEMATICAL PROPERTIES OF RANDOM VARIABLES*

It is often desirable to have a procedure by which one can evaluate the output records of an experiment. A suitable way would be to measure quantities associated with the outcome of an experiment. Such a quantity is called *a random variable*. Strictly speaking, a random variable X is a real valued function from the outcome ω in a sample space Ω to the set of real numbers. A value of the random variable X is the number $X(\omega)$ when the outcome ω occurs. When X takes on a discrete value, it is called a discrete random variable.

Since by definition a random variable is a function on a probability space, it is often of interest to be able to know the probability that a certain value of the random variable $X(.)$ occurs in a given set. We adopt the notation $P(X = \alpha)$ to mean $P([\omega|X(\omega) = \alpha])$, that is the probability of the subset of Ω consisting of those outcomes ω for which $X(\omega) = \alpha$. In a similar way, $P(X > 0)$ means $P([\omega|X(\omega) > 0])$. It is required for X to be a random variable that

1. $P(X = -\infty) = P(X \quad +\infty) = 0$
2. For all real β, the quantity $[\omega|X(\omega) \leq \beta]$ is an event which implies that

$$P([\omega|X(\omega) \leq \beta]) = P(X \leq \beta)$$

A. Distribution Functions

One way of describing random variables is in terms of their distribution functions. Given a random variable X, *the distribution function* $F(x)$ is a mapping from the reals to the interval $[0,1]$:

$$F(x) = P(X \leq x) \tag{6.8}$$

where the argument x is a typical value. The distribution function is monotonicly increasing in the sense that

$$\lim_{x\to\infty} F(x) = 1 \quad \text{and} \quad \lim_{x\to-\infty} F(x) = 0$$

Another way of describing random variables is in terms of their density functions. When $F(x)$ is continuous and differentiable everywhere, the *probability density function* $P(x)$ associated with the random variable X is

$$p(x) = dF(x)/dx \tag{6.9}$$

From (6.8) and (6.9), it is readily seen that $p(x)\,dx$ to first order is $P(x < X < x + dx)$.

A random vector $\underline{X}$ of order n consists of n random variables $X_1, X_2, \dots X_n$ with distribution and probability density functions defined by:

$$F(\underline{x}) = P[(X_1 \leq x_1) \cap \dots \cap (X_n \leq x_n)] \tag{6.10}$$

$$p(\underline{x}) = (\partial^n/\partial x_1 \dots \partial x_n) F(\underline{x}) \tag{6.11}$$

Consider the events $[X \leq x]$ and $[Y \leq y]$ associated with the random variables X and Y respectively. If these events are independent for all x and y, then it follows from (6.3) and (6.8) that the joint distribution function is

$$F(x,y) = F(x)F(y) \tag{6.12}$$

and correspondingly the joint probability density function is:

$$p(x,y) = p(x)p(y) \tag{6.13}$$

Let $h(.)$ be a well-behaved scalar valued function of a scalar variable and X a random variable. If X and Z are independent random variables, so are $h(X)$ and $g(Z)$.

B. Mathematical Expectations

We now move to define mathematical expectation. The *mathematical expectation* or *mean* of a random variable X, written as $E[X]$, is the number defined by:

$$E[X] = \int_{-\infty}^{+\infty} xp(x)dx \tag{6.14}$$

where the integral is assumed absolutely convergent. In the same way, a function $g(X)$ of the random variable X will have the mathematical expectation

$$E[g(X)] = \int_{-\infty}^{+\infty} g(x)p(x)dx \tag{6.15}$$

As an operator, the mathematical expectation has the following properties:

1. For a constant β, $E[\beta] = \beta$.
2. It is a linear operator. More precisely, if $g_1(X)$ and $g_2(X)$ are two functions of the random variable X and α and β are two constants, then

$$E[\alpha g_1(X) + \beta g_2(X)] = \alpha E[g_1(X)] + \beta E[g_2(X)]$$

3. If $X_1,\ldots,X_n$ denote mutually independent random variables, then

$$E[X_1 X_2 \ldots X_n] = E[X_1]E[X_2]\ldots E[X_n] \tag{6.16}$$

The *variance* σ^2 of a random variable provides a measure of the dispersion around the mean value and is defined by:

$$\sigma^2(X) = E[(X-E[X])^2] = \int_{-\infty}^{+\infty} (x-E[X])^2 p(x)dx \tag{6.17a}$$

An alternative form of (6.17a) is

$$\begin{aligned} \sigma^2(X) &= E[X^2-2E[X]X + (E[X])^2] \\ &= E[X^2]-2(E[X])^2 + (E[X])^2 \\ &= E[X^2]-(E[X])^2 \end{aligned} \tag{6.17b}$$

where we have made use of the properties of the expectation operator. Form (6.17b) is easy to remember. We note that the definition of the mean generalizes in an obvious way to a vector. Let $\underline{X} = [X_1 \; X_2 \; \dots \; X_n]^t$. thus:

$$E[\underline{X}] = [E[X_1] \; E[X_2] \; \dots \; E[X_n]]^t \tag{6.18a}$$

For random n-vector $\underline{X}$, the variance is now replaced by the (nxn) *covariance matrix* $\mathrm{Cov}(\underline{X})$ given by:

$$\mathrm{Cov}(\underline{X}) = E[(\underline{X}-E[\underline{X}])(\underline{X}-E[\underline{X}])^t]$$

$$= \begin{bmatrix} \sigma^2(X_1) & \mathrm{Cov}(X_1,X_2) & \dots & \mathrm{Cov}(X_1,X_n) \\ \mathrm{Cov}(X_1,X_2) & \sigma^2(X_2) & \dots & \cdot \\ \cdot & & & \cdot \\ \cdot & & & \cdot \\ \cdot & & & \cdot \\ \mathrm{Cov}(X_1,X_n) & & \dots & \sigma^2(X_n) \end{bmatrix} \tag{6.18b}$$

where the superscript t denotes matrix transpose. From (6.17b) and (6.18b) we note that the variance is always nonnegative, and the covariance matrix is nonnegative definite and symmetric.

C. *Two Random Variables*

For two random variables X and Y we summarize some important relations:

1. The *conditional probability density of X given Y* $p(x|y)$ is given by Bayes' Rule *[7]*,

$$p(x|y) = p(x,y)/p(y) \tag{6.19}$$

and from which one obtains the important formula

$$p(x) = \int_{-\infty}^{+\infty} p(x|y)p(y)\,dy \tag{6.20}$$

Also, if X and Y are independent then using (6.13) in (6.19) it reduces to:

$$p(x|y) = p(x) \tag{6.21}$$

2. By definition:

$$E[X] = \int_{-\infty}^{+\infty}\int_{-\infty}^{+\infty} x\, p(x,y)dxdy \tag{6.22}$$

$$E[Y] = \int_{-\infty}^{+\infty}\int_{-\infty}^{+\infty} y\, p(x,y)dxdy \tag{6.23}$$

$$E[X^2] = \int_{-\infty}^{+\infty}\int_{-\infty}^{+\infty} x^2 p(x,y)dxdy \tag{6.24}$$

$$E[Y^2] = \int_{-\infty}^{+\infty}\int_{-\infty}^{+\infty} y^2 p(x,y)dxdy \tag{6.25}$$

$$E[XY] = \int_{-\infty}^{+\infty}\int_{-\infty}^{+\infty} xy\, p(x,y)dxdy \tag{6.26}$$

and

$$\mathrm{Cov}(X,Y) = E[(X-E[X])(Y-E[Y])] \tag{6.27}$$

The quantity $E[XY]$, defined by (6.26), is often called

the correlation of X *and* Y. As a consequence, we define the *coefficient of correlation* between X and Y by:

$$\rho(X,Y) = \mathrm{Cov}(X,Y)/\sqrt{\sigma^2(X)}\ \sqrt{\sigma^2(Y)} \qquad (6.28)$$

provided that the variances of X and Y are finite and strictly positive.

3. The random variables X and Y are said to be *uncorrelated* if $E[X^2]$ and $E[Y^2]$, as defined by (6.24) and (6.25) respectively, are finite and

$$\mathrm{Cov}(X,Y) = 0 \qquad (6.29a)$$

From (6.28) this implies that

$$\rho(X,Y) = 0 \qquad (6.29b)$$

4. Suppose X and Y are two independent random variables. Then it is easy to show that they are uncorrelated. Starting from (6.27), expanding and using (6.16) we get:

$$\begin{aligned}\mathrm{Cov}(X,Y) &= E[XY]-E[X]E[Y]-E[X]E[Y]+E[X]E[Y]\\ &= E[X]E[Y] - E[X]E[Y]\\ &= 0\end{aligned}$$

which agrees with (6.29a). Therefore, an alternative way to define two uncorrelated random variables is when $E[XY] = E[X]E[Y]$. If $E[XY] = 0$, the random variables X and Y are termed *orthogonal*.

We caution the reader that two uncorrelated random variables need not necessarily be independent. The absence of correlation implies that the general condition

$$E[h(X)g(Y)] = E[h(X)]E[g(Y)]$$

is only satisfied for $h(X) = X$, whilst independence

requires that this condition be satisfied for all functions $h(\cdot)$ and $g(\cdot)$.

The conditionally expected value of a random variable X, given that Y has taken the value y, is

$$E[X|Y = y] = E[X|Y] = \int_{-\infty}^{+\infty} x\ p(x|y)dx \tag{6.30}$$

Note that the result of integration will be a number, depending on y. But since y is the outcome of a random experiment, the conditional expectation is a random variable. To calculate its expected value, we proceed as follows: from (6.14) and (6.30) we get:

$$E[E[X|Y]] = \int_{-\infty}^{+\infty} p(y) \int_{-\infty}^{+\infty} x\ p(x|y)dxdy$$
$$= \int_{-\infty}^{+\infty} x\{ \int_{-\infty}^{+\infty} p(x|y)p(y)dy\}dx$$

using (6.20) it simplifies to

$$= \int_{-\infty}^{+\infty} x\ p(x)dx$$
$$= E[X] \tag{6.31}$$

If the random variables X and Y are independent, then it follows from (6.21) and (6.30) that

$$E[X|Y] = E[X] \tag{6.32a}$$

and more generally

$$E[h(X)|Y) = E[h(X)] \tag{6.32b}$$

for any function $h(\cdot)$. We can generalize (6.30) by using $g(X,Y)$ in place of X to obtain

$$E[g(X,Y)|Y] = \int_{-\infty}^{+\infty} g(x,y)p(x,y)dx \tag{6.33}$$

and again the result is a random variable which is a function of the random variable Y. To emphasize this point, let $g(X,Y) = g_1(X)g_2(Y)$, then (6.33) becomes:

$$E[g(X,Y)|Y] = g_2(Y)E[g_1(X)|Y] \tag{6.34}$$

which represents a useful formula.

Since the various notions of random variables can be easily extended from random scalars to random vectors, with the notation (6.18) in mind, we next go on to consider Gaussian random vectors since most of our analysis for state and parameter estimation will assume that the probability distributions are Gaussian.

6.2.3 *STOCHASTIC PROCESSES*

A. Definition and Properties

Hitherto, most of our discussions have been centered around an experiment with a number (or an n-tuple of numbers) of possible outcomes and the time factor has been set aside. In this section we extend the previous analysis to the case where the outcome is a function mapping an underlying time set (commonly nonnegative integers) into the reals. Thus, we will deal with a random process rather than a random variable. More precisely, *a discrete-time random process* results in a function mapping from $\omega \in \Omega$ to a set of values $x_\omega(k)$ for $k \in I_t \triangleq \{0,1,2,...\}$, the discrete-time set. Looked at in this light, a scalar discrete-time random process behaves like an infinite-dimensional random vector. We adopt the notation $\{\underline{x}(k)\}$ to denote $\{(\underline{x}_\omega(k),k)|k \geq 0, \omega \in \Omega\}$, that is a particular sequence of vectors taken as a result of an experiment. The quantity $\underline{x}(m)$ will then denote the random vector obtained by looking at the process at time m, as well as the value taken by that vector.

From the above discussion, it is readily seen that a random

process is just a generalization of the concept of a random variable. Hence, most of the properties presented in Section 6.2.1 will carry over here. As an example, let m be an arbitrary integer and $(k_1, k_2, \ldots, k_m)$ be arbitrary instants in the underlying time set I_t. Then the set of all probability densities

$$p\{x(k_1), x(k_2), \ldots, x(k_m)\}$$

or the corresponding distribution functions can serve to define the probability structure of the random process. In what follows, we will provide some fundamental properties of random processes.

The *mean* $\underline{m}(k)$ of a random process is simply the time function $E[\underline{x}(k)]$. Given two discrete-time instants j and $r \in I_t$ and let

$$\underline{x}(\cdot) \triangleq [x_1(\cdot) \; x_2(\cdot) \; \ldots \; x_n(\cdot)]^t$$

then the *autocorrelation matrix* $R(j,r)$ is the set of quantities $E[\underline{x}(j)\underline{x}(r)^t]$, written in full as

$$R(j,r) = \begin{bmatrix} E[x_1(j)x_1(r)] & E[x_1(j)x_2(r)] \ldots & E[x_1(j)x_n(r)] \\ E[x_2(j)x_1(r)] & E[x_2(j)x_2(r)] \ldots & \cdot \\ \cdot & \cdot & \cdot \\ \cdot & \cdot & \cdot \\ \cdot & \cdot & \cdot \\ E[x_n(j)x_1(r)] & \ldots\ldots\ldots & E[x_n(j)x_n(r)] \end{bmatrix} \tag{6.35}$$

In a similar way, the *covariance matrix* $W(j,r)$ is the set of quantities $E[\{\underline{x}(j)-\underline{m}(j)\}\{\underline{x}(r)-\underline{m}(r)\}^t]$ for all j and r. Its full description takes a form similar to (6.35) with appropriate changes. When $j = r$, the covariance matrix $W(j,j)$ becomes a nonnegative definite symmetric matrix. Thus, we see that a random process is entirely characterised by the properties of the random variable (or vector) at different discrete-time instants.

The *first order densities* of a process are the set of densities $p\{\underline{x}(j)\}$ for all $j \in I_t$. The *second order densities* of a process are the set $p\{\underline{x}(j),\underline{x}(r)\}$ for all $j,r \in I_t$. Given these densities, we can apply the rules of the previous section like (6.14), (6.17a), to obtain the mean and variance of a process.

Define $[j_m, j_{m+1}]$ as a set of nonintersecting intervals in the discrete-time set I_t. Then a process is said to have *uncorrelated (orthogonal or independent)* increments if the quantity $[\underline{x}(j_m) - \underline{x}(j_{m+1})]$ is a sequence of *uncorrelated (orthogonal or independent)* random vectors.

A process $\{\underline{x}(k)\}$ is said to be *strict-sense stationary,* or simply *stationary* if its associated probability densities are unaffected by time translation; that is, for arbitrary integer m and discrete times $j_1,\ldots j_m$ and s,

$$p\{\underline{x}(j_1),\underline{x}(j_2),\ldots,\underline{x}(j_m)\} = p\{\underline{x}(j_1+s),\underline{x}(j_2+s)\ \ldots,\underline{x}(j_m+s)\} \tag{6.36}$$

If we consider two first order densities, we have

$$p\{\underline{x}(j)\} = p\{\underline{x}(j+s)\} \tag{6.37}$$

which implies that the first order probability density is, in this case, independent of j. Consequently, the mean $\underline{m}(j)$ of the process $\{\underline{x}(j)\}$ is a constant,

$$E[\underline{x}(j)] = \underline{m}(j) = \underline{m} \tag{6.38}$$

For the second order density we have

$$R(j,r) = E[\{\underline{x}(j)-\underline{m}\}\{\underline{x}(r)-\underline{m}\}^t]$$
$$= R(j-r) \tag{6.39}$$

that is the autocorrelation function depends only on the difference $(j-r)$.

We now move a step forward and consider pairs of random processes. In view of the above discussions, we summarize the important properties:

1. Two random processes $\{\underline{x}(k)\}$ and $\{\underline{y}(k)\}$ are said to be *uncorrelated* if

$$E[\underline{x}(j)\underline{y}^t(r)] = E[\underline{x}(j)]E[\underline{y}^t(r)] \tag{6.40}$$

   for all $j,r \in I_t$.

2. Two random processes $\{\underline{x}(k)\}$ and $\{\underline{y}(k)\}$ are said to be *orthogonal* if

$$E[\underline{x}(j)\underline{y}^t(r)] = [0] \tag{6.41}$$

   for all $j,r \in I_t$.

3. Two random processes $\{\underline{x}(k)\}$ and $\{\underline{y}(k)\}$ are said to be *independent* if for any sets $\{j_i\}$ and $\{r_i\}$ the vector random variable $[\underline{x}^t(j_1)\ \underline{x}^t(j_2)\ \dots\ \underline{x}^t(j_n)]^t$ is *independent* of the vector random variable $[\underline{y}^t(r_1)\ \underline{y}^t(r_2)\dots\underline{y}^t(r_m)]^t$.

4. Two random processes $\{\underline{x}(k)\}$ and $\{\underline{y}(k)\}$ are *jointly stationary* if the combined process $\{[\underline{x}^t(k) y^t(k)]^t\}$ is *stationary*.

B. *Gauss and Markov Processes*

Having presented the description and mathematical properties of a random process, we now direct attention to a particular class of stochastic processes called *Markov processes*. Consider a set of ordered parameters $j_0 < j_1 < j_2 < \dots < j_n$. A stochastic process $\{\underline{x}(j)\}$ is called a *Markov process* if we can write

$$p\{\underline{x}(j_n)|\underline{x}(j_{n-1}),\ldots,\underline{x}(j_0)\} = p\{\underline{x}(j_n)|\underline{x}(j_{n-1})\} \quad (6.41)$$

which means that the entire past history of the process is contained in the last state.

We now develop an expression for the joint probability density function for a Markov process. Using Bayes' theorem, similarly to (6.19), we have

$$p\{\underline{x}(j_n),x(j_{n-1}),\ldots,\underline{x}(j_0)\} = p\{\underline{x}(j_n)|\underline{x}(j_{n-1}),\ldots,\underline{x}(j_0)\} \times p\{\underline{x}(j_{n-1}),\ldots,\underline{x}(j_0)\} \quad (6.42)$$

If the process is Markovian, then from (6.41) and (6.42) we get

$$p\{\underline{x}(j_n),\underline{x}(j_{n-1}),\ldots,\underline{x}(j_0)\} = p\{\underline{x}(j_n)|\underline{x}(j_{n-1})\}p\{\underline{x}(j_{n-1}),\ldots,\underline{x}(j_0)\} \quad (6.43)$$

Doing the same operation on $p\ \underline{x}(j_{n-1}),\ldots\underline{x}(j_0)$ and repeating, we finally obtain

$$p\{\underline{x}(j_n),\underline{x}(j_{n-1}),\ldots,\underline{x}(j_0)\}= p\{\underline{x}(j_n)|\underline{x}(j_{n-1})\}p\{\underline{x}(j_{n-1})|\underline{x}(j_{n-2})\} \ldots p\{\underline{x}(j_1)|\underline{x}(j_0)\}p\{\underline{x}(j_0)\} \quad (6.44)$$

This means that we can describe completely a Markov process in terms of its *transition probability densities* $p\{\underline{x}(j_m)|\underline{x}(j_{m-1})\}$ and the distribution of the initial state.

Another important class of stochastic processes is *white noise*. Recall that a stationary discrete-time stochastic process $\{\underline{x}(k)\}$ with zero mean is one whose autocorrelation function is

$$R(s) = E[\{\underline{x}(j+s)\}\{\underline{x}(j)\}^t] \quad (6.45)$$

The *power spectrum* of this random process is given by

$$\Phi(z) = \sum_{k=-\infty}^{+\infty} z^{-k} R(k) \quad (6.46)$$

that is, the power spectrum is the $\mathcal{Z}$-transform of the autocorrelation function of a stationary random process. In the important special case, where $\Phi(z)$ is constant, we obtain white noise. Note the analogy with white light which contains all frequencies in its spectrum. Constancy of the power spectrum is equivalent to:

$$E[\underline{x}(j)\underline{x}^t(r)] = C\delta_{jr} \tag{6.47}$$

for some constant matrix C. The discrete-time (Kronecker) delta function δ_{jr} is 0 for $j \neq r$ and 1 for $j = r$. In this case, the vectors $\underline{x}(j_n)$, $\underline{x}(j_{n-1}),\ldots,\underline{x}(j_0)$ constitute a sequence of uncorrelated random variables.

Next, we consider Gaussian random variables, vectors and processes. A random variable X is a *Gaussian random variable* if its probability density is of the form

$$p(x) = (1/\sqrt{2\pi\sigma^2})\exp[-(x-m)^2/2\sigma^2] \tag{6.48}$$

Simple evaluation of $E[X]$ in (6.14) using (6.48) shows that

$$E[X] = m \tag{6.49}$$

and the variance $E[(X-E[X])^2]$ is σ^2.

Sometimes the notation "X is $N(m,\sigma^2)$" is used to denote that X is Gaussian (normal) with mean m and variance σ^2.

Let $\underline{X}$ be a random n-vector. If its covariance matrix $\mathrm{Cov}(\underline{X})$ in (6.18b) is nonsingular then we say that $\underline{X}$ is *Gaussian* if and only if its probability density is of the form

$$p(\underline{x}) = (1/[2\pi]^{n/2})(1/\det[W]^{\frac{1}{2}})\exp[-\tfrac{1}{2}(\underline{x}-\underline{m})^t W^{-1}(\underline{x}-\underline{m})] \tag{6.50}$$

for some vector m and matrix W. By similarity to the scalar case, we can easily show that

$$E[\underline{X}] = \underline{m} \qquad (6.51a)$$

and

$$E[(\underline{X}-\underline{m})(\underline{X}-\underline{m})^t] = W \qquad (6.51b)$$

Likewise, we write "$\underline{X}$ is $N(\underline{m},W)$" to denote that $\underline{X}$ is Gaussian with mean $\underline{m}$ and covariance W.

From these definitions we see why Gaussian distributions are so attractive. The Gaussian probability density function can be described uniquely in terms of the two quantities: mean and covariance so that we do not have to worry about higher order moments. This is fortunate, since many physical phenomena can be described in terms of Gaussian distributions.

Due to the importance of Gaussian random variables in our work on estimation, we provide below some of its basic properties:

1. Given a Gaussian random vector $\underline{X} = [X_1 \ \ldots \ X_n]^t$, any vector formed by some of its components is also Gaussian, that is $\underline{Y} = [X_j \ \ldots \ X_m]^t$ is Gaussian for all $j \neq m \leq n$.
2. When the random variables $X_1,\ldots,X_n$, which comprise a Gaussian random vector $\underline{X}$, are uncorrelated then they are independent. This is easy to see, since in this case the covariance matrix W and its inverse are diagonal implying that

$$p(\underline{x}) = p(x_1)p(x_2)\ldots p(x_n)$$

3. Gaussian random vectors retain their Gaussian character under linear transformation. To demonstrate this fact, let $\underline{X}$ be $N(\underline{m},W)$ and consider

$$\underline{Y} = A\underline{X} + \underline{b} \qquad (6.52)$$

where A is a constant matrix and $\underline{b}$ is a vector. Then by simple evaluation we can show that

$$E[\underline{Y}] = A\,\underline{m} + \underline{b} \qquad (6.53a)$$

and

$$\mathrm{Cov}[\underline{Y}] = AWA^t \tag{6.53b}$$

that is $\underline{Y}$ is $N(A\underline{m} + \underline{b},\ AWA^t)$.

Finally, we conclude this section by defining Gaussian processes. Let $j_1, \dots j_n$ be any selection of points in the discrete-time set I_t. A random process is a *Gaussian random process* if the random variables $x(j_1), x(j_2), \dots, x(j_n)$ are jointly Gaussian, that is

$$p\{x(j_1), \dots, x(j_n)\} = (1/[2\pi]^{n/2})(1/\det[W]^{\frac{1}{2}})\exp[-\tfrac{1}{2}(\underline{x}-\underline{m})^t W^{-1}(\underline{x}-\underline{m})] \tag{6.54a}$$

where
$$\left.\begin{aligned}
\underline{x} &= [x(j_1)\ x(j_2)\ \dots\ x(j_n)]^t \\
m_i &= E[x(j_i)] \\
\underline{m} &= [m_1\ m_2\ \dots\ m_n]^t \\
w_{ir} &= E[\{x(j_i) - m_i\}\{x(j_r) - m_r\}] \\
W &= [w_{ir}]
\end{aligned}\right\} \tag{6.54b}$$

A complete probabilistic description of the process is thus provided by $E[x(j_i)]$ and $\mathrm{Cov}[x(j_i), x(j_r)]$ for all j_i and j_r.

As indicated before, the material provided so far gives the basic background on what is needed in probability theory for the understanding of state and parameter estimation. However, some other ideas are left to be dealt with in the next sections.

6.3 Linear Discrete Models with Random Inputs

We shall restrict attention in this chapter to linear discrete-time dynamical systems with random inputs. Examples of such systems are found in the fields of economics and industrial control. In economic studies, certain statistics (or economic indices) may be compiled monthly or quarterly whilst control strategies (like budget controls) may be applied yearly.

6.3.1 *MODEL DESCRIPTION*

We shall consider linear discrete-time dynamical models described by

$$\underline{x}(k+1) = A(k)\underline{x}(k) + G(k)\underline{w}(k) \tag{6.55}$$

$$y(k) = H(k)\underline{x}(k) \tag{6.56}$$

$$\underline{z}(k) = \underline{y}(k) + \underline{v}(k) \tag{6.57}$$

where the integer k denotes the discrete-time instant, and

$\underline{x}(k)$ is the system state
$\underline{y}(k)$ is the system output when the output disturbance is absent
$\underline{z}(k)$ is the measured system output
$\underline{w}(k)$ is the random input to the system
$\underline{v}(k)$ is the output disturbance.

Fig. 6.1 gives a block diagram of the system. Note that the system model (6.55)-(6.57) holds for $k \geq 0$. As mentioned in Section 6.2.2, the symbol $\{\underline{x}(k)\}$ will be used to denote the set $\{(\underline{x}(k),k) | k \geq 0\}$. Thus, we have $\{\underline{x}(k)\}$, $\{\underline{y}(k)\}$, $\{\underline{z}(k)\}$, $\{\underline{w}(k)\}$ and $\{\underline{v}(k)\}$ representing the system process, output process, measurement process, input noise process and output noise process, respectively.

We begin our study of linear discrete-time system (6.55)-(6.57) by making certain assumptions. These are:

1. The processes $\{\underline{v}(k)\}$ and $\{\underline{w}(k)\}$ are each white noise processes.
   This means that the random vectors $\underline{v}(j)$, $\underline{v}(r)$ are independent for any j and r with $j \neq r$. Similarly, $\underline{w}(j)$, $\underline{w}(r)$ are independent random vectors for any j and r with $j \neq r$.

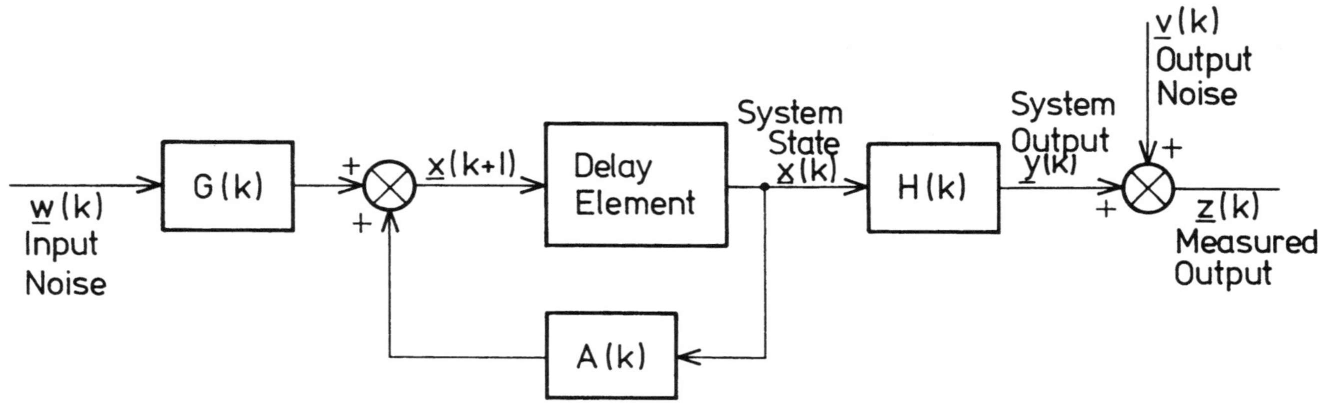

Fig. (6.1) Linear discrete-time model with random disturbances

2. The processes $\{\underline{v}(k)\}$ and $\{\underline{w}(k)\}$ are individually zero mean, Gaussian random processes with known covariances. This implies that

$$E[\underline{v}(k)] = \underline{0} \tag{6.58a}$$

$$E[\underline{w}(k)] = \underline{0} \tag{6.58b}$$

$$E[\underline{v}(k)\ \underline{v}^t(j)] = R(k)\delta_{kj} \tag{6.59a}$$

$$E[\underline{w}(k)\ \underline{w}^t(j)] = Q(k)\delta_{kj} \tag{6.59b}$$

with $R(k)$ and $Q(k)$ being nonnegative definite matrices for all k.

3. The processes $\{\underline{v}(k)\}$ and $\{\underline{w}(k)\}$ are independent processes.
In view of the above assumption and the zero mean assumption, we have

$$E[\underline{v}(k)\ \underline{w}^t(j)] = 0 \tag{6.60}$$

for all k and j.

4. The initial state $\underline{x}(0)$ is a Gaussian random vector with a known mean $\underline{m}_0$ and known covariance W_0, that is

$$E[\underline{x}(0)] = \underline{m}_0 \tag{6.61a}$$

$$E[(\underline{x}(0)-\underline{m}_0)(\underline{x}(0)-\underline{m}_0)^t] = W_0 \tag{6.61b}$$

5. The noise processes $\{\underline{v}(k)\}$ and $\{\underline{w}(k)\}$ are independent of $\underline{x}(0)$, that is

$$E[\underline{x}(0)\ \underline{v}^t(k)] = 0 \quad \text{for all } k \tag{6.62a}$$

$$E[\underline{x}(0)\ \underline{w}^t(k)] = 0 \quad \text{for all } k \tag{6.62b}$$

For convenience, we can sum up the above assumptions as follows

The noise processes $\{\underline{v}(k)\}$ and $\{\underline{w}(k)\}$ are zero mean, independent Gaussian processes with covariances given by (6.59a) and (6.59b). The initial state is $N(\underline{m}_0, W_0)$

which is independent of the processes $\{\underline{v}(k)\}$ *and* $\{\underline{w}(k)\}$.

The above assumptions are justified in many practical cases:

(i) The white noise assumption enables us to represent disturbances having 'short' correlation times compared to the system.

(ii) Since it is impossible to measure $\underline{x}(k)$ exactly for arbitrary k, it is unlikely the $\underline{x}(0)$ will be available. This leads to the adoption of a random initial condition for the system which is not related to the disturbances acting on the system.

(iii) The Gaussian assumption on the random processes gives complete information about the statistical properties of the noise processes. Also, experiments have established that many naturally occurring processes are Gaussian. Moreover, the central limit theorem of probability theory *[1-5]* suggests an approximately Gaussian character for the sum of a number of individual, possibly nonGaussian processes.

6.3.2 *SOME USEFUL PROPERTIES*

With the standard assumptions (1) through (5) of the previous section, we now provide some important properties of the random process $\{\underline{x}(k)\}$ of the system (6.55)-(6.57).

The first property is that *$\underline{x}(k)$ is a Gaussian random vector*. To show this we use (6.55) iteratively to yield (see Chapter 2 for more details):

$$\underline{x}(k) = \Psi(k,0)\,\underline{x}(0) + \sum_{j=0}^{k-1} \Psi(k,j+1)\,G(j)\,\underline{w}(j) \qquad (6.63)$$

where the transition matrix $\Psi(s,r)$,

$$\left.\begin{aligned} \Psi(s,r) &= A(s-1)A(s-2)\ldots A(r);\ s > r,\quad \Psi(s,s) = I \\ \Psi(s,r)\Psi(r,m) &= \Psi(s,m) \quad \text{for all } s,\ r \text{ and } m \\ &\qquad\qquad \text{with } s \geq r \geq m \end{aligned}\right\} \qquad (6.64)$$

Since the random vectors $\underline{x}(0)$, $\underline{w}(0)$, $\underline{w}(1),\ldots,\underline{w}(k-1)$ are individually Gaussian and independent, then they are jointly Gaussian. By virtue of the fact that linear transformations of Gaussian random vectors preserve their Gaussian character, it follows that $\underline{x}(k)$ is a Gaussian random vector.

Recall the property that a random vector derived from the entries of a Gaussian random vector is also Gaussian. Then, for arbitrary s and j_i, $i = 1,\ldots,s$, the set of random vectors $\underline{x}(j_i)$ is jointly Gaussian. This leads to the second property: *$\{\underline{x}(k)\}$ is a Gaussian random process.*

A little reflection on (6.63) will show that $\underline{x}(m)$ depends on the first m values of the noise process $\{\underline{w}(k)\}$ and on the initial state $\underline{x}(0)$. Therefore, $\underline{x}(m)$ is independent of $\underline{w}(m)$ and we can show in the same way that $\underline{x}(m-1),\ldots,\underline{x}(0)$ are all independent of $\underline{w}(m)$. We could thus write:

$$p\{\underline{x}(k)\,|\,\underline{x}(k-1),\ldots,\underline{x}(0)\} \;=\; p\{\underline{x}(k)\,|\,\underline{x}(k-1)\} \qquad (6.65)$$

Because of (6.41) and (6.65), the process $\{\underline{x}(k)\}$ is *Markovian*. To summarize, we say that the state of the system (6.55) is a *Gauss-Markov process.*

For essentially the same reasons, one can show that the measurement process $\{\underline{z}(k)\}$ is *a Gaussian process.* In view of the assumptions made on the noise processes, we see that *$\{x(k)\}$ and $\{\underline{z}(k)\}$ are jointly Gaussian processes.* But, since the output process $\{\underline{y}(k)\}$ is not white and the random variables $\underline{y}(s)$ and $\underline{y}(r)$ for $|s-r| > 1$ may be correlated, we conclude that the process $\{\underline{z}(k)\}$ is *a non-Markov process.* To elaborate more, it may be possible that $\underline{y}(m-2)$ and $\underline{y}(m-1)$ jointly convey more information about $\underline{y}(m)$ than $\underline{y}(m-1)$ alone. Hence, the past history is not necessarily included in the last value and one may need more records to recover the present.

6.3.3 *PROPAGATION OF MEANS AND COVARIANCES*

Since the processes $\{\underline{x}(k)\}$ and $\{\underline{z}(k)\}$ are jointly Gaussian, their probabilistic properties are entirely determined by their means and covariances. Our purpose now is to determine the behaviour of these means and covariances along the discrete-time set. We start by evaluating the evolution of the means.

From (6.63), the linearity of the expectation operator, and making use of (6.58a), (6.61a), we obtain:

$$E[\underline{x}(k)] = \Psi(k,0)\underline{m}_0 \tag{6.66a}$$

Equivalently,

$$\begin{aligned} E[\underline{x}(k+1)] &= \underline{m}(k+1) \\ &= A(k)E[\underline{x}(k)] \\ &= A(k)\underline{m}(k) \end{aligned} \tag{6.66b}$$

Similarly, from (6.57) we have:

$$\begin{aligned} E[\underline{z}(k)] &= H(k)E[\underline{x}(k)] \\ &= H(k)\underline{m}(k) \end{aligned} \tag{6.66c}$$

Note that (6.66b) is a recursive relationship which enables us to compute the mean at each discrete-instant.

Next consider the covariance functions. The covariance matrix at the (k+1)th instant is defined by:

$$W(k+1) = E[\{\underline{x}(k+1) - \underline{m}(k+1)\}\{\underline{x}(k+1)-\underline{m}(k+1)\}^t] \tag{6.67}$$

To calculate the inner term in (6.67) we combine (6.55) and (6.66b) to obtain:

$$\underline{x}(k+1)-\underline{m}(k+1) = A(k)[\underline{x}(k)-\underline{m}(k)] + G(k)\underline{w}(k)$$

Substituting the above expression into (6.67), it becomes:

$$W(k+1) = E[\{A(k)[\underline{x}(k)-\underline{m}(k)]+G(k)\underline{w}(k)\}\{[\underline{x}(k)-\underline{m}(k)]^tA^t(k) + \underline{w}^t(k)G^t(k)\}] \quad (6.68)$$

We have already seen in Section 6.3.2 that the random sequences $\underline{x}(k)$ and $\underline{w}(k)$ are independent so that

$$E[\underline{x}(k)\underline{w}^t(k)] = E[\underline{w}(k)\underline{x}^t(k)] = 0$$

In addition

$$E[\underline{m}(k)\underline{w}^t(k)] = \underline{m}(k)E[\underline{w}^t(k)] = 0$$

and

$$E[\underline{w}(k)\underline{m}^t(k)] = 0$$

since the noise is assumed to have a zero mean, (see (6.58b)). The use of the above expressions in (6.68) with some manipulations, reduces it to

$$W(k+1) = A(k)W(k)A^t(k) + G(k)Q(k)G^t(k) \quad (6.69)$$

where we have used (6.59b) and the linearity property of the expectation operator. This constitutes a difference equation for $W(k)$, allowing computation of this matrix recursively, starting with the known matrix W_0. Given that the process is Gaussian, the two quantities $\underline{m}(k+1)$ and $W(k+1)$, defined by (6.66b) and (6.69), are sufficient to determine the probabilistic structure of the state $\underline{x}(k+1)$.

We can equally determine a general expression for $W(k)$. To accomplish this we start from the definition:

$$W(k,j) = E[\{\underline{x}(k)-\underline{m}(k)\}\{\underline{x}(j)-\underline{m}(j)\}^t] \quad \text{for } k \geq j \quad (6.70a)$$

Following a parallel development, it can be shown that

$$W(k,j) = \Psi(k,j)\{\Psi(j,0)W_0\Psi^t(j,0) + \sum_{m=0}^{j-1} \Psi(j,m+1)G(m)Q(m)G^t(m)\Psi^t(j,m+1)\} \quad (6.70b)$$

where $\Psi(s,r)$ is defined in (6.64). It should be noted that (6.66a) and (6.70b) together provide an alternative form for all the probabilistic information there is to know about the Gaussian process $\{\underline{x}(k)\}$. Specializing (6.70b) to the case $k = j$, we arrive at

$$\begin{aligned} W(k,k) &= W(k) \\ &= \Psi(k,0)W_0\Psi^t(k,0) \\ &\quad + \sum_{m=0}^{k-1} \Psi(k,m+1)G(m)Q(m)G^t(m)\Psi^t(k,m+1) \end{aligned} \tag{6.71}$$

Since (6.70b) is valid for $k \geq j$, then the comparison of (6.70b) and (6.71) reveals that

$$\begin{aligned} W(k,j) &= \Psi(k,j)W(j,j) \\ &= \Psi(k,j)W(j) \qquad k \geq j \end{aligned} \tag{6.72a}$$

To obtain the corresponding expression when $k \leq j$, we note from (6.70a) that $W(k,j) = W^t(j,k)$; therefore,

$$W(k,j) = W(j)\Psi^t(j,k) \qquad k \leq j \tag{6.72b}$$

Turning to the measurement process $\{\underline{z}(k)\}$ for which the mean has already been defined by (6.66c). The covariance essentially follows from that for the process $\{\underline{x}(k)\}$. Using (6.56), (6.57) and (6.66c), it follows that

$$\begin{aligned} \mathrm{Cov}[\underline{z}(k),\underline{z}(j)] &= E[\{\underline{z}(k)-H(k)\underline{m}(k)\}\{\underline{z}(j)-H(j)\underline{m}(j)\}^t] \\ &= E[H(k)\{\underline{x}(k)-\underline{m}(k)\}\{\underline{x}(j)-\underline{m}(j)\}^t H^t(j)] \\ &\quad + E[H(k)\{\underline{x}(k)-\underline{m}(k)\}\underline{v}^t(j)] \\ &\quad + E[\underline{v}(k)\{\underline{x}(j)-\underline{m}(j)\}^t H^t(j)] \\ &\quad + E[\underline{v}(k)\underline{v}^t(j)] \end{aligned}$$

which can be simplified into

$$\begin{aligned} \mathrm{Cov}[\underline{z}(k),\underline{z}(j)] &= H(k)W(k)\Psi^t(j,k)H^t(j) + R(k)\delta_{kj} \quad \text{for } k \le j \\ &= H(k)\Psi(k,j)W(j)H^t(j) + R(k)\delta_{kj} \quad \text{for } k \ge j \end{aligned} \tag{6.73}$$

where we have made use of (6.59a), (6.70a) and the fact that the process $\{\underline{v}(k)\}$ is independent of $\{\underline{x}(k)\}$. This completes our development of the Gauss-Markov discrete-time model. Next we illustrate the different concepts by examples.

6.3.4 *EXAMPLES*

Example 1

Consider a discrete-time process described by

$$X(k+1) = X(k) + u(k) \; ; \quad k = 0,1,\ldots.$$

where $\{u(k),u(k-1),\ldots,u(0)\}$ is a sequence of independent random variables. Assume that $X(0)$ is a random variable which is independent of the sequence $\{u(k),\ldots,u(0)\}$. Is the process Markovian?

To answer the question posed above, we must examine the conditional probability

$$p[x(k+1)\,|\,x(k),\ldots,x(0)]$$

By iterating the discrete model we get

$$X(k) = u(k-1) + u(k-1) + \ldots + u(0) + X(0)$$

In view of the independence assumption, it is clear that $X(k)$ is independent of $u(k)$. The same is true for $X(k-1)$, $X(k-2)$, ... $X(0)$. We could therefore write:

$$p\{x(k+1)\,|\,x(k),\ldots,x(0)\} = p\{x(k+1)\,|\,x(k)\}$$

and hence the process is Markovian.

Example 2

A scalar process is modelled by the difference equation

$$x(k+1) = a\,x(k) + v(k)\ ; \qquad |a| < 1$$

where $\{\underline{v}(k)\}$ is Gaussian white noise with zero mean and covariance q. The initial state $x(0)$ is $N(0,W_0)$ and it is uncorrelated with $\{\underline{v}(k)\}$ for all k. It is required to examine the asymptotic behaviour of the variance. What will happen when $E[x(0)] = m_0 \neq 0$? The variance of the process is given by (6.69). Setting $A(k) = a$, $G(k) = 1$, $Q(k) = 1$, we obtain:

$$W(k+1) = a^2 W(k) + q$$

Starting from W_0 at $k = 0$ and iterating, we get:

$$W(k+1) = a^{2k+2} W_0 + q(1+a^2+\ldots+a^{2k})$$

$$= a^{2k+2} W_0 + q(1-a^{2k+2}/1-a^2)$$

Since $|a| < 1$, then

$$W(\infty) = q/1-a^2$$

The assumption $x(0)$ is $N(0,W_0)$ implies that $m(k+1) = 0$ and we conclude that the process is a stationary Gaussian process. Suppose that $E[x(0)] = m_0 \neq 0$, then from (6.66b) we have:

$$m(k) = a^k\, m_0$$

which tends to zero as $k \to \infty$. Therefore, the condition $|a| < 1$ always yields an asymptotically stationary process (as $k \to \infty$) with a zero mean irrespective of the actual mean of the initial state.

Example 3

Here we derive the formulae that define a Gauss-Markov process of the form (6.55) in terms of the probability distribution of the initial state $p\{\underline{x}(0)\}$ and the transition probability density $p\{\underline{x}(k+1)|x(k)\}$.

Recall the assumptions made in Section 6.3.1. Since the random vectors $\underline{x}(k+1)$ and $\underline{x}(k)$ are jointly Gaussian, it is sufficient to calculate the conditional mathematical expectation $E[\underline{x}(k+1)|\underline{x}(k)]$ and the conditional covariance matrix.

For the conditional expectation we have:

$$E[\underline{x}(k+1)|\underline{x}(k)] = A(k)\underline{x}(k) + G(k)E[\underline{w}(k)|\underline{x}(k)]$$

But $\{\underline{w}(k)\}$ and $\{\underline{x}(k)\}$ are independent by assumption so that

$$E[\underline{w}(k)|\underline{x}(k)] = E[\underline{w}(k)] = \underline{0}$$

hence,

$$E[\underline{x}(k+1)|\underline{x}(k)] = A(k)\underline{x}(k)$$

For the conditional covariance matrix

$$E[\{\underline{x}(k+1)-E[\underline{x}(k+1)|\underline{x}(k)]\}\{\underline{x}(k+1)-E[\underline{x}(k+1)|\underline{x}(k)]\}^t|\underline{x}(k)]$$

it is readily simplified, using the conditional expectation derived above, into:

$$E[G(k)\underline{w}(k)\underline{w}^t(k)G^t(k)|\underline{x}(k)] =$$
$$E[G(k)\underline{w}(k)\underline{w}^t(k)G^t(k)] = G(k)Q(k)G^t(k)$$

When the matrix $G(k)Q(k)G^t(k) = D(k)$ is nonsingular, we will have:

$$p\{\underline{x}(k+1)|\underline{x}(k)\} = \{1/(2\pi)^{n/2}\}\{\det[D(k)]\}^{\frac{1}{2}}\exp\{-\tfrac{1}{2}[\underline{x}(k+1) - A(k)\underline{x}(k)]^t D^{-1}(k)[\underline{x}(k+1)-A(k)\underline{x}(k)]\}$$

Example 4

Consider the random process (6.55)-(6.57) subject to the noise processes $\{\underline{w}(k)\}$ and $\{\underline{v}(k)\}$ having known nonzero means. What will the expressions for the evolution of the mean and covariance of $\underline{x}(k)$?

Let the mean of the input noise process be $\underline{n}(k)$. By taking the expectation of (6.55), we get:

$$\begin{aligned} E[\underline{x}(k+1)] &= E[A(k)\underline{x}(k)] + E[G(k)\underline{w}(k)] \\ &= A(k)E[\underline{x}(k)] + G(k)E[\underline{w}(k)] \end{aligned}$$

which can be written as:

$$\underline{m}(k+1) = A(k)\underline{m}(k) + G(k)\underline{n}(k)$$

The solution of this equation, given $\underline{m}(0)$, can be expressed as:

$$\underline{m}(k) = \Psi(k,0)\underline{m}(0) + \sum_{j=0}^{k-1} \Psi(k,j+1)G(j)\underline{n}(j)$$

A comparison of this expression with (6.66b) shows that the mean $\underline{m}(k)$ now depends on the random sequence $\underline{n}(0),\ldots,\underline{n}(k-1)$.

To calculate the covariance matrix we write:

$$W(k+1) = E[\{\underline{x}(k+1)-\underline{m}(k+1)\}\{\underline{x}(k+1)-\underline{m}(k+1)\}^t]$$

It should be noted that

$$\underline{x}(k+1)-\underline{m}(k+1) = A(k)[\underline{x}(k)-\underline{m}(k)] + G(k)[\underline{w}(k)-\underline{n}(k)]$$

Combining the above two expressions, we arrive at:

$$W(k+1) = A(k)W(k)A^t(k) + G(k)Q(k)G^t(k)$$

where $Q(k)$ is the covariance of $\{\underline{w}(k)\}$. This shows that the covariance of the process $\{\underline{x}(k)\}$ is unaltered.

Example 5

Here we develop expressions for the joint probability function $p\{\underline{x}(k),\underline{y}(k)\}$ and the conditional probability density $p\{\underline{x}(k)|\underline{y}(k)\}$ of two jointly Gaussian random vectors.

Let $\underline{x}(k)$ be $N(\underline{m},W)$ and $\underline{y}(k)$ be $N(\underline{n},V)$, and define $\underline{z}(k) = [\underline{x}^t(k) \quad \underline{y}^t(k)]^t$. Then

$$E[\underline{z}(k)] = E[\underline{x}^t(k) \quad \underline{y}^t(k)]^t$$

$$\underline{r}(k) = \begin{pmatrix} \underline{m}(k) \\ \underline{n}(k) \end{pmatrix}$$

The covariance matrix $E[\{\underline{z}(k)-E[\underline{z}(k)]\}\{\underline{z}(k)-E[\underline{z}(k)]\}^t]$ can be written in the following form:

$$\mathrm{Cov}[\underline{z}(k),\underline{z}(k)] = L = \left(\begin{array}{c|c} W & S \\ \hline S^t & V \end{array}\right)$$

where the partition is conformable with the dimensions of $\underline{x}(k)$ and $\underline{y}(k)$, q_1 and q_2 respectively. The S matrix is defined by:

$$S = E[\{\underline{x}(k)-\underline{m}(k)\}\{\underline{y}(k)-\underline{n}(k)\}^t]$$

Therefore the random vector $\underline{z}(k)$ is Gaussian with the probability density function, or the joint density function of $\underline{x}(k)$ and $\underline{y}(k)$, of the form:

$$p\{\underline{z}(k)\} = p\{\underline{x}(k),\underline{y}(k)\}$$

$$= 1/(2\pi)^{q_1+q_2}\}\{1/(\det[L])^{\frac{1}{2}}\}\exp\{-\tfrac{1}{2}[\underline{z}(k) - \underline{r}(k)]^t L^{-1}[\underline{z}(k)-\underline{r}(k)]\}$$

For convenience, we will obtain an explicit expression for L^{-1}. Since $L^{-1}L = I$, it can be easily shown [3] that

$$L^{-1} = \begin{pmatrix} D_1 & D_2 \\ D_2^t & D_3 \end{pmatrix}$$

with

$$D_1 = [W - S\, V^{-1}\, S^t]^{-1}$$

$$D_2 = - D_1\, S\, V^{-1}$$

$$D_3 = [V - S^t\, W^{-1}\, S]^{-1}$$

Note that W, V and L are positive definite. In addition, L is symmetric and hence L^{-1} will be symmetric and positive definite. Recall that

$$p\{\underline{y}(k)\} = \{1/(2\pi)^{q_2}\}\{1/\det[V]^{\frac{1}{2}}\}\exp\{-\tfrac{1}{2}[\underline{y}(k) -\underline{m}(k)][\underline{y}(k)-\underline{n}(k)]^t\}$$

then from Bayes' rule (6.19) we can write the conditional probability density of $\underline{X}$ given $\underline{Y}$ as:

$$\begin{aligned} p\{\underline{x}(k)|\underline{y}(k)\} &= p\{\underline{x}(k),\underline{y}(k)\}/p\{\underline{y}(k)\} \\ &= \{1/(2\pi)^{q_1}\}\{\det[V]/\det[L]\}^{\frac{1}{2}} \\ &\quad \exp\left\{-\frac{1}{2}\begin{pmatrix}\underline{x}(k)-\underline{m}(k)\\ \underline{y}(k)-\underline{n}(k)\end{pmatrix}^t \begin{pmatrix} D_1 & D_2 \\ D_2^t & D_3-V^{-1}\end{pmatrix}\begin{pmatrix}\underline{x}(k)-\underline{m}(k)\\ \underline{y}(k)-\underline{n}(k)\end{pmatrix}\right\} \end{aligned}$$

By expanding the quadratic form in the exponential, using the expressions for D_2, D_3 and rearranging the terms, we have:

$$\begin{aligned} &\begin{pmatrix}\underline{x}(k)-\underline{m}(l)\\ \underline{y}(k)-\underline{n}(k)\end{pmatrix}^t \begin{pmatrix} D_1 & D_2 \\ D_2^t & D_3-V^{-1}\end{pmatrix}\begin{pmatrix}\underline{x}(k)-\underline{m}(k)\\ \underline{y}(k)-\underline{n}(k)\end{pmatrix} \\ &= [\underline{x}(k)-\underline{m}(k)]^t D_1 [\underline{x}(k)-\underline{m}(k)] \\ &\quad + 2[\underline{x}(k)-\underline{m}(k)]^t D_2 [\underline{y}(k)-n(k)] \\ &\quad + [\underline{y}(k)-\underline{n}(k)]^t [D_3-V^{-1}][\underline{y}(k)-\underline{n}(k)] \\ &= [\underline{x}(k)-\underline{m}(k)-SV^{-1}\{\underline{y}(k)-\underline{n}(k)\}]^t D_1 [\underline{x}(k) \\ &\quad -\underline{m}(k)-SV^{-1}\{\underline{y}(k)-\underline{n}(k)\}] \end{aligned}$$

From the determinants of partitioned matrices *[6]*, we can write:

$$\det[L] = \det[W-SV^{-1}S^t]\det[V]$$

or

$$\det[L]/\det[V] = \det[W-SV^{-1}S^t]$$
$$\equiv \det[D_1^{-1}]$$

Define

$$\underline{h}(k) = \underline{m}(k) + SV^{-1}\{\underline{y}(k)-\underline{n}(k)\}$$

Therefore, $p\{\underline{x}(k)|\underline{y}(k)\}$ can finally be simplified to

$$p\{\underline{x}(k)|\underline{y}(k)\} = \{1/(2\pi)^{q_1}\}\{1/\det[D_1^{-1}]^{\frac{1}{2}}\}\exp\{-\tfrac{1}{2}[\underline{x}(k) - \underline{h}(k)]^t D_1[\underline{x}(k)-\underline{h}(k)]\}$$

which, when compared with (6.50), shows that $p\{\underline{x}(k)|\underline{y}(k)\}$ is a Gaussian density function with mean $\underline{h}(k)$ and covariance D_1^{-1}. To conclude, the conditional mean and the covariance matrix are given by:

$$\underline{h}(k) = E[\underline{x}(k)|\underline{y}(k)]$$
$$= \underline{m}(k) + SV^{-1}[\underline{y}(k)-\underline{n}(k)]$$
$$D_1^{-1} = \mathrm{Cov}\{\underline{x}(k)|\underline{y}(k)\}$$
$$= W - SV^{-1}S^t$$

We note that the conditional mean depends on $\underline{y}(k)$, while the covariance matrix is independent of $\underline{y}(k)$. The usefulness of the results obtained here lies in their relevance to the development of the Kalman filter. It is of prime interest to examine the vector $\underline{r}(k) = [\underline{x}(k)-\underline{h}(k)]$. It is obvious that $\underline{r}(k)$ is a Gaussian random vector. From the definition of $\underline{h}(k)$, it is easy to see that

$$E[\underline{r}(k)] = E[\underline{x}(k)-E[\underline{x}(k)|\underline{y}(k)]]$$
$$= \underline{m}(k)-\underline{m}(k) = \underline{0}$$

Also,

$$
\begin{aligned}
W\{\underline{r}(k),\underline{y}(k)\} &= E[\{\underline{r}(k)-\underline{0}\}\{\underline{y}(k)-\underline{n}(k)\}^t] \\
&= E[\{\underline{x}(k)-\underline{h}(k)\}\{\underline{y}(k)-\underline{n}(k)\}^t] \\
&= E[\{\underline{x}(k)-\underline{m}(k)-SV^{-1}[\underline{y}(k)-\underline{n}(k)]\}\{\underline{y}(k)-\underline{n}(k)\}^t] \\
&= S - S = 0
\end{aligned}
$$

We then conclude that the random vector $\underline{r}(k) = \underline{x}(k)-E[x(k)|\underline{y}(k)]$ is independent of $\underline{y}(k)$ and has a zero mean.

6.4 The Kalman Filter

In this section, we consider the problem of estimating the state of linear dynamical systems which are subject to stochastic disturbances. We shall focus attention on the filtering problem and derive the Kalman filter.

6.4.1 *THE ESTIMATION PROBLEM*

Consider a dynamic system whose state as a function of time is an n-dimensional discrete-time stochastic process $\{\underline{x}(k)\}$; $k \in I_t = \{0,1,2,...\}$. Suppose that we have made a sequence of measurements $\underline{z}(0),\underline{z}(1),...,\underline{z}(m)$, at consecutive discrete instants, which are related to $\underline{x}(k)$ by means of an appropriate measurement system. We wish to utilize the measurement data in some way to infer the value of $\underline{x}(k)$. Let us assume that the sequence $\{\underline{z}(j),\ j = 0,...\ m\}$ is a discrete-time stochastic process.

Given the measurement records $\{\underline{z}(0),...,\underline{z}(m)\}$, we denote (at least for the time being) an estimate of $\underline{x}(k)$ based on these measurements by $\hat{\underline{x}}(k|m)$. As a function of the measurements, define the estimate of the state to be

$$\hat{\underline{x}}(k|m) = g_k[\underline{z}(j),\ j = 0,...,m] \tag{6.74}$$

We can then state the following:

The estimation problem is one of determining $\underline{g}_k[\cdot]$ *in an appropriate way.*

In essence, the solution of the estimation problem implies the development of a suitable algorithm by which one can produce an approximate form of the state of noisy systems. Three versions of the above problem are now discussed.

A. The Filtering Problem

The filtering problem is obtained by setting $k = m$ in (6.74). It therefore means the recovery at time k of some information about $\underline{x}(k)$, which corresponds to $\hat{\underline{x}}(k|k)$, using measurement data up till time k. One should note the following points:

1. We wish to obtain the approximate value of $\underline{x}(k)$ at time k.
2. The measurement records are available at time k and not at a later time, and
3. all the measurement records up to time k are used in estimating the state.

The reason for stressing these points is to distinguish the filtering problem from the prediction and smoothing problems to be defined below.

B. The Smoothing Problem

This problem results from the estimation problem stated previously when $k < m$. The smoothing problem thus differs from the filtering problem in that the information about $\underline{x}(k)$, in the form of measurement data, need not become available at time k and measurement records derived later than time k can be utilized in obtaining information about $\underline{x}(k)$. It should be noted that:

1. there is a delay in producing the estimate of the state, and
2. more data records are used than in the filtering problem.

C. The Prediction Problem

Here we use $k > m$ in (6.74). The aim of the prediction problem is to obtain at time k information about $\underline{x}(k+s)$ for some $s > 0$. It therefore represents a forecast, i.e. we wish to guess how $\underline{x}(\cdot)$ will behave after a specified period of time.

In the sequel we shall consider only the filtering problem because of its wide use in control and systems applications, see *[1-5]* for details.

6.4.2 *PRINCIPAL METHODS OF OBTAINING ESTIMATES*

From now onwards, we shall limit out discussions to linear discrete-time dynamical systems of the type (6.55)-(6.57) where $\{\underline{x}(k)\}$ and $\{\underline{z}(k)\}$ are Gaussian random processes. Our purpose is to indicate how knowledge of the value taken by one random variable, the measurement $\underline{z}(k)$ in our case, can provide information about the value taken by a second random variable, the state $\underline{x}(k)$. In particular, we wish to solve the following problem (filtering problem):

> *Find at time k some information about $\underline{x}(k)$ from $[\underline{z}(0), \underline{z}(1), \ldots, \underline{z}(k)]$*

Let the information that we are seeking to find be summarised by the vector $\hat{\underline{x}}(k)$, *the estimate of $\underline{x}(k)$.* Since both $\{\underline{x}(k)\}$ and $\{\underline{z}(k)\}$ are random processes, so is $\{\hat{\underline{x}}(k)\}$. In general, $\hat{\underline{x}}(k)$ will not be equal to $\underline{x}(k)$. It is thus reasonable to define

$$\tilde{\underline{x}}(k) = \underline{x}(k) - \hat{\underline{x}}(k) \tag{6.75}$$

as the *estimation error*. Since the estimate $\hat{\underline{x}}(k)$ can be derived in several ways, the problem of interest would then be to find an estimate, which is a function of $\underline{z}(k)$, such that it is *optimal* with respect to some criterion. In addition, it is necessary to ensure that the estimates possess certain convergence properties with respect to the real values of the state.

A. Minimum Variance Estimate

Define $\underline{Z}(k) = \{\underline{z}(0),\ldots,\underline{z}(k)\}$, then an average measure of the estimation error in (6.75) is provided by:

$$E[\{\underline{X}-\hat{\underline{x}}\}^t C\{\underline{X}-\hat{\underline{x}}\}|\underline{Z}(k)] \tag{6.76}$$

where C is a nonnegative definite symmetrical weighting matrix. We note in (6.76) that

(1) $\hat{\underline{x}}$ is a fixed vector which needs to be determined from a knowledge of $\underline{Z}(k)$;

(2) the average measure is a scalar quantity which is convenient for comparison;

(3) it has a zero value when the estimate is exact.

The *minimum variance estimate* $\hat{\underline{x}}$ is defined as one for which

$$E[||\underline{X}-\hat{\underline{x}}||^2_C \; |\underline{Z}(k)] \leq E[||\underline{X}-\underline{y}||^2_C \; |\underline{Z}(k)] \tag{6.77}$$

for all vectors $\underline{y}$, determined in some way from $\underline{z}(k)$ and where $||\underline{h}||^2_C = \underline{h}^t C\underline{h}$. It should be observed in (6.77) that $\underline{y}$ in general depends on $\underline{z}$ but it is independent of $\underline{x}$. The right-hand side of (6.77) can be written as

$$E[||\underline{X}-\underline{y}||^2_C|\underline{Z}(k)] =$$

$$\int_{-\infty}^{+\infty} (\underline{x}-\underline{y})^t C(\underline{x}-\underline{y})p(\underline{x}|\underline{z})d\underline{x}$$

which, by adding and subtracting appropriate terms, can be put in the form:

$$E[||\underline{X}-\underline{y}||_C^2|\underline{Z}(k)] =$$

$$[\underline{y}^t - \int_{-\infty}^{+\infty} \underline{x}^t p(\underline{x}|\underline{z})d\underline{x}]C[\underline{y} - \int_{-\infty}^{+\infty} \underline{x}\ p(\underline{x}|\underline{z})d\underline{x}]$$

$$+ \int_{-\infty}^{+\infty} \underline{x}^t C\underline{x}\ p(\underline{x}|\underline{z})d\underline{x} - \left|\left|\int_{-\infty}^{+\infty} \underline{x}\ p(\underline{x}|\underline{z})d\underline{x}\right|\right|_C^2 \qquad (6.78)$$

It is evident that the right-hand side of (6.78) has a unique minimum when $\underline{y} = E[\underline{X}|\underline{Z}(k)]$, which implies that

$$\hat{\underline{x}} = E[\underline{X}|\underline{Z}(k)]$$

$$= \int_{-\infty}^{+\infty} \underline{x}\ p(\underline{x}|\underline{z})d\underline{x} \qquad (6.79)$$

We now conclude that

> *The minimum variance estimate* $\hat{\underline{x}}$ *is the conditional mean estimate; that is, the conditional expectation of* $\underline{X}$ *given* $\underline{Z}(k)$.

The value of the average mean square error associated with the estimate $\hat{\underline{x}}$ can be obtained from (6.78) by substituting $\underline{y} = \hat{\underline{x}}$ to yield:

$$E[||\underline{X}-\hat{\underline{x}}||_C^2|\underline{Z}(k)] = E[||\underline{X}||_C^2|\underline{Z}(k)] - ||\hat{\underline{x}}||_C^2 \qquad (7.80)$$

The estimate $\hat{\underline{x}}$ is often called the least-squares estimate or the minimum mean-square estimate. It is interesting to observe that the above analysis is carried out for arbitrary probability densities. For a given configuration of stochastic processes, all that is needed is to evaluate the conditional probability density function.

In the light of our discussions in Section 6.2, we wish to emphasize that what we have established is a procedure by which one can compute an estimate (known vector) of a random process given a particular set of measurements (values of another random process). Such procedure is essentially a rule of association between the measurement values and the value of the estimate. We therefore define the random variable which has $\hat{\underline{x}}$ as a particular value as $\hat{\underline{X}}$. One should think of $\hat{\underline{X}}$ as a function which depends on $\underline{z}$ or $\underline{Z}(k)$ and generates $\hat{\underline{x}}$. It is frequently called an *estimator* of $\underline{X}$, and from (6.79) we obtain

$$\hat{\underline{X}} = E[\underline{X}|\underline{Z}] \tag{6.81}$$

as the *minimum variance estimator*.

A major property of the conditional mean estimate (minimum variance estimate) $\hat{\underline{x}}$ is that it is an *unbiased estimate*, that is,

$$\begin{aligned} E[\underline{X}-\hat{\underline{x}}|\underline{Z}(k)] &= E[\underline{X}|\underline{Z}(k)] - \hat{x} \\ &= \hat{\underline{x}} - \hat{\underline{x}} \\ &= 0 \end{aligned} \tag{6.82}$$

The above expression shows that the conditional expected error in using $\underline{x}$ as an estimate of $\underline{X}$, given $\underline{Z}(k)$, is zero.

B. Maximum Likelihood Estimate

We can define the conditional probability density function $p\{\underline{Z}(k)|\underline{x}(k)\}$ as *the Likelihood function.* This is a function of $\underline{x}(k)$ whose maximum indicates the most likely value of the sequence $\underline{Z}(k) = \{\underline{z}(0),\ldots,\underline{z}(k)\}$ that we obtain using the parameters $\underline{x}(k)$.

In many cases we will maximize the logarithm of this function [1,5]. If the logarithm has a continuous first derivative, then a necessary condition for *a maximum likelihood estimate* $\hat{\underline{x}}_m$ can be obtained by differentiating

$$\mathrm{Log}[p\{\underline{Z}(k)|\underline{x}(k)\}]$$

with respect to $\underline{x}(k)$, so that we have

$$\{\partial/\partial\underline{x}(k)\ \mathrm{Log}[p\{\underline{Z}(k)|\underline{x}(k)\}]\}_{\underline{x}(k)\ =\ \hat{\underline{x}}_m} = \underline{0} \tag{6.83}$$

From Bayes' rule (6.19), it is easy to see that $p\{\underline{Z}(k)|\underline{x}(k)\} = p\{\underline{Z}(k),\underline{x}(k)\}/p\{\underline{x}(k)\}$ so that the maximum likelihood estimate requires the prior data $p\{\underline{x}(k)\}$.

C. Maximum A Posteriori Estimate

This estimate is obtained by considering the *a posteriori* probability density function $p\{\underline{x}(k)|\underline{Z}(k)\}$. It thus deals with the inverse problem of the maximum likelihood. The *maximum a posteriori estimate* $\hat{\underline{x}}_a$ is the value of $\hat{\underline{x}}$ which maximizes the distribution $p\{\underline{x}(k)|\underline{Z}(k)\}$. Similarly to (6.83) it is given by:

$$\{\partial/\partial\underline{x}\ \mathrm{Log}[p\{\underline{x}(k)|\underline{Z}(k)\}]\}_{\underline{x}(k)\ =\ \hat{\underline{x}}_a} = \underline{0} \tag{6.84}$$

The maximum likelihood and maximum *a posteriori* estimates can be related through Bayes' rule:

$$\begin{aligned} p\{\underline{x}(k)|\underline{Z}(k)\} &= p\{\underline{x}(k),\underline{Z}(k)\}/p\{\underline{Z}(k)\} \\ &= p\{\underline{Z}(k)|\underline{x}(k)\}p\{\underline{x}(k)\}/p\{\underline{Z}(k)\} \end{aligned}$$

When the *a priori* distribution is uniform, both estimates are identical.

We note that the three estimates described thus far are presented for arbitrary probability distributions. It has been verified *[1,5]* that when the random vectors are normally distributed (Gaussian) the maximum *a posteriori* estimates are precisely the conditional mean estimates. For this reason, we will adopt the minimum variance as our criterion for determining the optimal estimator (filter).

6.4.3 *DEVELOPMENT OF THE KALMAN FILTER EQUATIONS*

Our objective here is to develop the mechanism by which the best estimate of $\underline{x}(k)$, given the measurement sequence $\underline{Z}(k) = \{\underline{z}(0),\ldots,\underline{z}(k)\}$, can be obtained. For the linear model (6.55)-(6.57) with Gaussian random processes, this mechanism is known as *the Kalman Filter [8,9]*. In his original derivation of the discrete filter, Kalman *[8]* used the concept of orthogonal projections. Subsequent to this work, several methods have been developed to derive the discrete Kalman filter (see *[1-5,10-15]* for details). We will follow here the procedure of *[10]*. First, we present the optimal filtering problem and state the associated assumptions.

A. The Optimal Filtering Problem

The problem of interest is to estimate, at each discrete instant, in an optimal way, the state of a linear dynamical system using noisy measurements of the output records. The general model of the system is of the form:

$$\left.\begin{aligned} \underline{x}(k+1) &= A(k)\underline{x}(k) + G(k)\underline{w}(k) \\ \underline{z}(k) &= H(k)\underline{x}(k) + \underline{v}(k) \end{aligned}\right\} \qquad (6.85)$$

which is similar to the model (6.55)-(6.57). The assumptions concerning the model and disturbances are summed up below.

Assumption 1

$\{\underline{w}(k)\}$ and $\{\underline{v}(k)\}$ are Gaussian white noise sequences such that

$$E[\underline{w}(k)] = 0\,, \quad E[\underline{v}(k)] = \underline{0}$$

$$E[\underline{w}(k)\,\underline{w}^t(j)] = Q\delta_{kj}$$

$$E[\underline{v}(k)\,\underline{v}^t(j)] = R\delta_{kj}$$

Assumption 2

The random processes $\{\underline{w}(k)\}$ and $\{\underline{v}(k)\}$ are uncorrelated, that is

$$E[\underline{w}(k)\ \underline{v}^t(j)] = 0 \quad \text{for all } k \text{ and } j$$

Assumption 3

The initial state $\underline{x}(0)$ is a Gaussian random vector with mean $E[\underline{x}(0)] = \underline{m}(0)$ and covariance

$$E[\{\underline{x}(0)-\underline{m}(0)\}\{\underline{x}(0)-\underline{m}(0)\}^t] = W(0)$$

Assumption 4

The initial state $\underline{x}(0)$ and the noise processes $\{\underline{w}(k)\}$ and $\{\underline{v}(k)\}$ are uncorrelated, that is

$$E[\underline{x}(0)\ \underline{w}^t(k)] = 0 \quad \text{for all } k$$

$$E[\underline{x}(0)\ \underline{v}^t(k)] = 0 \quad \text{for all } k$$

Assumption 5

The elements of the system matrices $A(k)$, $G(k)$ and $H(k)$ are known.

It should be noted that Assumption 2 is not strictly necessary. It is convenient though, since the final expression for the filter would be much simpler in this case. We adopt the minimization of the conditional variance (6.76) as our criterion for determining the best (optimal) estimate.

We saw in Section 6.4.2 that the optimal estimate which minimizes the conditional variance is the conditional mean estimate,

$$\begin{aligned}\hat{\underline{x}}(k|k) &= E[\underline{x}(k)|\underline{z}(0),\ldots,\underline{z}(k)] \\ &= E[\underline{x}(k)|\underline{Z}(k)] \qquad (6.86)\end{aligned}$$

Note that the argument k is used twice in defining the optimal estimate. The k to the left of the conditioning bar denotes the discrete instant at which the estimate is required whereas the other k denotes the discrete instant up to which the output records are available. Therefore, the estimate $\hat{\underline{x}}(k|k-1)$ is the estimate of the state $\underline{x}(k)$ at the instant k given the sequence of measurements up to (k-1), that is

$$\begin{aligned}\hat{\underline{x}}(k|k-1) &= E[\underline{x}(k)|\underline{z}(0),\ldots,\underline{z}(k-1)] \\ &= E[\underline{x}(k)|\underline{Z}(k-1)] \qquad (6.87)\end{aligned}$$

which is actually a one-step predictor (see part C of Section 6.4.1). By convention, we will define $\hat{\underline{x}}(k-1|k-1)$ for k = 0 (that is $\hat{\underline{x}}(-1|-1)$ to be $\underline{m}(0)$, that is, the expected value of $\underline{x}(0)$ given no measurements. For the same reason, the initial value of the associated error covariance matrix P(-1,-1) is taken to be W(0). We can now state the basic optimal filtering problem

For the linear, discrete-time system of (6.85) under Assumptions (1)-(5), determine the estimates

$$\hat{\underline{x}}(k|k-1) \quad \text{and} \quad \hat{\underline{x}}(k|k)$$

defined by (6.86) and (6.87), and the associated error covariance matrices $P(k|k-1)$ and $P(k|k)$.

B. Solution Procedure

The approach to the development of the Kalman filter equations can be divided into a number of distinct steps.

Step 1 (Transition of the state $\underline{x}(k-1)$ to $\underline{x}(k)$)

We assume that the estimate $\hat{\underline{x}}(k-1|k-1)$ is known and we wish to determine the one-step predictor $\hat{\underline{x}}(k|k-1)$. Rewrite the dynamic model in the form

$$\underline{x}(k) = A(k-1)\underline{x}(k-1) + G(k-1)\underline{w}(k-1) \qquad (6.88)$$

On taking the conditional mean of (6.88), we obtain:

$$E[\underline{x}(k)|\underline{Z}(k-1)] = A(k-1)E[\underline{x}(k-1)|\underline{Z}(k-1)] + G(k-1)E[\underline{w}(k-1)|\underline{Z}(k-1)] \quad (6.89)$$

In view of Assumptions (1), (2) and (4), the random vector $w(k-1)$ is independent of $\underline{Z}(k-1)$ so that

$$E[\underline{w}(k-1)|\underline{Z}(k-1)] = E[\underline{w}(k-1)] = 0$$

and hence (6.89) reduces to:

$$\hat{\underline{x}}(k|k-1) = A(k-1)\hat{\underline{x}}(k-1|k-1) \quad (6.90)$$

Next we determine the error covariance matrix associated with the one-step predictor, that is

$$P(k|k-1) = E[\{\underline{x}(k)-\hat{\underline{x}}(k|k-1)\}\{\underline{x}(k)-\hat{\underline{x}}(k|k-1\}^t|\underline{Z}(k-1)] \quad (6.91)$$

We note that the vector $[\underline{x}(k)-\hat{\underline{x}}(k|k-1)]$ is independent of the sequence $\underline{Z}(k-1)$, (see the result of Example 5 in Section 6.3). This simplifies (6.91) to:

$$P(k|k-1) = E[\{\underline{x}(k)-\hat{x}(k|k-1)\}\{\underline{x}(k)-\hat{x}(k|k-1)\}^t] \quad (6.92)$$

Using (6.88) and (6.90), the one-step prediction error can be written as:

$$\underline{x}(k)-\hat{\underline{x}}(k|k-1) = A(k-1)[\underline{x}(k-1)-\hat{\underline{x}}(k-1|k-1)] + G(k-1)\underline{w}(k-1) \quad (6.93)$$

On expanding (6.92) with the aid of (6.93), we obtain:

$$P(k|k-1) = A(k-1)P(k-1|k-1)A^t(k-1) + A(k-1)E[\{\underline{x}(k-1) -\hat{\underline{x}}(k-1|k-1)\underline{w}^t(k-1)|\underline{Z}(k-1)] + G(k-1)QG^t(k-1) + G(k-1)E[\underline{w}(k-1)\{\underline{x}(k-1)-\hat{x}(k-1|k-1)\}^t|\underline{Z}(k-1)]G^t(k-1)$$

Since $\underline{w}(k-1)$ has zero mean and $\underline{x}(k-1)$ is a function of $\underline{x}(0)$ and $\{\underline{w}(0),\ldots,\underline{w}(k-2)\}$ but not of $\underline{w}(k-1)$, then

$$E[\underline{x}(k-1)\ \underline{w}^t(k-1)] = 0$$

Also,

$$E[\hat{\underline{x}}(k-1|k-1)\underline{w}^t(k-1)|\underline{Z}(k-1)] = \hat{\underline{x}}(k-1|k-1)E[\underline{w}^t(k-1)]$$
$$= 0$$

so that the error covariance matrix could be determined from the expression

$$P(k|k-1) = A(k-1)P(k-1|k-1)A^t(k-1) + G(k-1)QG^t(k-1) \tag{6.94}$$

Step 2 (One-Step Prediction of the Filtered Estimate)

Next, we wish to express the estimate of $\underline{x}(k)$ given the measurement sequence $\underline{Z}(k)$. To accomplish this, we use the maximum *a posteriori*. First, the conditional density function $p\{\underline{x}(k)|\underline{Z}(k)\}$ can be written as

$$p\{\underline{x}(k)|\underline{Z}(k)\} = p\{\underline{x}(k)|\underline{Z}(k-1),\underline{z}(k)\}$$

where we separate out the last measurement from the previous measurement records $\underline{Z}(k-1)$. Then, applying Bayes' theorem to the above expression leads to

$$\begin{aligned} p\{\underline{x}(k)|\underline{Z}(k)\} &= p\{\underline{x}(k)|\underline{Z}(k-1),\underline{z}(k)\} \\ &= p\{\underline{x}(k),\underline{Z}(k-1),\underline{z}(k)\}/p\{\underline{Z}(k-1),\underline{z}(k)\} \\ &= p\{\underline{x}(k),\underline{Z}(k-1)\}p\{\underline{z}(k)|\underline{x}(k),\underline{Z}(k-1)\}/ \\ &\qquad p\{\underline{Z}(k-1),\underline{z}(k)\} \\ &= p\{\underline{z}(k)|\underline{x}(k),\underline{Z}(k-1)\}p\{\underline{x}(k)|\underline{Z}(k-1)\}/ \\ &\qquad p\{\underline{z}(k)|\underline{Z}(k-1)\} \end{aligned} \tag{6.95}$$

To compute (6.95) we consider the observation equation

$$\underline{z}(k) = H(k)\underline{x}(k) + \underline{v}(k) \tag{6.96}$$

We see that knowledge of $\underline{x}(k)$ implies that the only random vector left is $\underline{v}(k)$, which is independent of $\underline{Z}(k)$ in view of assumptions (1), (2) and (4). We can thus write:

$$p\{\underline{z}(k)|\underline{x}(k),\underline{Z}(k-1)\} = p\{\underline{z}(k)|\underline{Z}(k-1)\}$$

which, when substituted into (6.95), gives:

$$p\{\underline{x}(k)|\underline{Z}(k)\} = p\{\underline{z}(k)|\underline{x}(k)\}p\{\underline{x}(k)|\underline{Z}(k-1)\}/p\{\underline{z}(k)|\underline{Z}(k-1)\} \tag{6.97}$$

In order to determine the maximum *a posteriori* estimate using (6.97), it will only be necessary to evaluate the probability densities of the numerator since the denominator is not an explicit function of $\underline{x}(k)$.

For a given $\underline{x}(k)$, $\underline{z}(k)$ is a Gaussian random vector with mean

$$E[\underline{z}(k)|\underline{x}(k)] = H(k)\underline{x}(k)$$

since $E[\underline{v}(k)|\underline{x}(k)] = E[\underline{v}(k)] = \underline{0}$

The covariance matrix is given by:

$$\begin{aligned} E[\{\underline{z}(k)-H(k)\underline{x}(k)\}\{\underline{z}(k)-H(k)\underline{x}(k)\}^t] &= E[\underline{v}(k)\ \underline{v}^t(k)] \\ &= R(k) \end{aligned}$$

Thus, $p\{\underline{z}(k)|\underline{x}(k)\}$ is $N(H(k)\underline{x}(k),R(k))$.

Turning to the *a priori* density function $p\{\underline{x}(k)|\underline{Z}(k-1)\}$, it is easy to show that this function is actually a Gaussian of mean $\hat{\underline{x}}(k|k-1)$ and covariance $P(k|k-1)$; again see Example 5 of Section 6.3 . Hence, $p\{\underline{x}(k)|\underline{Z}(k-1)\}$ is $N(\hat{\underline{x}}(k|k-1), P(k|k-1))$. To sum up, we can now write the *a posteriori* density function $p\{\underline{x}(k)|\underline{Z}(k)\}$ in (6.97) as:

$$\begin{aligned} p\{\underline{x}(k)|\underline{Z}(k)\} = K \exp[-\tfrac{1}{2}\{\underline{z}(k)-H(k)\underline{x}(k)\}^t R^{-1}(k)\{\underline{z}(k)-H(k)\underline{x}(k)\} \\ + \{\underline{x}(k)-\hat{\underline{x}}(k|k-1)\}^t P^{-1}(k|k-1)\{\underline{x}(k)-\hat{\underline{x}}(k|k-1)\}] \end{aligned} \tag{6.98}$$

where the factor K takes into account the denominator $p\{\underline{z}(k)|\underline{Z}(k-1)\}$ in (6.97).

In order to develop the maximum *a posteriori* estimate (which is identical with the conditional mean estimate in this case), we can differentiate the logarithm of (6.97) with respect to $\underline{x}(k)$ and set the result to zero to obtain the estimate $\hat{\underline{x}}(k|k)$. If we do this, we obtain:

$$H^t(k)R^{-1}(k)\{\underline{z}(k)-H(k)\underline{x}(k)\} - P^{-1}(k|k-1)\{\underline{x}(k)-\hat{\underline{x}}(k|k-1)\} = 0$$

For $\underline{x}(k) = \hat{\underline{x}}(k|k)$ and rearranging, we get:

$$\hat{\underline{x}}(k|k) = \hat{\underline{x}}(k|k-1) + [H^t(k)R^{-1}(k)H(k) + P^{-1}(k|k-1)]^{-1}H^t(k)R^{-1}(k)\{\underline{z}(k)-H(k)\hat{\underline{x}}(k|k-1)\} \tag{6.99a}$$

which is the new value of the state estimate given a new observation.

Finally, to calculate the variance of the estimation error we use the matrix identity [6]

$$[I-(M+N)^{-1}M] = (M+N)^{-1}N$$

with $M = H^t(k)R^{-1}H(k)$, $N = P^{-1}(k|k-1)$ in (6.99) to arrive at

$$\hat{\underline{x}}(k|k) = [H^t(k)R^{-1}(k)H(k) + P^{-1}(k|k-1)]^{-1}[H^t(k)R^{-1}(k)\underline{z}(k) + P^{-1}(k|k-1)\hat{\underline{x}}(k|k-1)] \tag{6.99b}$$

Using (6.96) in (6.99b) and after some algebraic manipulation, the result is

$$\underline{x}(k)-\hat{\underline{x}}(k|k) = -[H^t(k)R^{-1}(k)H(k)+P^{-1}(k\ k-1)][H^t(k)R^{-1}(k)\underline{v}(k) - P^{-1}(k\ k-1)\{\underline{x}(k)-\hat{\underline{x}}(k|k-1\}]$$

In the above expression we note that $\underline{v}(k)$ and $\underline{x}(k)$ are independent, $\underline{v}(k)$ and $\hat{\underline{x}}(k|k-1)$ are also independent and $\underline{v}(k)$ has zero mean. By virtue of these facts, it can be readily shown that

$$P(k|k) = E[\{\underline{x}(k)-\hat{\underline{x}}(k|k)\}\{\underline{x}(k)-\hat{\underline{x}}(k|k)\}^t]$$
$$= [H^t(k)R^{-1}(k)H(k)+P^{-1}(k|k-1)]^{-1}$$

which can be alternatively written, using the well-known matrix inversion lemma *[1,10]*

$$[F_1+F_2F_3F_2^t]^{-1} = F_1^{-1}-F_1^{-1}F_2[F_3^{-1}+F_2^tF_1^{-1}F_2]^{-1}F_2^tF_1^{-1} \qquad (6.100)$$

with $F_1 = P^{-1}(k|k-1)$, $F_2 = H^t(k)$, $F_3 = R^{-1}(k)$ and the indicated inverses exist:

$$P(k|k) = P(k|k-1)-P(k|k-1)H^t(k)[H(k)P(k|k-1)H^t(k) + R(k)]^{-1}H(k)P(k|k-1) \qquad (6.101)$$

and subsequently we write (6.99) as:

$$\hat{\underline{x}}(k|k) = \hat{\underline{x}}(k|k-1) + P(k|k-1)H^t(k)[H(k)P(k|k-1)H^t(k) + R(k)]^{-1}\{\underline{z}(k)-H(k)\hat{\underline{x}}(k|k-1)\} \qquad (6.102)$$

In summary, (6.90), (6.94), (6.101) and (6.102) constitute the equations of the optimal minimum variance filter.

Close examination of (6.102) will reveal that the optimal estimate $\hat{\underline{x}}(k|k)$ is the sum of the one-step predictor $\hat{\underline{x}}(k|k-1)$ and the difference between *the actual output* $\underline{z}(k)$ and *the predicted output* $H(k)\hat{\underline{x}}(k\ k-1)$, weighted by the term

$$K(k) = P(k|k-1)H^t(k)[R(k)+H(k)P(k|k-1)H^t(k)]^{-1} \qquad (6.103)$$

which is often called *the filter gain*. The quantity $\tilde{\underline{z}}(k|k-1) = \{\underline{z}(k)-H(k)\hat{\underline{x}}(k|k-1)\}$ under assumptions (1) to (5) and the minimum error variance criterion turns out to be a white noise stochastic process frequently called *the innovations process*. The reason for this is that it contains all of the new information in the measurement $\underline{z}(k)$. The procedure of computing the Kalman filter is carried out recursively in the following order:

Given $P(0|-1) = W(0)$ and $\underline{x}(0|-1) = \underline{m}(0)$

(a) Compute the filter gain using

$$K(k) = P(k|k-1)H^t(k)[R(k)+H(k)P(k|k-1)H^t(k)]^{-1}$$

(b) Compute the state estimate vector

$$\hat{\underline{x}}(k|k) = \hat{\underline{x}}(k|k-1)+K(k)\{\underline{z}(k)-H(k)\hat{\underline{x}}(k|k-1)\} \qquad (6.104)$$

(c) Compute the error covariance matrix

$$P(k|k) = [I-K(k)H(k)]P(k|k-1) \qquad (6.105)$$

These equations can be represented as shown in Fig. 6.2 . It is interesting to note that (6.101), which enables us to compute the error covariance, is a matrix equation of the Riccati type (see Chapter 8).

Occasionally, it is required to determine the function $\hat{\underline{x}}(k+1|k)$ in the mean square error sense. This can be obtained directly from (6.90), (6.102) and (6.103) as:

$$\hat{\underline{x}}(k+1|k) = [A(k)-K^+(k)H(k)]\hat{\underline{x}}(k|k-1)+K^+(k)\underline{z}(k) \qquad (6.106)$$

where $K^+(k)$ is defined by

$$K^+(k) = A(k)P(k|k-1)H^t(k)[R(k)H(k)P(k|k-1)H^t(k)]^{-1}$$

$$= A(k)K(k) \qquad (6.107)$$

and is sometimes called *the Kalman filter gain*.

It is remarked that the estimate $\hat{\underline{x}}(k+1|k)$ is actually the

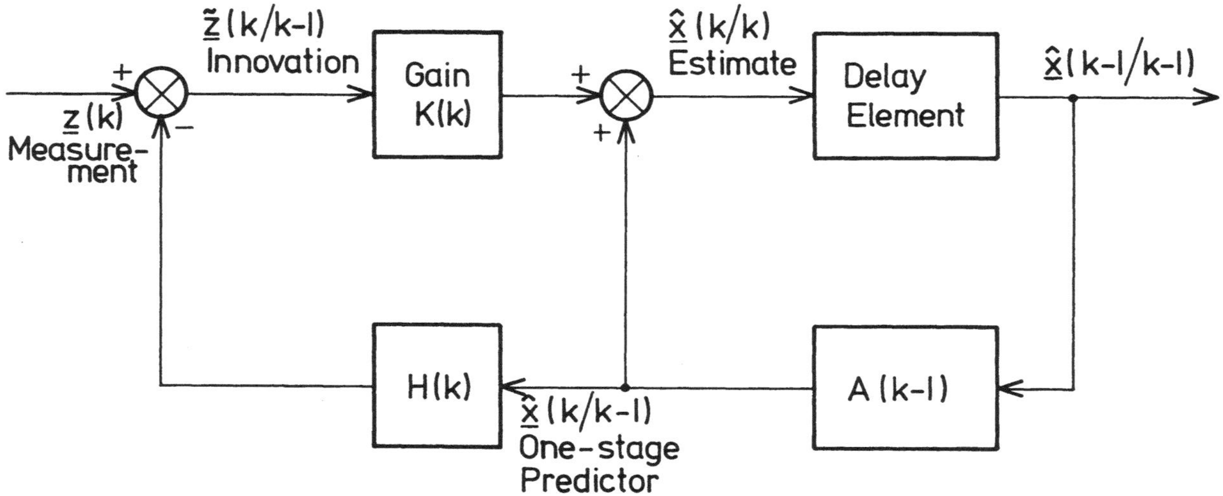

Fig. (6.2) Block-diagram of the Kalman filter

one-stage predictor at the discrete instant k given the measurement records $\underline{Z}(k)$. The reader should distinguish between this quantity and $\hat{\underline{x}}(k|k)$ which is the filtered estimate at k. For this reason, we have used different symbols for the corresponding gains. Fig. 6.3 gives the structure of the one-step predictor.

Next, we present a number of properties of the Kalman filter (or Kalman one-stage predictor).

C. Some Important Properties

With reference to the development of the celebrated Kalman filter we now list some of its important properties:

(1) Consideration of (6.104) or (6.107) shows that the Kalman filter has the same structure as the process. In fact, the Kalman filter is *linear, discrete-time, finite-dimensional system.* Sometimes the estimate $\hat{\underline{x}}(k|k)$ is called *the linear, unbiased, minimum variance estimate.* Note that all the computations are performed recursively.

(2) The input to the filter, or the one-step predictor, is the noise process $\{\underline{z}(k)\}$. The output is $\{\hat{\underline{x}}(k|k)\}$ for the filter and $\{\hat{\underline{x}}(k|k-1)\}$ for the predictor. Obviously, the output sequence (estimate) depends on the input sequence (measurement); but the interesting thing is that the conditional error covariance matrix is actually independent of $\underline{Z}(k)$, see (6.101), and of $\underline{Z}(k-1)$ as in (6.94). More importantly, the gain $K(k)$, or $K^{+}(k)$, is also independent of $\underline{Z}(k)$. Because of these, both the error covariance matrices $P(k|k-1)$, $P(k|k)$ and the gains $K(k)$, $K^{+}(k)$ can be pre-computed.

(3) As discussed earlier, the processes $\{\underline{x}(k)\}$ and $\{\underline{z}(k)\}$ are jointly Gaussian, which in turn implies that $\{\underline{x}(k)|\underline{Z}(k-1)$ is Gaussian. We also saw that the conditional density $p\{\underline{x}(k)|\underline{Z}(k-1)\}$ has mean of $\{\hat{\underline{x}}(k|k-1)$ and error covariance of $P(k|k-1)$. It follows that the Kalman filter equations provide an updating procedure for the entire

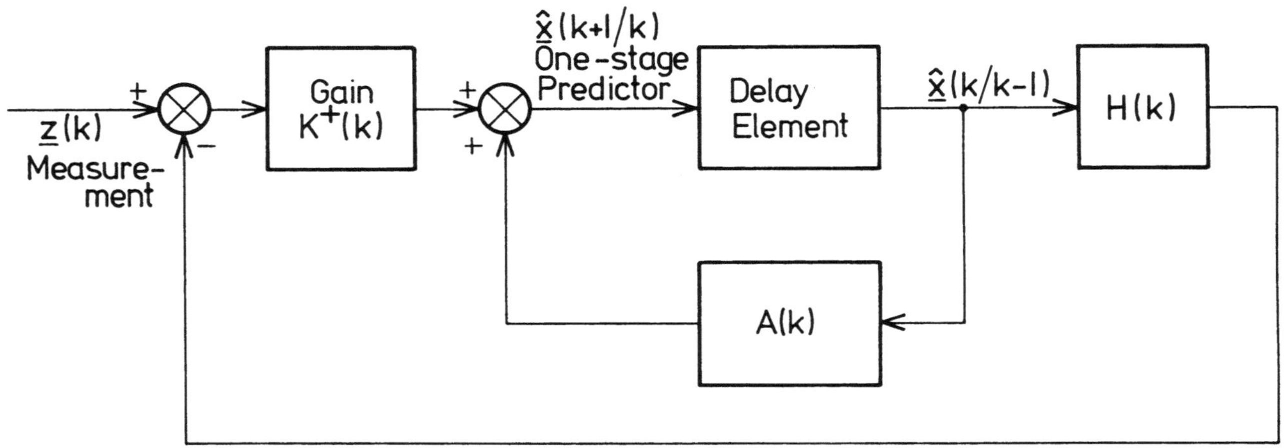

Fig. (6.3) Structure of the one-stage Kalman predictor

conditional probability density function of $\underline{x}(k)$.

(4) Consider the case of linear, shift-invariant discrete systems of the form

$$\left.\begin{aligned} x(k+1) &= A\underline{x}(k) + G\underline{w}(k) \\ z(k) &= H\underline{x}(k) + \underline{v}(k) \end{aligned}\right\} \qquad (6.108)$$

where the matrices (A, G, H) are constant and, in addition, the noise processes are white and stationary. In general, $P(k|k)$ and therefore $K(k)$ will not be constant. This means that the Kalman filter will normally be time varying despite time invariance and stationarity in the process.

In fact, time-invariance of a linear process driven by stationary white noise is necessary but not sufficient for stationarity of the state and output processes *[3]*. Normally, asymptotic stability of the noise-free system is also required. This corresponds to the condition that all of the eigenvalues of the system matrix be strictly less than one in absolute value (see Chapter 3 for related results).

We can now state the following:

For a linear process of the form (6.108), which is time-invariant and driven by stationary white noise, time-invariant filters (constant error covariance and gain) are obtained when either the process is asymptotically stable ($|\lambda_j(A)| < 1$) or the pair [A,H] is completely detectable and the pair [A,GD] is completely stabilizable where $DD^t = Q$.

Further accounts of the Kalman filter can be found in *[1,3]*.

Next we consider some illustrative examples.

6.4.4 *EXAMPLES*

Example 1

Let us consider discrete state estimation for the scalar process

$$x(k+1) = a\,x(k) + w(k)$$

$$z(k) = x(k) + v(k)$$

where w and v are N(0,r) and N(0,q), respectively. Assume that $W(0) = \alpha$ and $m(0) = \beta$. We wish to show one cycle of computing the Kalman filter equations.

Given that $\hat{x}(-1|-1) = \alpha$, then (6.90) yields at $k = 0$

$$\hat{x}(0|-1) = \beta$$

From (6.94), with $P(-1|-1) = \alpha$, we get

$$P(0|-1) = \alpha + q$$

Then, using (6.103)-(6.105), we obtain at $k = 0$

$$\begin{aligned} K(0) &= P(0|-1)[r+P(0|-1)]^{-1} \\ &= (\alpha+q)/(r+\alpha+q) \end{aligned}$$

$$\begin{aligned} \hat{x}(0|0) &= \hat{x}(0|-1)+K(0)\{z(0)-\hat{x}(0|-1)\} \\ &= [1-K(0)]\hat{x}(0|-1)+K(0)z(0) \\ &= \beta r/(r+\alpha+q) + (\alpha+q)z(0)/(r+\alpha+q) \end{aligned}$$

$$\begin{aligned} P(0|0) &= [1-K(0)]P(0|-1) \\ &= r(\alpha+q)/(r+\alpha+q) \end{aligned}$$

Example 2

We consider a constant x of which we record n successive measurements z(k), each one of which is N(0,r). Let the initial estimate $x(0)$ be $N(m,W)$. We want to show that for arbitrary s

$$\hat{x}(s|s) = [rm + W \sum_{j=1}^{s} z(j)]/(r+sW)$$

$$P(s|s) = rW/(r+sW)$$

To solve this problem, we note that the model is

$$\underline{x}(k+1) = x(k) \ ; \quad z(k) = x + v(k)$$

thus $A(k) = 1, \ G(k) = 0, \ Q = 0.$
Also $H(k) = 1$.

For this case, (6.90) and (6.94) become

$$\hat{x}(k|k-1) = \hat{x}(k-1|k-1)$$

$$P(k|k-1) = P(k-1|k-1)$$

Also from (6.103)-(6.105) we get

$$K(k) = P(k|k-1)[r+P(k|k-1)]^{-1}$$
$$= P(k-1|k-1)/[r+P(k-1|k-1)]$$

$$\hat{x}(k|k) = \hat{x}(k|k-1) + K(k)[z(k)-\hat{x}(k|k-1)]$$
$$= [1-K(k)]\hat{x}(k|k-1) + K(k)z(k)$$
$$= [r \ \hat{x}(k-1|k-1)+P(k-1|k-1)z(k)]/[r+P(k-1|k-1)]$$

and

$$P(k|k) = [1-K(k)]P(k-1|k-1)$$
$$= r \ P(k-1|k-1)/[r+P(k-1|k-1)]$$

It is clear that the new estimate $\hat{x}(k|k)$ is a linear combination of

(1) the old estimate $\hat{x}(k-1|k-1)$ weighted by the variance of the new measurement, and

(2) the new measurement $z(k)$ weighted by the variance of the old estimate.

This is a consequence of the trade-off between the confidence we have in the old estimates and those in the new measurements.

On using the above relations in conjunction with the data $\hat{x}(-1|-1) = m$, $P(-1|-1) = W$, we obtain the desired results:

$$\hat{x}(s|s) = [rm + W \sum_{j=1}^{s} z(j)]/(r+sW)$$

$$P(s|s) = rW/(r+sW)$$

These results show that

(i) as we use new measurement records, the variance $P(s|s)$ of the estimation error decreases,

(ii) in the limit when $s \to \infty$, all traces of the initial conditions disappear, and we have

$$\lim_{s\to\infty} \hat{x}(s|s) = \sum_{j=1}^{s} z(j)/s$$

$$\lim_{s\to\infty} P(s|s) = \lim_{s\to\infty} 0/W = 0$$

which means that the estimate asymptotically approaches the arithmetic mean of the measurement records.

Example 3

In our third example, we consider the scalar process

$$x(k+1) = a\, x(k) + w(k)$$

$$z(k) = x(k) + v(k)$$

with the standard assumptions : $\{w(k)\}$ is a zero mean Gaussian white sequence with constant variance q, $\{v(k)\}$ is a zero mean Gaussian white sequence with constant variance r, $x(0)$ is a zero mean Gaussian random variable with variance $W(0)$

and a is a constant. We also assume that the two Gaussian white sequences and $x(0)$ are independent.

The linear minimum variance filter equation is

$$\hat{x}(k|k) = \hat{x}(k|k-1) + K(k)[z(k)-\hat{x}(k|k-1)]$$

since $H(k) = 1$ for all k.

From (6.94) and (6.103), we obtain the results:

$$P(k|k-1) = a^2P(k-1|k-1) + q$$

and

$$K(k) = [a^2P(k-1|k-1)+q][a^2P(k-1|k-1)+q+r]^{-1}$$
$$= [a^2P(k-1|k-1)+q]/[a^2P(k-1|k-1)+q+r]$$

respectively. The filtering error variance equation is then seen to be

$$P(k|k) = [1-K(k)]P(k|k-1)$$
$$= r[a^2P(k-1|k-1)+q]/[a^2P(k-1|k-1)+q+r]$$

subject to the initial condition $P(-1|-1) = W(0)$.

Since $P(k-1|k-1) \geq 0$ by definition, then we can see that $P(k|k-1) \geq q$. This means that the variance of the system disturbance sets the performance limit on the prediction accuracy.

It is readily seen from the gain equation that $0 \leq K(k) \leq 1$ for most cases. Combining the gain equation with the error covariance equation, we arrive at

$$P(k|k) = r\,K(k)$$

which entails that $0 \leq P(k|k) \leq r$ for $k \geq 0$.

Suppose that $W(0)$ is very large ($>>r$). Then the first measurement record $z(0)$ will reduce the filtering error variance from $W(0)$ to $P(0|0) \leq r << W(0)$.

Another point to note is that when $q >> r$. In this case we see that $K(k) \approx 1$ and $P(k\ k)\quad r$ for all k. The interpretation of this is that the performance limit on filtering accuracy is now set by the measurement error variance.

On examining the asymptotic behaviour of the error variance $P(k|k)$ with $q = 0$, we find that

$$P = r\ a^2P/a^2P+r$$

where $$P \equiv P(k|k) = P(k-1|k-1)$$

The above expression possesses two solutions:

$$P = 0 \quad \text{and} \quad P = (a^2-1)r/a^2$$

We note that $P = 0$ is allowed only when $a^2 < 1$ since P is a variance. To study the nature of the steady-state value P we define

$$\delta P(k|k) = P(k|k) - P$$

$$\delta P(k-1|k-1) = P(k-1|k-1) - P$$

By direct manipulation we obtain:

$$\delta P(k|k) = [a^2r/(a^2P+r)][r/r+a^2P(k-1|k-1)]\delta P(k-1|k-1)$$

Now for $a^2 < 1$, it is readily seen that

$$\delta P(k|k) < \delta P(k-1|k-1) \quad \text{for all } k$$

and for $P = 0$,

$$P(k|k) < P(k-1|k-1)$$

so that we conclude that $P = 0$ is a stable equilibrium point for the filtering error variance whenever $a^2 < 1$.

Let us consider the case when $a^2 > 1$. Here we must consider both solution $P = 0$ and $P = (a^2-1)r/a^2$. For $P = 0$, it is easy to show that

$$\delta P(k|k) = a^2 r/[r+a^2 P(k-1|k-1)]\,\delta P(k-1|k-1)$$

In the case when $\delta P(k-1|k-1) = 0$ for some k, we see that $\delta P(k|k) = a^2 \delta P(k-1|k-1)$. This means that even if the filtering error variance becomes zero, it will not remain zero. Hence, $P = 0$ is an unstable equilibrium point. For the second solution, $P = (a^2-1)r/a^2$,

$$\delta P(k|k) = [r/r+a^2 P(k-1|k-1)]\,\delta P(k-1|k-1)$$

Since $a^2 > 1$, the above expression implies that $\delta P(k|k) < \delta P(k-1|k-1)$. Consequently, $P = (a^2-1)r/a^2$ is a stable equilibrium point when $a^2 > 1$.

To summarize, the filtering error variance will converge to zero if $a^2 < 1$ and to $(a^2-1)r/a^2$ if $a^2 > 1$. Therefore, for sufficiently long filtering times, the state of the linear discrete time process can be determined exactly when $-1 < a < 1$, but can only be specified to within an error variance of $(a^2-1)r/a^2$ when $q = 0$.

Next we consider a decentralized computational algorithm for the linear optimal filtering problem.

6.5 Decentralised Computation of the Kalman Filter

Although estimation algorithms have been widely applied to problems in the aerospace field, there have been rather fewer applications for the case of industrial systems, socioeconomic systems, etc. The main reason for this is that the relevant models in the case of industrial and other systems are usually of much higher dimension so that practical utilization of estimation techniques is hindered by computational problems. In this section, we consider the problem of state estimation using decomposition and a multilevel structure.

In the literature, there have been several approaches to state estimation in large-scale systems *[16-20]*. The technique developed in *[17,19]* yield suboptimal estimates. Although the technique of *[16]* gives optimal estimates, it requires excessive *a priori* data information. The optimal Kalman filter algorithm suggested in *[18]* is not readily extendable to a system comprising more than two subsystems. We therefore consider the two-level, decentralized computational structure developed in *[20]*, which provides optimal state estimation and which is applicable to systems comprising N subsystems.

6.5.1 *LINEAR INTERCONNECTED DYNAMICAL SYSTEMS*

Consider a linear discrete-time system comprising N_s interconnected dynamical subsystems defined by:

$$\underline{x}_j(k+1) = A_{jj}(k)\underline{x}_j(k) + \sum_{\substack{m=1\\ j\neq m}}^{N_s} A_{jm}(k)\underline{x}_m(k) + \underline{w}_j(k) \qquad j = 1,2,\ldots,N_s \tag{6.109}$$

with the outputs given by

$$\underline{z}_j(k+1) = H_j(k)\underline{x}_j(k+1) + \underline{v}_j(k+1) \tag{6.110}$$

where $\underline{x}_j(k)$ is an n_j state vector, $\underline{y}_j(k)$ is an r_j observation vector. The noise processes $\{\underline{w}_j(k)\}$ and $\{\underline{v}_j(k)\}$ are uncorrelated zero mean Gaussian white noise sequences with covariances Q_j and R_j respectively. The objective is to develop the equations for the optimal filtered estimate $\hat{\underline{x}}(k+1|k+1)$ of the overall system such that the computations are carried out on a subsystem by subsystem basis; that is, to develop a decentralized computational structure for the optimal Kalman filter.

6.5.2 *THE BASIS OF THE DECENTRALIZED FILTER STRUCTURE*

We saw in the previous section that one of the appealing properties of the Kalman filter from a practical point of view is its recursive nature. This recursive property of the filter arises essentially from the fact that if an estimate is computed, based on measurement records up to that discrete instant, then when receiving another set of measurements one could subtract out from these measurements that part which could be anticipated from the results of the first measurement records. This means that the updating is based on that part of the new records which is orthogonal to the old records. The procedure is repeated up to the desired discrete instant at which we wish to obtain the filtered estimate. We see that a successive orthogonalization procedure constitutes the bulk of the filter computation.

The decentralized filter computation structure we will develop below *[20]* exploits the orthogonalization procedure in an efficient way. In the decentralized filter for systems of the form (6.109) and (6.110), the orthogonalization procedure is performed subsystem by subsystem. This entails that the optimal estimate of the state of subsystem j is obtained by successively orthogonalizing the estimation error based on new measurement records for subsystems $1,\dots,N_s$ with respect to an appropriate space formed by all measurement records of all the subsystems up to that instant. It will be shown later that much computational saving results using this *successive orthogonalization procedure* since at each stage only low order subspaces are manipulated.

The actual orthogonalization procedure that is performed in the Kalman filter is based on the following important geometrical result *[21]*:

Let $\underline{\beta}$ be a member of space H^* of random variables which is a closed subspace of L_2, and let $\hat{\underline{\beta}}$ denote its orthogonal projection on a closed subspace $\mathcal{Z}_1$ of H^* (thus $\hat{\underline{\beta}}_1$ is the best estimate of $\underline{\beta}$ in $\mathcal{Z}_1$). Let $\underline{z}_2$ be an m-dimensional vector of the projections of $\underline{z}_2$ onto $\mathcal{Z}_1$ (thus $\hat{\underline{z}}_2$ is the vector of the best estimates of $\underline{z}_2$ in $\mathcal{Z}_1$). Let $\tilde{\underline{z}}_2 = \underline{z}_2 - \hat{\underline{z}}_2$. Then the projection of β onto the subspace $\mathcal{Z}_1 \oplus \mathcal{Z}_2$, denoted $\hat{\underline{\beta}}$, is

$$\hat{\underline{\beta}} = \hat{\underline{\beta}}_1 + E[\underline{\beta}\tilde{\underline{z}}_2^t]\{E[\tilde{\underline{z}}_2\, \tilde{\underline{z}}_2^t]\}^{-1}\tilde{\underline{z}}_2$$

The above result can be interpreted as:

> $\hat{\underline{\beta}}$ is $\hat{\underline{\beta}}_1$ plus the best estimate of $\underline{\beta}$ in the subspace $\tilde{\mathcal{Z}}$ generated by $\tilde{\underline{z}}_2$.

To see the implication of using this result in our problem, we consider the Hilbert space $\mathcal{Z}$ formed by the measurement records of the overall system. At the discrete instant (k+1), this space is denoted by $\mathcal{Z}(k+1)$. The optimal minimum variance estimate is given by

$$\begin{aligned}\hat{\underline{x}}(k+1|k+1) &= E[\underline{x}(k+1)|\mathcal{Z}(k+1)] \\ &= E[\underline{x}(k+1)|\mathcal{Z}(k)] + E[\underline{x}(k+1)|\tilde{\underline{y}}(k+1|k)]\end{aligned} \tag{6.111}$$

This expression provides an algebraic statement of the geometrical result given above.

We note in (6.111) that the first term can be evaluated directly. The basic idea of the decentralized filter is to decompose the second term such that the optimal estimate $\hat{x}(k+1|k+1)$ is derived using the two terms by considering the estimate as the orthogonal projection of $\underline{x}_j(k+1)$ taken on the Hilbert space generated by:

$$\mathcal{Z}(k) \oplus \tilde{\underline{z}}_1(k+1|k) \oplus \tilde{\underline{z}}_2^1(k+1|k+1) \oplus \tilde{\underline{z}}_3^2(k+1|k+1) \oplus \ldots \ldots \oplus \tilde{\underline{z}}_{N_s}^{N_s-1}(k+1|k+1)$$

where $\tilde{\underline{z}}_j^{j-1}(k+1|k+1)$ is the subspace generated by the subspace of measurement records $\underline{z}_j(k+1)$ and its projection on the subspaces generated by $Z(k) + Z_1(k+1) + \ldots + Z_{j-1}(k+1)$.

To apply the above idea we rewrite (6.111) as

$$\hat{\underline{x}}_j(k+1|k+1) = E[\underline{x}_j(k+1)|Z(k),\underline{z}_1(k+1),\ldots,\underline{z}_{N_s}(k+1)]$$

Manipulation of this expression shows that [20]

$$\begin{aligned}\hat{\underline{x}}_j(k+1|k+1) &= E[\underline{x}_j(k+1)|Z(k),\underline{z}_1(k+1),\underline{z}_2(k+1)+\ldots+\underline{z}_j(k+1),\\ &\qquad \underline{z}_{j+1}(k+1),\ldots,\underline{z}_{N_s-1}(k+1)]\\ &\quad + E[\underline{x}_j(k+1)|\tilde{\underline{z}}_{N_s}^{N_s-1}(k+1|k+1)]\\ &= E[\underline{x}_j(k+1)|Z(k)] + E[\underline{x}_j(k+1)|\tilde{\underline{z}}_1(k+1|k)]\\ &\quad + \sum_{m=2}^{N_s} E[\underline{x}_j(k+1)|\tilde{\underline{z}}_m^{m-1}(k+1|k+1)] \end{aligned} \tag{6.112}$$

where

$$\tilde{\underline{z}}_{N_s}^{N_s-1}(k+1|k+1) = \underline{z}_{N_s}(k+1)-E[\underline{z}_{N_s}(k+1)|Z(k),\underline{z}_1(k+1),\ldots,\underline{z}_{N_s-1}(k+1)] \tag{6.113}$$

On utilizing the idea of successive orthogonalization of the spaces defined above, we obtain the algebraic structure of the decentralized filter.

6.5.3 *THE RECURSIVE EQUATIONS OF THE FILTER*

We now develop the recursive equations of the decentralized filter for state estimation in interconnected systems of the type (6.109) and (6.110). First, we write (6.109) in the compact form

$$\underline{x}_j(k+1) = \sum_{m=1}^{N_s} A_{jm}(k)\underline{x}_m(k) + \underline{w}_j(k) \tag{6.114}$$

$$j = 1,\ldots,N_s$$

In the light of the analysis of Section 6.4.3, the optimal state prediction for the jth subsystem is given by:

$$\hat{\underline{x}}_j(k+1|k) = \sum_{m=1}^{N_s} A_{jm}(k)\hat{\underline{x}}_m(k|k) \tag{6.115}$$

The associated prediction error is defined by:

$$\tilde{\underline{x}}_j(k+1|k) = \underline{x}_j(k+1|k) - \hat{\underline{x}}_j(k+1|k) \tag{6.116}$$

A recursive expression for the covariance of the prediction error can be written as:

$$\begin{aligned} P_{jj}(k+1|k) &= \sum_{m=1}^{N_s}\sum_{r=1}^{N_s} A_{jm}(k)P_{mr}(k|k)A_{jr}^t(k) + Q_j(k) \\ &= \sum_{m=1}^{N_s} A_{jm}(k)\left\{\sum_{r=1}^{N_s} P_{mr}(k|k)A_{jr}^t(k)\right\} + Q_j(k) \end{aligned} \tag{6.117}$$

Also,

$$\begin{aligned} P_{jm}(k+1|k) &= \sum_{r=1}^{N_s}\sum_{s=1}^{N_s} A_{jr}(k)P_{rs}(k|k)A_{ms}^t(k) \\ &= \sum_{r=1}^{N_s} A_{jr}(k)\left\{\sum_{s=1}^{N_s} P_{rs}(k|k)A_{ms}^t(k)\right\} \end{aligned} \tag{6.118}$$

Proceeding in parallel to the development which leads to (6.112) and (6.113), it can be shown *[20]* that

$$\begin{aligned} \hat{\underline{x}}_j(k+1|k+1) &= \hat{\underline{x}}_j(k+1|k+1)_s \\ &+ \sum_{m=s+1}^{N_s} P_{x_j\tilde{z}_m^{m-1}}(k+1|k+1)P^{-1}_{\tilde{z}_m^{m-1} z_m^{m-1}}(k+1|k+1)\tilde{\underline{z}}_m^{m-1}(k+1|k+1) \end{aligned} \tag{6.119}$$

where

$$\hat{\underline{x}}_j(k+1|k+1)_s = \hat{\underline{x}}_j(k+1|k+1)_{s-1} + K_{j_s}^{s-1}(k+1)\tilde{\underline{z}}_s^{s-1}(k+1|k+1) \tag{6.120}$$

and

$$P_{jj}(k+1|k+1)_s = P_{jj}(k+1|k+1)_{s-1} - K_{j_s}^{s-1}(k+1)P_{\tilde{z}_s^{s-1}\tilde{x}_{j_{s-1}}}(k+1|k+1) \quad (6.121)$$

where

$$K_{j_s}^{s-1}(k+1) = P^{s-1}_{\tilde{x}_j\tilde{z}_s^{s-1}}(k+1|k+1)P^{-1}_{\tilde{z}_s^{s-1}\tilde{z}_s^{s-1}}(k+1|k+1) \quad (6.122)$$

$$\underline{\tilde{z}}_s^{s-1}(k+1|k+1) = \underline{\tilde{z}}_s^{s-2}(k+1|k+1) - K_{s_{s-1}}^{s-2}(k+1)\underline{\tilde{z}}_{s-1}^{s-2}(k+1|k+1) \quad (6.123)$$

$$K_{s_{s-1}}^{s-2}(k+1) = P_{\tilde{z}_s^{s-2}\tilde{z}_s^{s-2}}(k+1|k+1)P_{\tilde{z}_{s-1}^{s-2}\tilde{z}_{s-1}^{s-2}}(k+1|k+1) \quad (6.124)$$

$$P_{\tilde{z}_s^{s-1}\tilde{z}_s^{s-1}}(k+1|k+1) = P_{\tilde{z}_s^{s-2}\tilde{z}_s^{s-2}}(k+1|k+1) - K_{s_{s-1}}^{s-2}(k+1)P_{\tilde{z}_{s-1}^{s-2}\tilde{z}_{s-1}^{s-2}}(k+1|k+1) \quad (6.125)$$

$$P_{\tilde{z}_s^{s-1}\tilde{z}_s^{s-1}}(k+1|k+1) = H_s(k+1)P_{\tilde{x}_s^{s-1}\tilde{x}_s^{s-1}}(k+1|k+1)H_s^t(k+1) + R_s(k+1) \quad (6.126)$$

$$P_{\tilde{x}_j^{s-1}\tilde{z}_s^{s-1}}(k+1|k+1) = P_{\tilde{x}_j^{s-1}\tilde{x}_s^{s-1}}(k+1|k+1)H_s^t(k+1) \quad (6.127)$$

and

$$P_{jm}(k+1|k+1)_s = P_{jm}(k+1|k+1)_{s-1} - K_{j_s}^{s-1}(k+1)P_{\tilde{z}_s^{s-1}\tilde{z}_m^{s-1}}(k+1|k+1) \quad (6.128)$$

In summary, the relations (6.116)-(6.118) and (6.120)-(6.128) give the recursive equations of the decentralized filter.

The mechanization of the algorithm for one computational cycle of the filter is:

(1) From (6.116)-(6.118) we compute the prediction estimate as well as its error covariance matrix.

(2) Put $s = 1$ and use the conditions

$$\hat{\underline{x}}_j(k+1|k+1)_0 = \hat{\underline{x}}_j(k+1|k)$$

$$P_{jm}(k+1|k+1)_0 = P_{jm}(k+1|k)$$

for $j,m = 1,\ldots,N_s$, together with (6.120) through (6.128) to compute the filtered estimate $\hat{\underline{x}}_j(k+1|k+1)_s$ and the corresponding error covariance matrix.

(3) If $s = N_s$, the resulting estimate is the optimal Kalman estimate and the associated covariance matrix is the minimum error covariance matrix.
For $s < N_s$ go to Step (2).

6.5.4 *A COMPUTATIONAL COMPARISON*

A close examination of the decentralized filter algorithm and the global Kalman filter will reveal that:

(i) Both the decentralized filter and the Kalman filter are algebraically equivalent.

(ii) The computer storage requirements for the decentralized filter are roughly similar to those of the global Kalman filter. However, if the processing is carried out on a multiple-processor configuration then the decentralized filter will be more convenient, since the storage can be distributed between the processors.

(iii) To make a comparison based on the computational time requirements, we adopt as a good measure the number of elementary multiplication operations involved. We first let n be the dimension of state $\underline{x}(k)$, m be the dimension of observation $\underline{z}(k)$ and assume that $H(k)$ is block-diagonal and each subsystem has the same number of states and outputs.

Then it can be shown [20] that the number of multiplications required for the global Kalman filter is given by:

$$M_k = 1.5(n^2+n^3) + mn[1/N_s+(2m+1)/2N_s + m+1 + (n+1)/2] + m^2(3m+1)/2$$

where N_s is the number of subsystems.

Given that all subsystems have equal numbers of states n/N_s and equal number of observations m/N_s, the number of multiplications required for the decentralized filter is given by [20]:

$$\begin{aligned} M_d = {} & 1.5(n^2+n^2) + N_s\{mn/2+mn(m+N_s)/2N_s^3 \\ & + m^2(1+3m/N_s)/2N_s^2 + N_s(N_s-1)n^2m/2N_s^3 \\ & + N_s[n^2m/N_s^3+nm^2/N_s^3 + nm/N_s^2 + nm(n+N_s)/2N_s^3]\} \end{aligned}$$

It is easy to show that for high order systems, the decentralized filter will give substantial savings in computation time.

6.5.5 *Example*

A linearized discrete-time model of a power system comprising 11 coupled synchronous machines [22] can be put in the form:

$$\underline{x}(k+1) = A\underline{x}(k) + \underline{w}(k)$$

where the nth machine is taken as the reference to ensure the complete reachability and observability of the system. Each machine is represented by a second-order model.

Here A is a (20x20) matrix given in Table 6.1). The vector

TABLE 6.1 The A Matrix

1	.00625	0	0	0	0	0	0	0	0	0	0	0	0	0	0	0	0	0	0
-.587	1	.0546	0	.03359	0	.0164	0	.02535	0	-.0103	0	-.0169	0	-.03	0	-.0098	0	-.0706	0
0	0	1	.00625	0	0	0	0	0	0	0	0	0	0	0	0	0	0	-.059	0
.099	0	-.8772	1	0	0	.0875	0	.0417	0	.0107	0	.00949	0	-.0125	0	-.044	0	0	0
0	0	0	0	1	.00025	0	0	0	0	0	0	0	0	0	0	0	0	-.072	0
.0668	0	.0641	0	-.725	1	.0389	0	.0269	0	.0026	0	-.0066	0	-.0293	0	-.0734	0	0	0
0	0	0	0	0	0	1	.00625	0	0	0	0	0	0	0	0	0	0	-.064	0
.0532	0	.1479	0	.0478	0	.8248	1	.0254	0	.0166	0	.0072	0	-.019	0	-.0478	0	0	0
0	0	0	0	0	0	0	0	1	.00625	0	0	0	0	0	0	0	0	-.056	0
.07059	0	.0537	0	.0374	0	.0269	0	-.779	1	.0162	0	-.0004	0	-.008	0	-.069	0	0	0
0	0	0	0	0	0	0	0	0	0	1	.00625	0	0	0	0	0	0	-.062	0
.0123	0	.0122	0	.0105	0	.0075	0	.0107	0	-.623	1	.0121	0	-.0058	0	-.057	0	0	0
0	0	0	0	0	0	0	0	0	0	0	0	1	.00625	0	0	0	0	-.275	0
.044	0	.0538	0	.03359	0	.0474	0	.0344	0	.0736	0	-1.042	1	.035	0	.045	0	0	0
0	0	0	0	0	0	0	0	0	0	0	0	0	0	1	.00625	0	0	-.0636	0
.0042	0	-.0019	0	.00035	0	.0024	0	.00258	0	.00784	0	.0043	0	-.558	1	-.042	0	0	0
0	0	0	0	0	0	0	0	0	0	0	0	0	0	0	0	1	.00625	0	0
.0317	0	.0467	0	.0248	0	.035	0	.0251	0	.038	0	.0155	0	.05	0	-.882	1	-.0029	0
0	0	0	0	0	0	0	0	0	0	0	0	0	0	0	0	0	0	1	.00625
.0099	0	.0087	0	.006	0	.00503	0	.01126	0	.00985	0	.0033	0	.01	0	-.0063	0	.888	1

$\underline{w}(k)$ is a zero mean Gaussian white noise. For the j-th machine the observation is given by:

$$y_j(k) = [0 \quad 1]\underline{x}_j(k) + v_j(k)$$

where $y_j(k)$ is the speed and $v_j(k)$ is a zero mean Gaussian white noise sequence. The associated covariance matrices are:

$$R = I_{20}, \qquad Q = 5I_{11}$$

$$W(0) = 25I_{20}$$

where I_m is the (mxm) identity matrix. The initial estimate of the states was taken to be zero, whilst the initial states were all taken to be 10.

With the *a priori* data given above, both the global and the decentralized filters were simulated over a time horizon of 80 discrete points.

Figures (6.4) to (6.6) show the first three states and the corresponding estimates using the global Kalman filter. Figures (6.7) to (6.9) give the states and the corresponding estimates using the decentralized filter structure.

It is worth noting that the global Kalman filter shows numerical unstability towards the end of the horizon, whilst the decentralized filter is stable. Essentially, numerical errors build up to make the global filter unstable. In the case of the decentralized algorithm, only second-order subsystems are used at each stage and thus avoid numerical inaccuracies so that the resulting filter remains stable.

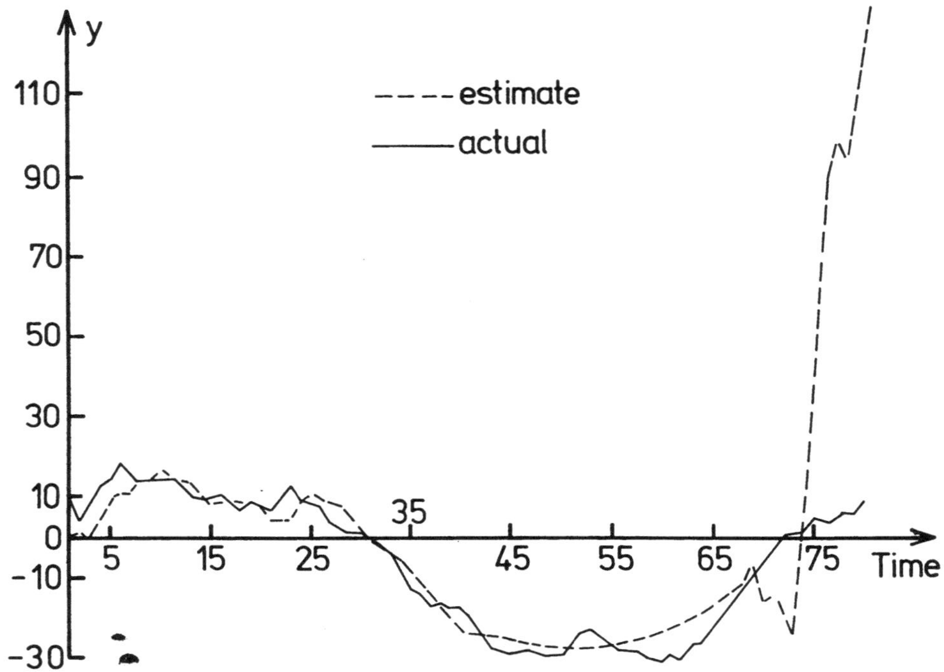

Fig. (6.4) Simulation of the first state and its estimate using the global Kalman filter

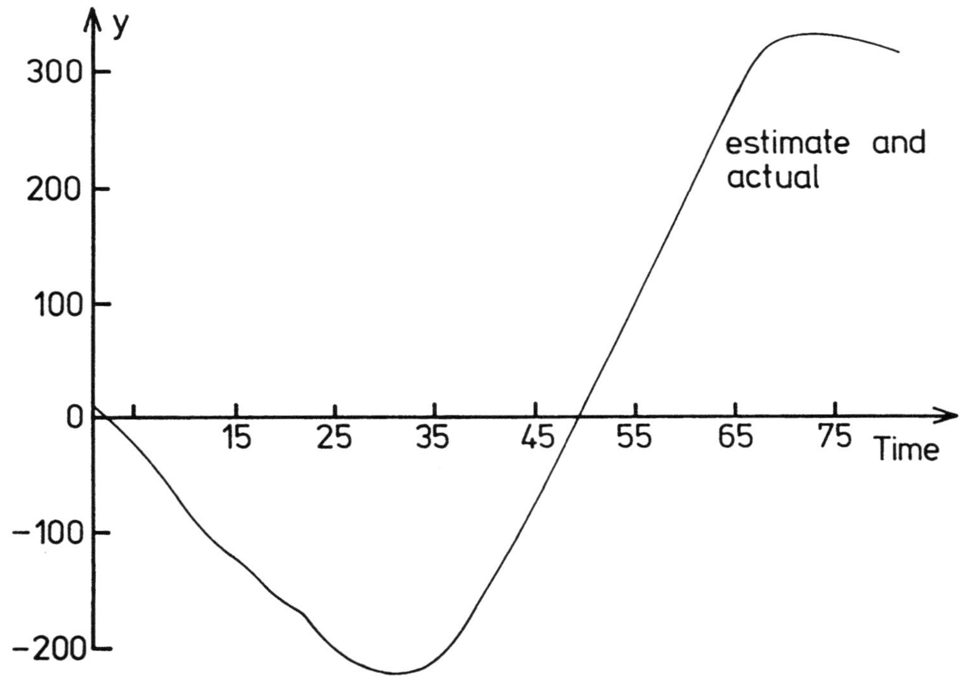

Fig. (6.5) Simulation of the second state and its estimate using the global Kalman filter

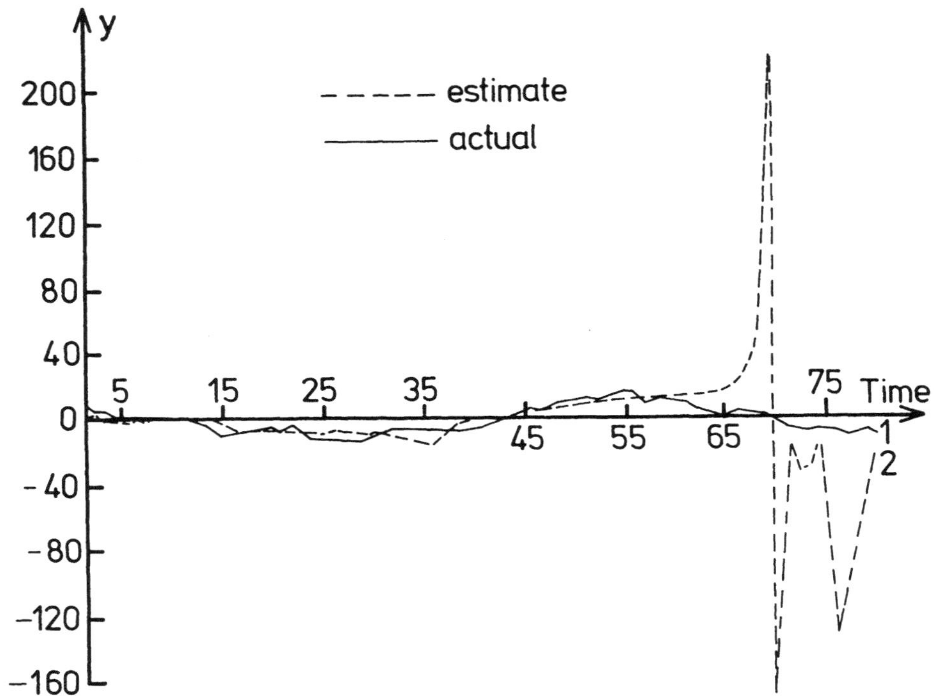

Fig. (6.6) Simulation of the third state and its estimate using the global Kalman filter

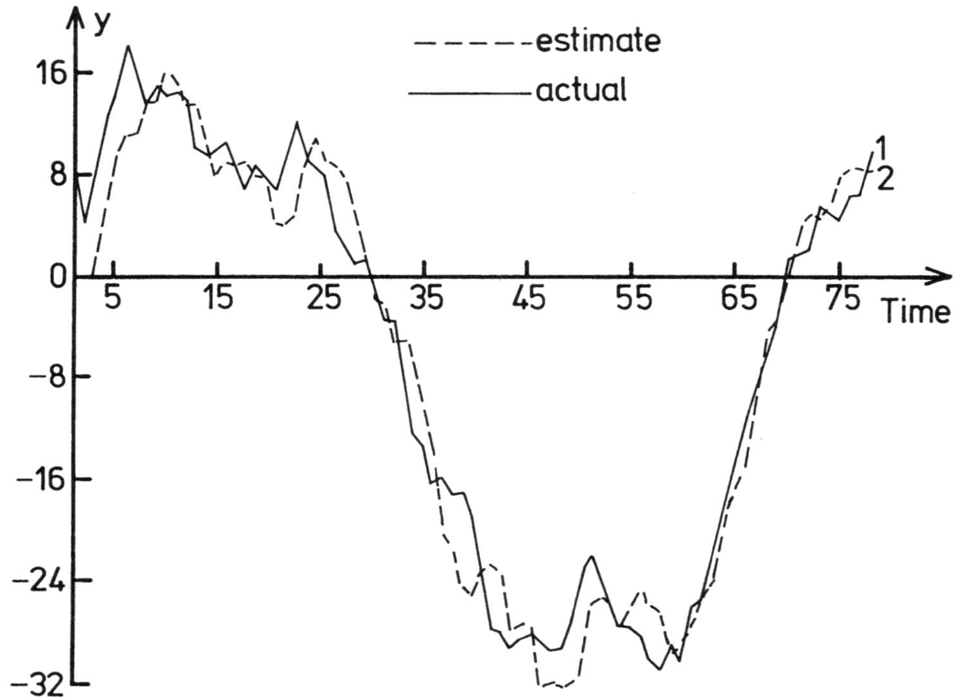

Fig. (6.7) Simulation of the first state and its estimate using the decentralized filter

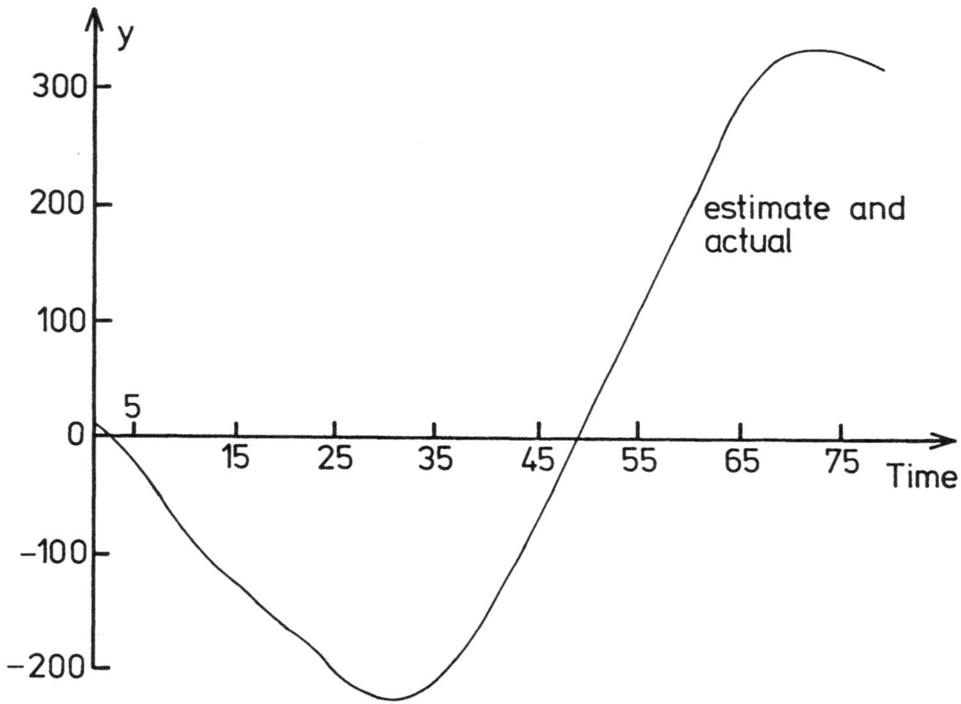

Fig. (6.8) Simulation of the second state and its estimate using the decentralized filter

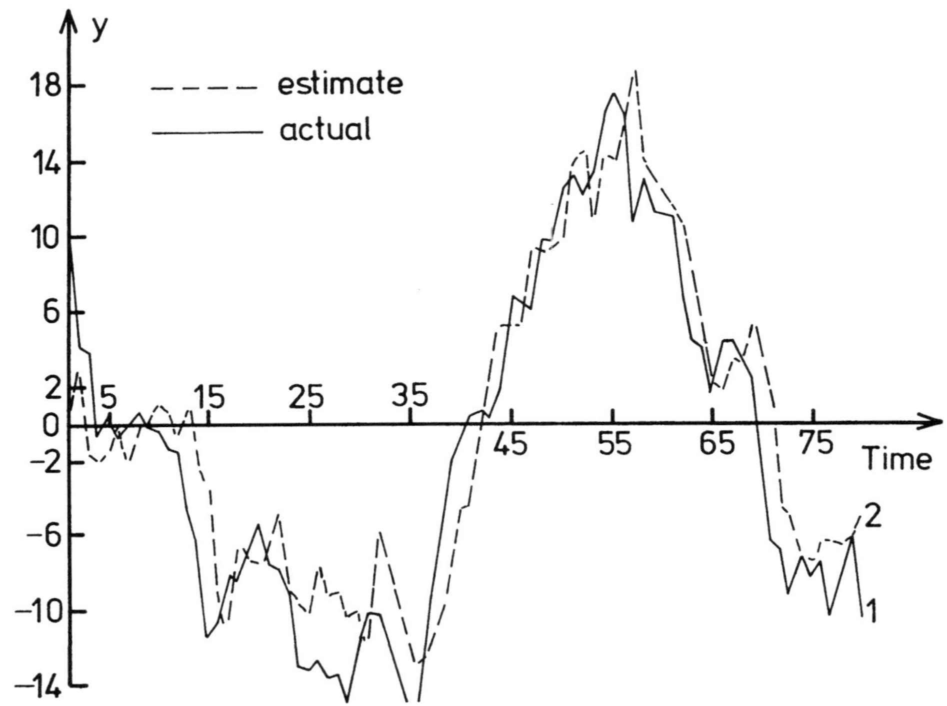

Fig. (6.9) Simulation of the third state and its estimate using the decentralized filter

6.6 Parameter Estimation

In this section, we will consider the problem of parameter estimation in static and dynamic systems. We will consider the least squares method and its variants. We will begin by describing the linear static model, and then we will go on to consider the least squares method for parameter estimation.

6.6.1 *LEAST SQUARES ESTIMATION*

Essentially, the least squares method provides a procedure for estimating the unknown parameters of static models through the minimization of the sum of the squares of the errors. An alternative approach is to apply the maximum likelihood method which yields identical results. This approach is followed here. First, we present the parameter estimation problem.

A. *Linear Static Models*

A model for parameter estimation in linear static systems can be cast into the form:

$$z = H \underline{\Theta} + \underline{e} \tag{6.129}$$

where $\underline{z}$ represents an n-dimensional vector of measurement records, and $\underline{\Theta}$ represents an m-dimensional vector of unknown parameters. The quantity $\underline{e}$ is a random vector which enables us to take into account the errors of measurements. The elements of the transformation matrix H are usually known. Sometimes H is called the observation matrix. H is assumed to be of full rank; that is, rank[H] = m. This assumption is valid in almost all practical situations since it means that we have m independent measurement units to record information about the unknown parameters.

For convenience, we assume that

(1) the error $\underline{e}$ is a zero mean Gaussian vector whose elements have the same variance and which are uncorrelated

with each other; that is,

$$E[\underline{e}] = \underline{0} \tag{6.130a}$$

$$E[\underline{e}\ e^t] = \sigma^2 I_n \tag{6.130b}$$

(2) the unknown vector $\underline{\theta}$ is independent of $\underline{e}$.

B. Standard Least Squares Method and Properties

As mentioned earlier, we will adopt the maximum likelihood estimation approach to determine the estimate $\hat{\underline{\theta}}$ of the unknown parameters by minimizing the sum of the squares of the errors. Recall from Section 6.4 that a possible candidate of the likelihood function is the conditional probability density $p\{\underline{z}|\underline{\theta}\}$. To calculate this function, we note from (6.129) that

$$E[\underline{z}|\underline{\theta}] = H\underline{\theta} \tag{6.131}$$

in view of assumptions (1) and (2) above. Also, note that the covariance matrix is given by:

$$E[(\underline{z}-H\underline{\theta})(\underline{z}-H\underline{\theta})^t|\underline{\theta}]$$

and using (6.129), it reduces to:

$$E[\underline{e}\ \underline{e}^t|\underline{\theta}]$$

In the light of assumption (2), the above expression, with the help of (6.130b), becomes:

$$E[\underline{e}\ \underline{e}^t] = \sigma^2 I_n$$

Thus, we obtain:

$$p\{\underline{z}|\underline{\theta}\} = K \exp\{-\tfrac{1}{2}(\underline{z}-H\underline{\theta})^t(\sigma^2 I_n)^{-1}(\underline{z}-H\underline{\theta}) \tag{6.132}$$

where K is an appropriate constant. We wish to find the

value of $\underline{\Theta}$ which maximizes the conditional probability density defined by (6.132). We saw earlier that the same value maximizes the logarithm of the density function. Thus,

$$\partial/\partial\underline{\Theta}[\text{Log } p\{\underline{z}|\underline{\Theta}\}] = \underline{0}$$

which gives

$$\partial/\partial\underline{\Theta}[(\underline{z}-H\underline{\Theta})^t\sigma^{-2}(\underline{z}-H\underline{\Theta})] = \underline{0}$$

or

$$H\hat{\underline{\Theta}} = \underline{z}$$

Since H is assumed to be of full rank, it possesses a pseudo inverse *[21]* and from the above expression we obtain the least squares estimate as

$$\hat{\underline{\Theta}} = (H^tH)^{-1}H^t\underline{z} \tag{6.133}$$

In view of (6.130a), it is straightforward to show that the least squares estimate (6.133) is the *unbiased estimate*, that is $E[\hat{\underline{\Theta}}] = E[\underline{\Theta}]$.

The estimation error is defined by $\tilde{\underline{\Theta}} = \underline{\Theta}-\hat{\underline{\Theta}}$ and its variance is given by

$$E[\tilde{\underline{\Theta}}\ \tilde{\underline{\Theta}}^t] = E[(\underline{\Theta}-\hat{\underline{\Theta}})(\underline{\Theta}-\hat{\underline{\Theta}})^t]$$

which, upon using (6.129) and (6.133), can be put in the form:

$$\begin{aligned} E[\tilde{\underline{\Theta}}\ \tilde{\underline{\Theta}}^t] &= E[(H^tH)^{-1}H^t\underline{e}\underline{e}^tH(H^tH)^{-1}] \\ &= \sigma^2(H^tH)^{-1} \end{aligned} \tag{6.134}$$

in view of (6.130b). This expression is valid when H is deterministic or when H is a matrix of random variables which are independent of $\underline{\Theta}$.

Now consider the case when the noise variance σ^2 is not known.

To obtain an estimate of σ^2, called $\hat{\sigma}^2$, we will analyse the residual term $[\underline{z}-H\hat{\underline{\theta}}]$. This term can be written using (6.129) and (6.133) as:

$$\begin{aligned}[\underline{z}-H\hat{\underline{\theta}}] &= H\underline{\theta}+\underline{e}-H(H^tH)^{-1}H^t(H\underline{\theta}+\underline{e}) \\ &= [I_n-H(H^tH)^{-1}H^t]\underline{e} \qquad (6.135)\end{aligned}$$

We note $E[\underline{e}^t\underline{e}] = \text{Tr}\, E[\underline{e}\ \underline{e}^t] = n\sigma^2$ where $\text{Tr}[B]$ is the trace of the matrix B. On examining the matrix $S = [I_n-H(H^tH)^{-1}H^t]$ we find that $S = S^t$; that is, the matrix is symmetric and, more importantly,

$$\begin{aligned}S^2 &= [I_n-H(H^tH)^{-1}H^t][I_n-H(H^tH)^{-1}H^t] \\ &= [I_n-H(H^tH)^{-1}H^t] \\ &= S \qquad (6.136)\end{aligned}$$

which means that the matrix S is idempotent. Using the above results, we proceed to examine the variance matrix associated with the residual term in (6.135). Thus,

$$\begin{aligned}E[(\underline{z}-H\hat{\underline{\theta}})^t(\underline{z}-H\hat{\underline{\theta}})] &= E[\underline{e}^tS^2\underline{e}] \\ &= E[\underline{e}^tS\underline{e}] \\ &= E[\text{Tr}[\underline{e}^tS\underline{e}]] \qquad (6.137)\end{aligned}$$

where we have used the fact that $(\underline{e}^tS\underline{e})$ is a scalar and hence can be replaced by its trace. By using the cyclic property of trace operators [6], we have

$$\begin{aligned}E[\text{Tr}[\underline{e}^tS\underline{e}]] &= E[\text{Tr}[S\underline{e}^t\underline{e}]] \\ &= \text{Tr}[S\sigma^2 n] \qquad (6.138)\end{aligned}$$

The substitution of (6.136) into (6.138) yields the variance matrix in the form:

$$E[(\underline{z}-H\hat{\underline{\Theta}})^t(\underline{z}-H\hat{\underline{\Theta}})] = \sigma^2\{n-Tr[H(H^tH)^{-1}H^t]\}$$
$$= \sigma^2\{n-Tr[H^tH(H^tH)^{-1}]\}$$
$$= \sigma^2(n-m) \quad (6.139)$$

so that the estimate $\hat{\sigma}^2$ will be given by:

$$\hat{\sigma}^2 = (1/n-m)(\underline{z}-H\hat{\underline{\Theta}})^t(\underline{z}-H\underline{\Theta}) \quad (6.140)$$

Next, we consider the use of the least squares method in estimating the parameters of dynamic models.

C. Application to Parameter Estimation of Dynamic Models

In Chapter 2, we have shown that a single input-single output discrete model is described by a transfer function of the form:

$$Y(z)/U(z) = (b_0+b_1z^{-1}+\ldots+b_nz^{-n})/(1+a_0z^{-1}+\ldots+a_nz^{-n}) \quad (6.141a)$$

One possible form of the corresponding difference equation can be written as:

$$y(k) = -a_1y(k-1)-\ldots-a_ny(k-n)+b_ou(k)+\ldots+b_nu(k-n)+e(k) \quad (6.141b)$$

Here, $Y(z) = Z[y(k)]$, $U(z) = Z[u(k)]$ and $e(k)$ is a random variable which takes into account the noise or uncertainty in the model.

We assume that $\{e(k)\}$ is a sequence of zero mean, independent random variables having the same probabiliity distribution. Suppose that we study the discrete system (6.141b) for a large number of samples N $(N >> n)$. By iterating (6.141b) N times we obtain:

$$y(k+1) = -a_1 y(k) - \dots - a_n y(k-n+1) + b_0 u(k+1) - \dots + b_n u(k-n+1) + e(k+1)$$
$$\vdots$$
$$y(k+N) = -a_1 y(k+N-1) - \dots - a_n y(k+N-n) + b_0 u(k+N) + \dots + b_n u(k-n+N) + e(k+N)$$

These N difference equations can be put in the compact form:

$$\underline{y}_N = H\underline{\theta}_N + \underline{e}_N \tag{6.142a}$$

with

$$\underline{y}_N = [-y(k+1) \ \dots \ -y(k+N)]^t \tag{6.142b}$$

$$\underline{\theta}_N = [a_1 \dots a_n \ b_0 \ \dots b_n]^t \tag{6.142c}$$

$$\underline{e}_N = [e(k+1) \ \dots \ e(k+N)] \tag{6.142d}$$

and

$$H = \begin{bmatrix} -y(k) & \dots & -y(k-n+1) & u(k+1) & \dots & u(k-n+1) \\ -y(k+1) & \dots & -y(k-n+2) & u(k+2) & & \cdot \\ \vdots & & \vdots & \vdots & & \vdots \\ -y(k+N-1) & \dots & -y(k+N-n) & u(k+N) & \dots & u(k+N-n) \end{bmatrix} \tag{6.142e}$$

Since the model (6.142a) is similar to (6.129), it is thus quite straightforward to write an expression for the least squares estimate $\hat{\underline{\theta}}_N$ of $\underline{\theta}_N$. In view of (6.142c), $\underline{\theta}_N$ will give us the estimates of the parameters of the transfer function (6.141a). Using (6.133), the estimate $\hat{\underline{\theta}}_N$ is given by:

$$\hat{\underline{\theta}}_N = (H^t H)^{-1} H^t \underline{y}_N \tag{6.143}$$

We note by virtue of (6.142e) that the (H^tH) is a $(2n+1) \times (2n+1)$ symmetrical matrix and can be written as:

$$(H^t H) = \begin{bmatrix} C & F \\ F^t & D \end{bmatrix} \tag{6.144a}$$

where the matrix blocks C, F and D have dimensions $(n \times n)$, $n \times (n+1)$ and $(n+1) \times (n+1)$ respectively, and their respective elements take the form:

$$c_{js} = \sum_{m=k-j+1}^{k+N-j} y(m)y(m-s+j) \qquad (6.144b)$$

$$f_{js} = \sum_{m=k-j+1}^{k+N-j} y(m)u(m+j-s+\ ) \qquad (6.144c)$$

$$d_{js} = \sum_{m=k-j+2}^{k+N-j+1} u(m)u(m-s+j-1) \qquad (6.144d)$$

Similarly, the (2n+1) vector $(H^t \underline{y}_N)$ could be written as:

$$(H^t \underline{y}_N) = [\underline{q}^t \quad \underline{g}^t]^t \qquad (6.145a)$$

where the vectors $\underline{q}$ and $\underline{g}$ have dimensions n and (n+1) with respective elements of the form:

$$q_j = \sum_{m=k+1}^{k+N} y(m)y(m-j) \qquad (6.145b)$$

$$g_j = \sum_{m=k+1}^{k+N} y(m)u(m-j+1) \qquad (6.145c)$$

We see from (6.144) and (6.145) that the main computations of the least squares method are of the form of sums of products, which can be easily programmed. It has been shown in *[2]* that the estimate (6.143), under the randomness properties of $\{\underline{e}(k)\}$ and hypotheses concerning $u(k)$, is asumptotically unbiased. Further discussions on the use of the above method can be found in advanced works on parameter estimation theory *[23]*.

D. Recursive Least Squares

In implementing the least squares algorithm (6.143)-(6.145), the data information is first grouped in batches and then processed. When an additional piece of data comes into the system, the entire procedure has to be repeated. This is somewhat unsatisfactory in practice. It would be highly desirable to compute the parameters recursively as the new data records become available.

We now consider the problem of determining the estimate $\hat{\underline{\theta}}_{N+1}$ after (N+1) measurement records, given that the model with N measurements is in the form (6.142). The new observation equation can be obtained from (6.141b) by advancing the arguments (N+1) discrete step, that is:

$$\begin{aligned} y(k+N+1) &= -a_1 y(k+N)-\ldots-a_n y(k+N-n+1) \\ &\quad + b_0 u(k+N+1)+\ldots+u(k+N-n+1)+e(k+N+1) \\ &= \underline{h}^t_{N+1}\,\underline{\theta} + e(k+N+1) \end{aligned} \tag{6.146a}$$

where

$$h^t_{N+1} = [-y(k+N)\ldots y(k+N-N+1)\;\; u(k+N+1)\ldots u(k+N-n+1)] \tag{6.146b}$$

By augmenting (6.142a) and (6.146a) we arrive at

$$\begin{bmatrix} \underline{y}_N \\ \hline y(k+N+1) \end{bmatrix} = \begin{bmatrix} H \\ \hline \underline{h}^t_{N+1} \end{bmatrix} \underline{\theta} + \begin{bmatrix} \underline{e} \\ \hline e(k+N+1) \end{bmatrix} \tag{6.147}$$

which again is in the form (6.129). Hence, the estimate $\hat{\underline{\theta}}_{N+1}$ can be written as

$$\hat{\underline{\theta}}_{N+1} = [H^t H+\underline{h}^t_{N+1}\underline{h}^t_{N+1}]^{-1}[H^t\underline{y}_N+\underline{h}_{N+1}y(k+N+1)] \tag{6.148}$$

Our purpose now is to simplify (6.148) and hopefully put it in a more appropriate computable form. To accomplish this, we will use the well-known matrix inversion lemma *[1]* as given by (6.100). In the present case we obtain:

$$\begin{aligned} [H^t H+\underline{h}_{N+1}\underline{h}^t_{N+1}]^{-1} &= (H^t H)^{-1} \\ &\quad - (H^t H)^{-1}\underline{h}_{N+1}(1+\underline{h}^t_{N+1}(H^t H)^{-1}\underline{h}_{N+1})^{-1}\underline{h}^t_{N+1}(H^t H)^{-1} \end{aligned} \tag{6.149}$$

The substitution of (6.149) into (6.148) using (6.143) yields:

$$\underline{\hat{\theta}}_{N+1} = \underline{\hat{\theta}}_N + (H^tH)^{-1}\underline{h}_{N+1}y(k+N+1)$$

$$- (H^tH)^{-1}\underline{h}_{N+1}(1+\underline{h}^t_{N+1}(H^tH)^{-1}\underline{h}_{n+1})^{-1}\underline{h}^t_{N+1}(H^tH)^{-1}[H^t\underline{y}_N$$

$$+ \underline{h}_{N+1}y(k+N+1)] \qquad (6.150)$$

On using the matrix identity $[I+M]^{-1} = I-[I+M]^{-1}M$ with $M = \underline{h}^t_{N+1}(H^tH)^{-1}\underline{h}_{N+1}$ in (6.150), utilizing (6.143) and rearranging, we obtain:

$$\underline{\hat{\theta}}_{N+1} = \underline{\hat{\theta}}_N + (H^tH)^{-1}\underline{h}_{N+1}[1+\underline{h}^t_{N+1}(H^tH)^{-1}\underline{h}_{N+1}]^{-1}[y(k+N+1) - \underline{h}^t_{N+1}\underline{\hat{\theta}}_N] \qquad (6.151)$$

Let us put

$$K_N = (H^tH)^{-1}\underline{h}_{N+1}[1+\underline{h}^t_{N+1}(H^tH)^{-1}\underline{h}_{N+1}]^{-1} \qquad (6.152a)$$

which leads to the final expression of the estimate after (N+1) measurements

$$\underline{\hat{\theta}}_{N+1} = \underline{\hat{\theta}}_N + K_N[y(k+N+1)-\underline{h}^t_{N+1}\underline{\hat{\theta}}_N] \qquad (6.152b)$$

This expression shows that we can compute the estimate after (N+1) measurement records by using the previous value of the estimate (after N measurements), plus a corrective term which is proportional to the difference between the *predicted value* $(\underline{h}^t_{N+1}\underline{\hat{\theta}}_N)$ and the value of the *measured output* $y(k+N+1)$. The factor K_N could be considered as a "gain" for the corrective term.

We now determine a recursive relation to update the gain K_N for every new observation. By analogy with the expression (6.134) for the error covariance matrix, we write

$$P_N = \beta\,(H^tH)^{-1} \qquad (6.153a)$$

so that

$$K_N = P_N \underline{h}_{N+1} [\beta + \underline{h}^t_{N+1} P_N \underline{h}_{N+1}]^{-1} \quad (6.153b)$$

At the next discrete instant we need to compute

$$P_{N+1} = \beta [H^t H + \underline{h}_{N+1} \underline{h}^t_{N+1}]^{-1}$$

which can be expanded using the matrix inversion lemma into

$$P_{N+1} = P_N - P_N \underline{h}_{N+1} [\beta + \underline{h}^t_{N+1} P_N \underline{h}_{N+1}]^{-1} \underline{h}^t_{N+1} P_N \quad (6.153c)$$

We note that the term $[\beta + \underline{h}^t_{N+1} P_N \underline{h}_{N+1}]$ is actually a scalar so that no matrix inversion is required. Essentially, we use (6.153) in conjunction with (6.152b) to compute the estimate $\hat{\underline{\theta}}_{N+1}$ by storing only the estimate $\hat{\underline{\theta}}_N$ and the symmetric matrix P_N at each iteration.

So far we have assumed in our development of the least squares method that the observation noise $\{\underline{e}(k)\}$ is an independent random sequence. We will next study different models for linear dynamical systems in which the noise sequence is correlated and we will develop methods to handle such cases.

E. The Generalized Least Squares Method

Consider a noisy dynamical system described by the state model

$$\underline{x}(k+1) = A\underline{x}(k) + G\underline{w}(k) \quad (6.154a)$$

$$\underline{y}(k) = H\underline{x}(k) \quad (6.154b)$$

where we assume that the vector $\underline{w}(k)$ is a white nois vector. Combining (6.154a) and (6.154b) and iterating we get:

$$\left.\begin{aligned} \underline{y}(k+1) &= HA\underline{x}(k) + G\underline{w}(k) \\ \underline{y}(k+2) &= HA^2\underline{x}(k) + HAG\underline{w}(k) + HG\underline{w}(k+1) \\ &\vdots \\ \underline{y}(k+m) &= HA^m\underline{x}(k) + \sum_{j=1}^{m} HA^{m-j}\underline{w}(k+j-1) \end{aligned}\right\} \quad (6.155)$$

By the Cayley-Hamilton theorem *[6]*, we have

$$A^m = -\alpha_{m-1}A^{m-1}-\alpha_{m-2}A^{m-2}-\ldots-\alpha_1 A-\alpha_0 I \tag{6.156}$$

Substituting (6.156) into the last relation of (6.155), we arrive at:

$$\underline{y}(k+m) = -H\left[\sum_{s=0}^{m-1} \alpha_s A^s\right]\underline{x}(k)-\sum_{j=1}^{m} H\left[\sum_{s=0}^{m-j} \alpha_s A^s\right]\underline{w}(k+j-1)$$

which, when expanded using the previous relations of (6.155), yields:

$$\underline{y}(k+m) = -\alpha_{m-1}\underline{y}(k+m-1)-\alpha_{m-2}\underline{y}(k+m-2)-\ldots-\alpha_0\underline{y}(k)+\underline{\psi}(k+m) \tag{6.157}$$

In (6.157), the term $\underline{\psi}(k+m)$ is a *correlated* noise sequence. We see thus that the input-output relationship of a linear discrete system, whose dynamics are subjected to white noise, contains a correlated noise sequence. This implies that the least squares method, as developed earlier, would yield a *biased* estimate. Hence, a modification of the present least squares method is required which leads to the *generalized least squares method*. To outline this method we consider the linear discrete model

$$\underline{x}(k+1) = A\underline{x}(k) + Bu(k) + Gw(k) \tag{6.158a}$$

$$y(k) = H\underline{x}(k) \tag{6.158b}$$

whose single input-single output description can be put in the general form (see Chapter 2):

$$\Phi(z^{-1})Y(z) = \Psi(z^{-1})U(z)+\Omega(z^{-1})W(z) \tag{6.159a}$$

where

$$\left.\begin{aligned} \Phi(z^{-1}) &= 1+a_1z^{-1}+\ldots+a_nz^{-n} \\ \Psi(z^{-1}) &= b_0+b_1z^{-1}+\ldots+b_nz^{-n} \\ \Omega(z^{-1}) &= 1+c_1z^{-1}+\ldots+c_nz^{-n} \end{aligned}\right\} \tag{6.159b}$$

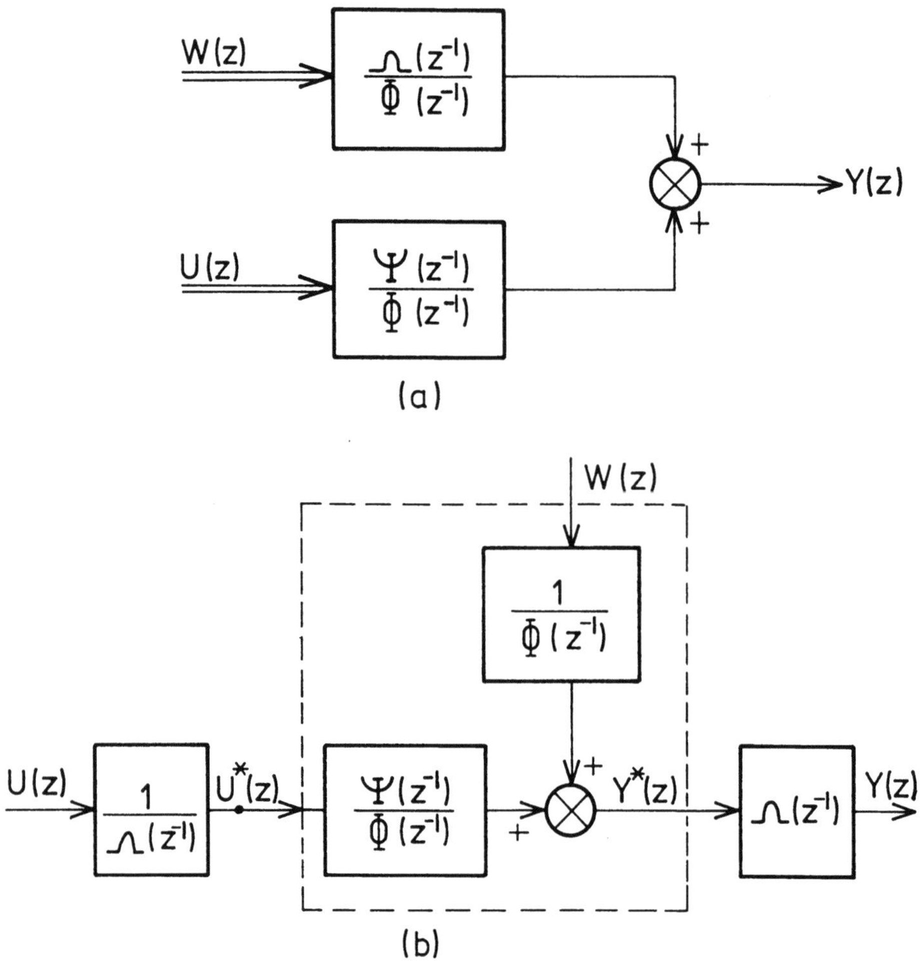

Fig. (6.10) Input-output description of a noisy dynamic system

We assume that

(1) the system (6.158a), or equivalently (6.159a), is stable,

(2) the system (6.158a) is completely reachable and completely observable. This ensures that the model (6.159a) has no common factors amongst the three polynomials $\Phi(z^{-1})$, $\Psi(z^{-1})$ and $\Omega(z^{-1})$.

In the development of the generalized least squares method we begin by writing (6.159a) in the form:

$$\Phi(z^{-1})Y(z)/\Omega(z^{-1}) = \Psi(z^{-1})U(z)/\Omega(z^{-1})+W(z) \qquad (6.160)$$

Block diagram representations of (6.159a) and (6.160) are displayed in Figures (6.10a) and (6.10b).

When $w(k)$ is a white noise sequence we can apply the least squares method to the block diagram inside the dotted lines in Fig. (6.10b). In the case that the polynomial $\Omega(z^{-1})$ is known, we could then design a filter for $1/\Omega(z^{-1})$ and we could operate on the intermediate quantities $U^{*}(z)$ and $Y^{*}(z)$.

In the light of (6.157), it is easy to see that knowledge of $\Omega(z^{-1})$ involves knowing the covariance matrix of the correlated noise which acts on the system. This is rarely possible in practice. The generalized least squares method suggests an iterative procedure to update the noise covariance matrix while monitoring the residual term (difference between the measured and predicted values of the output sequence). The different steps involved are summed up below *[23,25]:*

Step 1

Select an *a priori* filter which initially realizes $1/\Omega(z^{-1})$, call it $\Omega_j(z^{-1})$, such that

$$\Omega_j(z^{-1} = 1+\omega_{j1}z^{-1}+\ldots+\omega_{jn}z^{-n}$$

and set $j = 1$.

Step 2

Compute the filtered values

$$U_j^*(z) = \Omega_j(z^{-1})U(z)$$
$$Y_j^*(z) = \Omega_j(z^{-1})Y(z)$$

Step 3

Evaluate the estimate $\hat{\underline{\Theta}}_j$ by $\hat{\underline{\Theta}}_j$ by applying the least squares method to the model

$$\Phi(z^{-1})Y_j^*(z) = \Psi(z^{-1})U_j^*(z) + W(z)$$

Step 4

Compute the noise sequence

$$w(k) = \phi(k)y(k) - \psi(k)u(k)$$

where $\phi(k) = Z^{-1}[\Phi(z^{-1})]$ and $\psi(k) = Z^{-1}[\Psi(z^{-1})]$

Step 5

From the residual relation

$$\varepsilon_j(k) = w(k)/Z^{-1}[\Omega_j(z^{-1})]$$

construct the recurrence relationships

$$\varepsilon_j(k) = -\omega_{j1}\varepsilon(k-1)-\ldots-\omega_{jn}\varepsilon_j(k-n)+w(k)$$
$$\vdots \qquad\qquad \vdots$$
$$\varepsilon_j(k+m) = -\omega_{j1}(k+m-1)-\ldots-\omega_{jn}\varepsilon(k+m-n)+w(k+m)$$

and apply the standard least squares method to determine the coefficients of the filter which will give us the new filter $\Omega_{j+1}(z^{-1})$. Note that $w(k), w(k+1), \ldots.$ are independent random variables.

Step 6

Repeat the procedure with $\Omega_{j+1}(z^{-1}), \Omega_{j+2}(z^{-1}), \ldots$ until the estimates converge.

Convergence properties of the above procedure have not been studied theoretically, but on applications the simulation results were satisfactory. Other variants of the least squares method can be found in *[23,26]* which are advanced works in parameter estimation techniques.

Next, we consider parameter estimation techniques for large-scale systems using two-level computational algorithms.

6.6.2 *TWO-LEVEL COMPUTATIONAL ALGORITHMS*

Many problems arising in systems engineering applications are described by high dimensional models. The practical utilization of parameter estimation techniques to determine the parameters of the model is hindered by computational difficulties. It is thus desirable to seek estimation techniques which distribute the computational effort. In this subsection we consider the problem of parameter estimation using decomposition and a two-level structure. We will then begin by examining linear static models and then go on to dynamical models.

A. Linear Static Models

It is assumed that the following linear static model is given by

$$\underline{z} = H\underline{\Theta} + \underline{e} \tag{6.161}$$

where $\underline{z}$ is the measurement vector of dimension n, $\underline{\Theta}$ is the random parameter vector (to be estimated) of dimension m, $\underline{e}$ is the measurement error vector of dimension n and H is the (nxm) transformation matrix with known elements. We assume that $\{\underline{\Theta}\}$ and $\{\underline{e}\}$ are independent Gaussian random vectors with the statistical parameters:

$$\left.\begin{aligned} E[\underline{\theta}] &= \underline{m}, & \text{var}\{\underline{\theta}\} &= P \\ E[\underline{e}] &= \underline{0}, & \text{var}\{\underline{e}\} &= Q \end{aligned}\right\} \tag{6.161}$$

and Q is a full matrix. We know from subsection 6.6.1 that the least-squares estimate (minimum variance estimate) is given by $\hat{\underline{\theta}} = E[\underline{\theta}|\underline{z}]$ or equivalently (6.133). For large values of n, we would like to compute this estimate within a two-level structure. To accomplish this, we decompose the measurement vector $\underline{z}$ and the parameter vector $\underline{\theta}$ into N_s subvectors such that

$$\underline{z} = \begin{bmatrix} \underline{z}_1 \\ \underline{z}_2 \\ \vdots \\ \underline{z}_{N_s} \end{bmatrix} ; \quad \underline{\theta} = \begin{bmatrix} \underline{\theta}_1 \\ \underline{\theta}_2 \\ \vdots \\ \underline{\theta}_{N_s} \end{bmatrix} \tag{6.162a}$$

where $\underline{z}_j$ and $\underline{\theta}_j$ are n_j- and m_j-vectors respectively, with

$$\sum_{j=1}^{N_s} n_j = n ; \quad \sum_{j=1}^{N_s} m_j = m \tag{6.162b}$$

In terms of the decomposition defined by (6.162), the jth submodel can be expressed as:

$$\underline{z}_j = H_{jj}\underline{\theta}_j + \sum_{\substack{k=1 \\ k\neq j}}^{N_s} H_{jk}\underline{\theta}_k + \underline{e}_j \tag{6.163}$$

The development of the two-level algorithm to determine the minimum variance estimate for the collection of submodels (6.163), $j = 1,\ldots,N_s$ is based on the multiple projection approach *[20]* which has been presented in Section 6.5. As discussed there, this algorithm incorporates a series of successive orthogonalization on the measurement subspace (as parts of the whole space) for each subsystem (which constitutes a portion of the overall system) within a two-level structure in order to provide the optimal estimate.

B. *A Two-Level Multiple Projection Algorithm*

A detailed derivation of the two-level algorithm is given in *[24]*. The main results are summarised below:

(1) The optimal minimum variance estimate $\hat{\underline{\theta}}_j^{N_s}$ is given by:

$$\hat{\underline{\theta}}_j^{N_s} = \underline{m}_j + P_{\theta_j z_1} P^{-1}_{z_1 z_1}\left[\underline{z}_1 - \sum_{k=1}^{N_s} H_{1k}\underline{m}_k\right] + \sum_{k=2}^{N_s} P_{\tilde{\theta}_j^{k-1}\tilde{z}_k^{k-1}} P^{-1}_{\tilde{z}_k^{k-1}\tilde{z}_k^{k-1}} \tilde{\underline{z}}_k^{k-1}$$

$$= \hat{\underline{\theta}}_j^{r} + \sum_{k=r+1}^{N_s} P_{\tilde{\theta}_j^{k-1}\tilde{z}_k^{k-1}} P^{-1}_{\tilde{z}_k^{k-1}\tilde{z}_k^{k-1}} \tilde{\underline{z}}_k^{k-1} \qquad (6.164a)$$

where

$$\hat{\underline{\theta}}_j^{r} = \underline{m}_j + P_{\theta_j z_1} P^{-1}_{z_1 z_1}\left[\underline{z}_1 - \sum_{k=1}^{N_s} H_{1k}\underline{m}_k\right] + \sum_{s=2}^{r} P_{\tilde{\theta}_j^{s-1}\tilde{z}_s^{s-1}} P^{-1}_{\tilde{z}_s^{s-1}\tilde{z}_s^{s-1}} \tilde{\underline{z}}_s^{s-1} \qquad (6.164b)$$

is the estimate after r iterations between the two-levels.

(2) The different covariance matrices are:

$$P_{\theta_j z_k} = \sum_{r=1}^{N_s} P_{\theta_j \theta_r} H^t_{kr} \qquad (6.165a)$$

$$P_{z_j z_j} = H_{jj}\left[P_{\theta_j\theta_j} H^t_{jj} + \sum_{\substack{k=1\\k\neq j}}^{N_s} P_{\theta_j\theta_k} H^t_{jk}\right] + \left[\sum_{\substack{k=1\\k\neq j}}^{N_s} H_{jk} P_{\theta_k\theta_j}\right] H^t_{jj} + \sum_{\substack{r=1\\r\neq j}}^{N_s} H_{jr} \sum_{\substack{k=1\\k\neq j}}^{N_s} P_{\theta_r\theta_k} H^t_{jk} + Q_{jj} \qquad (6.165b)$$

$$P_{z_j z_k} = H_{jj}[P_{\Theta_j\Theta_k} H_{kk}^t + \sum_{\substack{s=1\\s\neq j}}^{N_s} P_{\Theta_j\Theta_s} H_{ks}^t]$$

$$+ \left[\sum_{\substack{r=1\\r\neq j}}^{N_s} H_{jr} P_{\Theta_r\Theta_k}\right] H_{kk}^t$$

$$+ \sum_{\substack{r=1\\j\neq r}}^{N_s} H_{jr} \sum_{\substack{s=1\\s\neq j}}^{N_s} P_{\Theta_r\Theta_s} H_{ks}^t + Q_{jk} \qquad (6.165c)$$

We note that (6.164) provides the description and the associated procedure for computing the optimal linear least-squares estimate. The expressions (6.165a)-(6.165c) give the propagation of the covariance matrices. When taken together, (6.164a) -(6.165c) establish the trend of the average patterns. Evaluation of the estimation errors is given next.

(3) The measurement estimation error $\underline{\tilde{z}}_j^r$ (for the jth submodel after r iterations) and the associated covariance matrices are given by:

$$\underline{\tilde{z}}_j^r = \underline{\tilde{z}}_j^{r-1} - P_{\tilde{z}_j^{r-1}\tilde{z}_r^{r-1}} P^{-1}_{\tilde{z}_r^{r-1}\tilde{z}_r^{r-1}} \underline{\tilde{z}}_r^{r-1} \qquad (6.166a)$$

$$P_{\tilde{z}_r^{r-1}\tilde{z}_r^{r-1}} = P_{\tilde{z}_r^{r-2}\tilde{z}_r^{r-2}} - P_{\tilde{z}_r^{r-2}\tilde{z}_{r-1}^{r-2}} P^{-1}_{\tilde{z}_{r-1}^{r-2}\tilde{z}_{r-1}^{r-2}} P_{\tilde{z}_{r-1}^{r-2}\tilde{z}_r^{r-2}} \qquad (6.166b)$$

and in general,

$$P_{\tilde{z}_j^{r-1}\tilde{z}_j^{r-1}} = P_{\tilde{z}_j^{r-2}\tilde{z}_j^{r-2}} - P_{\tilde{z}_j^{r-2}\tilde{z}_{r-1}^{r-2}} P^{-1}_{\tilde{z}_{r-1}^{r-2}\tilde{z}_{r-1}^{r-2}} P_{\tilde{z}_{r-1}^{r-2}\tilde{z}_j^{r-2}} \qquad (6.166c)$$

$$P_{\tilde{z}_j^{r-2}\tilde{z}_{j-1}^{r-2}} = P_{\tilde{z}_j^{r-3}\tilde{z}_{j-1}^{r-3}} - P_{\tilde{z}_j^{r-3}\tilde{z}_{r-2}^{r-3}} P^{-1}_{\tilde{z}_{r-2}^{r-3}\tilde{z}_{r-2}^{r-3}} P_{\tilde{z}_{r-2}^{r-3}\tilde{z}_{j-1}^{r-3}} \qquad (6.166d)$$

(4) The variance and covariance matrices of the parameter

estimation error propagate according to:

$$P_{\tilde{\theta}_j^r \tilde{\theta}_j^r} = P_{\tilde{\theta}_j^{r-1} \tilde{\theta}_j^{r-1}} - P_{\tilde{\theta}_j^{r-1} \tilde{z}_r^{r-1}} P^{-1}_{\tilde{z}_r^{r-1} \tilde{z}_r^{r-1}} P_{\tilde{z}_r^{r-1} \tilde{r}_j^{r-1}} \qquad (6.167a)$$

$$P_{\tilde{\theta}_j^r \tilde{\theta}_k^r} = P_{\tilde{\theta}_j^{r-1} \tilde{\theta}_k^{r-1}} - P_{\tilde{\theta}_j^{r-1} \tilde{z}_r^{r-1}} P^{-1}_{\tilde{z}_r^{r-1} \tilde{z}_r^{r-1}} P_{\tilde{z}_r^{r-1} \tilde{r}_j^{r-1}} \qquad (6.167b)$$

$$P_{\tilde{\theta}_j^{r-1} \tilde{z}_r^{r-1}} = P_{\tilde{\theta}_j^{r-2} \tilde{z}_r^{r-2}} - P_{\tilde{\theta}_j^{r-2} \tilde{z}_{r-1}^{r-2}} P^{-1}_{\tilde{z}_{r-1}^{r-2} \tilde{z}_{r-1}^{r-2}} P_{\tilde{z}_{r-1}^{r-2} \tilde{z}_r^{r-2}} \qquad (6.168c)$$

It should be emphasized that the optimal estimate is obtained after N_s iterations. We can easily show that when Q is block-diagonal (uncorrelated Gaussian sequence), the optimal estimate is also obtained after N_s iterations. We note in this case that

$$E[\underline{e}_k \, \underline{z}_{k-j}^t] = 0 \qquad \text{for all } j = 1,\ldots,k-1$$

C. The Recursive Version

Following the development of the recursive least-squares algorithm, we now consider a recursive version of the two-level multiple projection algorithm. Let us assume that the model (6.161) is obtained after k measurement records. For emphasis we rewrite it as:

$$\underline{z}_k = H_k \underline{\theta} + \underline{e}_k \qquad (6.168a)$$

where the subscript k denotes discrete instants and not vector components. Now, if we received a new measurement at k+1 given by:

$$\underline{\alpha}_{k+1} = h_{k+1} \underline{\theta} + \underline{v}_{k+1} \qquad (6.168b)$$

then by adjoining (6.168b) to (6.168a) we can obtain:

$$\underline{z}_{k+1} = H_{k+1} \underline{\theta} + \underline{e}_{k+1} \qquad (6.169a)$$

where

$$\underline{z}_{k+1} = \begin{bmatrix} \underline{z}_k \\ --- \\ {}_{k+1} \end{bmatrix} ; \quad H_{k+1} = \begin{bmatrix} H_k \\ --- \\ h_{k+1} \end{bmatrix} ; \quad \underline{e}_{k+1} = \begin{bmatrix} \underline{e}_k \\ --- \\ \underline{v}_{k+1} \end{bmatrix} \tag{6.169b}$$

The minimum variance estimate based on (k+1) measurement records, $\hat{\underline{\theta}}_{k+1}$, is given by

$$\begin{aligned} \hat{\underline{\theta}}_{k+1} &= E[\underline{\Theta} | \underline{z}_{k+1}] \\ &= E[\underline{\Theta} | \underline{z}_k, \underline{\alpha}_{k+1}] \\ &= E[\underline{\Theta} | \underline{z}_k] + E[\underline{\Theta} | \tilde{\underline{\alpha}}_{k+1}] - \underline{m} \end{aligned} \tag{6.170}$$

which can be put in the form:

$$\hat{\underline{\Theta}}_{k+1} = \hat{\underline{\theta}}_k + P_{\Theta\tilde{\alpha}^k_{k+1}} P^{-1}_{\tilde{\alpha}^k_{k+1}\tilde{\alpha}^k_{k+1}} \tilde{\underline{\alpha}}^k_{k+1} \tag{6.171}$$

where the innovations vector $\tilde{\underline{\alpha}}^k_{k+1}$ is given by:

$$\begin{aligned} \tilde{\underline{\alpha}}^k_{k+1} &= \underline{\alpha}_{k+1} - E[\underline{\alpha}_{k+1} | \underline{z}_k] \\ &= \underline{\alpha}_{k+1} - h_{k+1} E[\underline{\Theta} | \underline{z}_k] - E[\underline{v}_{k+1} | \underline{z}_k] \end{aligned}$$

In the case that $E[\underline{v}_{k+1}\ \underline{e}_k^t] = 0$, it follows that $E[\underline{v}_{k+1}\ \underline{z}_k^t] = 0$ and the above expression reduces to:

$$\begin{aligned} \tilde{\underline{\alpha}}^k_{k+1} &= \underline{\alpha}_{k+1} - h_{k+1}\ \hat{\underline{\theta}}_k \\ &= h_{k+1}\ \tilde{\underline{\theta}}_k + \underline{v}_{k+1} \end{aligned} \tag{6.172}$$

for which the associated error covariance matrix takes the form:

$$\begin{aligned} P_{\tilde{\alpha}^k_{k+1}\tilde{\alpha}^k_{k+1}} &= E[(h_{k+1}\tilde{\underline{\theta}}_k + \underline{v}_{k+1})(\underline{h}_{k+1}\tilde{\underline{\theta}}_k + \underline{v}_{k+1})^t] \\ &= h_{k+1} P_{\tilde{\theta}_k\tilde{\theta}_k} h^t_{k+1} + Q_{k+1} \end{aligned} \tag{6.173}$$

In a similar way, it is easy to show [24] that

$$P_{\theta\tilde{\alpha}^{k}_{k+1}} = P_{\tilde{\theta}_k\tilde{\theta}_k} h^t_{k+1} = P^t_{\tilde{\alpha}^{k}_{k+1}\underline{\theta}} \tag{6.174}$$

and the estimation error covariance matrix (after k+1 measurements) is given by:

$$P_{\tilde{\theta}_{k+1}\tilde{\theta}_{k+1}} = P_{\tilde{\theta}_k\tilde{\theta}_k} - P_{\theta\tilde{\alpha}^{k}_{k+1}} P^{-1}_{\tilde{\alpha}^{k}_{k+1}\tilde{\alpha}^{k}_{k+1}} P_{\tilde{\alpha}^{k}_{k+1}\theta} \tag{6.175}$$

It should be noted that the above recursive algorithm has not considered the decomposition of $\underline{\theta}$ and $\underline{\alpha}_{k+1}$. In principle, this can be easily done using the multiple projection approach, see [24] for details.

Since the two-level multiple projection algorithm is based on the same idea as the decentralized filter developed in Section 6.4, one should expect that the implementation of the two-level algorithm would require less computational effort than the centralized least squares algorithm. This is particularly noticeable in cases where a large number of parameters are to be estimated.

Next, we consider parameter estimation of linear dynamical systems using the maximum *a posteriori* approach.

D. Linear Dynamical Models

In the previous subsection, we discussed the application of the least squares method in estimating the parameters of linear dynamical models which are derived from single-input single-output transfer functions. Here, we consider linear dynamical models with multi-inputs and multi-outputs. Our objective is to develop methods to determine the unknown parameters of the models. Let the state space model of a linear dynamical model be described by:

$$\underline{X}(k+1) = A\underline{X}(k) + \Gamma\underline{w}(k) \tag{6.176a}$$

$$\underline{Y}(k) = H\underline{X}(k) + \underline{v}(k) \tag{6.176b}$$

in which both the state and measurement equations are contaminated by the noise processes $\{\underline{w}(k)\}$ and $\{\underline{v}(k)\}$ respectively. We are interested in the cases where the coefficients of the system matrix A and/or the observation matrix H are unknown constant parameters. One possible way to deal with such cases is to collect these constant parameters into one vector $\underline{\alpha}$ and define it by the simple difference equation

$$\underline{\alpha}(k+1) = \underline{\alpha}(k) \qquad \text{for all } k \tag{6.177}$$

We can now treat $\underline{\alpha}(k)$ as new state variables. By adjoining (6.177) to (6.176) and defining

$$\underline{x}(k) = [\underline{X}^t(k) \quad \underline{\alpha}^t(k)]^t$$

as the overall state vector, it follows that we can express the dynamics of $\underline{x}(k)$ as:

$$\underline{x}(k+1) = \underline{\Phi}[\underline{x}(k),k] + G\underline{w}(k) \tag{6.178a}$$

$$\underline{y}(k) = \underline{h}[\underline{x}(k),k] + \underline{v}(k) \tag{6.178b}$$

where $\underline{x}(k)$ is the n-dimensional vector of the states and parameters to be estimated, $\Phi[.,.]$ is the n-dimensional nonlinear, vector-valued function which describes the structure of the system and includes any deterministic inputs, $y(k)$ is the m-dimensional measurement vector and $\underline{h}[.,.]$ is the m dimensional nonlinear, vector-valued observation function. The noise vectors $\underline{w}(k)$ and $\underline{v}(k)$ are uncorrelated white Gaussian sequences with means and variances given by:

$$E[\underline{w}(k)] = \underline{q}(k) \; ; \qquad E[\underline{w}(k)\underline{w}^t(j)] = Q(k)\delta_{kj}$$

$$E[\underline{v}(k)] = \underline{r}(k) \; ; \qquad E[\underline{v}(k)\underline{v}^t(j)] = R(k)\delta_{kj}$$

where δ_{kj} is the Kronecker delta. The matrices $Q(k)$ and

R(k) are assumed to be positive definite and block-diagonal. The initial state is also assumed to be a Gaussian random vector with

$$E[\underline{x}(0)] = \underline{m}(0) \; ; \qquad E[\underline{x}(0)\underline{x}^t(0)] = P(0)$$

where P(0) is also assumed to be block diagonal and positive definite.

The important thing to note is that the introduction of (6.177) converts the linear model (6.176) to the nonlinear model (6.178). However, the benefit gained by this formulation is that we could estimate both the state and parameters of the dynamical model.

E. The Maximum A Posteriori Approach

To determine the estimate of the state $\underline{x}(k)$ of the dynamical model (6.178), we will use the maximum *a posteriori* (MAP approach. In subsection 6.4.2, it is shown that the MAP estimate can be obtained by maximizing the conditional distribution $p\{\underline{x}(k)|\underline{Y}(k)\}$ where $\underline{Y}(k)$ corresponds to $\underline{Z}(k)$ in (6.84). It turns out *[1,7]* that this maximization is equivalent to the minimization of the cost function

$$J = \frac{1}{2}||\underline{x}(0)-\underline{m}(0)||^2_{P^{-1}(0)} + \frac{1}{2}\sum_{k=0}^{N-1}\{||\underline{z}(k+1) - \underline{h}[\underline{x}(k+1),k+1]-\underline{r}(k+1)||^2_{R^{-1}(k+1)} + ||\underline{w}(k)-\underline{q}(k)||^2_{Q^{-1}(k)}\} \tag{6.179}$$

with respect to $\underline{w}(k)$ and subject to the dynamical constraints (6.178). Note that $||\underline{d}||^2_C = \underline{d}^t C\underline{d}$ and N is the terminal discrete instant. This problem is a standard dynamic optimization one (see Chapter 8). We will tackle the problem by a decomposition/coordination method. To do this, we introduce an interconnection vector $\underline{\pi}_j(k)$ defined by:

$$\underline{\pi}_j(k) = \underline{g}_j[\underline{x}_m(k),k] \qquad j \neq m \tag{6.180}$$

which when used in (6.178) will decompose the overall model into a group of lower-order submodels *[27-30]*. Thus we can write the jth submodel of order n_j as

$$\underline{x}_j(k+1) = [\underline{\Phi}_j\ \underline{x}_j,\underline{\pi}_j,k] + G_j\underline{w}_j(k) \tag{6.181a}$$

$$\underline{y}_j(k) = \underline{h}_j[\underline{x}_j,\underline{\pi}_j,k] + \underline{v}_j(k) \tag{6.181b}$$

Note that we used $\underline{f}[\underline{x},\underline{y},\underline{z},k] \triangleq \underline{f}[\underline{x}(k),\underline{y}(k),\underline{z}(k),k]$ in (6.181) for convenience. Also,

$$\sum_{j=1}^{N_s} n_j = n$$

where N_s is the number of submodels. The cost functional (6.179) can also be decompose as:

$$J = \sum_{j=1}^{N_s} J_j$$

$$= \sum_{j=1}^{N_s} \frac{1}{2}\{||\underline{x}_j(0)-\underline{m}_j(0)||^2_{P_j^{-1}(0)}$$

$$+ \sum_{k=0}^{N-1} ||\underline{z}_j(k+1)-\underline{h}_j[\underline{x}_j,\underline{\pi}_j,k+1]-\underline{r}_j(k+1)||^2_{R_j^{-1}(k+1)}$$

$$+ ||\underline{w}_j(k) - \underline{q}_j(k)||^2_{Q_j^{-1}(k)}\} \tag{6.182}$$

where J_j is the cost functional associated with the jth sub-problem. Continuing this effort, we write the corresponding Hamiltonian of the overall problem H and of the jth subproblem H_j as follows *[11]*:

$$H = \sum_{j=1}^{N_s} H_j =$$

$$= \sum_{j=1}^{N_s} \tfrac{1}{2}\{||\underline{z}_j(k+1)-\underline{h}_j[\underline{x}_j,\underline{\pi}_j,k+1]-\underline{r}_j(k+1)||^2_{R_j^{-1}(k+1)}$$

$$+ ||\underline{w}_j(k)-\underline{q}_j(k)||^2_{Q_j^{-1}(k)} + \underline{\beta}_j^t(k)[\underline{\pi}_j(k)-\underline{g}_j(\underline{x}_j,k)]$$

$$+ \underline{\lambda}_j^t(k+1)[\underline{\Phi}_j[\underline{x}_j,\underline{\pi}_j,k] + G_j\underline{w}_j(k)]\} \tag{6.183}$$

where $\underline{\beta}_j^t(k)$ and $\underline{\lambda}_j^t(k)$ are Lagrange multiplier vectors introduced to append the equality constraints (6.180) and (6.181) to the cost functional (6.182).

Direct application of the discrete maximum principle [11] to the Hamiltonian (6.183), for given $\underline{\beta}(k)$ and $\underline{\pi}(k)$ sequences, yields the following conditions:

$$\underline{x}_j(k+1) = \underline{\Phi}_j[\underline{x}_j,\underline{\pi}_j,k] + \underline{q}_j(k) - G_jQ_j(k)$$

$$- G_jQ_j(k)G_j^t\underline{\Psi}_j^{-t}[\underline{\lambda}_j(k)-\underline{\rho}_j(k)] \tag{6.184a}$$

$$\underline{\lambda}_j(k+1) = \underline{\Psi}_j^{-t}(k)[\underline{\lambda}_j(k)-\underline{\rho}_j(k)]$$

$$+ H_j(k+1)R_j^{-1}(k+1)\underline{\nu}_j(k+1) \tag{6.184b}$$

with the boundary conditions

$$\underline{\lambda}_j(0) = P_j^{-1}(0)[\underline{x}_j(0)-\underline{m}_j(0)] \tag{6.185a}$$

$$\underline{\lambda}_j(N) = \underline{0} \tag{6.185b}$$

where (for $j \neq m$)

$$\underline{\nu}_j(k+1) = \underline{z}_j(k+1)-\underline{h}_j[\underline{x}_j,\underline{\pi}_j,k]-\underline{r}_j(k+1) \tag{6.186a}$$

$$\underline{\rho}_j(k) = \partial/\partial\underline{x}_j(k)\{\underline{\beta}_m^t[\underline{\pi}_m(k)-\underline{g}_m(\underline{x}_j,k)]\} \tag{6.186b}$$

$$\underline{\Psi}_j^t = \partial\underline{\Phi}_j[\underline{x}_j,\underline{\pi}_m,k]/\partial\underline{x}_j(k) \tag{6.186c}$$

$$H_j(k+1) = \partial\underline{h}_j[\underline{x}_j,\underline{\pi}_j,k+1]/\partial\underline{x}_j(k) \tag{6.186d}$$

It must be emphasized that (6.184)-(6.186) have to be solved for fixed sequences of $\underline{\beta}(k)$ and $\underline{\pi}(k)$. Next, we will see how this can be implemented within a two-level structure.

F. A Two-Level Structure

We seek a computational algorithm which will

(i) determine the filtered estimate of the state vector at (k+1) from that at k; that is, possess a sequential feature,

(ii) update the multiplier sequences $\underline{\beta}(k)$ and $\underline{\pi}(k)$ iteratively to obtain the optimal solution.

Sage and coworkers *[27-30]* suggested a two-level computational algorithm where, at the first level, the technique of discrete invariant imbedding *[31]* is used to obtain the sequential state estimate. The basic idea is to treat the final discrete-time instant as a running variable. This means that the two-point boundary problem under consideration is now generalized to (imbedded into) a more general problem, described essentially by a nonlinear partial differential equation. A suitable approximation scheme *[31]* is then used to solve the problem in a sequential manner. For the jth subsystem, (6.184)-(6.186), the following sequential estimation algorithm is obtained *[27-30]*:

$$\hat{\underline{x}}_j(k+1) = \underline{\Phi}_j[\hat{\underline{x}}_j,\underline{\pi}_j,k]+\underline{q}_j(k)+K_j(k+1)\hat{\underline{\nu}}_j(k+1)+S_j(k+1)\underline{\rho}_j(k) \tag{6.187a}$$

$$K_j(k+1) = P_j(k+1)H_j^t(k+1)R_j^{-1}(k+1) \tag{6.187b}$$

$$P_j(k+1) = [P_j(k+1|k) + D_j(k)]E_j^{-1}(k+1) \tag{6.187c}$$

$$E_j(k+1) = [I-\{\partial M_j(k+1)/\partial\hat{\underline{x}}_j(k+1|k)\}P_j(k+1|k)+C_j(k+1)] \tag{6.187d}$$

$$P_j(k+1|k) = \underline{\psi}_j(k)P(k)\underline{\psi}_j^t(k)+G_jQ_j(k)G_j^t \tag{6.187e}$$

where the different variables are defined as:

$$S_k(k+1) = -\{P_j(k+1)[I-\{\partial M_j(k+1)/\partial \underline{\hat{x}}_j(k+1|k)\}G_jQ_j(k)G_j^t]$$
$$- G_jQ_j(k)G_j^t\}\underline{\psi}_j^{-t}(k) \qquad (6.188a)$$

$$C_j(k+1) = \partial/\partial\underline{\hat{x}}_j(k)\{\underline{\psi}_j^{-t}(k)\underline{\rho}_j(k)$$
$$- [\partial M_j(k+1)/\partial\underline{\hat{x}}_j(k+1|k)]G_jQ_j(k)G_j^t\underline{\psi}_j^{-t}(k)\underline{\psi}_j(k)\}P_j(k)\underline{\psi}_j^t(k)$$
$$(6.188b)$$

$$D_j(k) = \partial/\partial\underline{\hat{x}}_j(k)\{G_jQ_j(k)G_j^t\underline{\psi}_j^{-t}(k)\underline{\psi}_j(k)\}P_j(k)\underline{\psi}_j^t(k) \qquad (6.188c)$$

$$M_j(k+1) = H_j^t(k+1)R_j^{-1}(k+1)\underline{\nu}_j(k+1) \qquad (6.188d)$$

$$\underline{\hat{\nu}}_j(k+1) = \underline{z}_j(k+1)-\underline{r}_j(k+1)-\underline{h}_j[\underline{\hat{x}}_j(k+1|k),\underline{\pi}_j(k+1),k+1] \qquad (6.188e)$$

$$\underline{\hat{x}}_j(k+1|k) = \underline{\Phi}_j[\underline{\hat{x}}_j,\underline{\pi}_j,k] + G_j\underline{q}_j(k) \qquad (6.188f)$$

Having solved the first level problem (or collection of subproblems), it is then necessary to update the multiplier sequences $\underline{\beta}(k)$ and $\underline{\pi}(k)$. This is accomplished at the second level. Instead of updating (coordinating) optimally the subsystems filters, a sequential coordination procedure [27] is suggested to gradually improve the sequences $\underline{\pi}(k)$ and $\underline{\beta}(k)$ with the passing of time.

For this purpose, a sequential predictor-corrector scheme can be implemented. First, $\underline{\pi}_j(k)$ and $\underline{\beta}_j(k)$ are predicted by linear extrapolation:

$$\underline{\beta}_j(k+1|k) = 2\underline{\beta}_j(k) - \underline{\beta}_j(k-1) \qquad (6.189a)$$
$$\underline{\pi}_j(k+1|k) = 2\underline{\pi}_j(k) - \underline{\pi}_j(k-1) \qquad (6.189b)$$

Then the predicted values of $\underline{\beta}_j$ are corrected using a gradient procedure:

$$\underline{\beta}_j(k+1) = \underline{\beta}_j(k+1|k) + \zeta_j[\underline{\pi}_j(k)-\underline{g}_j(\underline{\hat{x}}_m,k)] \qquad (6.190a)$$

where ζ_j is an appropriate step length. The sequence $\underline{\pi}_j(k)$ is corrected using the predicted values of the states as

$$\underline{\pi}_j(k+1) = \underline{g}_j[\hat{\underline{x}}_m(k+1|k), k+1] \qquad (6.190b)$$

or by the averaging formula

$$\underline{\pi}_j(k+1) = \frac{1}{2}\{\underline{\pi}_j(k+1|k) + \underline{g}_j[\hat{\underline{x}}_m(k), k]\} \qquad (6.190c)$$

The implementation of the two-level structure can be summarized below:

Step 1

Initialize $\underline{\pi}_j(0) = \underline{g}_j(\underline{x}_m, 0)$ and $\underline{\beta}_j(0) = \underline{0}$ for $j = 1,\ldots,N_s$

Step 2

Compute $\underline{\pi}_j(k+1)$, $\underline{\beta}_j(k+1)$, $\hat{\underline{x}}_j(k+1|k)$ and $P_k(k+1|k)$, for $j = 1,\ldots,N_s$ from (6.189), ($.190), (6.188f) and (6.187e), respectively.

Step 3

Compute $P_j(k+1)$ and $\hat{\underline{\nu}}_j(k+1)$ from (6.186), (6.188), (6.187c) and (6.187d).

Step 4

Compute $\hat{\underline{x}}_j(k+1)$ from (6.187a), (6.187b) and (6.188a). Record it as the extimate.

Step 5

Change k to k+1. If $k > N$ stop, otherwise go to Step 2.

From the above analysis, we note the following:

(1) The optimality condition for the sequence $\underline{\pi}(k)$,

$\partial H/\partial \underline{\pi}(k) = 0$, is not satisfied explicitly. Therefore, the final result would give *a suboptimal estimate of the parameters.*

(2) It is shown *[29,30]* that when the ratio $||Q||/||R||$ is large, good convergence behaviour is obtained.

(3) The method essentially provides the filtered estimate of the states and parameters. It seems unlikely that one can get an optimal filtered estimate using the MAP approach since any other coordination scheme will eventually destroy the sequential feature and yield a smoothing solution; see *[32]* for further discussions.

6.6.3 *EXAMPLES*

To illustrate the various notions developed in this subsection, we present two examples.

Example 1

This example is concerned with the implementation of the two-level multiple projection algorithm. A single-input single-output discrete transfer function of the type (6.141a) is considered. Note that it can be easily put in the form (6.142a) which is equivalent to (6.161). Two specific models were used in simulating the two-level algorithm:

Model 1

Had the parameter values

$$b_0 = 0, \quad b_1 = 1, \quad b_2 = .5$$
$$a_1 = -1.5, \quad a_2 = .7$$

Model 2

Had the parameter values

$$b_0 = 1, \quad b_1 = .1, \quad b_2 = -.8, \quad b_4 = -.45$$
$$a_1 = -2.3, \quad a_2 = 1.93, \quad a_3 = -.713, \quad a_4 = .1102,$$
$$a_5 = -.0056$$

A Gaussian, zero mean, unit variance random signal was used in exciting both models during simulations on a PDP-11 computer facility available at UMIST. The measurement vector $\underline{z}$ was decomposed into subvectors $\underline{z}_1$ of order n_1, $\underline{z}_2$ of order n_2 and $\underline{z}_3$ of order n_3 such that $n_1+n_2+n_3 = L$ (number of samples). The parameter vector $\underline{\theta} = [-a_1 \ldots -a_n \; b_0 \ldots b_m]^t$ was decomposed into $\underline{\theta}_1$ of order m_1, $\underline{\theta}_2$ of order m_2 and $\underline{\theta}_3$ of order m_3 such that $m_1 < n$, $m_2 = n-m_1$ and $m_3 = m+1$.

The simulation results for the non-recursive algorithm are given in Tables 6.2 and 6.3 for Models 1 and 2, respectively.

The recursive (non-decomposed) algorithm was then tested, since the decomposed (extended) algorithm can be simply considered as a combination of the recursive (non-decomposed) and non-recursive algorithm. Figs. (6.11)-(6.15) show the simulation results of Model 1 for $Q = .1$, $P = 1$, $\underline{m} = \underline{0}$, and using 100 recursions. Other operating conditions have been considered in *[24]*. It has been concluded that the two-level algorithms have good performance in estimating the unknown parameters.

Example 2

With reference to (6.176), an eighth-order power system has the following data:

$$A = \begin{bmatrix} a_1 & 0 & 0 & 0 & 0 & 0 & 0 & 0 \\ .096 & a_2 & 0 & 0 & 0 & 0 & 0 & 0 \\ -.002 & -.005 & a_3 & -.253 & .041 & -.003 & -.025 & -.001 \\ .007 & .014 & -.029 & a_4 & 0 & .006 & .059 & .002 \\ -.03 & -.061 & 2.028 & -2.303 & a_5 & -.021 & -.224 & -.008 \\ .048 & .758 & 0 & 0 & 0 & a_6 & 0 & .023 \\ -.012 & -.027 & 1.209 & -1.4 & .161 & -.013 & a_7 & .006 \\ .815 & 0 & 0 & 0 & 0 & 0 & 0 & a_8 \end{bmatrix}$$

$$\Gamma = \text{diag}\{.001 \;\; .001 \;\; .001 \;\; .001 \;\; .003 \;\; .003 \;\; .003\}$$

$$H = I_8$$

TABLE 6.2 Simulation Results of Model 1 (all figures are rounded to four decimals)

Dimensions of Subsystems	Initial Data	Estimation Results: After One Iteration	After Two Iterations	After Three Iterations
$n_1=20, m_1=1$ $n_2=10, m_2=1$ $n_3=10, m_3=3$	$\underline{m} = \underline{0}$ $P = I$ $Q = 0.1$	$\theta_1 = 1.5086$, $\theta_2 = -.7421$, $\theta_3 = .0856$, $\theta_4 = .9904$, $\theta_5 = .4990$ $P = \text{diag}\{.1822, .2743, .8294, .8250, 1.0749\} \times 10^{-2}$	$\theta_1 = 1.5085$, $\theta_2 = -.7264$, $\theta_3 = .0853$, $\theta_4 = .9828$, $\theta_5 = .4974$ $P = \text{diag}\{.1334, .1417, .3866, .3947, .4972\} \times 10^{-2}$	$\theta_1 = 1.5121$, $\theta_2 = -.7435$ $\theta_3 = .0339$, $\theta_4 = .9358$ $\theta_5 = .4388$ $P=\text{diag}\{.1103, .1103, .2596, .2668, .3190\} \times 10^{-2}$
$n_1=30, m_1=1$ $n_2=40, m_2=1$ $n_3=30, m_3=3$	$\underline{m} = \underline{0}$ $P = I$ $Q = 0.1$	$\theta_1 = 1.5057$, $\theta_2 = -.7320$, $\theta_3 = .0576$, $\theta_4 = .9540$, $\theta_5 = .4737$ $P = \text{diag}\{.1452, .1481, .3656, .3665, .4849\} \times 10^{-2}$	$\theta_1 = 1.4967$, $\theta_2 = -.7214$, $\theta_3 = .0078$, $\theta_4 = .9348$, $\theta_5 = .4565$ $P = \text{diag}\{.8448, .8423, 1.7422, 1.8365, 2.3793\} \times 10^{-3}$	$\theta_1 = 1.4934$, $\theta_2 = -.7198$ $\theta_3 = -.0268$, $\theta_4 = .9281$ $\theta_5 = .4253$ $P=\text{diag}\{.6767, .6601, 1.1310 1.1585, 1.4644\} \times 10^{-3}$
$n_1=30, m_1=1$ $n_2=40, m_2=1$ $n_3=30, m_3=3$	$\underline{m} = \underline{0}$ $P = 10I$ $Q = 0.1$	$\theta_1 = 1.5079$, $\theta_2 = -.7340$, $\theta_3 = .0589$, $\theta_4 = .9578$, $\theta_5 = .4740$ $P = \text{diag}\{.1458, .1484, .3670, .3680, .4876\} \times 10^{-2}$	$\theta_1 = 1.4983$, $\theta_2 = -.7228$, $\theta_3 = -.7159$, $\theta_4 = .9369$, $\theta_5 = .4564$ $P = \text{diag}\{.8474, .8428, 1.7454, 1.8404, 2.3854\} \times 10^{-3}$	$\theta_1 = 1.4946$, $\theta_2 = -.7210$ $\theta_3 = -.0265$, $\theta_4 = .9294$ $\theta_5 = .4250$ $P=\text{diag}\{.6784, .6609, 1.1330 1.1600, 1.4670\} \times 10^{-3}$

TABLE 6.3 Simulation Results of Model 2 (all figures are rounded to four decimals)

Dimensions of Subsystems	Initial Data	Estimation Results: After One Iteration	After Two Iterations	After Three Iterations
$n_1=20, m_1=2$ $n_2=40, m_2=3$ $n_3=40, m_3=5$	$\underline{m} = \underline{0}$ $P = I$ $Q = 0.1$	$\Theta_1 = 2.1780,\ \Theta_2 = -1.6558$ $\Theta_3 = .4870,\ \Theta_4 = -.0659$ $\Theta_5 = .0134,\ \Theta_6 = 1.0387$ $\Theta_7 = .2029,\ \Theta_8 = -.7861$ $\Theta_9 = -.0542,\ \Theta_{10} = -.5073$ $P = \text{diag}\{.1510, .8274, 1.0414, .6862, .1136, .0687, .1758, .1743, .3015, .1811\} \times 10^{-1}$	$\Theta_1 = 2.1533,\ \Theta_2 = -1.6383$ $\Theta_3 = .5714,\ \Theta_4 = -.1768$ $\Theta_5 = .0534,\ \Theta_6 = 1.0087$ $\Theta_7 = .1967,\ \Theta_8 = -.7817$ $\Theta_9 = -.1899,\ \Theta_{10} = -.5656$ $P = \text{diag}\{.6493, 3.6310, 4.9130, 3.0405, .4154, .2122, .8379, .8442, 1.1726, .8735\} \times 10^{-2}$	$\Theta_1 = 2.2282, \Theta_2 = -1.7853$ $\Theta_3 = .6781, \Theta_4 = -.2064$ $\Theta_5 = .0626, \Theta_6 = .9648$ $\Theta_7 = .1009, \Theta_8 = -.8501$ $\Theta_9 = -.1402, \Theta_{10} = -.5589$ $P = \text{diag}\{.4222, 2.3881, 3.0568, 1.8290, .2533, .1155, .5113, .4680, .7562, .5391\} \times 10^{-2}$
$n_1=20, m_1=2$ $n_2=40, m_2=3$ $n_3=40, m_3=5$	$\underline{m} = \underline{0}$ $P = I$ $Q = 0.01$	$\Theta_1 = 2.2655,\ \Theta_2 = -1.8489$ $\Theta_3 = .6465,\ \Theta_4 = -\ .1014$ $\Theta_5 = .011,\ \Theta_6 = 1.0115$ $\Theta_7 = .1281,\ \Theta_8 = -\ .7966$ $\Theta_9 = -.0177,\ \Theta_{10} = -\ .4678$ $P = \text{diag}\{.2443, 1.3832, 1.6871, 1.1154, .1852, .0803, .2234, .2276, .5001, .2629 \times 10^{-2}\}$	$\Theta_1 = 2.2427,\ \Theta_2 = -1.8152$ $\Theta_3 = .6589,\ \Theta_4 = -\ .1367$ $\Theta_5 = .0241,\ \Theta_6 = 1.0041$ $\Theta_7 = .1420,\ \Theta_8 = -\ .7866$ $\Theta_9 = -.0679,\ \Theta_{10} = -\ .4919$ $P = \text{diag}\{1.0762, 6.0539, 8.0247, 4.7447, .6323, .1871, 1.2016, 1.1868\ 1.6162, 1.2096 \times 10^{-3}\}$	$\Theta_1 = 2.2738, \Theta_2 = -1.8852$ $\Theta_3 = .7187, \Theta_4 = -\ .1564$ $\Theta_5 = .0268, \Theta_6 = .9928$ $\Theta_7 = .1057, \Theta_8 = -\ .8025$ $\Theta_9 = -.0401, \Theta_{10} = -.4918$ $P = \text{diag}\{.7428, 4.0369, 4.9504, 2.9496, .4079, .0967, .8290, .7416, 1.0988, .7702 \times 10^{-3}\}$

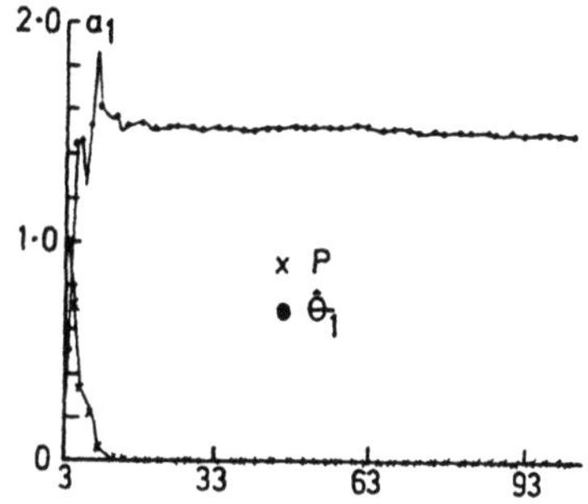

Fig. (6.11) 100 recursions for model 1, parameter θ_1

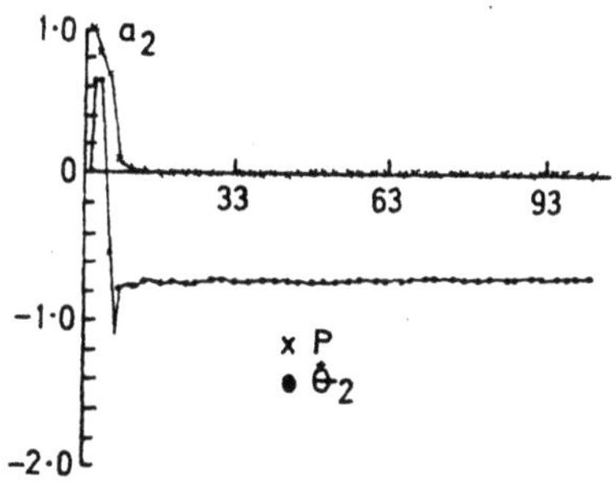

Fig. (6.12) 100 recursions for model 1, parameter θ_2

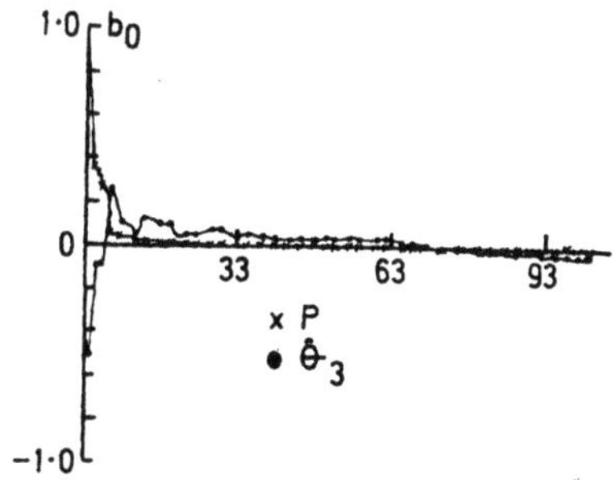

Fig. (6.13) 100 recursions for model 1, parameter θ_3

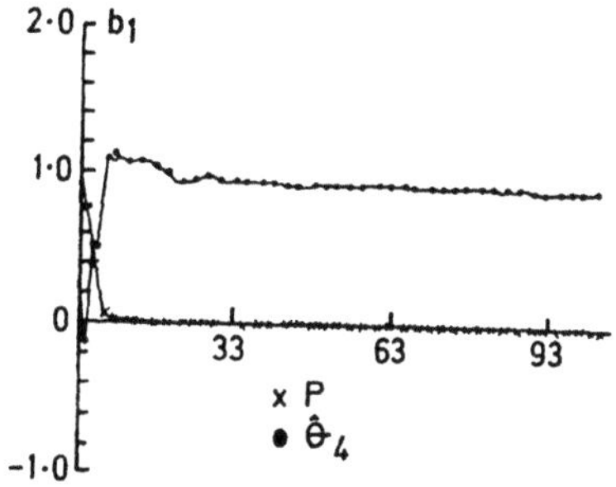

Fig. (6.14) 100 recursions for model 1, parameter θ_4

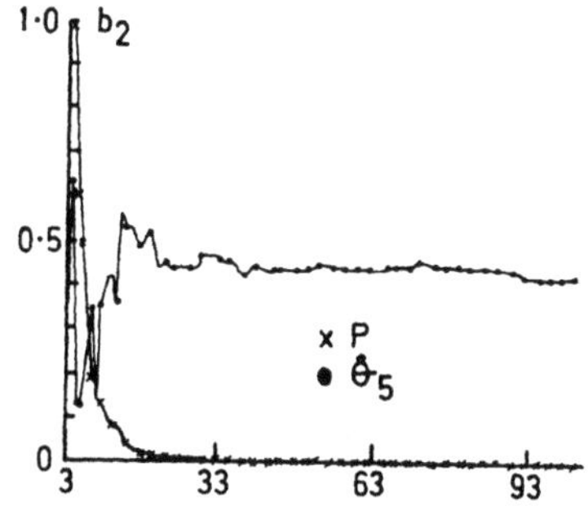

Fig. (6.15) 100 recursions for model 1, parameter θ_5

where the real values of the parameters are:

$$a_1 = .835, \quad a_2 = .861, \quad a_3 = .882, \quad a_4 = .928,$$
$$a_5 = .088, \quad a_6 = .165, \quad a_7 = .156, \quad a_8 = .011$$

The noise vectors are zero mean Gaussian sequences. In one simulation run, the noise covariances were taken to be

$$Q = I_8 \quad \text{and} \quad R = .1I_8$$

while in a second run, they were

$$Q = 10I_8 \quad \text{and} \quad R = .1I_8$$

In both simulation runs, the initial state was chosen to be $\underline{x}(0) = [1 \;\; .1 \;\; .5 \;\; 2 \;\; .2 \;\; .4 \;\; 1 \;\; 1]$ with initial covariance $P(0) = \text{diag}\{2 \;\; 2 \;\; 2 \;\; 2 \;\; 4 \;\; 4 \;\; 4 \;\; 4\}$. The final time N was taken to be 100.

The eighth-order system with eight unknown parameters (overall order = 16) has been decomposed into four subsystems as follows:

Subsystem 1 (order 2)

has one state variable and one unknown parameter.

subsystem 2 (order 2)

has one state variable and one unknown parameter.

subsystem 3 (order 10)

has five state variables and five unknown parameters.

subsystem 4 (order 2)

has one state variable and one unknown parameter.

All the initial parameters were taken to be zero.

The simulation results of the MAP approach, using UNIVAC 1108 available at the Computer Centre of Kuwait University, gave the estimated parameters as:

First run

$a_1 = .751, \quad a_2 = .778, \quad a_3 = .853, \quad a_4 = .937,$

$a_5 = .111, \quad a_6 = .169, \quad a_7 = .183, \quad a_8 = .0089$

Second run

$a_1 = .809, \quad a_2 = .842, \quad a_3 = .871, \quad a_4 = .929,$

$a_5 = .094, \quad a_6 = .166, \quad a_7 = .170, \quad a_8 = .010$

From these results, it is clear that the true values were not obtained. However, the maximum deviation was 20% in the first run and 8% in the second run. This shows that increasing $||Q||/||R||$ gives improved estimation results.

6.7 Problems

1. Let $\underline{x}, \underline{y}, \underline{z}$ be jointly distributed random vectors, and $g(.)$ a scalar-valued function. Show that

   (a) $E[g(\underline{y})\underline{x}|\underline{y}] = g(\underline{y})E[\underline{x}|\underline{y}]$

   (b) $E[g(\underline{y})\underline{x}] = E[g(\underline{y})E[\underline{x}|\underline{y}]]$

   (c) $E[g(\underline{y})|\underline{y}] = g(\underline{y})$

2. Consider the model of Example 3, Section 6.4, with the numerical values $a = 1$, $W(0) = 100$, $q = 25$ and $r = 15$.

   (a) determine the asymptotic value of $P(k|k)$

   (b) carry out a few computational cycles and hence detect the discrete instant at which the filter can be considered to be in a steady state

   (c) write the equation of the filtered estimate in a steady-state.

3. Suppose that for the linear discrete model

$$\underline{x}(k+1) = A(k)\,\underline{x}(k) + G(k)\,\underline{w}(k)$$
$$\underline{z}(k) = H(k)\,\underline{x}(k) + \underline{v}(k)$$

where $k = 0,1,\ldots,$ the noise processes $\{\underline{w}(k)\}$ and $\{\underline{v}(k)\}$ are zero mean Gaussian white sequences which are independent of $\underline{x}(0)$ and have covariance matrices $Q(k)$ and $R(k)$, respectively. Assume that these two sequences are correlated with each other with the cross-covariance matrix

$$E[\underline{w}(k)\,\underline{v}^t(j)] = S(k)\delta_{kj}$$

Show that one possible estimation algorithm for this case is given by

$$\hat{\underline{x}}(k+1|k) = A(k)\hat{\underline{x}}(k|k-1) + K(k)[z(k)-G(k)\hat{x}(k|k-1)]$$

Derive the expressions for the gain $K(k)$ and the error covariance matrix $P(k+1|k)$.

4. A discrete model of a noise-free second-order system representing a falling body in a constant field is described by

$$\underline{x}(k+1) = \begin{bmatrix} 1 & 1 \\ 0 & 1 \end{bmatrix}\underline{x} + \begin{bmatrix} -.5 \\ -1 \end{bmatrix}$$

where $\underline{x}^t = [x_1 \quad x_2]$; x_1 is the position and x_2 is the velocity. The observation equation is given by

$$z(k) = [1 \quad 0]\underline{x}(k) + v(k)$$

where the measurement noise $v(k)$ is a zero mean Gaussian process with covariance $r = 1$.
Let $\underline{x}(0)$ be $N(\underline{m},W)$ where

$$\underline{m} = [95 \quad 1]^t, \quad W = \begin{bmatrix} 5 & 0 \\ 0 & 1 \end{bmatrix}$$

The measurement records of the position are:

$z(0) = 102$, $z(1) = 100$, $z(2) = 97.9$, $z(3) = 94.4$,

$z(4) = 92.7$, $z(5) = 87.3$, $z(6) = 82.1$

It is required to estimate the position and the velocity of the falling body at $k = 6$. Obtain the corresponding error covariance matrix.

Consider the discrete system

$$\underline{x}(k+1) = \begin{bmatrix} 2 & 3 \\ 3 & -2 \end{bmatrix} \underline{x}(k) + \underline{w}(k)$$

$$\underline{y}(k) = \begin{bmatrix} 4 & 0 \\ 0 & 1 \end{bmatrix} \underline{x}(k) + \underline{v}(k)$$

where $\underline{w}(k)$ and $\underline{v}(k)$ are zero mean, independent Gaussian random sequences with

$$Q = I_2 \quad \text{and} \quad R = .5I_2$$

The initial state $\underline{x}(0)$ is a zero mean Gaussian random vector with $W(0) = 1.5I_2$. The following observations were recorded:

k	0	1	2	3	4	5
y_1	0	2	3	5	6	8
y_2	0	2	3	5	6	8

Use the decentralized filter to compute the estimate $\hat{\underline{x}}(5|5)$.

6. For the scalar process

$$x(k+1) = \alpha\, x(k) + w(k)$$

$$z(k) = x(k) + v(k)$$

where $w(k)$ is $N(0,q)$ and $v(k)$ is $N(0,r)$. Assume $\underline{x}(0)$ to be $N(0,s_0)$. Develop the filtered estimate equation by maximizing (i) $p\{x(k+1)|Z(k+1)\}$
(ii) $p\{\underline{x}(k+1)|Z(k)\}$
and compare the results.

7. Given a second-order system

$$\underline{x}(k+1) = \begin{bmatrix} -.5 & -.5 \\ 2 & 0 \end{bmatrix} \underline{x}(k) + \begin{bmatrix} 1 \\ 0 \end{bmatrix} w(k)$$

where $E[w(k)] = 0, \quad E[\underline{x}(0)] = \underline{0}$

$\mathrm{Cov}\{w(k)\} = 3, \quad \mathrm{Cov}\{\underline{x}(0)\} = 0$

Find the variance of the output state vector.

8. Let the innovation be defined by:

$$\tilde{\underline{z}}(k) = \underline{z}(k) - \hat{\underline{z}}(k|k-1)$$

Determine $\mathrm{Cov}[\tilde{\underline{z}}(k), \tilde{\underline{z}}(j)]$. Show that $\tilde{\underline{z}}(k)$ is orthogonal to $\{\underline{Z}(k-1)\}$.

6.8 References

[1] Sage, A.P. and J.L. Melsa
"Estimation Theory with Applications to Communications and Control", McGraw-Hill, N.Y., 1971.

[2] Astrom, K.J.
"Introduction to Stochastic Control Theory", Academic Press, N.Y., 1970.

[3] Anderson, B.D.O. and J.B. Moore
"Optimal Filtering", Prentice Hall, N.J., 1979.

[4] Meditch, J.S.
"Stochastic Optimal Linear Estimation and Control", McGraw-Hill, N.Y., 1969.

[5] Jazwinski, A.H.
"Stochastic Processes and Filtering Theory", Academic Press, N.Y., 1970.

[6] Brogan, W.L.
"Modern Control Theory", Quantum Publishing Inc., N.Y., 1974.

[7] Cox, H.
"On the Estimation of State Variables and Parameters for Noisy Dynamic Systems", IEEE Trans. Autom. Contr., vol. AC-9, pp. 5-12, 1964.

[8] Kalman, R.E.
"A New Approach to Linear Filtering and Prediction Problems", Trans. ASME, Ser. D: J. Basic Engineering, vol. 82, pp. 35-45, 1960.

[9] Kalman, R.E. and R.S. Bucy
"New Results in Linear Filtering and Prediction Theory", Trans. ASME, Ser. D: J. Basic Engineering, vol. 83, pp. 95-108, 1961.

[10] Singh, M.G. and A. Titli
"Systems: Decomposition, Control and Optimization", Pergamon Press, Oxford, 1978.

[11] Bryson, A.E. and Y.C. Ho
"Applied Optimal Control", Ginn and Co., Massachusetts, 1969.

[12] Athans, M. and E. Tse
"A Direct Derivation of the Optimal Linear Filter Using the Maximum Principle", IEEE Trans. Autom. Contr., vol. AC-12, pp. 690-698, 1967.

[13] Sorenson, H.W.
"Least-Squares Estimation : from Gauss to Kalman", IEEE Spectrum, vol. 7, pp. 63-68, 1970.

[14] Balakrishnan, A.V.
"A Martingale Approach to Linear Recursive State Estimation", SIAM J. Control, vol. 10, pp. 754-766, 1972.

[15] Bierman, G.J.
"A Comparison of Discrete Linear Filtering Algorithms" IEEE Trans. Aerospace and Electronic Systems, vol. AES-9, pp. 28-37, 1973.

[16] Pearson, J.D.
"Dynamic Optimization Techniques", in Optimization Methods for Large Scale Systems, edited by D.A. Wismer, McGraw-Hill, N.Y., 1971.

[17] Singh, M.G.
"Multi-level State Estimation", Int. J. Systems Sciences, vol. 6, pp. 535-555, 1975.

[18] Hassan, M.F.
"Optimal Kalman Filter for Large Scale Systems Using the Prediction Approach", IEEE Trans. Systems, Man and Cybern., vol. SMC-6, pp. , 1976.

[19] Shah, M.
"Suboptimal Filtering Theory for Interacting Control Systems", Ph.D. Thesis, Cambridge University, 1971.

[20] Hassan, M.F., G. Salut, M.G. Singh and A. Titli
"A Decentralized Computational Algorithm for the Global Kalman Filter", IEEE Trans. Autom. Contr., vol. AC-23, pp. 262-267, 1978.

[21] Luenberger, D.G.
"Optimization by Vector Space Methods", J. Wiley, N.Y., 1969.

[22] Darwish, M.G. and J. Fantin
"An Approach for Decomposition and Reduction of Dynamical Models for Large Scale Power Systems", Int. J. Systems Science, vol. 7, pp. 1101-1112, 1976.

[23] Eykoff, P.
"System Identification", J. Wiley and Sons, N.Y., 1974.

[24] Hassan, M.F., M.S. Mahmoud, M.G. Singh and M.P. Spathopolous
"A Two-Level Parameter Estimation Algorithm Using the Multiple Projection Approach", CSC Report No.518, UMIST, Manchester, UK, and also Automatica, vol. 18, pp. 621-630, 1982.

[25] Clarke, D.W.
"Generalized Least-Squares Estimation of the Parameters of a Dynamic Model", IFAC Symposium - Identification in Automatic Control Systems, Prague Paper # 3.17, 1967.

[26] Hasting-James, R. and M.W. Sage
"Recursive Generalized Least-Squares Procedure for On-Line Identification of Process Parameters", Proc. IEE, vol. 116, pp. 2057-2062, 1969.

[27] Arafeh, S. and A.P. Sage
"Multilevel Discrete-Time System Identification in Large Scale Systems", Int. J. Systems Science, vol. 8, pp. 753-791, 1974.

[28] Arafeh, S. and A.P. Sage
"Hierarchical System Identification of States and Parameters in Interconnected Power Systems", Int. J. Systems Science, vol. 9, pp. 817-846, 1975.

[29] Fry, C.M. and A.P. Sage
"Identification of Aircraft Stability and Control Parameters Using Hierarchical State Estimation", IEEE Trans. Aero. Elect. Syst., vol. AES-10, pp. 255-264, 1974.

[30] Guinzy, N.J. and A.P. Sage
"System Identification in Large Scale Systems with Hierarchical Structures", J. Computers and Elect. Eng., vol. 1, pp. 23-43, 1973.

[31] Lee, E.S.
"Quasilinearisation and Invariant Imbedding", Academic Press, N.Y., 1968.

[32] Chemoul, P., M.R. Katebi, D. Sastry and M.G. Singh
"Maximum *a posteriori* Parameter Estimation in Large-Scale Systems", Automatica, vol. 17, pp. 845-851, 1981.

Chapter 7
Adaptive Control Systems

7.1 Introduction

In many industrial and process control applications, the use of high-performance control systems is very desirable. Usually the plant parameters are poorly known or vary during normal operation, and this must be taken into consideration when designing and implementing feedback controllers. The problem of self-adjusting the parameters of a controller in order to stabilize the dynamic characteristics of a feedback control system when the plant parameters undergo large and unpredictable variations, has led to the development of adaptive control techniques *[1-6]*. Many solutions have been proposed over the past two decades in order to make a control system *"adaptive"* (that is, to assure high performance when large and unpredictable variations of the plant dynamic characteristics take place). In some sense, the problem can be viewed as *"combined identification and control"* of a particular system configuration.

Two important classes of adaptive systems are *model reference adaptive systems (MRAS's)* and *self-tuning regulators (STR's)*. We shall examine each system separately and then present the similarities and differences amongst them. In the next section, an introduction to the model reference adaptive systems is given to familiarize the reader with the various concepts which will be used in later sections.

7.2 Basic Concepts of Model Reference Adaptive Systems

We shall briefly summarize the concepts and block diagrams that are frequently used in model reference adaptive systems.

7.2.1 *THE REFERENCE MODEL*

A basic ingredient of model reference adaptive systems is a reference model which, in simple terms, specifies the desired performance of the control system. One of the main reasons for having a reference model is to avoid the difficulties encountered when attempting to translate the performance indices, such as rise time, damping, overshoot, bounds on state or control signals into either desired positions of eigenmodes (then using state or output feedback schemes; see Chapters 4 and 5) or appropriate weighting matrices to derive optimal control schemes (see Chapter 8). In this respect, the reference model is an idealized control system whose performance provides the desired performance requirements. Design of adaptive systems utilizes the reference model as a target to be followed with some precision.

The reference model can be used implicitly or explicitly. In the first case, it is used only for the computation of the control law, a situation which will be dealt with later on. In the second case, the reference model represents the main portion of the control. A preliminary block-diagram is shown in Fig. 7.1 to illustrate this case. We note that the error signal $\underline{e}(k)$, which is the difference between the performance of the reference model and that of the controlled plant, is used to derive a model-following control. As it stands, this configuration does not incorporate a block or device that performs the adaptation. More specifically, the model-following control block has a fixed configuration and hence it will not, in general, be capable of regulating the plant against drift variations in the parameters.

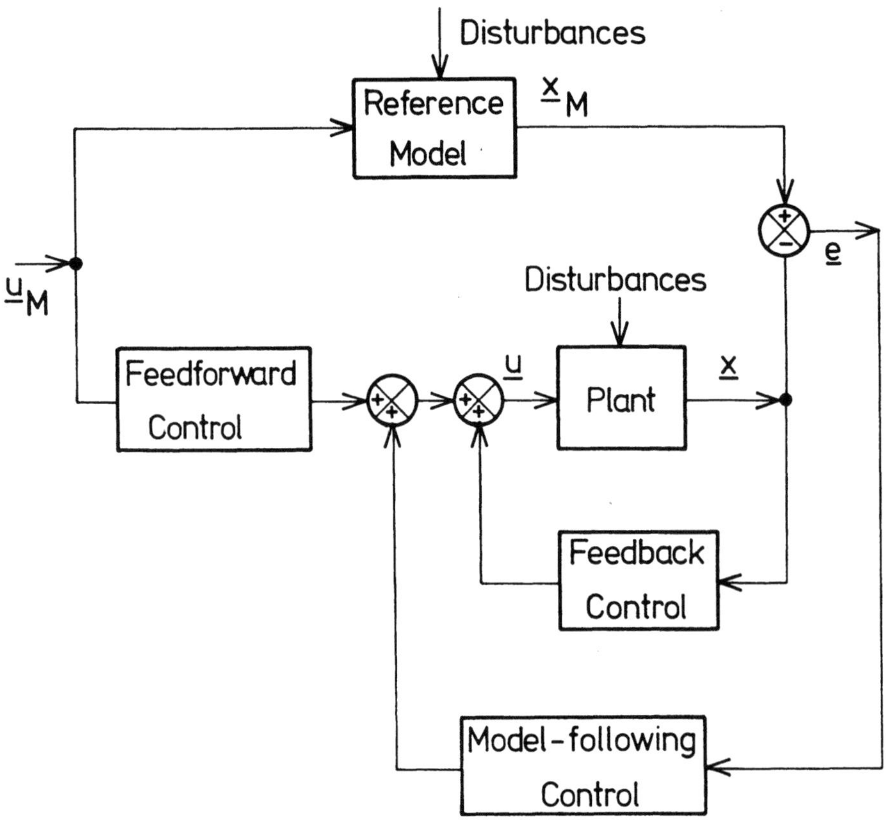

Fig. (7.1) Model-following control system

7.2.2 *THE ADAPTATION MECHANISM*

From a practical point of view, it would be preferable if a control system design has the provision that the desired performance requirements (set forth by a reference model for example) can be compared with the real performance as produced by the plant, and from this information to modify or upgrade a feedback control law. The realization of this design objective constitutes the second main ingredient of adaptive systems, that is, the adaptation mechanism. With reference to Fig. 7.2, which is an appropriate modification of Fig. 7.1, the adaptive control system incorporates an adaptation mechanism which adjusts the parameters of the feedback control loop and/or the input to the controlled plant in order to keep the real index of performance close to the desired index of performance. It is the task of the designer to determine the adaptation law or procedure which will maintain the index of performance as close as possible to the desired one in the presence of environmental variations. At this stage, we should emphasize that any adaptive system must have a closed-loop control on the performance indices, otherwise the adaptivity feature will be lost.

Going back to the block diagram in Figure 7.2, we note that in general the adaptation mechanism can perform parametric adaptation (modifying the parameters of the controller) or signal synthesis adaptation through adjusting a supplementary signal to the output of the controller. Both schemes can be operated in a deterministic or stochastic environment *[4]*. Sometimes, the plant, plus the feedback control blocks, is termed the simulated plant or adjustable system *[4,6]*. Due to the relative position of the reference model and the adjustable system, the configuration of Fig. 7.2 is often called *a parallel model reference adaptive system*. Other configurations are drawn in Fig. 7.3 for the purpose of comparison. These structures are frequently used in system identification *[7]* based on equation error method (series-parallel structures) and input error method (series structure). A unified approach to their analy-

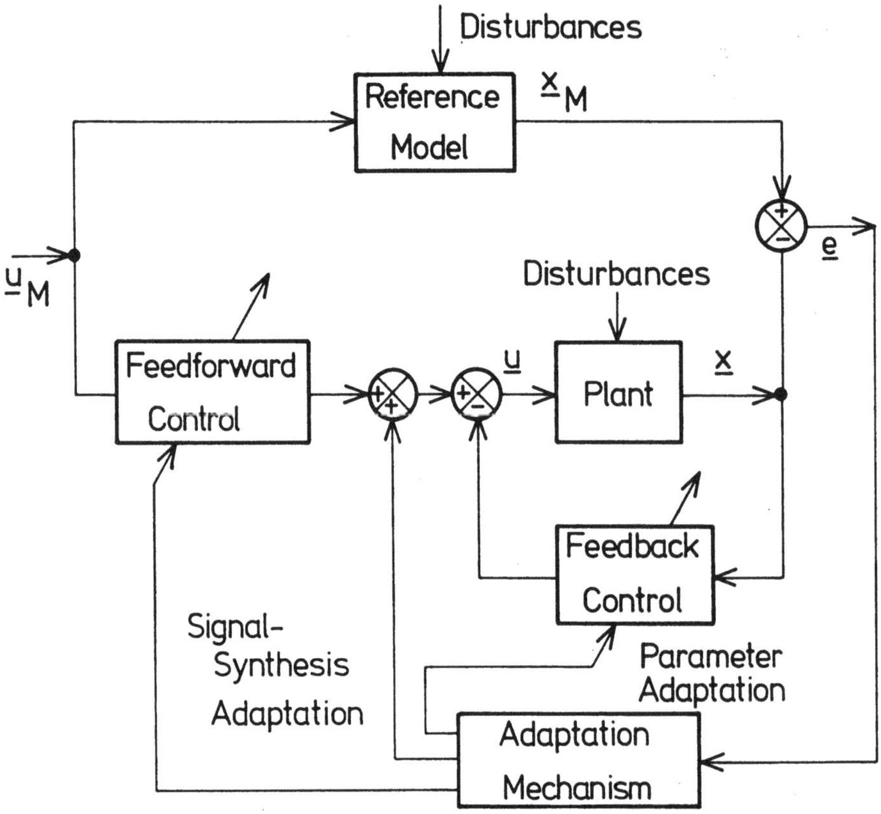

Fig. (7.2) Adaptive model-following control system

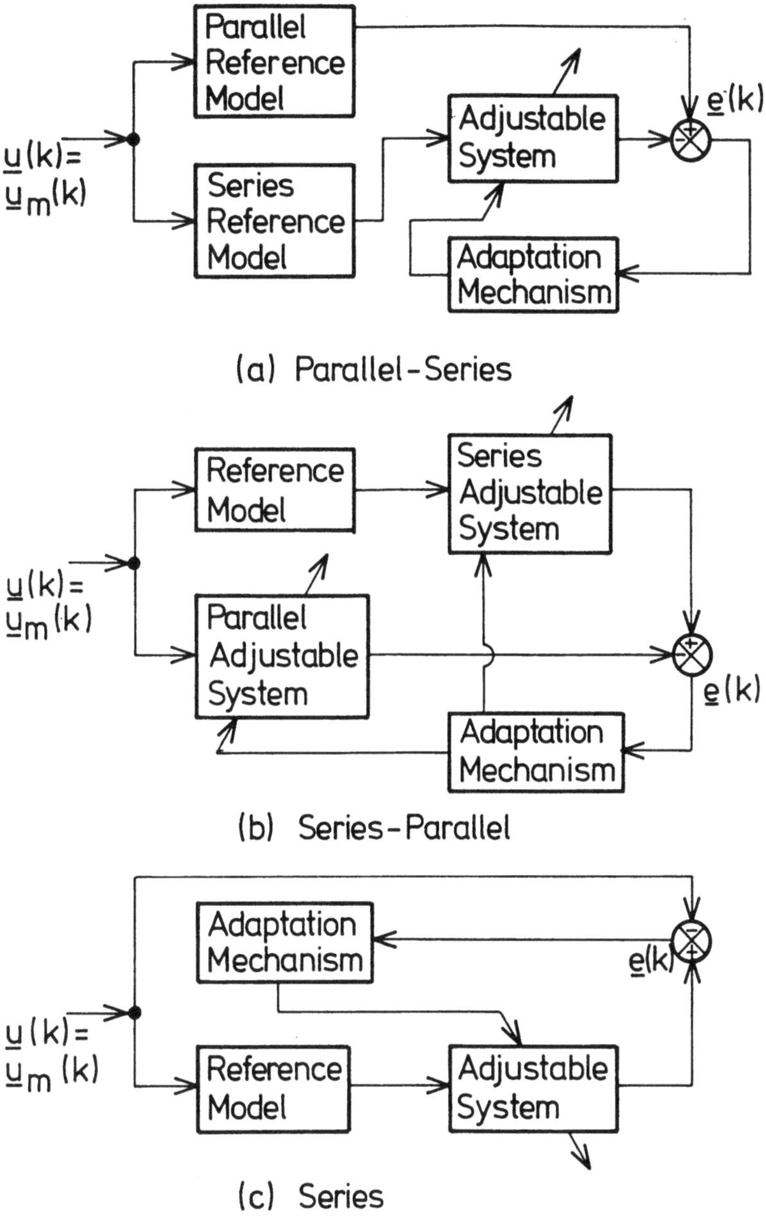

Fig. (7.3) Different MRACS configurations

sis and synthesis is presented in *[8]* based on the positivity concept.

For the class of applications in which a linear system representation may be acceptable, the adaptive system shown in Fig. 7.4 is quite convenient in implementation. The control matrices K_v, K_c and K_p are designed so as to assure good model following in the presence of parameter variations.

7.2.3 *NOTATIONS AND SOME DEFINITIONS*

We now introduce some notations and definitions pertinent to model reference adaptive systems. For simplicity in exposition, we consider first single-input single-output (SISO) systems. The adjustable system (simulated plant) is described by the difference equation:

$$y(k) = \sum_{j=1}^{n} a_j y(k-j) + \sum_{j=0}^{r} b_j u(k-j-d) + \sum_{j=0}^{n} c_j \xi(k-j) \quad (7.1)$$

where $y(k)$, $u(k)$, $\xi(k)$ are the output, input and unmeasurable disturbance, respectively, at instant k. In terms of the delay (shift) operator z^{-s} and the following polynomial delay operators

$$A(z^{-1}) = 1 - \sum_{j=1}^{n} a_j z^{-j} \quad (7.2a)$$

$$B(z^{-1}) = \sum_{j=0}^{r} b_j z^{-j} \qquad b_0 \neq 0 \quad (7.2b)$$

$$C(z^{-1}) = \sum_{j=0}^{n} c_j z^{-j} \quad (7.2c)$$

the system model (7.1) can be put in the form

$$A(z^{-1})Y(z) = z^{-d}B(z^{-1})U(z) + C(z^{-1})E(z) \quad (7.3)$$

where $Y(z)$, $U(z)$ and $E(z)$ are the Z-transform output, input

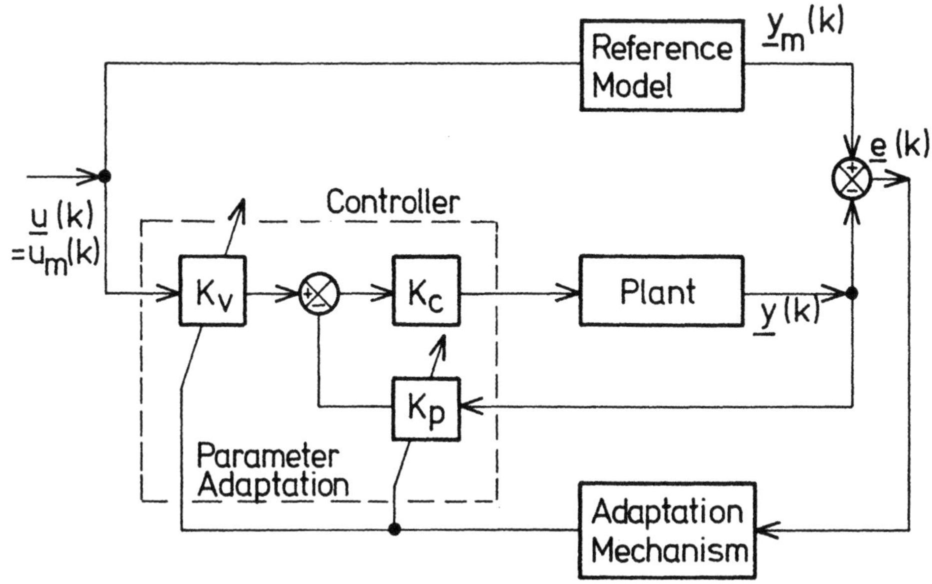

Fig. (7.4) Adaptive model-following control system for linear plants

and disturbance, respectively. The integer d in (7.1) and (7.3) is the plant time delay. It is assumed that coefficients a_j, b_j and c_j are unknown and constant or slowly varying. For the case when $d = 0$ (no delay between input and output sequences) and $\xi(n) = 0$ for all n (no disturbance), we obtain a deterministic plant with unknown parameters whose transfer function takes the form:

$$\begin{aligned} H_a(z) &= Y(z)/U(z) \\ &= B(z^{-1})/A(z^{-1}) \end{aligned} \tag{7.4}$$

Most of the time, we consider that the roots of the polynomial $B(z^{-1}) = 0$ in (7.2b), which corresponds to the zeros of $H_a(z)$ in (7.4), are all in the region $|z| < 1$; that is, within the unit circle. Therefore, these roots can be cancelled without leading to an unbounded control input. For stable plants, the above condition implies that only minimum phase systems *[9]* can be considered. In later developments, we shall see how this condition can be relaxed to deal with nonminimum phase systems.

The reference model is specified in terms of the difference equation:

$$y^m(k) = \sum_{j=1}^{n} a_j^m y^m(k-j) + \sum_{j=0}^{r} b_j^m u^m(k-j) \tag{7.5}$$

where $y^m(k)$ and $u^m(k)$ are the model output and reference input respectively. Define the polynomial

$$A^m(z^{-1}) = 1 - \sum_{j=1}^{n} a_j^m z^{-j} \tag{7.6a}$$

$$B^m(z^{-1}) = \sum_{j=0}^{r} b_j^m z^{-j} \qquad b_0 \neq 0 \tag{7.6b}$$

then the transfer function of the reference model can be expressed as

$$H_m(z) = Y^m(z)/U^m(z)$$
$$= B^m(z^{-1})/A^m(z^{-1}) \qquad (7.7)$$

Here $Y^m(z) = Z\{y^m(k)\}$ and $U^m(z) = Z\{u^m(k)\}$.
For practical purposes, it will be assumed that the poles of $H_m(z)$ in (7.7), or the roots of $A^m(z^{-1}) = 0$ in (7.6a), are inside the unit circle corresponding to stable reference model. Moreover, the reference model is taken to be completely reachable, (see Chapter 3).

An alternative formulation of (7.1) and (7.5), for $\xi(n) = 0$ for all n, can be obtained by defining the following quantities:

$$\underline{p}^t = [a_1,\ldots,a_n,b_0,\ldots,b_r] \qquad (7.8a)$$

$$\underline{p}_m^t = [a_1^m,\ldots,a_n^m,b_0^m,\ldots,b_r^m] \qquad (7.8b)$$

$$\underline{f}^t(k-1) = [y(k-1),\ldots,y(k-n),u(k-d),\ldots,u(k-r-d)] \qquad (7.8c)$$

$$\underline{g}^t(k-1) = [y^m(k-1),\ldots,y^m(k-n),u^m(k),\ldots,u^m(k-r)] \qquad (7.8d)$$

Thus,

$$y(k) = \underline{p}^t(k)\underline{f}(k-1) \qquad (7.9)$$

$$y^m(k) = \underline{p}_m^t\,\underline{g}(k-1) \qquad (7.10)$$

characterises the adjustable system and reference model, respectively. In (7.9), the parameter vector $\underline{p}(k)$ is shown explicitly to depend on k to account for slowly varying plants.

Another representation of MRAC systems could be obtained by using the state-space approach. In this case, the reference model is described by:

$$\underline{x}_m(k+1) = A_m\underline{x}_m(k) + B_m\underline{u}_m(k) \qquad (7.11)$$

where $\underline{x}_m(k)$, $\underline{u}_m(k)$ are the state and control input vectors at discrete instants k, with dimensions n and r, respectively. The matrices A_m and B_m are known and constant with appropriate dimensions. The adjustable system can generally be put in the form

$$\underline{x}(k+1) = A(k)\underline{x}(k) + B(k)\underline{u}(k) \tag{7.12}$$

where $\underline{x}(k)$ is the state vector (n-dimensional) and $\underline{u}(k)$ is the input vector (r-dimensional). $A(k)$ and $B(k)$ are time-varying matrices of dimensions (nxn) and (nxr), respectively. Depending on the nature of the adaptation mechanism, the input vectors $\underline{u}(k)$ and $\underline{u}_m(k)$ could be the same or different. It must be emphasized that the preceding representations of MRAC systems can be derived from each other (see Chapter 2).

It is convenient at this stage to define error quantities to represent the difference in response between the reference and adjustable models. The first quantity is the generalized state error $\underline{e}(k)$:

$$\underline{e}(k) = \underline{x}_m(k) - \underline{x}(k) \tag{7.13}$$

Let C be an output matrix of both models, then the generalized output error $\underline{\varepsilon}(k)$ is

$$\begin{aligned}\underline{\varepsilon}(k) &= C\underline{x}_m(k) - C\underline{x}(k) \\ &= C\underline{e}(k)\end{aligned} \tag{7.14}$$

When the reference and adjustable models are (7.1) and (7.5), the generalized output error $\varepsilon(k)$ becomes:

$$\varepsilon(k) = y^m(k) - y(k) \tag{7.15}$$

In processing the equations of the MRAC system, with a suitable adaptation loop, it turns out that some delay may occur. To avoid the impact of this delay, we distinguish between two cases

(1) information is available at the instant $(j-1)$ and it is used to compute the output at the j-th instant. This gives the *a priori* value of the output;

(2) information is available at the instant j and is used to compute the output at the j-th instant, that is, the adaptation has acted. This gives the *a posteriori* value of the output.

In view of the above discussions, the generalized output error $\varepsilon(k)$, as given by (7.15), is in fact the *a posteriori* value of the error. The *a priori* value would be

$$\varepsilon^a(k) = y^m(k) - y^a(k) \tag{7.16}$$

where $y^a(k)$ is the *a priori* value of the output.

By virtue of the fact that the reference model represents a target to be followed, the instant at which the generalized error dies out corresponds to perfect matching between the reference model and adjustable system. To achieve this condition in finite time would lead to a complicated design problem for the adaptation mechanism. Hence, it would be desirable to meet perfect matching over an infinite time, that is, one should seek asymptotic adaptation. Intuitively, this can be characterised by

$$\lim_{k\to\infty} [\underline{x}_m(k) - \underline{x}(k)] = \lim_{k\to\infty} \underline{e}(k) = \underline{0} \tag{7.17}$$

in the case of generalized state error, or

$$\lim_{k\to\infty} [y^m(k) - y(k)] = \lim_{k\to\infty} \varepsilon(k) = 0 \tag{7.18}$$

when the generalized output error is used. Next, we turn our attention to the design problem of MRAC systems.

7.2.4 *DESIGN CONSIDERATIONS*

It has been mentioned earlier that the fundamental problem in the synthesis of MRAC systems is the design of the adaptation mechanism for a given structure of the reference model and of the adjustable system. Before looking into this problem, we first present some assumptions that will be used in later developments. They are:

(a) Both the reference model and the adjustable system are linear and of the same dimension. In most cases, the reference model is shift-invariant.

(b) During the adaptation process, the parameters of the adjustable system depend only on the adaption mechanism.

(c) The initial difference between the parameters of the reference model and of the adjustable system is unknown.

(d) There should be some provision to measure the error quantities (state or output).

We note that assumption (a) above has been taken into consideration when defining the MRAC system elements. The usefulness of assumptions (b)-(d) is that they allow for an analytical treatment to be carried out.

Recall that the role of the adaptation mechanism can either be

(i) a parametric adaptation, by adjusting the parameters of the simulated plant, or

(ii) signal-synthesis adaptation, by applying an appropriate signal to the input of the plant.

In both cases, the adaptation process depends on the generalized error vector. To emphasize this point, we consider the case (i) of parametric adaptation and rewrite (7.12), using Fig. 7.2, as:

$$\underline{x}(k+1) = A(\underline{e},k)\underline{x}(k) + B(\underline{e},k)\underline{u}(k) \qquad (7.19)$$

where

$$\left.\begin{aligned} \underline{x}(0) &= \underline{x}_0 \\ A(0) &= A_0 \\ B(0) &= B_0 \end{aligned}\right\} \tag{7.20}$$

from which it is clear that $A(\cdot)$ and $B(\cdot)$ are time-varying matrices whose entries depend on the generalized state error vector $\underline{e}(k)$, defined by (7.13).

In the case (ii) of signal-synthesis adaptation, we write (7.12) in the form:

$$\underline{x}(k+1) = A\underline{x}(k) + B\underline{u}(k) + \underline{u}_a(\underline{e},k) \tag{7.21}$$

where

$$\left.\begin{aligned} \underline{x}(0) &= \underline{x}_0 \\ \underline{u}_a(0) &= \underline{u}_{a_0} \end{aligned}\right\} \tag{7.22}$$

from which we see that A and B are now constant matrices and the adaptation signal $\underline{u}_a$ depends on the error vector $\underline{e}(k)$ through the adaptation law. Similar formulas can be written for other MRAC configurations.

The design objective now becomes that of finding an adaptation law such that the parameter matrices $A(\underline{e},k)$ and $B(\underline{e},k)$ in (7.19), or the auxiliary signal $\underline{u}_a(\underline{e},k)$ in (7.21), are modified so that the error vector $\underline{e}(k)$, as defined by (7.13), tends toward the null vector $\underline{0}$ for any input $\underline{u}(k)$. Looked at in this light, this can intuitively be interpreted as a stability problem. In this respect, we can define the error vector $\underline{e}(k)$ as a new state vector for an equivalent system. This system must be globally asymptotically stable in order to satisfy condition (7.17). We must subsequently specify the conditions under which

$$\lim_{k\to\infty} A(\underline{e},k) = A_m \; ; \; \lim_{k\to\infty} B(\underline{e},k) = B_m \tag{7.23}$$

are obtained.

The stability approach to model reference adaptive system design appears to be a useful tool and has been used extensively in the literature [4,9-18]. Before considering this approach, we examine the construction of an equivalent system for a parallel model MRAC system as described by (7.11) and (7.19). Subtracting (7.19) from (7.11) using (7.13) and letting $\underline{u}(k) = u_m(k)$, results in:

$$\begin{aligned}
\underline{e}(k+1) &= A_m\underline{x}_m(k) - A(\underline{e},k)\underline{x}(k) + B_m\underline{u}(k) - B(\underline{e},k)\underline{u}(k) \\
&= A_m x_m(k) - A_m\underline{x}(k) + A_m\underline{x}(k) - A(\underline{e},k)\underline{x}(k) \\
&\quad + B_m\underline{u}(k) - B(\underline{e},k)\underline{u}(k) \\
&= A_m\underline{e}(k) + [A_m - A(\underline{e},k)]\underline{x}(k)\underline{x}(k) \\
&\quad + [B_m - B(\underline{e},k)\ \underline{u}(k)]
\end{aligned} \tag{7.24}$$

which characterises the MRAC system. It is easy to see that (7.24) represents a nonlinear, time-varying feedback system. This is depicted in Fig. 7.5, where it is decomposed into a linear part and a nonlinear time-varying part. We remark that:

(1) The condition $\underline{u}(k) = u_m(k)$ is not a restrictive one, since it complies with practical situations in which a common input signal is used to excite both the adjustable system and the reference model. In the sequel, this condition will be used unless otherwise stated.

(2) Stability results related to nonlinear time-varying feedback systems, which consist of a linear time-invariant part and a nonlinear time-varying part, have shown [10] that the stability of the feedback system is determined only by the characteristics of the linear part in addition to certain conditions placed on the nonlinear part.

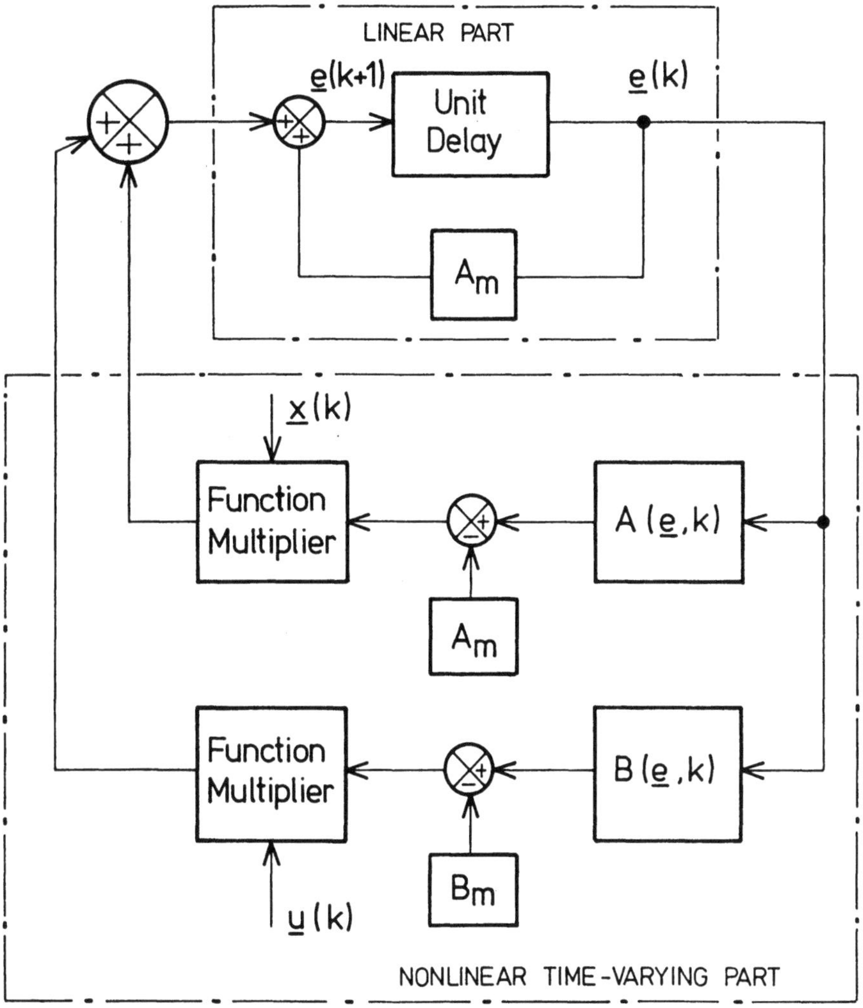

Fig. (7.5) Equivalent system representation

By considering (7.24), we see that the linear part is represented by the matrix A_m which is prespecified. Therefore, in order to meet the performance specifications, it would seem reasonable to consider a modified error vector of the form

$$\underline{\rho}(k) = E\,\underline{e}(k) \tag{7.25}$$

where E is an (n×n) matrix of design parameters. One way of looking into (7.25) is that it implies manipulating the generalized error vector through an appropriate gain (frequently called a *linear compensator*), to guarantee the overall stability of the MRAC system. Hence, the gain matrix E constitutes a part of the adaption mechanism.

Next, we consider the development of suitable methods for the design of model reference adaptive systems.

7.3 Design Techniques

We saw in the last section that the design of MRA systems is essentially a stability problem. There are two main approaches to tackling this problem: the first is based on Lyapunov analysis and the second uses hyperstability and positivity concepts *[4,6]*.

7.3.1 *TECHNIQUES BASED ON LYAPUNOV ANALYSIS*

Recall that a state-space representation of a parallel-type MRAS is given in terms of a reference model of the form

$$\underline{x}_m(k+1) = A_m\underline{x}_m(k) + B_m\underline{u}(k) \tag{7.26}$$

together with an adjustable system

$$\underline{x}(k+1) = A(\underline{e},k)\underline{x}(k) + B(\underline{e},k)\underline{u}(k) \tag{7.27}$$

where both systems are excited by the same input and a

parametric adaptation scheme is employed.

For the system described by (7.26) and (7.27), the dynamics of the error vector $\underline{e}(k)$, as defined by (7.13), are given by:

$$\underline{e}(k+1) = A_m\underline{e}(k) + \Gamma(k+1)\underline{x}(k) + \Omega(k+1)\underline{u}(k) \qquad (7.28)$$

where

$$\Gamma(k) = A_m - A(\underline{e},k) \qquad (7.29a)$$

$$\Omega(k) = B_m - B(\underline{e},k) \qquad (7.29b)$$

represents parameter alignment error terms. The subscript (k+1) is used on the r.h.s. of (7.28) to avoid delay in processing the adaptation mechanism.

At this stage we introduce the adaption mechanism to adjust the parameters of the model (7.27) as:

$$A(\underline{e},k+1) = A(\underline{e},k) + \alpha(k)P[\underline{e}(k)-A_m\underline{e}(k+1)]\underline{x}^t(k-1) \qquad (7.30a)$$

$$B(\underline{e},k+1) = B(\underline{e},k) + \alpha(k)P[\underline{e}(k)-A_m\underline{e}(k-1)]\underline{u}^t(k-1) \qquad (7.30b)$$

where $\alpha(k)$ is a suitable positive scalar for all k and P is an $(n\times n)$ symmetric, positive definite gain matrix. The recursive scheme (7.30) is usually initialized by:

$$A(\underline{e},0) = A_p \qquad (7.31a)$$

$$B(\underline{e},0) = B_p \qquad (7.31b)$$

where A_p and B_p are arbitrary selected nominal matrices of the adjustable system. We now define the quantities:

$$\underline{z}(k) = \begin{bmatrix} \underline{x}(k) \\ \underline{u}(k) \end{bmatrix} \qquad (7.32a)$$

$$\Phi(k) = \left[\Gamma(k) \;\vdots\; \Omega(k)\right] \qquad (7.32b)$$

as (n+m)-vector and $n\times(n+m)$ matrix representing an augmented state vector and an augmented parameter matrix, respectively. Using (7.32) in (7.28) results in

$$\underline{e}(k+1) = A_m\underline{e}(k) + \Phi(k+1)\underline{z}(k) \tag{7.33}$$

and from (7.29), (7.30) and (7.32) we obtain:

$$\Phi(k+1) = \Phi(k) + \Psi(k) \tag{7.34a}$$

where

$$\Psi(k) = -\alpha(k)P[\underline{e}(k)-A_m\underline{e}(k-1)]\underline{z}^t(k-1) \tag{7.34b}$$

We then proceed to define a candidate Lyapunov function L in the augmented state space formed by the product of the generalized error space and the parameter error space, that is, the $(\underline{e},\Phi)$ space. The function chosen is of the form:

$$L(k) = \beta[\underline{e}^t(k-1)W\underline{e}(k-1)] + \pi\,\mathrm{Tr}[\Phi^t(k)P^{-1}\Phi(k)] \tag{7.35}$$

where W is a symmetric, positive definite matrix, $\mathrm{Tr}[\cdot]$ is the trace operator and the scalars β, π are appropriate positive penality terms. It is important to note that $L(k) > 0$ by construction. For $\Delta L(k)$, the first forward difference, one uses (7.33) through (7.35) to obtain the expression:

$$\begin{aligned}\Delta L(k) &= L(k+1) - L(k)\\ &= \beta[\underline{q}^tW\underline{q} + \underline{q}^tWA_m\underline{p} + \underline{p}^tA_m^tW\underline{q} - \underline{p}^tQ\underline{p}]\\ &\qquad + \pi\,\mathrm{Tr}[\alpha^2(k)\underline{s}\,\underline{q}^t\,P\,\underline{q}\,\underline{s}^t\\ &\qquad - \underline{\alpha}(k)\Phi^t(k)\,\underline{q}\,\underline{s}^t\\ &\qquad - \underline{\alpha}(k)\,\underline{s}\,\underline{q}^t\,\Phi(k)]\end{aligned} \tag{7.36}$$

where

$$\left.\begin{aligned}\underline{p} &= \underline{e}(k+1) \quad ; \quad \underline{s} = \underline{z}(k-1)\\ \underline{q} &= \Phi\underline{s} \qquad\quad ; \quad \xi = \underline{s}^t\,\underline{s}\end{aligned}\right\} \tag{7.37}$$

and $$A_m^t W A_m - W = -Q \tag{7.38}$$

with $Q = Q^t > 0$. Making use of the cyclic property of trace operators [20], that is, $Tr[G_1G_2G_3] = Tr[G_2G_3G_1] = Tr[G_3G_1G_2]$ we can put (7.36) in the compact form:

$$L(k) = -[\underline{p}^t \quad \underline{q}^t]\, H \begin{bmatrix} \underline{p} \\ \underline{q} \end{bmatrix} \tag{7.39a}$$

where

$$H = \left[\begin{array}{c|c} \beta Q & -\beta A_m^t W \\ \hline -\beta W A_m & 2\alpha(k)\pi I_n - \alpha^2(k)\xi\pi P - \beta W \end{array}\right] \tag{7.39b}$$

The augmented error system (7.33) and (7.34) will be asymptotically stable if we can make $\Delta L(k) < 0$. From (7.39), this corresponds to the requirement that H, given by (7.39b), is positive definite. In view of the facts that $Q > 0$ and H is symmetric, then it is only required to ensure that the matrix

$$2\alpha(k)\pi I_n - \alpha^2(k)\pi\xi P - \beta[W + WA_m Q^{-1} A_m^t W] > 0 \tag{7.40}$$

Bearing in mind that Q and P are not related to each other, it is suggested that we choose

$$\alpha(k) = (\delta/\hat{\lambda}\ \xi) \quad ; \quad 0 < \delta \leq 1 \tag{7.41a}$$

$$\pi > \beta \tag{7.41b}$$

where $\hat{\lambda}$ is the maximum eigenvalue of P, so that (7.40) is satisfied for all $\underline{p}, \underline{q} \neq \underline{0}$. Consequently, $L(k)$ of (7.35) is a Lyapunov function and thus the adaptive scheme (7.30) is asymptotically stable in the $(\underline{p}, \underline{q})$ space. This corresponds to

$$\lim_{k\to\infty} \underline{e}(k) = \underline{0} \tag{7.42a}$$

$$\lim_{k\to\infty} \Phi(k)\underline{z}(k-1) = \underline{0} \tag{7.42b}$$

Consider that $\underline{u}(k)$ is bounded for all k, then (7.42a) implies that

$$\underline{x}(k) \longrightarrow \underline{x}_m(k)$$

which, on using (7.42b), becomes:

$$\Gamma(k+1)\underline{x}_m(k) + \Omega(k+1)\underline{u}(k) \longrightarrow \underline{0} \qquad (7.43)$$

When the reference model is reachable, it has been shown *[14]* that (7.43) reduces to:

$$\begin{aligned} \Gamma(k) &\longrightarrow \underline{0} \\ \Omega(k) &\longrightarrow \underline{0} \end{aligned} \qquad (7.44)$$

By virtue of (7.29), it is readily evident that the discrete system (7.26), (7.27) and (7.30) is capable of performing asymptotic adaptation.

It should be remarked that:

(1) The satisfaction of inequality (7.40) hinges upon a suitable selection of the scalars δ, β and π. In general, a search procedure is used to choose values of these scalars so as to make $\Delta L(k) < 0$.

(2) From (7.38), the matrix W is symmetric and satisfies a Lyapunov equation. To determine this matrix, one can use the methods described in Chapter 3.

Lyapunov functions have been successfully used for designing stable MRAS's, for a survey of various methods, see *[13,14]*. The only limitation on their use is the lack of a procedure to enlarge the class of suitable Lyapunov functions which would lead to adaptation rules other than (7.30). This is an important problem because in most cases we will be interested in generating several adaptation rules, which ensure the overall stability of an MRAS, from which we can then choose the pertinent one to suit a specific application.

As we shall see in the next section, the above problem can be resolved to some extent through the use of hyperstability theory and positivity concept.

7.3.2 *TECHNIQUES BASED ON HYPERSTABILITY AND POSITIVITY CONCEPTS*

The hyperstability problem was introduced by Popov *[10]* as a generalization of the absolute stability problem for multivariable, nonlinear, time-varying feedback systems of the type depicted in Fig. 7.6 . We note that this system is in a standard form, a typical example of which is the equivalent system shown in Fig. 7.5 . The basic feature of the standard system is that it is split into two main blocks: a feedforward block represents a linear time-invariant system, and a feedback block belongs to the family of nonlinear, time-varying systems.

A. Popov Inequality and Related Results

With reference to Fig. 7.6, we consider that the input and output vectors $\underline{y}(k)$ and $\underline{w}(k)$ of the feedback block are both m-dimensional. Popov considered the global asymptotic stability of a system of the form given in Fig. 7.6, but for the class of feedback blocks satisfying the inequality

$$\eta(k_0,n) = \sum_{j=k_0}^{n} \underline{w}^t(j)\underline{y}(j) \geq -\omega_0^2 , \quad \text{for all } n \geq k_0 \tag{7.45}$$

where ω_U^2 is a finite positive constant, independent of n . It is important to note that (7.45) expresses a relation on the input-output inner product. A globally (asymptotically) stable standard system, with feedback blocks satisfying (7.45), is said to be (asymptotically) *hyperstable*. To discuss the properties of hyperstable systems, we need to review some results related to positive dynamic systems *[10,21-23]*.

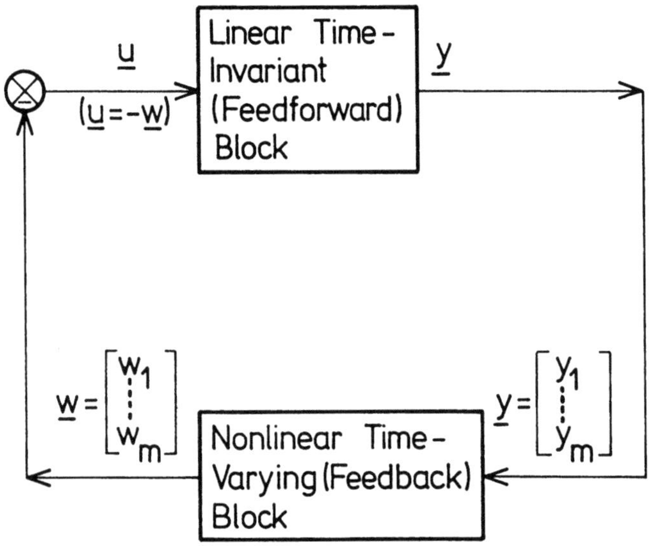

Fig. (7.6) Standard multivariable nonlinear, time-varying feedback system

In the sequel we consider a linear, time-invariant discrete system of the form

$$\underline{x}(k+1) = A\underline{x}(k) + B\underline{u}(k) \tag{7.46a}$$

$$\underline{y}(k) = C\underline{x}(k) + D\underline{u}(k) \tag{7.46b}$$

where $\underline{x}(k)$ is an n-dimensional state vector; $\underline{u}(k)$ and $\underline{y}(k)$ are m-dimensional vectors representing the input and output, respectively; and A,B,C, and D are constant matrices of appropriate dimensions. We assume that the pair (A,B) is completely reachable and that the pair (C,A) is completely observable. The system (7.46) is also characterised by the discrete square transfer matrix

$$H(z) = D + C(zI_n - A)^{-1}B \tag{7.47}$$

An (mxm) discrete matrix $H(z)$ of real rational functions is *positive real* if [22]:

(a) all elements of $H(z)$ are analytic outside the unit circle, that is, they do not have poles in the region $|z| > 1$;

(b) the eventual poles of any element of $H(z)$ on the unit circle $|z| = 1$ are simple, the associated residue matrix is positive semidefinite Hermitian ($H(z) = H^t(z^*) \geq 0$ where the asterisk means conjugate);

(c) the matrix $H(z) + H^t(z^*) = H(e^{j\omega}) + H^t(e^{-j\omega})$ is positive semidefinite Hermitian for all real values of ω which are not poles for any element of $H(e^{j\omega})$.

A discrete matrix $F(k,\ell)$ is termed positive definite if for each interval $[k_0, k_1]$ and for all the discrete vectors $\underline{h}(k)$ bounded in $[k_0, n]$ the following inequality holds:

$$\sum_{k=k_0}^{n} \underline{h}^t(k) \left[\sum_{\ell=k_0}^{k} F(k,\ell)\underline{h}(\ell) \right] \geq 0 \quad \text{for all } n \geq k_0 \tag{7.48}$$

The term $\left[\sum_{\ell=k_0}^{k} F(k,\ell)\underline{h}(\ell)\right]$ can be interpreted as the output of a block whose input is $\underline{h}(k)$, and hence (7.48) can be interpreted as the sum of the input-output inner product. The matrix $F(k,\ell)$, which is equivalent to the impulse response matrix [6], is frequently called a *discrete matrix kernel* [22]. For the class of discrete kernels $F(k-\ell)$ for which the Z-transform exists, the necessary and sufficient condition for $F(k-\ell)$ to be a positive definite discrete matrix kernel is that [22] its Z-transform be a positive real discrete transfer matrix.

The discrete system (7.46) is said to be *positive* if the sum of the input-output scalar products over the interval $[k_0,n]$ can be expressed by

$$\sum_{j=k_0}^{n} \underline{y}^t(j)\underline{u}(j) = \Theta[\underline{x}(n+1)] - \Theta[\underline{x}(k_0] + \sum_{j=0}^{n} \lambda[\underline{x}(k),\underline{u}(k)] \; ; \qquad \text{for all } k \geq 0 \qquad (7.49)$$

with

$$\lambda[\underline{x}(k),\underline{u}(k)] \geq 0 \qquad \text{for all } \underline{x}(k) \text{ and } \underline{u}(k) \qquad (7.50)$$

An equivalent statement of the positivity of the system (7.46) is that its transfer matrix $H(z)$, given by (7.47), be a positive real. Still other equivalent statements can be obtained by analyzing the stability behaviour of (7.46) and have been derived in [10,22]. The main result is that there exists a symmetric positive definite matrix P, a symmetric positive semidefinite matrix Q and matrices S and R such that:

$$A^tPA - P = -Q \qquad (7.51)$$

$$B^tPA + S^t = C \qquad (7.52)$$

$$D+D^t - B^tPB = R \qquad (7.53)$$

$$\begin{pmatrix} Q & S \\ S^t & R \end{pmatrix} \geq 0 \qquad (7.54)$$

In the case where the matrices Q, S and R are factorizable in the form

$$\left.\begin{aligned} Q &= MM^t \\ R &= N^tN \\ S &= MN \end{aligned}\right\} \qquad (7.55)$$

we obtain *the discrete, positive real lemma [22]*:

$$A^tPA - P = -MM^t \qquad (7.56)$$

$$B^tPA + N^tM^t = C \qquad (7.57)$$

$$D + D^t - B^tPB = N^tN \qquad (7.58)$$

in which we note that condition (7.54) is no longer required since it is automatically satisfied. Of interest now is the linking of the conditions stated in (7.49) and (7.50) to those of (7.51) through (7.54). The result is that *[6]* every solution $\underline{x}(k)$ of the system (7.46), which depends on the initial state $\underline{x}_0$ and input sequence $\underline{u}(k)$, satisfies the following relation

$$\sum_{j=0}^{n} \underline{y}^t(j)\underline{u}(j) = (\tfrac{1}{2})\underline{x}^t(n+1)P\underline{x}(n+1) - (\tfrac{1}{2})\underline{x}_0^tP\underline{x}_0 + (\tfrac{1}{2}) \sum_{j=0}^{n} [\underline{x}^t(j)Q\underline{x}(j) + 2\underline{u}^t(j)S^t\underline{x}(j) + \underline{u}^t(j)R\underline{u}(j)] \qquad (7.59)$$

where P, Q, S and R satisfy (7.51) to (7.54).

For the class of discrete, linear, time-varying systems of the type

$$\underline{x}(k+1) = A(k)\underline{x}(k) + B(k)\underline{u}(k) \qquad (7.60a)$$

$$\underline{y}(k) = C(k)\underline{x}(k) + D(k)\underline{u}(k) \qquad (7.60b)$$

the positivity conditions are only sufficient, in contrast to the time-invariant case where they are necessary and sufficient. In the time-varying case, the discrete, positive real lemma becomes:

$$A^t(k)P(k+1)A(k) - P(k) = -M(k)M^t(k) \qquad (7.61)$$

$$B^t(k)P(k+1)A(k) + N^t(k)M^t(k) = C(k) \qquad (7.62)$$

$$D(k) + D^t(k) - B^t(k)P(k+1)B(k) = N^t(k)N(k) \qquad (7.63)$$

Corresponding to (7.59), we have the relation

$$\sum_{j=0}^{n} \underline{y}^t(j)\underline{u}(j) = (\tfrac{1}{2})\underline{x}^t(n+1)P(n+1)\underline{x}(n+1) - (\tfrac{1}{2})\underline{x}_0^t P(0)\underline{x}_0$$
$$+ (\tfrac{1}{2}) \sum_{j=0}^{n} [\underline{x}^t(j)Q(j)\underline{x}(j) + 2\underline{u}^t(j)S(j)\underline{u}(j)$$
$$+ \underline{u}^t(j)R(j)\underline{u}(j)] \qquad (7.64)$$

where $P(k)$ is a sequence of positive definite matrices.

Going back to the standard feedback system depicted in Fig. 7.6, where the feedforward block is given by (7.46). Consider that the feedback block is described by

$$\underline{w}(k) = \underline{g}[\underline{y},k,j] \quad , \quad j \leq k \qquad (7.65)$$

where the pair $[\underline{y}(k),\underline{w}(k)]$ satisfies the inequality (7.45). Hence we state that

> *The necessary and sufficient condition for the feedback system described by (7.45), (7.46) and (7.65) to be (asymptotically) hyperstable is that the discrete transfer matrix $H(z)$ given by (7.47) is a (strictly) positive real matrix or equivalently the positivity conditions (7.51) through (7.54) are satisfied.*

Our main objective now is to apply the analytical results presented above to derive suitable adaptation schemes.

B. Systematic Procedure

The application of the hyperstability and positivity concepts to the design of an MRAS is performed systematically as follows:

Step I

Convert the MRAS into an equivalent standard feedback

system (similar to that of Fig. 7.6 but including a linear compensator of the type (7.25)).

Step II

Use appropriate adaptation laws and then find solutions for the portion appearing in the feedback path such that the inequality (7.45) be satisfied.

Step III

Find solutions for the remaining portion of the adaptation law appearing in the feedforward path such that the feedforward block be hyperstable.

Step IV

Implement the adaptation law and if changes are needed, go to Step I.

The above procedure can be utilized with parametric-type or signal-synthesis adaptation schemes.

C. Parametric Adaptation Scheme

We now provide a detailed specification of a parametric adaptation when used for a parallel-type MRAS. The discrete system at hand is described by the following components in state space form:

(1) A reference model

$$\underline{x}_m(k+1) = A_m \underline{x}_m(k) + B_m \underline{u}(k) \tag{7.66}$$

(2) An adjustable "parallel" system

$$\underline{x}^a(k+1) = A_p(k)\underline{x}(k) + B_p(k)\underline{u}(k) \tag{7.67}$$

$$\underline{x}(k+1) = A_p(k+1)\underline{x}(k) + B_p(k+1)\underline{u}(k) \tag{7.68}$$

For this system, we define:

(i) The generalized state error:

$$\underline{e}^a(k) = \underline{x}_m(k) - \underline{x}^a(k) \qquad (7.69)$$

$$\underline{e}(k) = \underline{x}_m(k) - \underline{x}(k) \qquad (7.70)$$

(ii) The adaptation algorithm:

$$\underline{\rho}^a(k) = E\,\underline{e}^a(k) \qquad (7.71)$$

$$\underline{\rho}(k) = E\,\underline{e}(k) \qquad (7.72)$$

$$A_p(k+1) = \sum_{j=0}^{k} F_s(\underline{\rho},k,j) + F_p(\underline{\rho},k) + A_p(0) \qquad (7.73)$$

$$B_p(k+1) = \sum_{j=0}^{k} G_s(\underline{\rho},k,j) + G_p(\underline{\rho},k) + B_p(0) \qquad (7.74)$$

where $\underline{x}_m(k)$, $\underline{x}^a(k)$ and $\underline{x}(k)$ are n-dimensional state vectors, $\underline{u}(k)$ is an r-dimensional input vector, A_m and B_m are constant matrices, $A_p(k)$ and $B_p(k)$ are time-varying matrices, and E is a constant square matrix. The superscript 'a' on the variable denotes its *a priori* value. We note that when $\lim_{k\to\infty} \underline{e}(k) = 0$; the effect of $F_p(.)$ and $G_p(.)$ in (7.73) and (7.74) dies out, implying that these matrices represent transient terms appearing during the adaptation process before reaching the equilibrium. Note also from (7.73) and (7.74) that F_p and G_p define a 'proportional' adaptation, whereas F_s and G_s specify a 'summation' adaptation.

From (7.66), (7.68) and (7.70) we derive the dynamics of the generalized state error:

$$\begin{aligned}\underline{e}(k+1) = A_m\,\underline{e}(k) &+ [A_m - A_p(k+1)]\underline{x}(k) \\ &+ [B_m - B_p(k+1)]\underline{u}(k) \qquad (7.75)\end{aligned}$$

To complete Step I of the systematic procedure, we combine (7.72)-(7.75) together to obtain the equivalent feedback system:

$$\underline{e}(k+1) = A_m\,\underline{e}(k) + I_n\,\underline{f}(k+1) \qquad (7.76)$$

$$\underline{\rho}(k) = E\,\underline{e}(k) \qquad (7.77)$$

$$\underline{w}(k+1) = -\underline{f}(k+1)$$

$$= \left[\left\{\sum_{j=0}^{k} F_s(\underline{\rho},k,j)\right\} + F_p(\underline{\rho},k) + A_0\right]\underline{x}(k)$$

$$+ \left[\left\{\sum_{j=0}^{k} G_s(\underline{\rho},k,j)\right\} + G_p(\underline{\rho},k) + B_0\right]\underline{u}(k) \tag{7.78}$$

$$A_0 = A_p(0) - A_m \; ; \; B_0 = B_p(0) - B_m \tag{7.79}$$

It should be remarked that:

(1) The model description (7.76)-(7.79) can be put in the form of Fig. 7.5 by inserting a block E at the output of the linear part and replacing $A(\underline{e},k)$, $B(\underline{e},k)$ by $A_p(k+1)$, $B_p(k+1)$, respectively.

(2) It can also be put in the standard form of Fig. 7.6 by noting that (7.76) and (7.77) define the linear time-invariant part but with input $\underline{f}(k) = -\underline{w}(k)$ and output $\underline{\rho}(k)$. This part is characterised by the transfer matrix

$$H(z) = zE[zI_n - A_m]^{-1}$$

written more appropriately in the form

$$H(z) = E + E[zI_n - A_m]^{-1}A_m \tag{7.80}$$

(3) The nonlinear time-varying part of the equivalent feedback system is specified by (7.78) and (7.79) in which the input is $\underline{\rho}(k)$ and the output is $\underline{w}(k)$.

(4) The adaptation algorithm (7.71) through (7.74) is of proportional plus summation type; which is quite general and realistic for practical applications.

According to the hyperstability theorem stated earlier, $H(z)$ of (7.80) must be a (strictly) positive real matrix for the feedback system to be (asymptotically) hyperstable. As we shall see later, this condition allows us to compute the compensator matrix **E**.

Our next task is to find expressions for the quantities F_s, F_p, G_s and G_p in order that the nonlinear, time-varying part of the equivalent feedback system verifies the positivity condition rewritten in the appropriate form:

$$\eta(0,n) = \sum_{k=0}^{n} \underline{\rho}^t(k+1)\underline{w}(k+1) \geq -\omega_0^2 \; ; \quad n > 0 \qquad (7.81)$$

The use of (7.78) in (7.81) converts it into:

$$\eta(0,n) = \sum_{k=0}^{n} \underline{\rho}^t(k+1) \left\{ \left[\left\{ \sum_{j=0}^{k} F_s(\underline{\rho},k,j) \right\} + F_p(\underline{\rho},k) + A_0 \right] \underline{x}(k) + \left[\left\{ \sum_{j=0}^{k} G_s(\underline{\rho},k,j) \right\} + G_p(\underline{\rho},k) + B_0 \right] \underline{u}(k) \right\} \geq -\omega_0^2 \qquad (7.82)$$

Our approach to satisfy (7.82) is to decompose it into suitable terms, each of which must verify an inequality of the same type. For example, a typical term would be

$$\sum_{k=0}^{n} \underline{\rho}^t(k+1) \sum_{j=0}^{k} F_s(\underline{\rho},k,j)\underline{x}(k) \geq -\omega_{0a}^2 \quad \text{for all } n \geq 0 \qquad (7.83)$$

which is constructed so as to isolate the effects of F_s. Using the properties of positive discrete kernels *[6,8]*, it has been shown *[8]* that an inequality of the type (7.83) can be verified by

$$F_s(\underline{\rho},k,j) = M\underline{\rho}(j+1)[P(k-j)\underline{x}(j)]^t \qquad (7.84)$$

where $P(k-j)$ is a positive definite matrix kernel and M is an appropriate positive definite matrix. There are two important points to mention about (7.84). First, that a particular case of interest results when $P(k-j) = P$, $k = j$ and 0 otherwise. Second, since $\underline{\rho}(k+1)$ is not yet available for processing, it must be related to $\underline{\rho}(k)$. A possible form would be

$$\underline{\rho}(k+1) = T(k)\underline{\rho}(k) \tag{7.85}$$

where $T(k)$ is an appropriate gain matrix. By taking these points into consideration, (7.84) becomes (with $M = I_n$) :

$$F_s(\underline{\rho},k,j) = T(j)\underline{\rho}(j)[P\underline{x}(j)]^t \tag{7.86}$$

Applying similar reasoning, we can proceed to find expressions for other matrices such that the complete forms for $A_p(k)$ and $B_p(k)$ now become:

$$A_p(k+1) = \sum_{j=0}^{k} T(j)\underline{\rho}(j)[Px(j)]^t + T(k)\underline{\rho}(k)[\tilde{P}\underline{x}(k)]^t + A_p(0) \tag{7.87}$$

$$B_p(k+1) = \sum_{j=0}^{k} T(j)\underline{\rho}(j)[R\underline{u}(j)]^t + T(k)\underline{\rho}(k)[\tilde{R}\underline{u}(k)]^t + B_p(0) \tag{7.88}$$

Alternative expressions for $A_p(k)$ and $B_p(k)$ would be obtained by rearranging (7.87) and (7.88) to arrive at:

$$A_p(k+1) = \tilde{A}_p(k) + T(k)\underline{\rho}(k)[(P+\tilde{P})\underline{x}(k)]^t \tag{7.89a}$$

$$\tilde{A}_p(k) = A_p(0) + \sum_{j=0}^{k-1} T(j)\underline{\rho}(j)[P\underline{x}(j)] \tag{7.89b}$$

$$B_p(k+1) = \tilde{B}_p(k) + T(k)\underline{\rho}(k)[(R+\tilde{R})\underline{u}(k)]^t \tag{7.90a}$$

$$\tilde{B}_p(k) = B_p(0) + \sum_{j=0}^{k-1} T(j)\underline{\rho}(j)[R\underline{u}(j)]^t \tag{7.90b}$$

where $P,R > 0$ and $\tilde{P},\tilde{R} \geq 0$. It is easy to interpret the quantities $\tilde{A}_p(k)$ and $\tilde{B}_p(k)$ as summarizing the contributions of past records. We emphasize that $\underline{\rho}(k)$ of (7.72) is a function of the error $\underline{e}(k)$ to be nullified gradually. This completes Step II of the design procedure.

The substitution of (7.89) and (7.90) into (7.75) results in

$$\underline{e}(k+1) = A_m \underline{e}(k) + [A_m - \tilde{A}_p(k)]\underline{x}(k) + [B_m - \tilde{B}_p(k)]\underline{u}(k) - T(k)\underline{\rho}(k)\Delta_s \qquad (7.91)$$

where

$$\Delta_s = \{\underline{x}^t(k)[P+\tilde{P}]\underline{x}(k) + \underline{u}^t(k)[R+\tilde{R}]\underline{u}(k)\} \qquad (7.92)$$

is a suitable scalar quantity.

Let $T^a(k)$ be an *a priori* estimate of $T(k)$, then from (7.91) one obtains

$$\underline{e}(k+1) = \underline{e}^a(k+1) - T(k)\underline{\rho}(k)\Delta_s + T^a(k)\underline{\rho}(k)\Delta_s \qquad (7.93)$$

where we used $\underline{e}(k) = \underline{e}^a(k)$. Using (7.72) and (7.85) in (7.93) converts it to:

$$\underline{\rho}(k+1) = \mathbf{E}\underline{\rho}(k)$$

$$= [I_n + E\Delta_s]^{-1}[\underline{\rho}^a(k+1) + ET^a(k)\underline{\rho}(k)\Delta_s] \qquad (7.94)$$

In the particular case $T^a(k) = 0$, (7.94) reduces to

$$\underline{\rho}(k+1) = [I_n + E\Delta_s]^{-1}\, \underline{\rho}^a(k+1) \qquad (7.95)$$

where $\underline{\rho}^a(k)$ can be computed from (7.71). We note that (7.95) does not depend on $T(k)$.

Suppose that in (7.84) we let $M \neq I_n$ but the other assumptions remain unchanged. It is a straightforward task to show that (7.87) and (7.88) now take the form

$$A_p(k+1) = \sum_{j=0}^{k} MT(j)\underline{\rho}(j)[P(k-j)\underline{x}(j)]^t + \tilde{M}(k)T(k)\underline{\rho}(k)[\tilde{P}(k)\underline{x}(k)]^t + A_p(0) \quad (7.96a)$$

$$B_p(k+1) = \sum_{j=0}^{k} NT(j)\underline{\rho}(j)[R(k-j)\underline{u}(j)]^t + \tilde{N}(k)T(k)\underline{\rho}(k)[\tilde{R}(k)\underline{u}(k)]^t + B_p(0) \quad (7.96b)$$

where $P(k-j)$, $R(k-j)$ are positive definite discrete matrix kernels, $\tilde{P}(k)$, $\tilde{R}(k)$, $\tilde{M}(k)$, $\tilde{N}(k)$ are positive definite matrices and M,N are positive definite constant matrices. In this case, the formula (7.95) under the conditions $P(k-j) = \tilde{P}(k) = P$ and $R(k-j) = \tilde{R}(k) = R$ become:

$$\underline{\rho}(k+1) = [I_n+E\Delta_g]^{-1}\underline{\rho}^a(k+1) \quad (7.97a)$$

$$\Delta_g = (M+\tilde{M})\underline{x}^t(k)P\underline{x}(k) + (N+\tilde{N})\underline{u}^t(k)R\underline{u}(k) \quad (7.97b)$$

It remains, in either case, to compute the linear compensator matrix E such that $H(z)$ of (7.80) is strictly positive real. We restrict E to be positive definite. Note that $H(z)$ will be strictly positive real if

$$\tilde{H}(z) = \frac{1}{2}E + E[zI_n-A_m]^{-1}A_m \quad (7.98)$$

is strictly positive real since $H(z) = \tilde{H}(z) + \frac{1}{2}E$ and E is positive definite.

The comparison of (7.47) and (7.98) reveals that $\tilde{H}(z)$ can be realized in a state-space representation with

$$\left.\begin{array}{ll} A = A_M , & B = A_m \\ C = E , & D = \frac{1}{2}E \end{array}\right\} \quad (7.99a)$$

Recall from the discrete positive real lemma that $\tilde{H}(z)$ is strictly positive real with the qualification (7.99a) when conditions (7.56) through (7.58) are verified. It is readily evident that $E = P$, where

$$A_m^t P A_m - P = -Q \qquad (7.99b)$$

is an admissible solution provided that A_m is a stable matrix. Note that this is equivalent to saying that E satisfies a Lyapunov equation. In effect, this implies that the feedback system (7.66)-(7.74) is asymptotically hyperstable. It should be stressed that Step III is now completed, and since Step IV has been implicitly accomplished, the design task is fulfilled.

In the course of experimentation, the weighting matrices P, R, $\tilde{P}$ and $\tilde{R}$ are frequently selected with large values. Other MRAS configurations can be dealt with in the same way and can be found in *[6,8]*.

Had we considered the MRAS problem description of the type (7.5)-(7.10), instead of the state-space representation (7.11) -(7.14), or equivalently (7.46)-(7.47), we would have obtained similar adaptation schemes. The result of such an effort is summarized in Table 7.1, where we have distinguished between two classes of systems:

(a) Systems in which the parameters of either the reference model or the adjustable system can undergo frequent changes.

(b) Systems in which the initial difference between the parameters of the reference model and the adjustable system are unknown.

We note the following concerning the algorithms in Table 7.1:

(1) For class "a" algorithms, the case of $F(k) = F \neq 0$ corresponds to a proportional plus summation scheme with a constant adaptation gain. The special case of $F = 0$ becomes that of summation-type adaptation.

(2) In the summation-algorithm of class "b" it can be seen

TABLE 7.1 Adaptation Schemes for Parallel MRAS

*The reference model**

$$y^m(k) = \sum_{j=1}^{n} a_j^m y^m(k-j) + \sum_{j=0}^{r} b_j^m u(k-j)$$

$$= \underline{p}_m^t \, \underline{g}(k-1)$$

* $\underline{p}_m^t = [a_1^m, \ldots, a_n^m, b_0^m, \ldots, b_r^m]$

$\underline{g}^t(k-1) = [y^m(k-1), \ldots, y^m(k-n), u^m(k), \ldots, u^m(k-r)]$

The parallel adjustable system†

$$y(k) = \sum_{j=1}^{n} a_j(k) y_j(k-j) + \sum_{j=0}^{r} b_j(k) u_j(k-j)$$

$$= \underline{p}^t(k)\underline{f}(k-1)$$

$$= [\underline{p}^s(k) + \underline{p}^p(k)]^t \, \underline{f}(k-1)$$

$$\underline{y}^a(k) = [\underline{p}^s(k-1)]^t \, \underline{f}(k-1)$$

† $\underline{p}^t(k) = [a_1(k), \ldots, a_n(k), b_0(k), \ldots, b_r(k)]$

$\underline{f}^t(k-1) = [y(k-1), \ldots, y(k-n), u(k), \ldots, u(k-r)]$

$\underline{p}^s(k)$ summation-part (memory term)

$\underline{p}^p(k)$ proportional-part (memoryless term)

$\underline{y}^a(k)$ *a priori* value of $\underline{y}(k)$

The generalized error

a priori $e^a(k) = y^m(k) - y^a(k)$

a posteriori $e(k) = y^m(k) - y(k)$

cont....

TABLE 7.1 (contd.)

The adaptation algorithm

$$\nu^a(k) = e^a(k) + \sum_{j=1}^{n} d_j\, e(k-j)$$

$$\nu(k) = e(k) + \sum_{j=1}^{n} d_j (k-j)$$

$$\underline{p}(k) = \underline{p}^s(k) + \underline{p}^p(k)$$

Class "a"	Class "b": Summation adaptation	Class "b": Proportional plus summation adaptation
1. $\underline{p}^s(k)=\underline{p}^s(k-1)+\underline{h}(k-1)\nu^a(k)$	1. $p^s(k)=\underline{p}^s(k-1)+\underline{h}(k-1)\nu^a(k)$	1. $\underline{p}^s(k)=\underline{p}^s(k-1)+\underline{h}(k-1)\nu^a(k)$
2. $\underline{h}(k-1)=G\underline{f}(k-1)/[1 + \underline{f}^t(k-1)\{G+F(k-1)\}\underline{f}(k-1)]$	2. $\underline{h}(k-1)=G(k-1)\underline{f}(k-1)/[1 + \underline{f}^t(k-1)G(k-1)\underline{f}(k-1)]$	2. $\underline{h}(k-1)=G(k-1)\underline{f}(k-1)/[1 + \underline{f}^t(k-1)\{G(k-1)+F(k-1)\}\underline{f}(k-1)]$
3. $\underline{p}^p(k) = \underline{r}(k-1)\nu^a(k)$	3. $\underline{p}^p(k) = \underline{0}$	3. $\underline{p}^p(k) = \underline{r}(k-1)\nu^a(k)$
4. $\underline{r}(k-1) = F(k-1)\underline{f}(k-1)/[1 + \underline{f}^t(k-1)\{G+F(k-1)\}\underline{f}(k-1)]$	4. $G(k) = G(k-1)-(1/\lambda)\dfrac{G(k-1)\underline{f}(k-1)\underline{f}^t(k-1)G(k-1)}{1+(1/\lambda)\underline{f}^t(k-1)G(k-1)\underline{f}(k-1)}$ $G(0) > 0, \quad \lambda > 1/2$	4. $\underline{r}(k-1) = F(k-1)\underline{f}(k-1)/[1 + \underline{f}^t(k-1)\{G(k-1)+F(k-1)\}\underline{f}(k-1)]$

cont....

TABLE 7.1) (contd.)

Class "a"	Class "b": Summation adaptation	Class "b": Proportional plus summation adaptation
5. $F(k)+{}^{1}/2G \geq 0$ for all k $G > 0$	5. $\nu^{a}(k) = y^{m}(k)-y^{a}(k) + \sum_{j=1}^{n} d_j\, e(k-j)$	5. $G(k) = G(k-1) - (1/\lambda)\dfrac{G(k-1)\underline{f}(k-1)\underline{f}^{t}(k-1)G(k-1)}{1+(1/\lambda)\underline{f}^{t}(k-1)G(k-1)\underline{f}(k-1)}$ $G(0) > 0, \quad \lambda > 1/2$
6. $\nu^{a}(k) = y^{m}(k) - y^{a}(k) + \sum_{j=1}^{n} d_j\, e(k-j)$	6. The coefficients d_j are selected such that $\tilde{h}(z) = \dfrac{1+\sum_{j=1}^{n} d_j z^{-j}}{1-\sum_{j=1}^{n} a_j z^{-j}} - (1/2\lambda)$ is a strictly positive real transfer function	6. $F(k) = \alpha G(k)$, $\alpha \geq - 1/2$
7. The coefficients d_j are selected such that $h(z) = \dfrac{1+\sum_{j=1}^{n} d_j z^{-j}}{1-\sum_{j=1}^{n} a_j z^{-j}}$ is a strictly positive real transfer function.		7. $\nu^{a}(k) = y^{m}(k) - y^{a}(k) + \sum_{j=1}^{n} d_j e(k-j)$
		8. The coefficients d_j are selected such that $\tilde{h}(z) = \dfrac{1+\sum_{j=1}^{n} d_j z^{-j}}{1-\sum_{j=1}^{n} a_j z^{-j}} - ({}^{1}/2\lambda)$ is a strictly positive real transfer function

that

$$G(k) \leq G(k-1)$$

implying that the adaption gain of the summation part is gradually decreasing with a rate weighted by the factor λ.

(3) For large values of λ, $G(k)$ tends towards a constant matrix in which case class "b"-summation adaptation reduces to class "a" adaptation.

It should be emphasized again that the development of the adaptation algorithm cited in Table 7.1 is based on the systematic procedure outlined earlier using hyperstability and positivity concepts.

D. Adaptive Model-Following Schemes

In this section we present adaptive schemes which are based on explicit model-following capabilities. The particular structure we are going to analyse is shown in Fig. 7.7, which employs a signal-synthesis adaptation algorithm. The system is described as follows:

(1) the reference model

$$\underline{x}_m(k+1) = A_m\underline{x}_m(k) + B_m\underline{u}_m(k) \tag{7.100}$$

(2) the plant to be controlled

$$\underline{x}(k+1) = A_p\underline{x}(k) + B_p\underline{u}_p(k) \tag{7.101}$$

(3) the plant control input

$$u_p(k) = -K_p\underline{x}(k) + K_m\underline{x}_m(k) + K_u\underline{u}_m(k) \tag{7.102}$$

where $\underline{x}_m(k)$ is the model state vector (n-dimensional), $\underline{x}(k)$ is the plant state vector (n-dimensional), $\underline{u}_m(k)$ is the model input vector (m-dimensional) and $\underline{u}_p(k)$ is the plant input vector (q-dimensional). The model matrices A_m and B_m

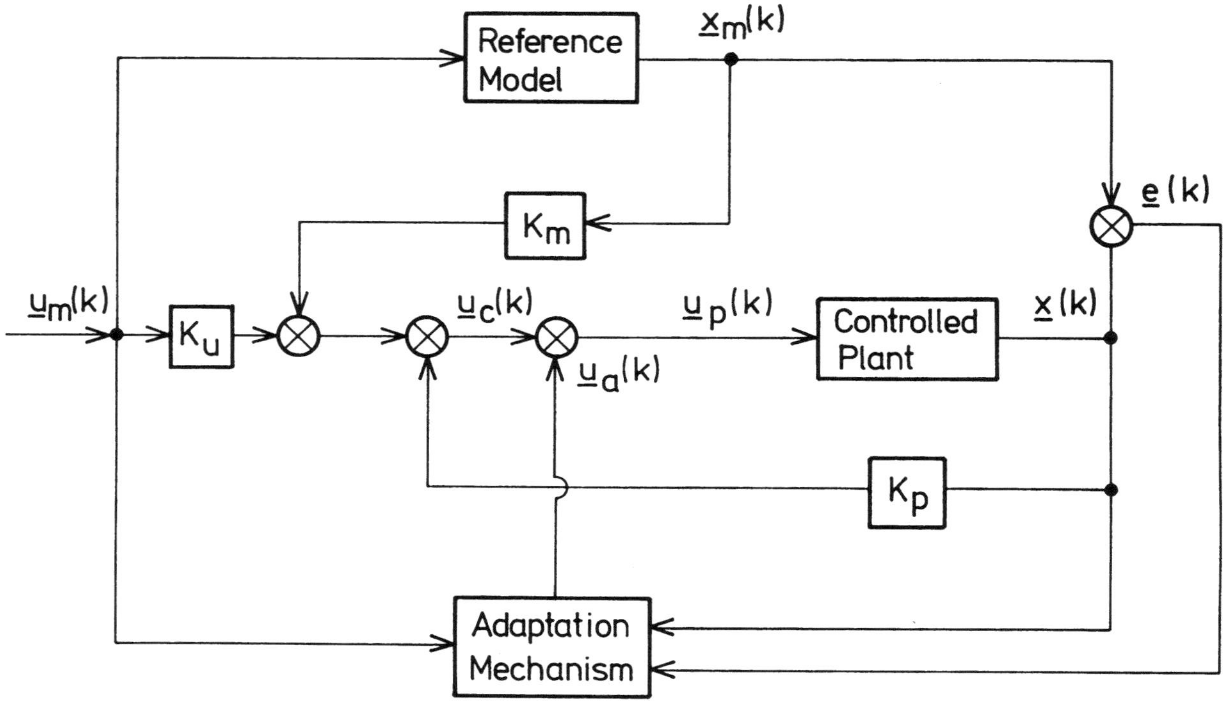

Fig. (7.7) Paraller AMFC system with signal-synthesis adaption

are constant with appropriate dimensions. K_u is the feedforward gain, K_p is the plant feedback gain and K_m is the model feedback gain. The objective is to determine the gain matrices K_p, K_m and K_u such that for null initial conditions perfect-model following occurs. In the literature, three basic approaches for perfect-model following have been considered:

(i) Algebraic conditions on the error equation *[24,25]*. This means that the generalized state error $\underline{e}(k) = \underline{x}_m(k) - \underline{x}(k)$ and its derivatives be null for any input $\underline{u}_m(k)$.

(ii) Algebraic conditions on the augmented system with $[\underline{x}_m^t(k), \underline{x}^t(k)]^t$, $\underline{u}_m(k)$, $\underline{e}(k)$ as its state, input and output, respectively. In effect, it is required to show *[26]* that the input $\underline{u}_m(k)$ has no influence on the output $\underline{e}(k)$.

(iii) Transfer matrix matching of both the model and the controlled plant *[27,28]*.

We consider only the first approach. From (6.100)-(6.102), the dynamics of the error $\underline{e}(k) = \underline{x}_m(k) - \underline{x}(k)$ can be expressed as:

$$\underline{e}(k+1) = (A_m - B_p K_m)\underline{e}(k) + [A_m - A_p + B_p(K_p - K_m)]\underline{x}(k) + [B_m - B_p K_u]\underline{u}_m(k) \qquad (7.103)$$

To have perfect-model following according to the first approach one must assure that for any $\underline{u}_m(k)$ and $\underline{e}(0) = \underline{0}$ we obtain $\underline{e}(k) = \underline{x}_m(k) - \underline{x}(k) = \underline{0}$ and $\Delta\underline{e}(k) = \underline{e}(k+1) - \underline{e}(k) = \underline{0}$. It can immediately be concluded that this can be achieved when

$$[A_m - A_p + B_p(K_p - K_m)]\underline{x}(k) + [B_m - B_p K_u]\underline{u}_m(k) = \underline{0} \qquad (7.104)$$

$$|\sigma[A_m - B_p K_m]| < 1 \qquad (7.105)$$

where $\sigma[.]$ is the set of eigenvalues. For (7.104) to hold for any $\underline{x}(k)$ and $\underline{u}_m(k)$, it is required that

$$A_m - A_p + B_p(K_p - K_m) = 0 \qquad (7.106)$$

$$B_p - B_p K_u = 0 \qquad (7.107)$$

The existence of solutions for K_m, K_p and K_u leading to verification of (7.106) and (7.107) is guaranteed if *[25]*:

$$\text{rank}[B_p] = \text{rank}[B_p \mid B_m] = \text{rank}[B_p \mid A_m - A_p] \qquad (7.108)$$

When $q < n$, a sufficient condition to satisfy (7.108) is given by *[24]*:

$$(I_n - B_p B_p^+)(A_m - A_p) = 0 \qquad (7.109a)$$

$$(I_n - B_p B_p^+) B_m = 0 \qquad (7.109b)$$

If (7.109) is satisfied, then the solutions of (7.106) and (7.107) are of the form:

$$K_m - K_p = B_p^+(A_m - A_p) \qquad (7.110a)$$

$$K_u = B_p^+ B_m \qquad (7.110b)$$

where B_p^+ is the Penrose pseudo-inverse of B_p *[29]*. The condition (7.109) was originally developed by Erzberger *[24]*. It is interesting to note that only K_u and the difference $(K_m - K_p)$ are required.

Despite the many attractive features of the linear model-following approach, it generally fails to overcome the problem of uncertainty in or variation of plant parameters *[6,30]*. This leads to the consideration of adaptive model-following control approach. It has the advantage of not requiring an explicit identification of the plant parameters while providing for an explicit form of the adaptation law. Hence the need for real-time numerical solutions is bypassed, and the design problem becomes that of asymptotically tending toward perfect-model following.

The starting point of adaptive model-following control (AMFC) systems is to generalize the plant control signal to the form:

$$\underline{u}_p(k) = -K_p(\underline{e},k)\underline{x}(k) + K_u(\underline{e},k)\underline{u}_m(k) + K_m\underline{x}_m(k) \qquad (7.111)$$

where

$$K_p(\underline{e},k) = K_p - \Delta K_p(\underline{e},k) \qquad (7.112a)$$

$$K_u(\underline{e},k) = K_u + \Delta K_u(\underline{e},k) \qquad (7.112b)$$

With this decomposition, one can write

$$\underline{u}_p(k) = \underline{u}_c(k) + \underline{u}_a(k) \qquad (7.113)$$

where

$$\underline{u}_c(k) = -K_p\underline{x}(k) + K_m\underline{x}_m(k) + K_u\underline{u}_m(k) \qquad (7.114a)$$

$$\underline{u}_a(k) = \Delta K_p(\underline{e},k)\underline{x}(k) + \Delta K_u(\underline{e},k)\underline{u}_m(k)$$

$$= \Delta K_p^s(k)\underline{x}(k) + \Delta K_u^s(k)\underline{u}_m(k) \qquad (7.114b)$$

The matrices K_p, K_m and K_u are constant and designed for some specific plant parameter values. One way is to compute K_p using linear optimal control and then use (7.110) to obtain the other matrices. The vector $\underline{u}_c(k)$ is a command signal and $\underline{u}_a(k)$ is an adaptation signal. The time-varying matrices $\Delta K_p(\underline{e},k)$ and $\Delta K_u(\underline{e},k)$ are to be designed in order to assure that $\underline{e}(k) \to \underline{0}$ as $k \to \infty$. To do this, we apply the procedure developed in the previous sections based on hyperstability theory. In deriving the equivalent feedback system, in the form of Fig. 7.6, it is found *[6,30,31]* that the standard form (7.76)-(7.79) cannot be easily obtained. One is thus led to modifying either the reference model *[6,30]* or the adaptation signal *[31]*. In the former case, the reference model is described by:

a priori

$$\underline{x}_m^a(k+1) = A_m\underline{x}_m(k) + B_m\underline{u}_m(k) \qquad (7.115a)$$

a posteriori

$$\begin{aligned}\underline{x}_m(k+1) &= A_m\underline{x}_m(k) + B_m\underline{u}_m(k)\\ &\quad - B_p[\Delta K_p(k+1) - \Delta K_p^s(k)]\underline{x}(k)\\ &\quad - B_p[\Delta K_u(k+1) - \Delta K_u^s(k)]\underline{u}_m(k) \qquad (7.115b)\end{aligned}$$

This model substitutes that of (7.100). The second and third terms in (7.115b) represent auxiliary transient signals which vanish when adaptation is achieved. The adaptive algorithm can be derived using parallel developments to that of parametric adaptation schemes. In Table 7.2 we provide a summary of the main steps.

To avoid the difficulties in selecting the matrix B_p, the adaptation signal $\underline{u}_a(k)$ can be modified to:

$$\underline{u}_a(k) = \Delta K_p(k+1)\underline{x}(k) + \Delta K_u(k+1)\underline{u}_m(k) \qquad (7.116)$$

and the reference model retains the form (7.100). When using (7.116), one replaces $\underline{\rho}^a(k+1)$, the *a priori* value at the (k+1)th instant, by $\underline{\rho}(k)$ and use similar results to that of part C of Section 7.3.2. The reader is referred to *[31]* for further details.

Next, we present some illustrative examples.

TABLE 7.2 Adaptive model-following control system

Component	Description
Reference Model	$\underline{x}_m^a(k+1) = A_m\underline{x}_m(k) + B_m\underline{u}_m(k)$ $\underline{x}_m(k+1) = \underline{x}_m^a(k+1) - B_p[\Delta K_p(k+1) - \Delta K_p^s(k)]\underline{x}(k) - B_p[\Delta K_u(k+1) - \Delta K_u^s(k)]\underline{u}_m(k)$
Controlled Plant	$\underline{x}(k+1) = A_p\underline{x}(k) + B_p\underline{u}_c(k) + B_p\underline{u}_a(k)$
Generalized State Error	$\underline{e}^a(k) = \underline{x}_m^a(k) - \underline{x}(k)$ $\underline{e}(k) = \underline{x}_m(k) - \underline{x}(k)$
Command Signal	$\underline{u}_c(k) = -K_p\underline{x}(k) + K_m\underline{x}_m(k) + K_u\underline{u}_m(k)$
Adaptation Signal	$\underline{u}_a(k) = \Delta K_p^s(k)\underline{x}(k) + \Delta K_u^s(k)\underline{u}_m(k)$
Adaptation Algorithm	(1) $\underline{\rho}^a(k+1) = \mathbf{E}\,\underline{e}^a(k+1)$ (2) $\underline{\rho}(k+1) = [I_n + E\{(M+\tilde{M})\underline{x}^t(k)P\underline{x}(k) + (N+\tilde{N})\underline{u}_m^t(k)R\underline{u}_m(k)\}]^{-1}\underline{\rho}^a(k+1)$ (3) $\Delta K_p(k+1) = \Delta K_p^s(k+1) + \Delta K_p^p(k+1)$ (4) $\Delta K_p^s(k+1) = \Delta K_p^s(k) + M\,\underline{\rho}(k+1)[P\underline{x}(k)]^t$ (5) $\Delta K_p^p(k+1) = \tilde{M}\,\underline{\rho}(k+1)[P\underline{x}(k)]^t \;;\; M,N > 0 \;;\; \tilde{M} \geq 0$

TABLE 7.2 (contd.)

Component	Description
Adaptation Algorithm (cont.)	(6) $\Delta K_u(k+1) = \Delta K_u^s(k+1) + \Delta K_u^p(k+1)$ (7) $\Delta K_u^s(k+1) = \Delta K_u^s(k) + N\,\underline{\rho}(k+1)[R\underline{u}_m(k)]^t$ (8) $\Delta K_u^p(k+1) = \tilde{N}\,\underline{\rho}(k+1)[R\underline{u}_m(k)]^t$
Perfect Model-Following Conditions	$(I_n-B_pB_p^+)(A_m-A_p) = 0$ $(I_n-B_pB_p^+)B_m = 0$
Stability Conditions	$E = P$ $(A_m-B_pK_m)^tP(A_m-B_pK_m) - P = -Q \;;\quad Q > 0$

7.3.3 *EXAMPLES*

Three examples will now be solved to illustrate design methods. The first two examples deal with Lyapunov design and the third example is concerned with MRAC systems.

Example 1

A first-order plant is described by

$$x(k+1) = a\,x(k) + b\,u(k)$$

The reference model is taken to be

$$x_m(k+1) = .5x_m(k) + u(k)$$

Application of (7.28) gives the error dynamics as

$$e(k+1) = .5e(k) + (.5-a)x(k) + (1-b)u(k)$$

The adaptation algorithm (7.30), using (7.37) and (7.41), can be expressed as

$$a(k+1) = a(k) + 2[e(k)-.5e(k-1)]x(k-1)/[x^2(k-1)+u^2(k-1)]$$

$$b(k+1) = b(k) + 2[e(k)-.5e(k-1)]u(k-1)/[x^2(k-1)+u^2(k-1)]$$

where we used the numerical values $P = 2$, $\delta = 1$ and an input $u(k)$ of the form of a sampled rectangular pulse of unit amplitude and a fundamental frequency $\omega_f = \pi$ rad/sec.

Computer simulations of the error and the unknown parameters are shown in Fig. 7.8 and Fig. 7.9. It is clear from the graphs that $a(k)$ and $b(k)$ converge to steady-state values of .50001 and .99995, respectively. We have examined the effect of the weighting terms δ and P. The result was that the smaller the product $(P.\delta)$ the slower the algorithm in approaching the steady-state values.

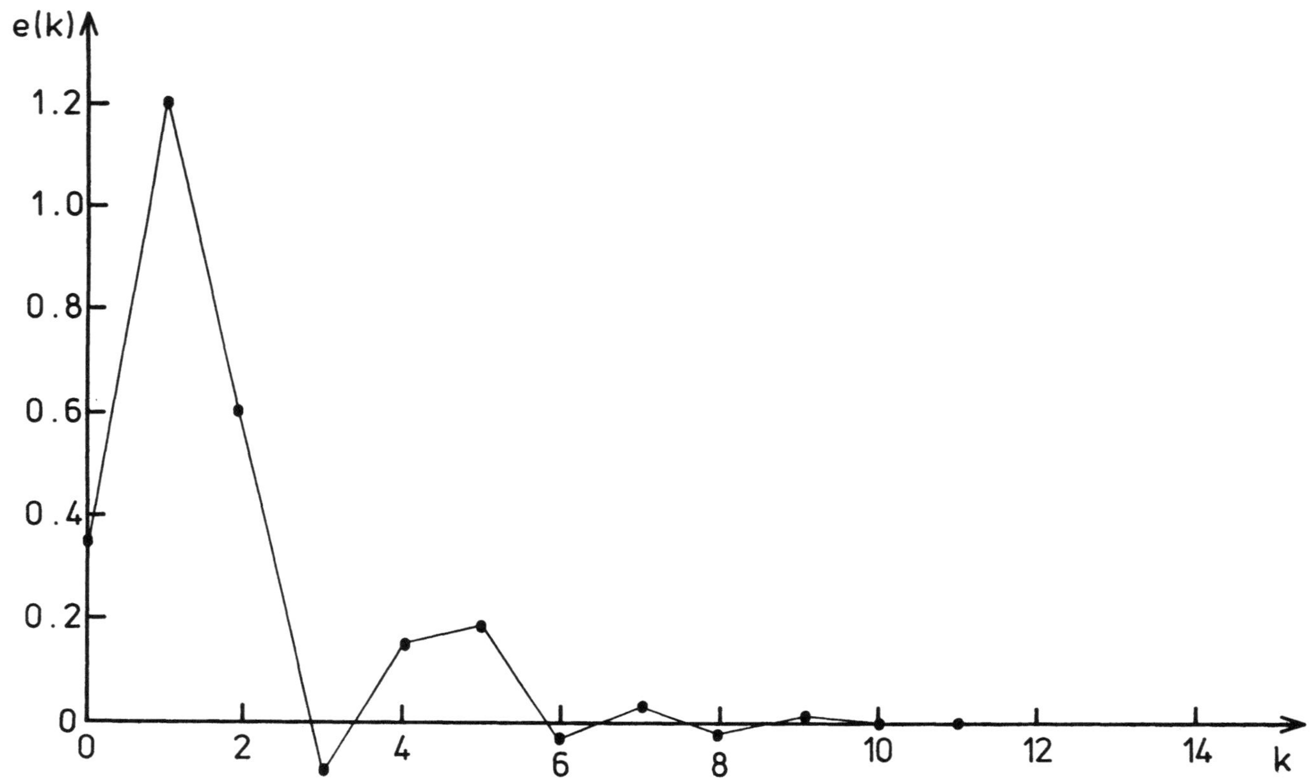

Fig. (7.8) Variation of error term

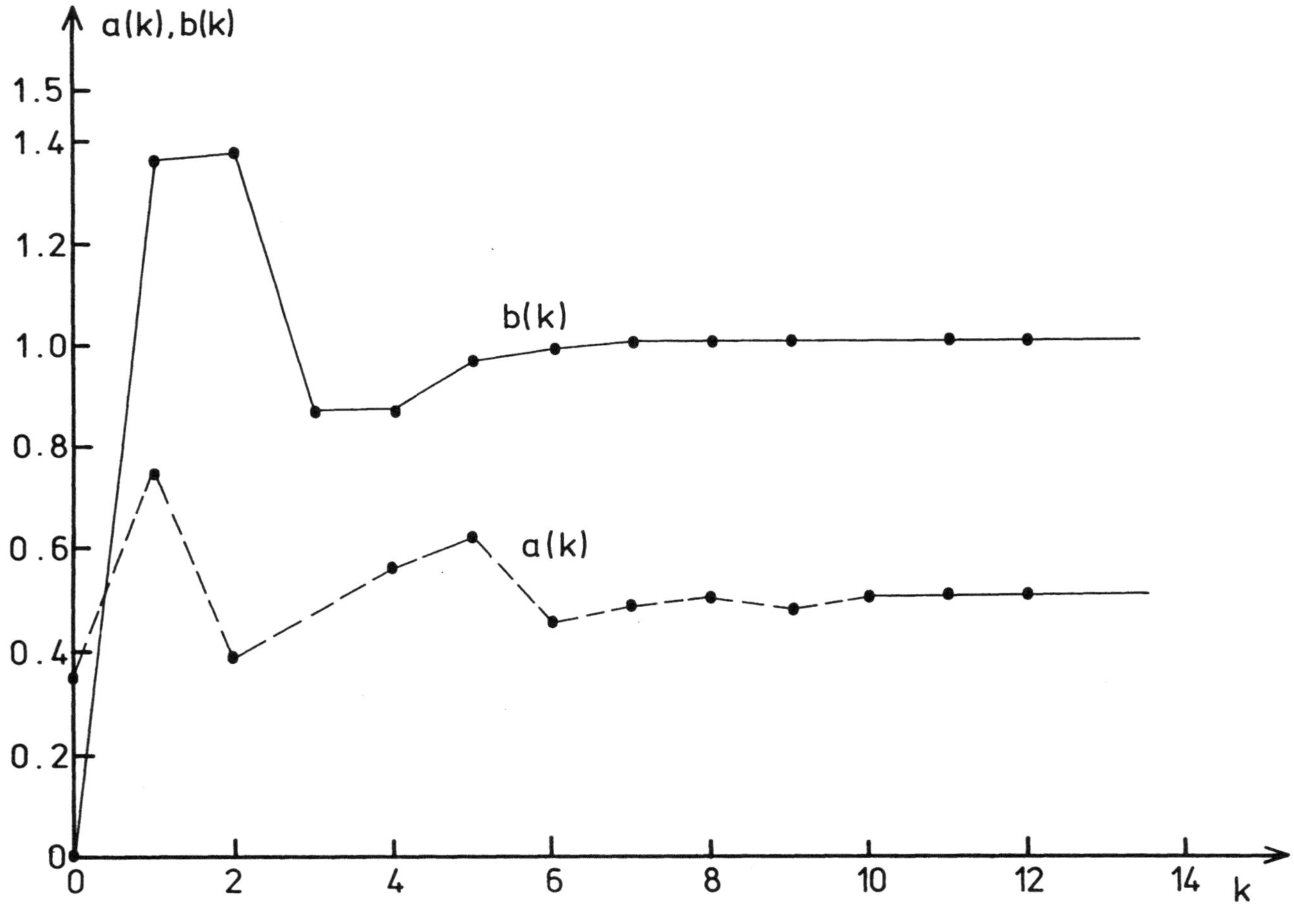

Fig. (7.9) Variation of unknown parameters

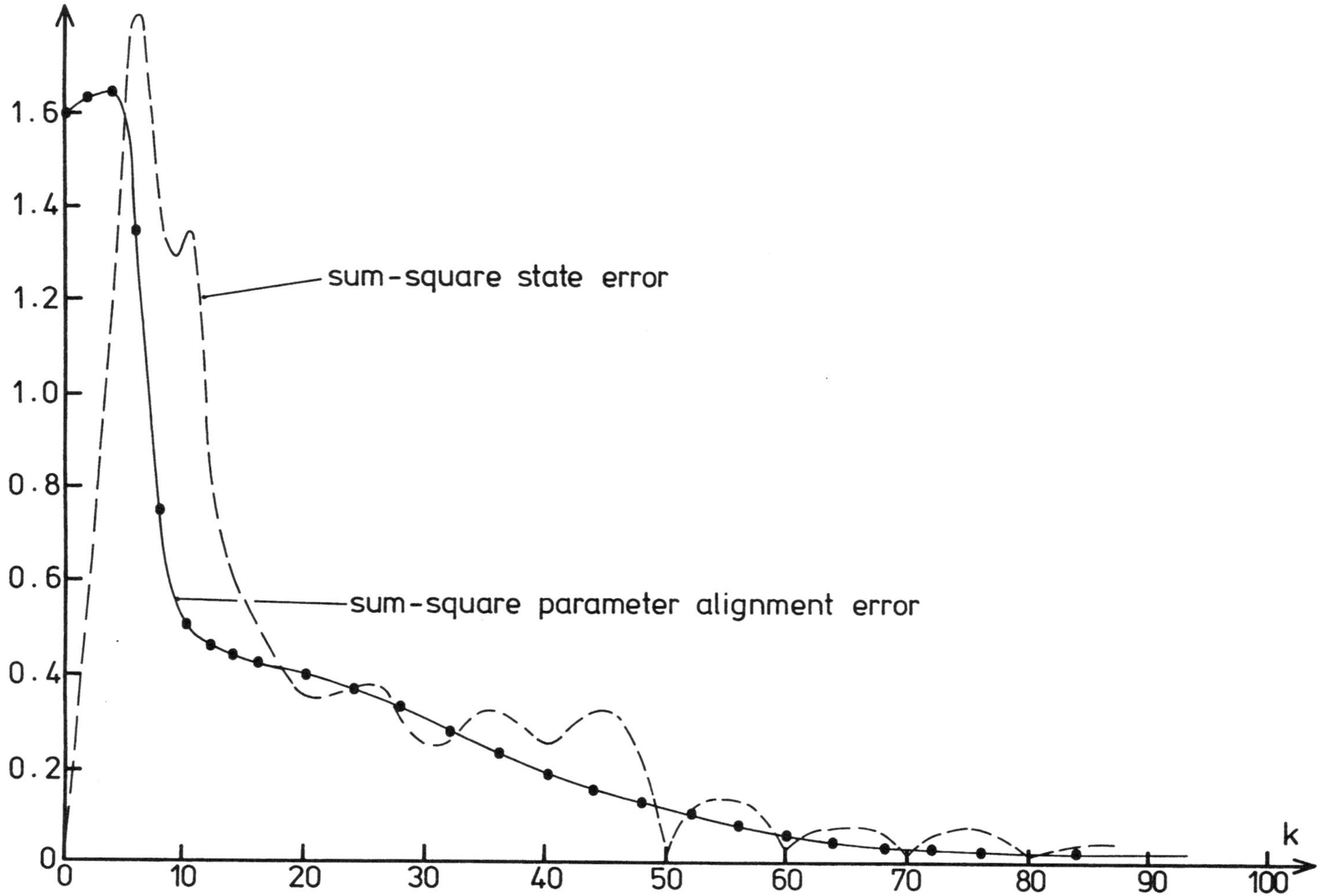

Fig. (7.10) Response of sum-square errors

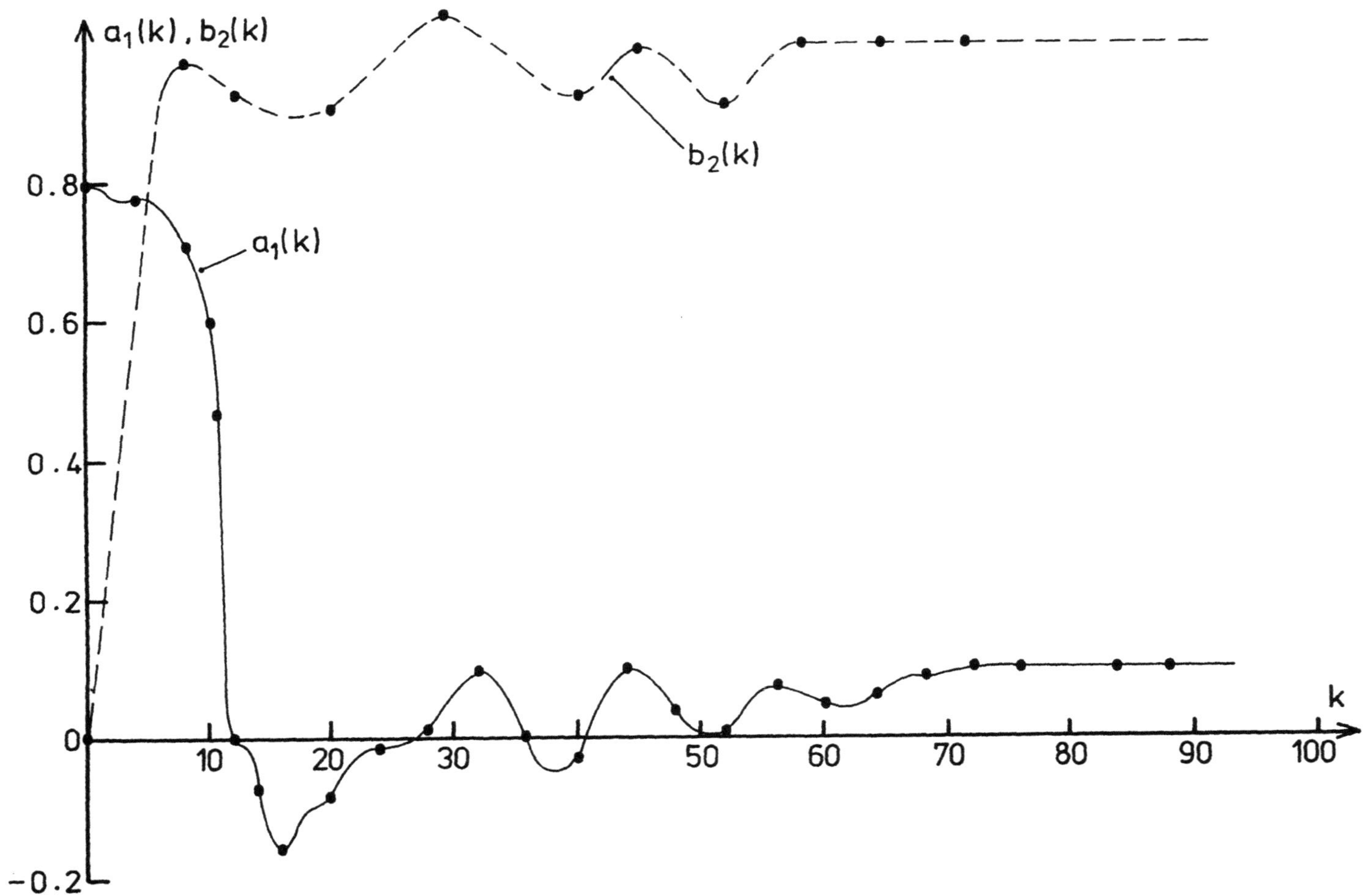

Fig. (7.11) Adaptive response of parameters $a_1(k)$ and $b_2(k)$

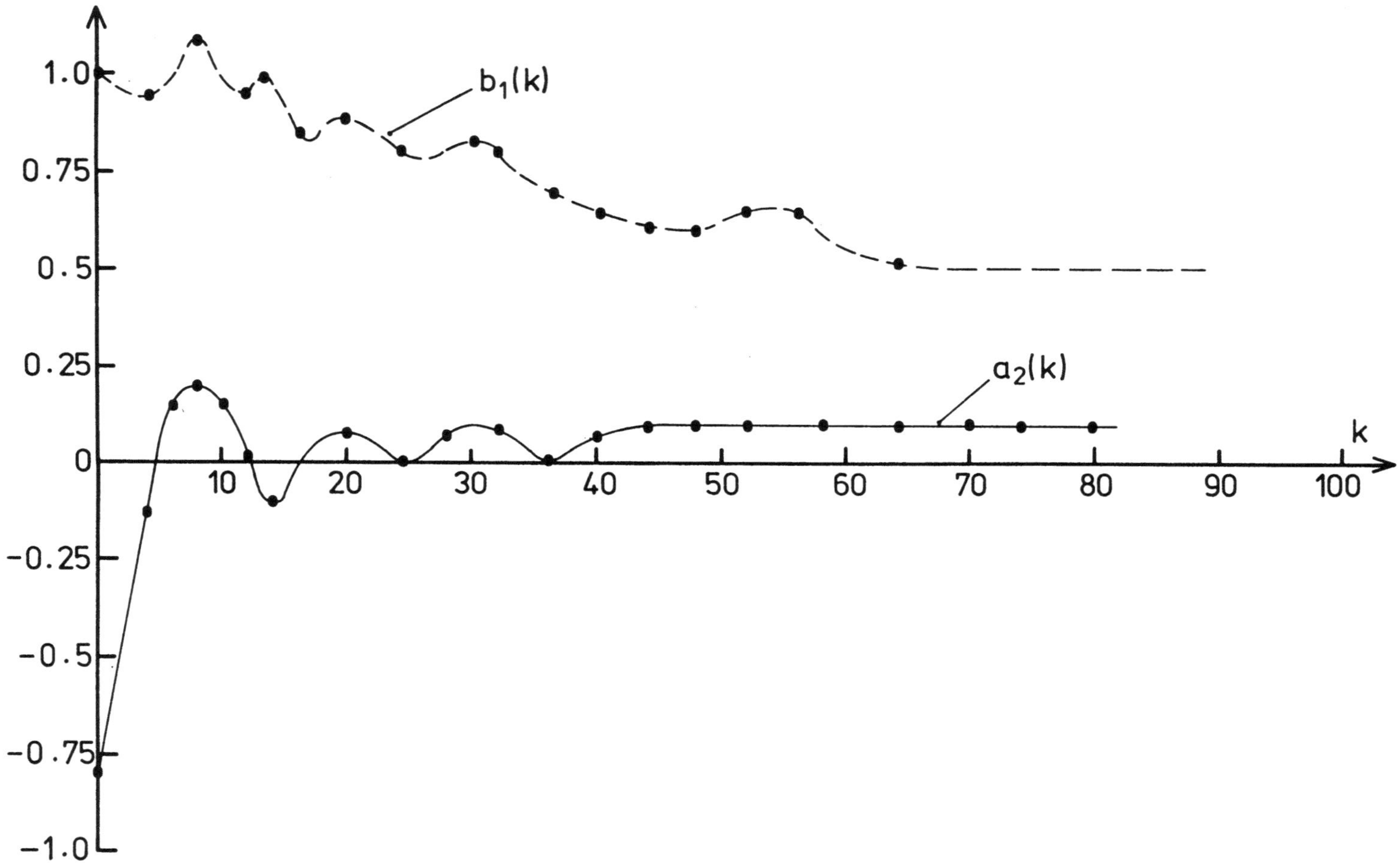

Fig. (7.12) Adaptive response of parameters $b_1(k)$ and $a_2(k)$

Example 2

A second-order MRAS of the form (7.26) and (7.27) has the following matrices:

$$A_m = \begin{pmatrix} .5 & .1 \\ .1 & .5 \end{pmatrix} \quad ; \quad \underline{b}_m = \begin{pmatrix} .5 \\ .1 \end{pmatrix}$$

$$A(\underline{e},k) = \begin{pmatrix} .5 & a_1(k) \\ a_2(k) & .5 \end{pmatrix} \quad ; \quad \underline{b}(\underline{e},k) = \begin{pmatrix} b_1(k) \\ b_2(k) \end{pmatrix}$$

For the purpose of simulation, the input $u(k)$ is taken to be a sine function of unit peak value and of frequency .3 r/sec with a phase shift of .3 rad. The weighting factor δ is set to 1 and the gain matrix P is chosen to be

$$P = \begin{pmatrix} 10 & 1 \\ 1 & 10 \end{pmatrix}$$

The response of sum-square parameter alignment and state errors are displayed in Fig. 7.10 . The behaviour of the unknown parameters $a_1(k)$, $a_2(k)$, $b_1(k)$ and $b_2(k)$ is shown in Figs. 7.11 and 7.12 . We note that the algorithm converged to the steady-state values of .10003, .09997, .5 and 1.0001, respectively, after 80 discrete steps. In this example, all attempts to change the values of δ or P result in slow convergence.

Both examples 1 and 2 show that Lyapunov adaptive design is simple and yields tractable results.

Example 3

Consider the second-order plant:

$$\underline{x}(k+1) = \begin{pmatrix} -a_1 & 1 \\ -a_2 & 0 \end{pmatrix} \underline{x}(k) + \begin{pmatrix} b_1 \\ b_2 \end{pmatrix} u(k)$$

where a_1, a_2, b_1 and b_2 are unknown parameters and for which the reference model is expressed by:

$$\underline{x}_m(k+1) = \begin{bmatrix} -.7 & 1 \\ -.12 & 0 \end{bmatrix} \underline{x}_m(k) + \begin{bmatrix} 1 \\ 1 \end{bmatrix} u(k)$$

The reference model is stable with eigenvalues -.3 and -.4 . A parametric adaptation scheme is implemented with $M = I_n$, $P = \tilde{P} = \mu_p I_n$, $R = \tilde{R} = \mu_r I_n$, where $0 < \mu_p, \mu_r < 10$. The algorithm is initialized with $a_1(0) = 0$, $a_2(0) = .2$, $b_1(0) = .5$, $b_2(0) = -.3$. From (7.99b), the compensator E has the form:

$$E = \begin{bmatrix} 1.69 & -1.056 \\ -1.056 & 2.69 \end{bmatrix}$$

Convergence behaviour of the unknown parameters is shown in Figs. 7.13 and 7.14 . It is seen from the graphs that the parameters experience some irregular changes, then quickly converge to the steady state values.

7.4 Self-Tuning Regulators

This section is devoted to the study of the second class of adaptive systems, that is, *Self-tuning regulators (STR's)*. We first provide a brief introduction to this type of system, and then move on to the analytical development.

7.4.1 *INTRODUCTION*

The purpose of self-tuning regulators is to control systems with unknown but constant (and sometimes slowly varying) parameters *[5]*. A block diagram representation of STR's is shown in Fig. 7.15. The regulator can be thought of as being composed of three parts: a parameter estimator, a controller and an interfacing part, which relates the controller parameters to the estimated parameters. We note that the parameter estimator acts on the system input-output records and produces estimates of certain process parameters. The controller is some sort of a digital filter whose coefficients are in general a nonlinear function of the estimated parameters. The way of

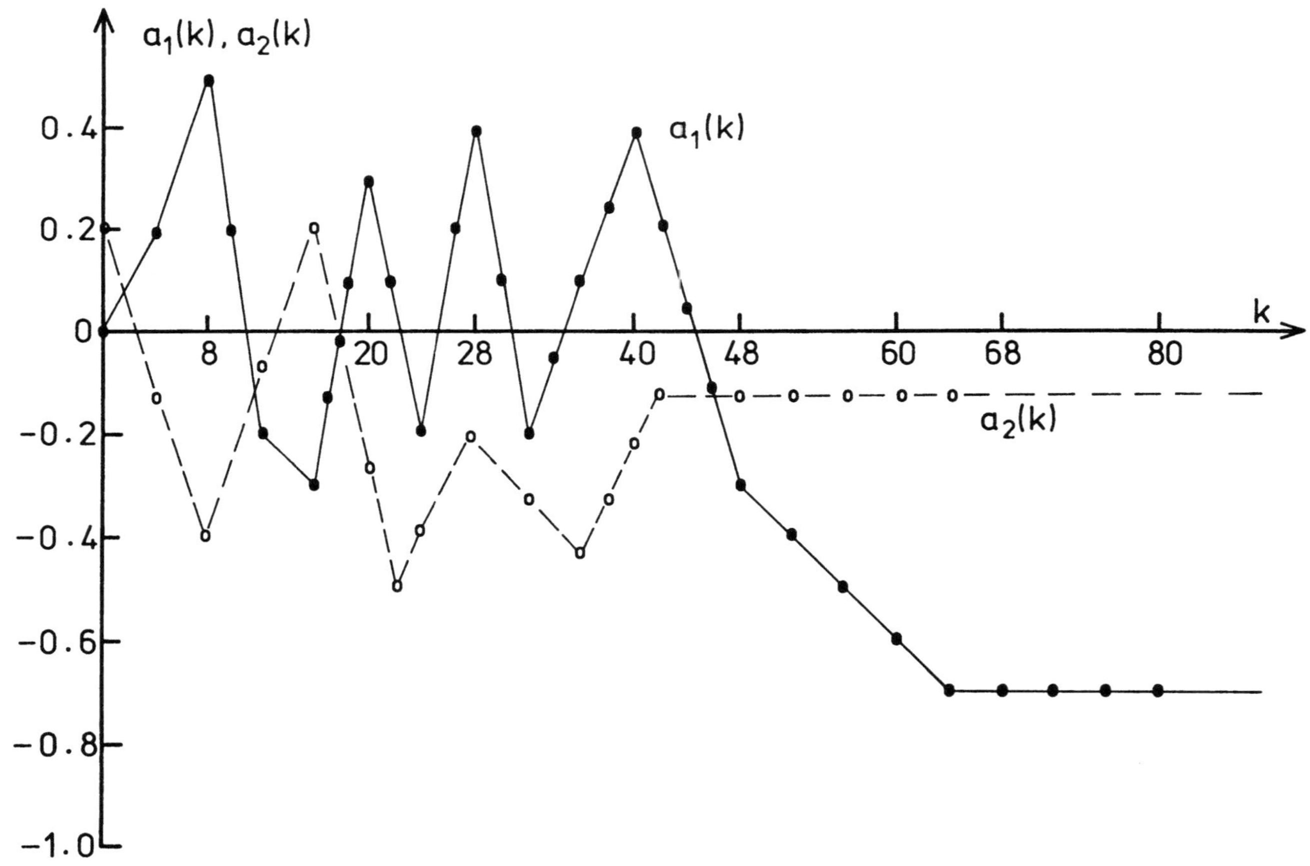

Fig. (7.13) Convergence of parameters $a_1(k)$ and $a_2(k)$

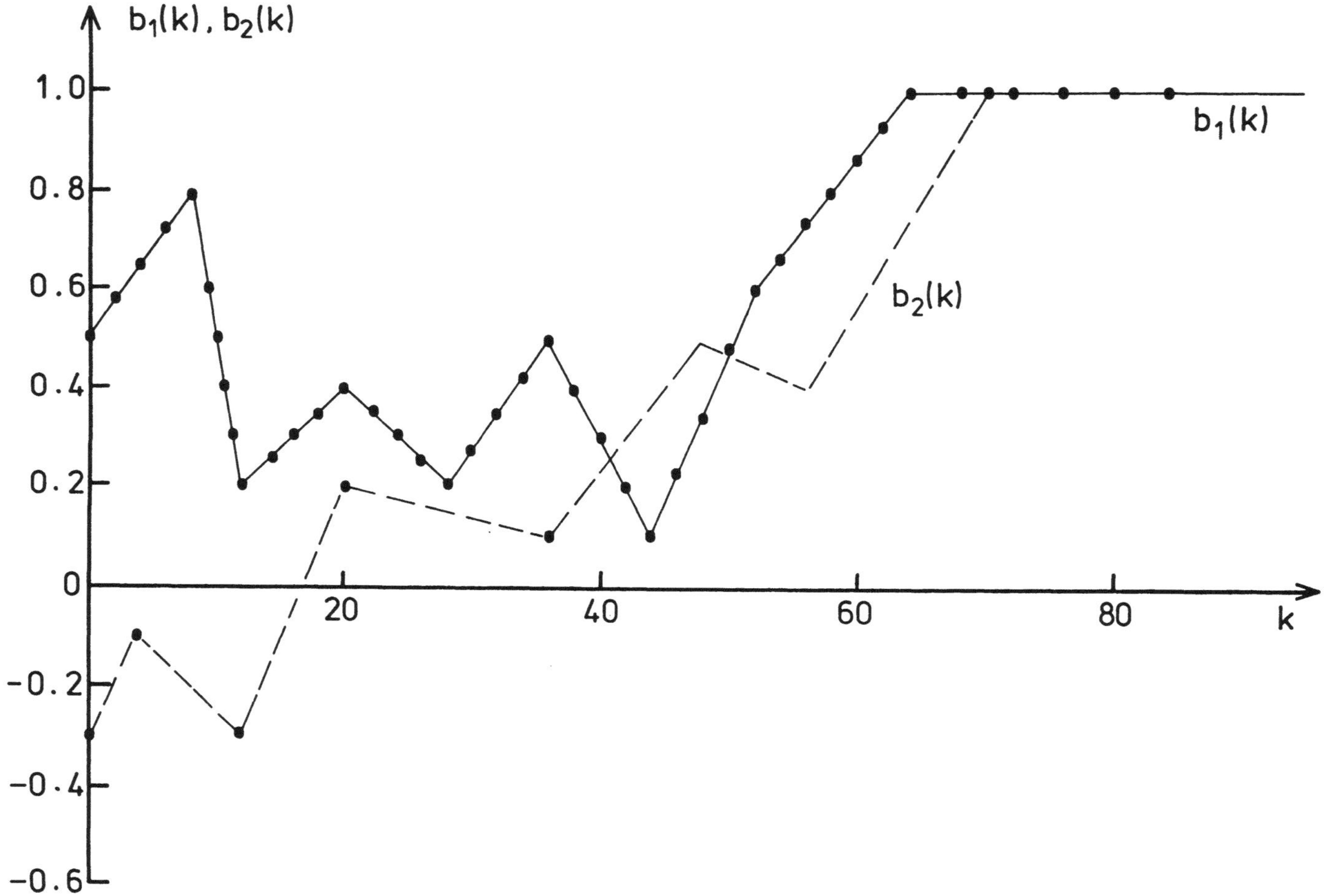

Fig. (7.14) Convergence of parameters $b_1(k)$ and $b_2(k)$

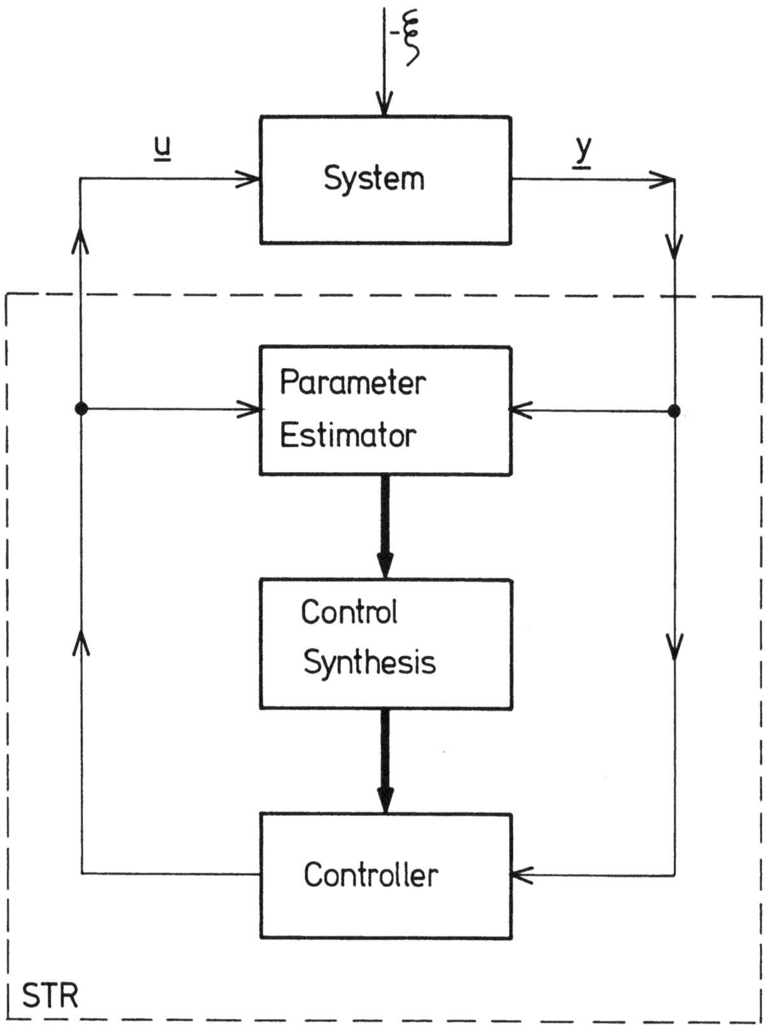

Fig. (7.15) Block-diagram of the self-tuning regulator

subdividing the self-tuning regulator is quite arbitrary; however, the above description is convenient for design and implementation purposes. In view of the fact that there are many different ways for parameter estimation and controller design, different types of self-tuning regulators can be obtained. Our purpose hereafter is to emphasize the main results. To this end, we start by describing the self-tuning problem of discrete systems.

7.4.2 *DESCRIPTION OF THE SYSTEM*

In terms of q^{-1}, the one-step backward shift operator, we consider systems described by the auto-regressive moving average (ARMA) model

$$A(q^{-1})\underline{y}(k) = B(q^{-1})\underline{u}(k-d-1) + C(q^{-1})\underline{\xi}(k) \qquad (7.117)$$

where $(d+1)$ is a time delay, $\underline{y}$ is the output vector, $\underline{u}$ is the input vector and $\{\underline{\xi}(k)\}$ is a sequence of independent, equally distributed random vectors with zero mean and covariance $E\{\underline{\xi}(k)\underline{\xi}^t(k)\} = R$. The vectors $\underline{y}$, $\underline{u}$ and $\underline{\xi}$ are all of dimension p. The polynomial matrices A, B and C are all of dimension $(p\mathrm{x}p)$ with unknown coefficients and they are given by:

$$\left.\begin{aligned} A(z) &= I_p+A_1 z +\ldots+ A_{n_a} z^{n_a} \\ B(z) &= B_0+B_1 z +\ldots+ B_{n_b-1} z^{n_b-1} \\ C(z) &= I_p+C_1 z +\ldots+ C_{n_c} z^{n_c} \end{aligned}\right\} \qquad (7.118)$$

where B_0 is nonsingular. This means that a transformation of the type $\underline{u}_t(k) = B_0\underline{u}(k)$ exists and consequently implies that we have one control variable for each loop which influences the corresponding output before other control variables.

In the sequel we consider that the output records $\{\underline{y}(k), \underline{y}(k-1),\ldots\}$ and the past control actions $\{\underline{u}(k-1,\underline{u}(k-2),\ldots\}$ are known at time k. Based on this information, we define an *admissible control strategy* $\underline{u}(k)$ to be a function of $\{\underline{y}(k), \underline{y}(k-1),\dots\}$ and $\{\underline{u}(k-1),\underline{u}(k-2),\dots\}$

In (7.118) the integers n_a, n_b and n_c are arbitrary; however, normally $n_a = n_b = n$. An equivalent representation of (7.117) is:

$$A(z^{-1})Y(z) = z^{-(d+1)}B(z^{-1})U(z) + C(z^{-1})E(z) \quad (7.119)$$

where $Y(z)$, $U(z)$, and $E(z)$ are the Z-transform output, input and disturbance, respectively. For single-input, single-output systems, the system model takes the form:

$$A(q^{-1})y(k) = q^{-(d+1)}B(q^{-1})u(k) + C(q^{-1})\xi(k) \quad (7.120)$$

where

$$\left.\begin{aligned} A(q^{-1}) &= 1+a_1q^{-1} + \dots + a_{n_a}q^{-n_a} \\ B(q^{-1}) &= b_0+b_1q^{-1} + \dots + b_{n_b-1}q^{-(n_b-1)} \\ C(q^{-1}) &= 1+c_1q^{-1} + \dots + c_{n_c}q^{-n_c} \end{aligned}\right\} \quad (7.121)$$

We now move on to discuss the other blocks in Fig. 7.15 .

7.4.3 *PARAMETER ESTIMATORS*

It has been shown *[5,33,35,39]* that self-tuning regulators should incorporate some sort of a recursive scheme for estimating the parameters of a 'prediction' model. This model represents the unknown system (7.117), or the corresponding SISO system (7.120), during on-line processing of the self-tuning regulator. Some recursive estimation schemes, frequently used in (STR's), are briefly described below, for SISO systems. Further details are found in Chapter 6.

A. The Least Squares Method

In this method, the prediction model is represented by the

difference equation:

$$\hat{y}(k) = -A(q^{-1})y(k-1) + B(q^{-1})u(k-1) \tag{7.122}$$

where $\hat{y}$ is the predicted output, and

$$\left.\begin{aligned} A(q^{-1}) &= \alpha_1 + \alpha_2 q^{-1} + \ldots + \alpha_s q^{-(s-1)} \\ B(q^{-1}) &= \beta_1 + \beta_2 q^{-1} + \ldots + \beta_r q^{-(r-1)} \end{aligned}\right\} \tag{7.123}$$

are polynomials in the backward shift operator q^{-1}. For convenience, we define

$$\underline{\Theta} = \begin{bmatrix} \alpha_1 \\ \vdots \\ \alpha_s \\ \beta_1 \\ \vdots \\ \beta_r \end{bmatrix} \quad ; \quad \underline{\Phi}(k) = \begin{bmatrix} -y(k-1) \\ \vdots \\ -y(k-s) \\ -u(k-1) \\ \vdots \\ -u(k-r) \end{bmatrix} \tag{7.124}$$

then from (7.122)-(7.124), we have

$$\hat{y}(k) = \underline{\Theta}^t \underline{\Phi}(k) \tag{7.125}$$

In the least squares method *[40]*, the criterion

$$\begin{aligned} J(k) &= \sum_{j=1}^{k} ||y(j)-\hat{y}(j)||^2 \\ &= \sum_{j=1}^{k} ||y(j)-\underline{\Theta}^t\underline{\Phi}(k)||^2 \end{aligned} \tag{7.126}$$

is minimized with respect to $\underline{\Theta}$ to yield the estimates $\hat{\underline{\Theta}}(k)$. The recurrence relations to produce the estimate are *[40]*:

$$\hat{\underline{\Theta}}(k) = \hat{\underline{\Theta}}(k-1)+(\tfrac{1}{k})\underline{s}(k)\Delta[k,\hat{\underline{\Theta}}(k-1)] \tag{7.127a}$$

$$\underline{s}(k) = R^{-1}(k)\underline{\Phi}(k)/1+(\tfrac{1}{k})[\underline{\Phi}^t(k)R^{-1}(k)\underline{\Phi}(k)-1] \tag{7.127b}$$

$$R(k) = R(k-1) + \left(\frac{1}{k}\right)[\underline{\Phi}(k)\underline{\Phi}^t(k) - R(k-1)] \qquad (7.127c)$$

$$\Delta[k,\hat{\underline{\theta}}(k-1)] = y(k) - \hat{\underline{\theta}}^t(k-1)\underline{\Phi}(k) \qquad (7.127d)$$

In practice, the factor $(1/k)$ is replaced by a sequence of positive scalars $\rho(k)$. A sequence $\rho(.)$ that decreases more slowly than $(1/k)$ corresponds to discounting old output records in the criterion (7.126). The case $\rho(.) = \rho_0$ corresponds to exponential forgetting of past records, with the base $1-\rho_0$. Sometimes we use $Q(k) = R^{-1}(k)$ so that (7.127c) can be written as:

$$Q(k) = \frac{[1/1-\rho(k)]\{Q(k-1)-Q(k-1)\underline{\Phi}(k)\underline{\Phi}^t(k)Q(k-1)\}}{\{[1-\rho(k)/\rho(k)] + \underline{\Phi}^t(k)Q(k-1)\underline{\Phi}(k)\}} \qquad (7.127e)$$

In applications the least squares method is known *[40]* to yield biased estimates.

B. The Extended Least Squares Method

To overcome the biasing problem, we consider an extended prediction model:

$$\hat{y}(k) = -A(q^{-1})y(k-1) + B(q^{-1})u(k-1) + C(q^{-1})\Delta(k-1) \qquad (7.128)$$

where
$$\Delta(k) = y(k) - \hat{y}(k) \qquad (7.129a)$$

$$C(q^{-1}) = \gamma_1 + \gamma_2 q^{-1} + \ldots + \gamma_t q^{-(t-1)} \qquad (7.129b)$$

are the prediction error and the associated distribution matrix. Similarly to (7.124), we define here

$$\underline{\theta} = \begin{bmatrix} \alpha_1 \\ \vdots \\ \alpha_s \\ \beta_1 \\ \vdots \\ \beta_r \\ \gamma_1 \\ \vdots \\ \gamma_t \end{bmatrix} ; \quad \underline{\Phi}(k) = \begin{bmatrix} -y(k-1) \\ \vdots \\ -y(k-s) \\ -u(k-1) \\ \vdots \\ -u(k-r) \\ -\Delta(k-1) \\ \vdots \\ -\Delta(k-t) \end{bmatrix} \qquad (7.130)$$

so that

$$\hat{y}(k) = \underline{\theta}^t(k-1)\underline{\Phi}(k) \tag{7.131a}$$

where

$$\Delta(k,\underline{\theta}) = y(k) - \underline{\theta}^t\underline{\Phi}(k) \tag{7.131b}$$

In (7.131a), the prediction is computed using the current estimate $\underline{\theta}(k-1)$. The recurrence relations for the extended least squares method are similar to those of the least squares with $\underline{\Phi}(k)$ now being defined by (7.130). Indeed, when t = 0, both methods become identical.

Other recursive estimation schemes can also be used; for example, stochastic approximations, maximum likelihood, generalized least squares, instrumental variables, etc., *[40,41]*. We now direct our attention to the derivation of control signals.

7.4.4 *CONTROL STRATEGIES*

It has been mentioned earlier that any admissible control strategy $\underline{u}(k)$ should be a function of all the previous input and output records. A suitable feedback control strategy would be

$$u(k) = [G(q^{-1})/F(q^{-1})]y(k) \tag{7.132}$$

for SISO systems, or

$$\underline{u}(k) = G(q^{-1})[F(q^{-1})]^{-1}\underline{y}(k) \tag{7.133}$$

for MIMO systems, where

$$G(q^{-1}) = g_0+g_1q^{-1}+\ldots+g_{n_g}q^{-n_g} \tag{7.134a}$$

$$F(q^{-1}) = I+f_1q^{-1}+\ldots+f_{n_f}q^{-n_f} \tag{7.134b}$$

In (7.134), the coefficients $f_1,\ldots,f_{n_f}, g_0,\ldots,g_{n_g}$ are scalar parameters for systems of the type (7.132) or matrices with appropriate dimensions for the multivariable systems

(7.133). Consider the SISO case. The controller (7.132) can be characterised by the vector of parameters:

$$\underline{\pi} = \begin{bmatrix} g_0 \\ \vdots \\ g_{n_g} \\ f_1 \\ \vdots \\ f_{n_f} \end{bmatrix} \tag{7.135}$$

The problem now is to relate the controller parameters $\underline{\pi}$ to the estimator parameters $\underline{\theta}$ and the matrix $R(k)$, which is related to the uncertainties of the parameter estimates. Many techniques have been suggested to solve this problem. We shall examine some of these, where $\underline{\pi}(k)$ depends on $\underline{\theta}(k)$ only.

A. Controllers Based on Linear Quadratic Theory

The linear-quadratic control theory (see *[41]* and Chapter 8) can be used to produce linear control laws in the following way. Rewriting the prediction model (7.128) in the form

$$[1+q^{-1}\,\mathcal{A}(q^{-1})]y(k) = \mathcal{B}(q^{-1})u(k) + [1+q^{-1}\,\mathcal{C}(q^{-1})]\xi(k) \tag{7.136}$$

where $\{\xi(k)\}$ is now treated as white noise; we can use it as a system model to replace (7.117). Note that (7.117) and (7.136) have the same structure. A measure of performance is taken to be *[5,41]*,

$$J = \lim_{N\to\infty} E\{(\frac{1}{N}) \sum_{j=k}^{N} [y^2(j)+\rho u^2(j)]\} \tag{7.137}$$

which is to be minimized subject to the model constraint (7.136). The optimization problem (7.136) and (7.137) is a standard one in control theory for which spectral factorization methods are directly applicable *[41]*. By converting

(7.136) into a difference equation of order s or into a state-space representation, one can apply steady state solutions of Riccati equations (see Chapter 8, or [41]). The end result is again a controller which is linearly related to the system output. If the relationship is in the frequency domain, the parameters of the associated transfer function depend on the model parameters (7.136). This kind of dependence is not simple and requires excessive computations.

B. Controllers Based on the Minimum Variance Criterion

In (7.137), the case of $\rho = 0$ means that there is no penalty on the control and which then reduces to the variance of the output. The corresponding controllers are called *minimum variance controllers (MV)*. It is readily evident that MV controllers are special cases of linear quadratic controllers. The advantage of these controllers is that their parameters can be determined rather easily from the model parameters. For a system governed by (7.136), the minimum variance controller has the form (7.132) where the polynomials $F(q^{-1})$ and $G(q^{-1})$ are obtained from the identity, see [5,32,42]:

$$B^{+}(q^{-1})\, B^{-*}(q^{-1})\,[1+q^{-1}\, C(q^{-1})]$$

$$= [1+q^{-1}\, A(q^{-1})]F(q^{-1})-q^{-1}\, B(q^{-1})G(q^{-1}) \quad (7.138)$$

where the polynomials $B^{+}(q^{-1})$ and $B^{-}(q^{-1})$ are factors of the polynomial $B(q^{-1})$ such that

(i) roots of $B^{+}(q^{-1})$ lie outside the unit disc or on the unit circle,

(ii) roots of $B^{-}(q^{-1})$ lie inside the unit disc.

This implies that

$$B(q^{-1}) = B^{+}(q^{-1})\, B^{-}(q^{-1}) \quad (7.139)$$

The polynomial $B^{-*}(q^{-1})$ is the reciprocal of $B^{-}(q^{-1})$. All polynomials are normalized in such a way that $B^{-}(0) = 1$.

Suppose that $B(q^{-1})$ can be expressed as a delayed stable polynomial, that is

$$B(q^{-1}) = q^{-d} B^{s}(q^{-1})$$

where $B^{s}(q^{-1})$ is stable. In this case the identity (7.138) reduces to

$$B^{s}(q^{-1})[1+q^{-1} C(q^{-1})]$$
$$= [1+q^{-1} A(q^{-1})]F(q^{-1}) - q^{-(d+1)} B^{s}(q^{-1})G(q^{-1}) \tag{7.140}$$

Using forms (7.123), (7.129b) and (7.134), it can be easily shown *[41]* that (7.140) can be converted into a set of simultaneous or recurrence relations. For these relations to have a solution, it is required that

$$\left.\begin{aligned} n_g &= s - 1 \\ n_f &= n_b + d-1 \end{aligned}\right\} \tag{7.141}$$

where n_g, n_f and n_b are the orders of $G(q^{-1})$, $F(q^{-1})$ and $B^{s}(q^{-1})$, respectively. The MV controller in this case can be interpreted as choosing the control signal such that the predicted value (d+1) steps ahead will be equal to zero.

A performance measure, which is more general than (7.137), takes the form *[33,34]*:

$$J = E\{[P_1 y(k+d) - P_2 r(k)]^2 + [P_3 u(k)]^2\} \tag{7.142}$$

where P_1, P_2 and P_3 are appropriate polynomials, and $r(k)$ is a desired sequence. In this case, a control law (controller) minimizing (7.142) subject to the model (7.117) in the steady state could be expressed in the form:

$$F^{*}y(k) + G^{*}u(k) + H^{*}r(k) = 0 \tag{7.143}$$

where the polynomials F^{*}, G^{*} and H^{*} are derived from the polynomials in (7.117) and (7.142).

To this end, it should be emphasized that the class of self-tuning regulators considered thus far are all characterised by the block diagram of Fig. 7.15 . The parameter estimates are frequently based on recursive estimation schemes. The controllers are linear systems whose parameters are generally obtained from the estimated parameters.

Next we discuss other approaches to self-tuning control of discrete systems with unknown parameters.

7.4.5 *OTHER APPROACHES*

The purpose of this section is to provide a concise treatment of some self-tuning regulators based on slightly different concepts than those of the preceding section. For the sake of clarity and simplicity in exposition, we will present each approach independently. However, at appropriate points we will give some interpretations and point out some interrelationships amongst these approaches.

A. *Pole/Zero Placement Approach*

We consider here the problem of designing closed-loop regulators for SISO discrete systems whose output is corrupted by a coloured, unobservable disturbance. The approach to tackling this problem has been developed in *[35-37]* where the authors specify the closed-loop pole and zero positions.

Consider the discrete system

$$[1+A(z^{-1})y(k)] = z^{-d}B(z^{-1})u(k) + C(z^{-1})\xi(k) \tag{7.144}$$

where

$$\left.\begin{aligned} A(z^{-1}) &= a_1 z^{-1} +\ldots+ a_{n_a} z^{-n_a} \\ B(z^{-1}) &= b_1 z^{-1} +\ldots+ b_{n_b} z^{-n_b} \\ C(z^{-1}) &= 1 + c_1 z^{-1} +\ldots+ c_{n_c} z^{-n_c} \end{aligned}\right\} \qquad (7.145)$$

and $\xi(k)$ is a white noise sequence of zero mean and variance σ^2. It has been suggested in *[35]* that we should multiply (7.144) by $[1+M(z^{-1})]$ where

$$M(z^{-1}) = m_1 z^{-1} + \ldots + m_d z^{-d} \qquad (7.146)$$

and the coefficients $(m_1,\ldots,m_d)$ are selected so as to cancel out the coefficients $(a_1,a_2,\ldots,a_d)$. With this choice, the system (7.144) becomes

$$[1+z^{-d}\tilde{A}(z^{-1})]y(k) = z^{-d}\tilde{B}(z^{-1})u(k) + C(z^{-1})[1+M(z^{-1})]\xi(k) \qquad (7.147)$$

where

$$\left.\begin{aligned} \tilde{A}(z^{-1}) &= \tilde{a}_1 z^{-1} +\ldots+ \tilde{a}_{n_a} z^{-n_a} \\ \tilde{B}(z^{-1}) &= \tilde{b}_1 z^{-1} +\ldots+ \tilde{b}_{n_b+d} z^{-(n_b+d)} \end{aligned}\right\} \qquad (7.148)$$

Suppose that an output feedback control of the form (7.132) and (7.134) is applied to (7.144) with $n_g = n_a-1$ and $n_f = n_b+d-1$. The control design objective *[35]* is to force the resulting closed-loop system to be

$$y(t) = [N(z^{-1})/D(z^{-1})]\xi(k) \qquad (7.149)$$

with corresponding output spectral density function $S_y(z^{-1})$ given by (see Chapter 6):

$$S_y(z^{-1}) = [N(z^{-1})N^*(z^{-1})/D(z^{-1})D^*(z^{-1})]\sigma^2 \qquad (7.150)$$

where the superscript * indicates complex conjugation and $N(z^{-1})$, $D(z^{-1})$ are pre-selected polynomials which specify the locations of the closed-loop zeros and poles.

To meet the design objective, we substitute (7.132) into (7.147) to yield

$$\{[1+z^{-d}\tilde{A}(z^{-1})]F(z^{-1}) - z^{-d}\tilde{B}(z^{-1})G(z^{-1})\}y(k)$$

$$= F(z^{-1})C(z^{-1})[1+M(z^{-1})]\xi(k) \qquad (7.151)$$

The comparison of (7.149) with (7.151) implies that the controller polynomials $F(z^{-1})$, $G(z^{-1})$ must be obtained as the solution of the polynomial identity

$$N(z^{-1})\{[1+z^{-d}\tilde{A}(z^{-1})]F(z^{-1})-z^{-d}\tilde{B}(z^{-1})G(z^{-1})\}$$

$$= D(z^{-1})F(z^{-1})C(z^{-1})[1+M(z^{-1})] \qquad (7.152)$$

Similarly to (7.140), the identity (7.152) can be converted into a set of simultaneous equations in the controller parameters $(g_0,\dots,g_{n_g},f_1,\dots,f_{n_f})$. Indeed, the existence of a solution depends on the orders (n_g,n_f) and the orders of $N(z^{-1})$ and $D(z^{-1})$, of the closed-loop system.

The case of interest *[35]* is obtained when $D(z^{-1}) = 1$ and $N(z^{-1}) = V(z^{-1})$ such that $[1+M(z^{-1})]C(z^{-1}) = V(z^{-1})+z^{-d}\tilde{C}(z^{-1})$ where $V(z^{-1})$ is the moving average of $y(k)$ of order d and whose coefficients are the first d terms in the quotient $C(z^{-1})/A(z^{-1})$. With these conditions, (7.152) reduces to

$$V(z^{-1})[\tilde{A}(z^{-1})F(z^{-1}) - \tilde{B}(z^{-1})G(z^{-1})]$$

$$= F(z^{-1})\tilde{C}(z^{-1}) \qquad (7.153)$$

which is of the type satisfying the MV property *[39]*. More importantly, the special case of $C(z^{-1}) = 1$, $\tilde{C}(z^{-1}) = 0$

yields the identity

$$\tilde{A}(z^{-1})F(z^{-1}) = \tilde{B}(z^{-1})G(z^{-1}) \qquad (7.154)$$

which has the solution *[39]*

$$G(z^{-1}) = z\tilde{A}(z^{-1}) \qquad (7.155)$$

$$F(z^{-1}) = z\tilde{B}(z^{-1})/\tilde{b}_1 \qquad (7.156)$$

subject to n_n (order of $N(z^{-1})$) = d, $n_g = n_a-1$ and $n_f = n_b+d-1$. Note that the case leading to (7.153) entails that the asymptotic control error takes the form

$$y(k) = V(z^{-1})\xi(k) \qquad (7.157a)$$

where the identity

$$V(z^{-1}) = [C(z^{-1})/A(z^{-1})]-z^{-(d+1)}W(z^{-1})/A(z^{-1}) \qquad (7.157b)$$

is enforced to meet the minimum variance control policy.

Consider now the design case where the closed-loop system poles are pre-specified only. The closed-loop zeros will be determined by the design procedure. In this case, if we apply the feedback control (7.132) to yield the closed-loop system

$$y(k) = [F(z^{-1})/D(z^{-1})]\xi(k) \qquad (7.158)$$

in which the closed-loop zeros correspond to the regulator poles, then we obtain the following identity

$$\{[1+z^{-d}\tilde{A}(z^{-1})]F(z^{-1})-z^{-d}\tilde{B}(z^{-1})G(z^{-1})\}$$

$$= D(z^{-1})C(z^{-1})[1+M(z^{-1})] \qquad (7.159)$$

for the pole-placement regulator. It should be clear that the case of $C(z^{-1})$, which corresponds to white noise, yields no appreciable simplification.

Recall that the pole/zero placement controller can be imbedded in the self-tuning regulator of Fig. 7.15, by using either the identity (7.152), (7.153) or (7.159) with $\tilde{A}(z^{-1})$ and $\tilde{B}(z^{-1})$ containing the estimated parameters. Other properties are discussed in *[35-37]*. These and other issues are demonstrated in the examples.

B. Implicit Identification Approach

This approach combines ideas from self-tuning and model reference adaptive systems. We illustrate this approach for SISO systems described by

$$A(q^{-1})y(k) = q^{-(d+1)} b_0 B(q^{-1})u(k) + w(k) \qquad (7.160)$$

where

$$\left.\begin{aligned} A(q^{-1}) &= 1 + a_1 q^{-1} + \ldots + a_n q^{-n} \\ B(q^{-1}) &= 1 + b_1 q^{-1} + \ldots + b_m q^{-m} \end{aligned}\right\} \qquad (7.161)$$

and $w(k)$ is a nonmeasurable disturbance. Our objective is to design a controller so as to make the difference between $y(k)$ and a reference model output $y^m(k)$ as small as possible in some sense. The reference model is given by

$$\begin{aligned} y^m(k) &= [\, q^{-(d+1)} B^m(q^{-1})/A^m(q^{-1}) \,] u^m(k) \\ &= [\, q^{(d+1)} \sum_{j=0}^{m} b_j^m q^{-j} / 1 + \sum_{j=1}^{n} a_j^m q^{-j} \,] u^m(k) \end{aligned} \qquad (7.162)$$

Two versions of the design problems are analysed. The first version results when $w(k) = 0$, that is the disturbances are suppressed leading to a pure servo-problem (deterministic design). In the second version, the disturbances are included as a part of the problem and the design is stochastic.

In *[9,43]* it is shown that when $w(k)$ is fixed, then the design proceeds as follows:

(1) Choose an asymptotically stable polynomial $F(q^{-1})$ of the form

$$F(q^{-1}) = 1 + f_1 q^{-1} + \ldots + f_{n_f} q^{-n_f} \tag{7.162a}$$

(2) Solve the polynomial equation

$$F(q^{-1})A^m(q^{-1}) = A(q^{-1})S(q^{-1}) + q^{-(d+1)}R(q^{-1}) \tag{7.162b}$$

where

$$R(q^{-1}) = r_0 + r_1 q^{-1} + \ldots + r_{n_r} q^{-n_r}$$

$$S(q^{-1}) = 1 + s_1 q^{-1} + \ldots + s_d q^{-d} \tag{7.162c}$$

$$n_r = \max(n_f + n - d - 1,\ n - 1)$$

(3) Use the control law

$$b_0 B(q^{-1})S(q^{-1})u(k) = F(q^{-1})B^m(q^{-1})u^m(k) - R(q^{-1})y(k) \tag{7.163}$$

We note that since the controller (7.163) cancels out the process zeros, only minimum phase systems should be considered. The $F(.)$ polynomial is interpreted as the characteristic polynomial of an observer *[9]*.

Define the error term $e(k) = y(k) - y^m(k)$ and consider that the controller parameters will be estimated instead of the parameters of the model (7.160). In comparison with Section 7.4.3, this procedure is often called implicit identification *[9]*. To implement such a procedure, it is first required that we obtain a model of the system expressed in terms of the unknown controller parameters. This can be done by combining (7.160)-(7.162) to yield:

$$F(q^{-1})A^m(q^{-1})e(k) = q^{-(d+1)}[b_0 B(q^{-1})S(q^{-1})u(k)$$

$$+ R(q^{-1})y(k)-F(q^{-1})B^m(q^{-1})u^m(k)] + S(q^{-1})w(k) \quad (7.164)$$

Instead of dealing with the output error, we will use a filtered version in the form

$$e_f(k) = [K(q^{-1})/P(q^{-1})]\underline{e}(k) \quad (7.165a)$$

where

$$\left.\begin{aligned} K(q^{-1}) &= 1 + k_1 q^{-1} +\ldots+ k_{n_k} q^{-n_k} \\ P(q^{-1}) &= P_1(q^{-1})P_2(q^{-1}) \\ &= 1 + p_1 q^{-1} +\ldots+ p_{n_p} q^{-n_p} \end{aligned}\right\} \quad (7.165b)$$

are stable polynomials and P_1, P_2 are factors of P of degree n_{p_1}, n_{p_2} with $P_1(0) = P_2(0) = 1$. The manipulation of (7.165) using (7.164) yields:

$$e_f(k) = [K(q)^{-1}/F(q^{-1})A^m(q^{-1})]q^{-(d+1)} \cdot \{b_0 u(k)/P_1(q^{-1})$$

$$+ b_0[B(q^{-1})S(q^{-1}) - P_2(q^{-1})] \cdot u(k)/P(q^{-1})$$

$$+ R(q^{-1})y(k)/P(k) - F(q^{-1})B^m(q^{-1})u^m(k)/P(k)\}$$

$$+ [K(q^{-1})S(q^{-1})/F(q^{-1})A^m(q^{-1})P(q^{-1}) \; w(k)] \quad (7.166)$$

In general, the choice of $K(.)$ and $P(.)$ will aid in classifying the different special cases. In terms of

$$\underline{\theta}^t = [\beta_1,\ldots,\beta_x, r_0/b_0,\ldots,r_{n_r}/b_0, 1/b_0] \quad (7.167a)$$

with $x = \max(m+d, n_{p_2})$,

$$\underline{\Phi} = \begin{bmatrix} u(k-1)/P(.) \\ \vdots \\ u(k-x)/P(.) \\ y(k)/P(.) \\ \vdots \\ y(k-n_r)/P(.) \\ -F(.)B^m(.)u^m(.)/P(.) \end{bmatrix} \tag{7.167b}$$

we can rewrite (7.166) as:

$$\begin{aligned} e_f(k) \quad = \quad & [K(q^{-1})/F(q^{-1})A^m(q^{-1})]q^{-(d+1)} \cdot \\ & [b_0 u(k)/P_1(q^{-1}) + b_0\, \underline{\Theta}^t \underline{\Phi}] \\ & + [K(q^{-1})S(q^{-1})/F(q^{-1})A^m(q^{-1})P(q^{-1})]w(k) \end{aligned} \tag{7.168}$$

Note that the model (7.168) involves the unknown parameters b_0 and $\underline{\Theta}$. Now the design approach consists of two phases - phase 1 estimates b_0 and $\underline{\Theta}$ and phase 2 constructs a control law based on the parameter estimates.

It has been shown *[9]* that the stochastic case can be dealt with by considering

$$\begin{aligned} w(k) &= C(q^{-1})v(k) \\ &= F(q^{-1})v(k) \end{aligned}$$

and estimating the coefficients of the polynomial $C(.)$ as a part of the vector $\underline{\Theta}$. The estimation of b_0 and $\underline{\Theta}$ could be accomplished via any of the methods discussed in Section 7.4.3 .

The control law (7.163) can be expressed using (7.167) as:

$$u(k) = - P_1(q^{-1})[\underline{\Theta}^t \underline{\Phi}(k)]$$

but in implementation, it takes the form

$$u(k) = - P_1(q^{-1}) [\underline{\hat{\theta}}^t \underline{\Phi}(k)] \tag{7.169}$$

where $\underline{\hat{\theta}}$ is the estimate of $\underline{\theta}$.

Several important special cases of the general formula (7.166) can be derived and summarized in Table (7.3) with cross referencing to the corresponding adaptive method.

Examination of this table leads us to conclude that:

(1) Both MRAC systems and ST regulators can be described using the implicit identification approach.

(2) Both types of adaptive schemes can be thought of as composed of two parts. The first is a parameter estimator based on a model obtained from an analysis of the known parameter case. The second part consists of a feedback law, which is identical to the known parameter case but with the estimates replacing the unknown parameters.

(3) The MRAC system can be derived from the basic structure of ST regulators.

C. State-Space Approach

Our discussion now is directed towards the development of a self-tuning approach for a system described by state-space equations. We start with a SISO system described in the state-space innovations form *[41,45]:*

$$\begin{aligned} \underline{x}(k) &= A\underline{x}(k-1) + \underline{b}u(k-1) + \underline{g}(k)e(k-1) \\ y(k) &= \underline{c}^t\underline{x}(k) + e(k) \end{aligned} \tag{7.170}$$

where $\underline{x}(k)$ stands for the state estimate (n-vector), $u(k)$ is the scalar input, $y(k)$ is the scalar output and $e(k)$

TABLE 7.3 Special cases of the implicit identification approach

Particular choice of matrices in (7.166)	Corresponding adaptive method
$P_1(.) = F(.)$; degree d $K(.) = P(.)$ $= P_1(.)P_2(.)$; degree (n+d-1) $P_2(.)$; degree (n-1) $w(k) = 0$	MRAC system with error signal augmentation
$P(.) = 1$ $K(.) = F(.)A^m(.)$ $w(k) = 0$	Self-tuning controller based on pole-placement *[44]*
$A^m(.) = 1$ $u^m(.) = y^m(.) = 0$ $K(.) = P(.) = 1$	Minimum-variance self-tuning regulator *[39]*
$K(.) = A^m(.)$ $P(.) = 1$ $F(.) = C(.)$	Self-tuning controller with generalized MV criteria *[33,34]*

represents the innovations process (a white process with zero mean and covariance $\sigma^2(k) = \underline{c}^t P(k)\underline{c} + \mu^2$).

The associated estimation error covariance is given by

$$P(k) = A[P(k-1) - P(k-1)\underline{c}\,\underline{c}^t P(k-1)/\sigma^2(k-1)]A^t + Q \; ; \quad Q > 0 \tag{7.171}$$

and the Kalman gain is

$$\underline{g}(k) = A\,P(k)\,\underline{c}\,/\,\sigma^2(k) \tag{7.172}$$

Introduce the linear transformation

$$\begin{aligned} \underline{x}_0(k) &= H\underline{x}(k) \\ &= H_2^{-1} H_1\, \underline{x}(k) \end{aligned} \tag{7.173}$$

where

$$H_1 = [\underline{c}^t \;\; A^t\underline{c}^t \;\; \dots \;\; (A^{n-1})^t\, \underline{c}]^t \tag{7.174a}$$

$$H_2^{-1} = [\underline{i}_0 \;\; A_0^t\underline{i}_0 \;\; \dots \;\; (A_0^{n-1})^t\, \underline{i}_0]^{-t} \tag{7.174b}$$

$$A_0 = \begin{bmatrix} -a_1 & 1 & 0 & \dots & 0 \\ -a_2 & 0 & 1 & \dots & 0 \\ \vdots & \vdots & \vdots & \dots & \vdots \\ \vdots & \vdots & \vdots & \dots & \vdots \\ -a_{n-1} & 0 & 0 & \dots & 1 \\ -a_n & 0 & 0 & \dots & 0 \end{bmatrix} = HAH^{-1} \tag{7.174c}$$

$$\underline{i}_0 = [1 \;\; 0 \;\; \dots \;\; 0] \tag{7.174d}$$

The form (7.174c) is called the observability canonical form, H_2^{-1} is a lower triangular Toeplitz matrix and the parameters $(a_1, \dots, a_n)$ are the coefficients of the characteristic polynomial of A *[45]*.

The use of (7.173) in (7.170) results in:

$$\left.\begin{aligned} \underline{x}_0(k) &= A_0\,\underline{x}_0(k-1) + \underline{b}_0\,u(k-1) + \underline{g}_0(k)\,e(k-1) \\ y(k) &= \underline{c}_0^t\,\underline{x}_0(k) + e(k) \end{aligned}\right\} \quad (7.175)$$

with $\underline{b}_0 = H\,\underline{b}$; $\underline{g}_0(k) = H\,\underline{g}(k)$

By taking the Z-transform of (7.175), we obtain *[38]*:

$$y(k) = [\underline{c}_0^t(I_n - A_0 z^{-1})^{-1}\underline{b}_0 z^{-1}]u(k) + [1 + \underline{c}_0^t(I_n - A_0 z^{-1})^{-1}\underline{g}_0(k) z^{-1}]e(k) \quad (7.176a)$$

or equivalently

$$A(z^{-1})y(k) = B(z^{-1})u(k) + C(z^{-1})e(k) \quad (7.176b)$$

where

$$\left.\begin{aligned} A(z^{-1}) &= 1 + a_1 z^{-1} + \ldots + a_n z^{-n} \\ B(z^{-1}) &= b_1 z^{-1} + \ldots + b_n z^{-n} \\ C(z^{-1}) &= 1 + c_1 z^{-1} + \ldots + c_n z^{-n} \\ c_j &= a_j + g_j(k) \ ; \quad j = 1,\ldots,n \end{aligned}\right\} \quad (7.177)$$

are the standard polynomials associated with (7.170); compare these with (7.121). To this end, the autoregressive moving-average form (7.176) is obtained from (7.170) via the transformation (7.173) and the relations (7.177). Note that (7.177) contains the system parameters which have to be defined (known or identified) before the states can be estimated. Recall also that from the standpoint of parameter identification, both (7.175) and (7.176b) are equivalent.

For parameter identification, we write (7.176b) in the form

$$y(k) = \underline{\theta}^t\,\underline{\Phi}(k) + e(k) \quad (7.178)$$

where

$$\underline{\theta}^t = [a_1,\ldots,a_n,b_1,\ldots,b_n,c_1,\ldots,c_n] \quad (7.179a)$$

$$\underline{\Phi}^t(k) = [-y(k-1),\ldots,-y(k-n),u(k-1),\ldots,u(k-n),e(k-1),\ldots \ldots,e(k-n)] \quad (7.179b)$$

and use the extended least-squares method *[40,46]* to obtain the recursive estimation sequence *[38]*:

$$\underline{\hat{\theta}}(k) = \underline{\hat{\theta}}(k-1) + \underline{s}(k)\varepsilon(k) \tag{7.180a}$$

$$\underline{s}(k) = R(k-1)\underline{\psi}(k)/[\lambda(k)+\underline{\psi}^t(k)R(k-1)\underline{\psi}(k)] \tag{7.180b}$$

$$R(k) = [1/\lambda(k)]\{R(k-1)-R(k-1)\underline{\psi}(k)\underline{\psi}^t(k)R(k-1)/\ [\lambda(k) + \underline{\psi}^t(k)R(k-1)\underline{\psi}(k)]\} \tag{7.180c}$$

$$\underline{\psi}(k) = \hat{\Phi}(k)/\hat{C}(z^{-1}) \tag{7.180d}$$

where $\lambda(k)$ is the forgetting function to discount old measurements $(.9 < \lambda(0) < 1,\ \lambda(k) = \rho\lambda(k-1) + (1-\rho);\ 0 < \rho < 1)$, $\varepsilon(k) = y(k) - \underline{\hat{\theta}}^t(k-1)\underline{\hat{\Phi}}(k)$, $\underline{\hat{\Phi}}(k)$ is the vector $\underline{\Phi}(k)$ with $\varepsilon(k)$ replacing $e(k)$ and $\hat{C}(z^{-1})$ is $C(z^{-1})$ with the coefficients d_j are replaced by their estimates from the vector $\underline{\hat{\theta}}$. With the aid of (7.174c), (7.175), (7.179b) and (7.180a) the system states $\underline{x}_0(k)$ can be reconstructed quite easily. Note that

$$\underline{b}_0^t = [b_1 \quad b_2, \ldots, b_n]$$

$$\underline{g}_0^t = [c_1-a_1, \ldots, c_n-a_n]$$

This completes the joint scheme for system-parameter identification and state estimation. We now move to developing a control law based on pole assignment algorithms for systems of the type (7.175).

From the theory of linear systems, it is well known *[45]* that the discrete system

$$\left.\begin{aligned} \underline{x}_c(k+1) &= A_c\ \underline{x}_c(k) + \underline{b}_c\ u(k) \\ y(k) &= \underline{c}_c^t\ \underline{x}_c(k) \end{aligned}\right\} \tag{7.181}$$

under the linear feedback control law

$$u(k) = r(k) - \underline{f}^t\ \underline{x}_c(k) \tag{7.182}$$

has the closed-loop characteristic polynomial

$$\Delta_c(z) = z^n + \sum_{j=1}^{n} \alpha_j z^{n-j} = z^n + \sum_{j=1}^{n} (a_j+f_j) z^{n-j} \qquad (7.183)$$

where

$$A_c = \begin{bmatrix} -a_1 & -a_2 & \cdots & -a_{n-1} & -a_n \\ 1 & 0 & \cdots & 0 & 0 \\ \vdots & \vdots & \cdots & \vdots & \vdots \\ 0 & 0 & & 0 & 0 \\ 0 & 0 & & 1 & 0 \end{bmatrix} \qquad (7.184a)$$

$$\underline{b}_c = \underline{i}_0 \; ; \quad \underline{c}_c^t = [b_1 \; b_2, \ldots, b_n] \qquad (7.184b)$$

This means that by assigning n desired eigenvalues, we can construct (7.183) and from which the feedback gain $\underline{f} = [f_1 \; f_2 \; \ldots \; f_n]$ can be easily determined. In order to be able to utilize this method efficiently, we need to convert the observability canonical form (7.175) of which we estimated the parameters and states, to the controllability canonical form (7.181) for which we have designed the linear controller (7.182). The conversion is a standard procedure in linear systems analysis *[46]* and is defined by

$$\underline{x}_0(k) = T_1 T_2^{-1} \underline{x}_c(k) = T \, \underline{x}_c(k) \qquad (7.185a)$$

where

$$T_1 = [\underline{b}_0 \; A_0\underline{b}_0, \ldots, A_0^{n-1}\underline{b}_0] \qquad (7.185b)$$

$$T_2 = [\underline{b}_c \; A_c\underline{b}_c, \ldots, A_c^{n-1} b_c] \qquad (7.185c)$$

By similarity to H_2^{-1}, T_2^{-1} is an upper triangular Toeplitz matrix with the first row $[1 \; a_1 \; \ldots \; a_{n-1}]$. From (7.185a) in (7.182) we arrive at

$$u(k) = r(k) - \underline{f}^t T^{-1} \underline{x}_0(k) \tag{7.186}$$

The use of (7.186) in (7.175) and (7.176) after some algebraic manipulations, results in:

$$y(k) = [B(z^{-1})/E_c(z^{-1})]r(k) + [D_c(z^{-1})/E_c(z^{-1})]e(k) \tag{7.187a}$$

where

$$E_c(z^{-1}) = 1 + \alpha_1 z^{-1} + \ldots + \alpha_n z^{-n}$$

$$= 1 + (a_1+f_1)z^{-1} + \ldots + (a_n+f_n)z^{-n} \tag{7.187b}$$

$$D_c(z^{-1}) = 1 + \mu_1 z^{-1} + \ldots + \mu_n z^{-n}$$

$$= 1 + (c_1+h_1)z^{-1} + \ldots + (c_n+h_n)z^{-n} \tag{7.187c}$$

and

$$\underline{h}^t = [h_1\ h_2\ \ldots\ h_n]$$

$$= \underline{f}^t T^{-1} S_1 S_2^{-1} \tag{7.187d}$$

with

$$S_1 = [\underline{b}_0\ D\underline{b}_0\ \ldots\ D^{n-1}\underline{b}_0] \tag{7.187e}$$

$$S_2 = [\underline{b}_c\ D^t\underline{b}_c\ \ldots\ (D^t)^{n-1}\underline{b}_c\ ] \tag{7.187f}$$

$$D = \begin{bmatrix} -c_1 & 1 & 0 & \ldots & 0 \\ -c_2 & 0 & 1 & \ldots & 0 \\ \vdots & \vdots & \vdots & \ldots & \vdots \\ \vdots & \vdots & \vdots & \ldots & \vdots \\ \vdots & \vdots & \vdots & \ldots & \vdots \\ -c_{n-1} & 0 & 0 & \ldots & 1 \\ -c_n & 0 & 0 & \ldots & 0 \end{bmatrix} \tag{7.187g}$$

This completes the derivation of the state feedback control law, and hence completes the design of pole assignment self-tuning controllers. Some remarks are in order:

(1) From (7.187a), the mean value of the steady state output for constant reference input $r(k) = r$ is:

$$y_m = E[y(k)]$$

$$= [B(1)/E_c(1)]r$$

Since, for practical situations, tracking systems require that $y_m = r$, therefore the control law (7.186) must be modified to

$$u(k) = u_r\, r(k) - \underline{f}^t T^{-1} \underline{x}_0(k)$$

$$= [E_c(1)/B(1)]r(k) - \underline{f}^t T^{-1} \underline{x}_0(k)$$

$$= [(1 + \sum_{k=1}^{n} \alpha_j)/\sum_{j=1}^{n} b_j]r(k) - \underline{f}^t T^{-1} \underline{x}_0(k) \tag{7.188}$$

which now gives the general pole assignment self-tuning control law.

(2) Sometimes it is desirable to have a fast tracking system. This means that the closed-loop poles must be assigned very near to the origin $z = 0$. Recall from Chapter 4 that when all the closed-loop poles are assigned at $z = 0$ we will have a deadbeat response and the output will reach the steady state in, at most, n steps. In our case, this corresponds to setting

$$\alpha_j = 0 \quad ; \quad j = 1,\ldots,n$$

which results in

$$\underline{f} = [-a_1,\ldots,-a_n]^t \quad , \quad u_r = 1/b(1) \quad ;$$

$$b(1) \neq 0 \quad , \quad E_c(z^{-1}) = 1$$

It is evident that the design procedure will be simplified.

(3) The developed algorithm can be equally applied to control the nonminimum and/or unstable systems. It must be

emphasized that only the controllability and observability properties are needed here. In contrast to the solution of polynomial equations, as required by previous design approaches, the state-space approach requires only matrix operations.

D. MULTIVARIABLE APPROACH

Our aim is to provide a brief account of the generalization of adaptive control design to multivariable systems. The systems under consideration are of the type (7.117) and (7.118). An equivalent description is

$$[I+A(z^{-1})]\underline{y}(k) = z^{-d}B(z^{-1})\underline{u}(k) + [I+C(z^{-1})]\underline{\xi}(k) \quad (7.189)$$

where $\underline{u}(k)$ and $\underline{y}(k)$ are p-vectors defining the measurable system input and output, respectively, and $\underline{\xi}(t)$ is a p-vector representing a zero-mean white-noise process with covariance R. The quantities $A(z^{-1})$, $B(z^{-1})$ and $C(z^{-1})$ are polynomial matrices in the backward shift operator z^{-1}, and in our case are of the form

$$X(z^{-1}) = X_1 z^{-1} + \ldots + X_{n_x} z^{-n_x} \quad (7.190)$$

where X_j, $j = 1,\ldots,n_x$ are (pxp) matrix coefficients.

One class of adaptive methods is the minimum variance control, discussed in part B of Section 7.4.4 . We now present a multivariable version of this control method, using a slightly different treatment. First, we introduce the polynomial identity:

$$C(z) = A(z)F(z) + z^{d+1}G(z) \quad (7.191)$$

where

$$\left.\begin{aligned} F(z) &= I+F_1 z + \ldots + F_d z^d \\ G(z) &= G_0+G_1 z + \ldots + G_{n-1} z^{n-1} \end{aligned}\right\} \quad (7.192)$$

with $n_a = n_b = n_c = n$ and the model (7.117) and (7.118) is used, (the reader is advised to compare (7.191) with (7.152) or (7.153)). Introduce further $F^*(z)$ and $G^*(z)$ given by:

$$F^*(z)G(z) = G^*F(z) \tag{7.193}$$

such that $F^*(0) = I$ and $\det[F(z)] = \det[F^*(z)]$.

It has been shown [32] that the admissible control strategy minimizing the criterion

$$\min_{\underline{u}} E[\underline{y}^t(k+d+1)Q\underline{y}(k+d+1)] \quad ; \quad Q \geq 0 \tag{7.194}$$

is given by

$$G^*(z^{-1})\underline{y}(k) + F^*(z^{-1})B(z^{-1})\underline{u}(k) = 0 \tag{7.195}$$

and the asymptotic control error

$$\underline{y}(k) = F(z^{-1})\underline{\xi}(k) \tag{7.196}$$

An equivalent expression to (7.195) can be obtained using (7.193) as:

$$\underline{u}(k) = - B^{-1}(z)G(z)F^{-1}(z)y(k) \tag{7.197}$$

where the indicated inverses exist. Again, for constant multivariable systems with unknown parameters, an identification scheme should be implemented first. The estimated parameters are then used to compute the control signal.

Another class of adaptive methods is based on pole assignment, for which the model (7.189) and (7.190) is utilized. By analogy to part A of Section 7.4.5, we introduce a control law of the form:

$$\underline{u}(k) = N(z^{-1})[I+D(z^{-1})]^{-1}\underline{y}(k) \tag{7.198}$$

where:

$$\left.\begin{aligned} D(z^{-1}) &= D_1 z^{-1} + \ldots + D_{n_d} z^{-n_d} \\ N(z^{-1}) &= N_0 + N_1 z^{-1} + \ldots + N_{n_n} z^{-n_n} \end{aligned}\right\} \qquad (7.199)$$

Normally, $n_n = n_a - 1$ and $n_d = n_b + d - 1$. Substituting (7.198) into (7.189) the closed-loop system becomes

$$\underline{y}(k) = [I+D(z^{-1})][I+P(z^{-1})]^{-1}[I+C(z^{-1})]\underline{\xi}(k) \qquad (7.200a)$$

where

$$[I+P(z^{-1})] = [I+A(z^{-1})][I+D(z^{-1})] - z^{-d}B(z^{-1})N(z^{-1}) \qquad (7.200b)$$

At this stage we choose the coefficients of polynomials $N(z^{-1})$ and $D(z^{-1}$ so that *[37]*:

$$[I+P(z^{-1})] = [I+C(z^{-1})][I+T(z^{-1})] \qquad (7.201)$$

where $T(z^{-1})$ is a polynomial of order $n_t \leq n_a + n_b + d-1-n_c$ (which assures that the solution of (7.200b) exists) and $\det[I+T(z^{-1})]$ specifies the poles of the closed-loop system (design parameters). We point out that (7.200b) and (7.201) have to be rearranged in the form of simultaneous linear equations and solved for the control law parameters $(D_1, \ldots, D_{n_d}, N_0, \ldots, N_{n_n})$. In implementation, it has been found that a more convenient form of control law than (7.198) would be

$$\begin{aligned} \underline{u}(k) &= [I+\tilde{D}(z^{-1})]^{-1}\tilde{N}(z^{-1})\underline{y}(k) \\ &= -\tilde{D}(z^{-1})\underline{u}(k) + \tilde{N}(z^{-1})\underline{y}(k) \end{aligned} \qquad (7.202)$$

where the polynomials $\tilde{D}(z)$ and $\tilde{N}(z)$ satisfy certain qualifications similar to (7.193). We repeat here that a self-tuning regulator can then be constructed using the control law (7.202) but preceded by a recursive estimation scheme. It is important to note that nonminimum-phase systems can be dealt with directly using the pole/zero placement approach. However, the MV approach will suffer some difficulties and most of the time will yield highly-sensitive closed-loop systems which may turn out to be unstable in practice. Further discussions on this issue are found in *[33,37,39]*.

7.4.6 *DISCUSSION*

Thus far we have treated model-reference adaptive control (MRAC) systems and self-tuning (ST) regulators as separate design methodologies. The exception from this was part B of Section 7.4.4 concerning the implicit identification approach. Recall that the emphasis was to derive the MRAC system from the (ST) regulator structure. Here, we shall highlight some of the similarities and differences between the two adaptive design methodologies.

In Sections 7.2 and 7.3, most of the analytical treatment was centered around the use of "parallel" reference models. The design objective was *asymptotic tracking* rather than *control regulation*. Using a "series-parallel" reference model (see Fig. 7.3), it has been shown *[6,17,18,48]* that the resulting adaptive configuration allows the desired response for regulation to be specified. On the other hand, the same problem has been considered in *[34,44]* by using self-tuning concepts coupled with an implicit reference model. In the terminology of MRAC systems, an implicit reference model can be formed by

(a) an adaptive predictor,
(b) a controller.

Explicit MRAC schemes can then be made equivalent to implicit MRAC when the output of the predictor is designed to behave as the output of the explicit reference model.

To illustrate the above arguments, we consider shift-invariant systems of the type (7.144) and (7.145) with $\xi = 0$ and having the minimum phase property. The control design objectives are *[17,18]:*

(a) Tracking Objective

The control should be such that in tracking, the output $y(k)$ satisfies

$$T(z^{-1})y(k) = z^{-d}D(z^{-1})u_m(k)$$

$$\left[1 + \sum_{j=1}^{n_t} t_j z^{-j}\right] y(k) = z^{-d}\left[\sum_{j=0}^{n_d} d_j z^{-j}\right] u_m(k) \quad (7.203)$$

(b) Regulation Objective

The control should be such that in regulation $(u_m(k) \equiv 0)$, an initial disturbance $(y(0) \neq 0)$ is eliminated according to

$$R(z^{-1})y(k+d) = 0$$

$$\left[1 + \sum_{j=0}^{n_r} r_j z^{-j}\right] y(k+d) = 0 \quad (7.204)$$

where $u_m(k)$ is a bounded input and $R(z^{-1})$ is an asymptotically stable polynomial. One solution of the problem addressed above is shown in Fig. 7.16, where

$$u(k) = [1/B(z^{-1})V(z^{-1})][R(z^{-1})y_m(k+d)-W(z^{-1})y(k)] \quad (7.205a)$$

$$V(z^{-1}) = 1 + \sum_{j=1}^{n_v} v_j z^{-j} \quad (7.205b)$$

$$W(z^{-1}) = \sum_{j=0}^{n_w} w_j z^{-j} \quad (7.205c)$$

such that

$$R(z^{-1}) = A(z^{-1})V(z^{-1}) + z^{-d}W(z^{-1})$$

It has been shown *[17,18]* that in the $\xi \neq 0$, the controller (7.205a) satisfies the minimum variance property *[34,39]*

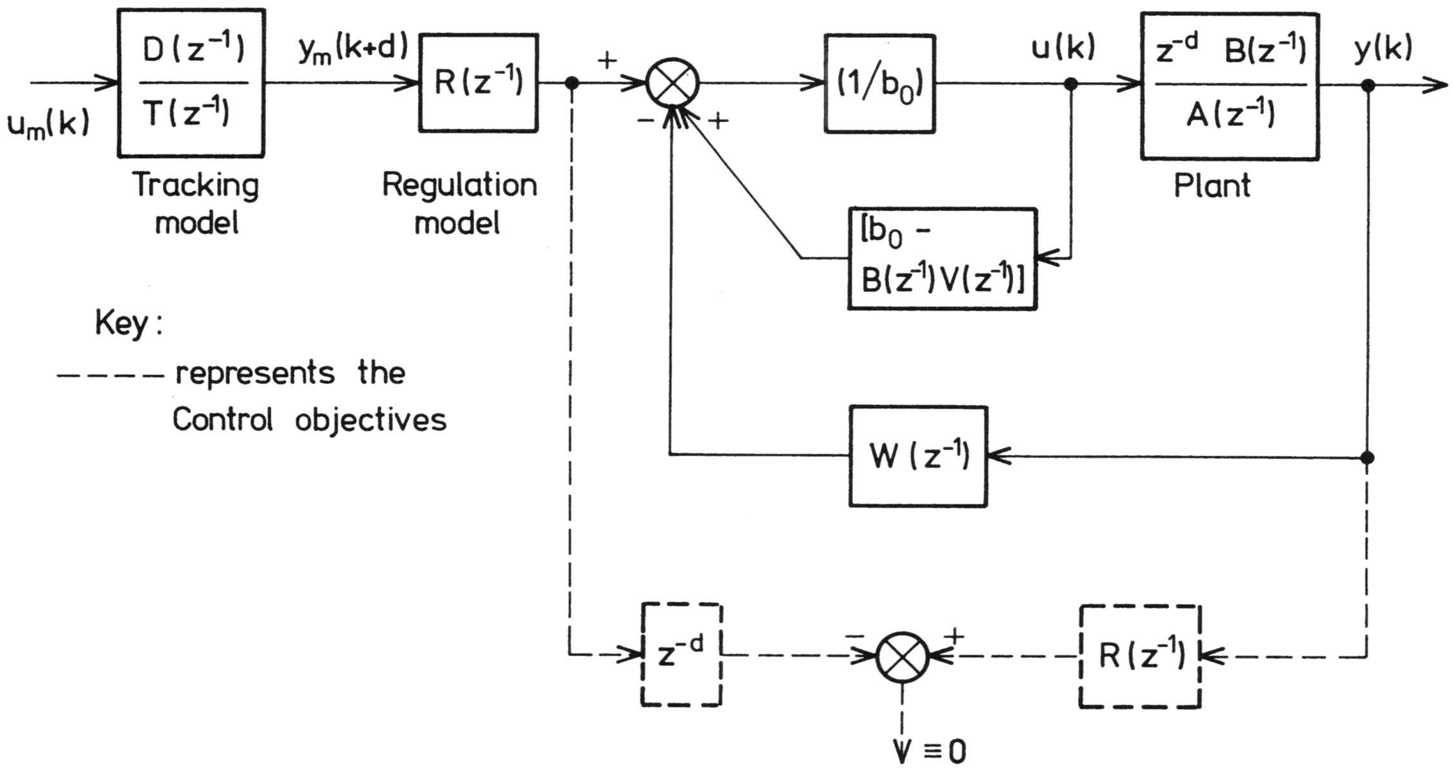

Fig. (7.16) Tracking and regulation control scheme for known plants

Further detailed discussions are found in *[48]*. Other methods are presented in *[47,49-51]*.

The various design methods are now illustrated by several examples.

7.4.7 *EXAMPLES*

As seen in the previous sections, numerous self-tuning regulator schemes are available. Here, we present the simulation results of experimenting with these schemes.

Example 1

A second order example of the type (7.117) and (7.118) with one-step time delay (d = 0) has the following values:

$n_a = 2$

$$A_1 = \begin{bmatrix} -1.5 & .3 \\ .2 & -1.5 \end{bmatrix}, \quad A_2 = \begin{bmatrix} .54 & -.1 \\ .1 & .56 \end{bmatrix}$$

$n_b = 2$

$$B_0 = \begin{bmatrix} 2 & -.3 \\ .1 & 1 \end{bmatrix}, \quad B_1 = \begin{bmatrix} -1.8 & .2 \\ -.2 & -.5 \end{bmatrix}$$

$n_c = 2$

$$C_1 = \begin{bmatrix} .2 & .1 \\ -.1 & .2 \end{bmatrix}, \quad C_2 = \begin{bmatrix} -.48 & -.2 \\ .2 & -.24 \end{bmatrix}$$

The disturbance vector $\underline{\xi}(k)$ is taken to be a sequence of normally distributed independent variables with zero mean value and unity covariance matrix. A minimum variance self-tuning regulator is considered with a performance criterion of the form (7.142) with $P_1 = P_2 = 1$, $P_3 = 0$ and $\underline{r}(k) = [1 \quad 1]^t$ for all k. The computer simulation was made over 9 periods each with 30 samples. In Figures 7.17 through 7.20, the

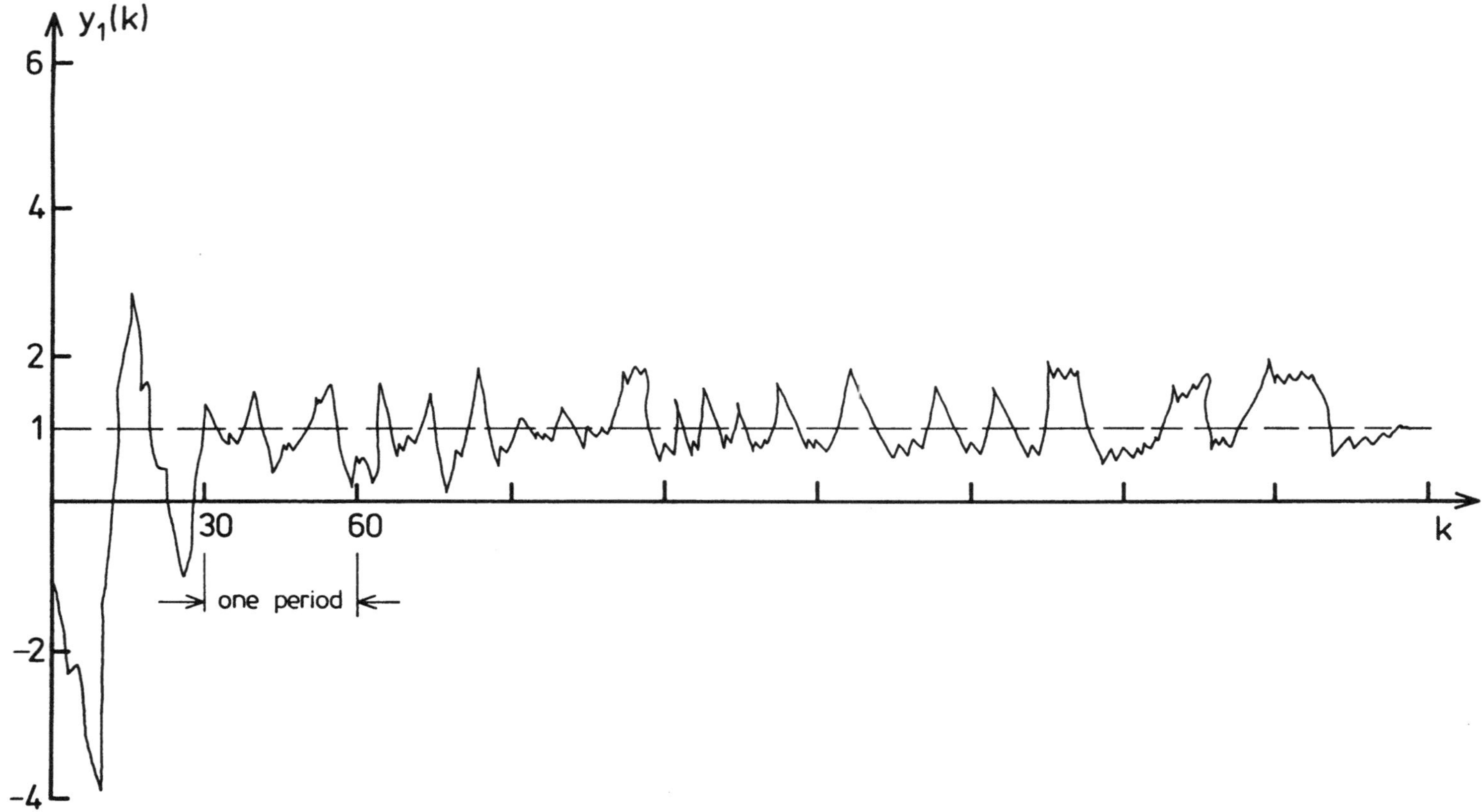

Fig. (7.17) Behaviour of output $y_1(k)$

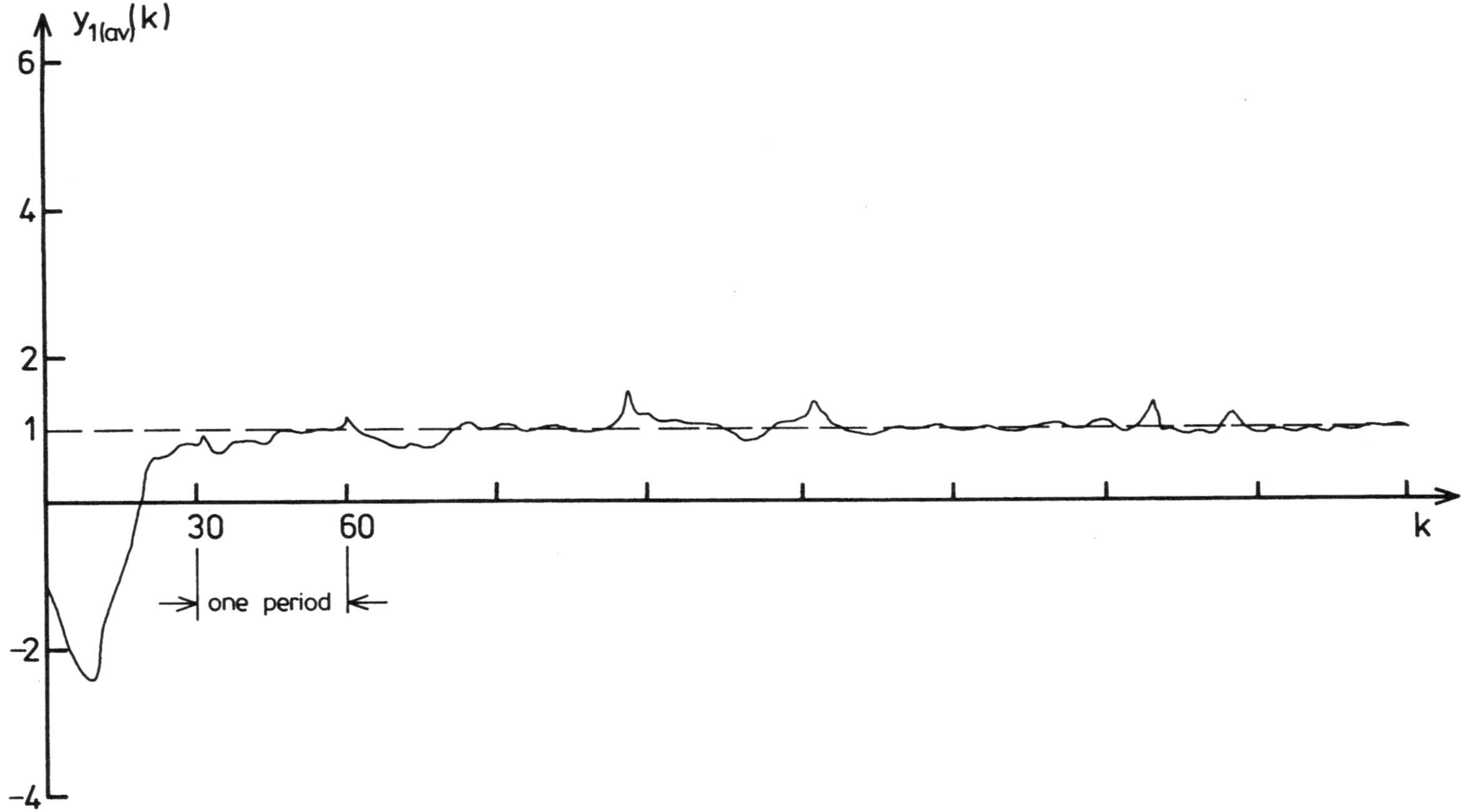

Fig. (7.18) Behaviour of average value of $y_1(k)$

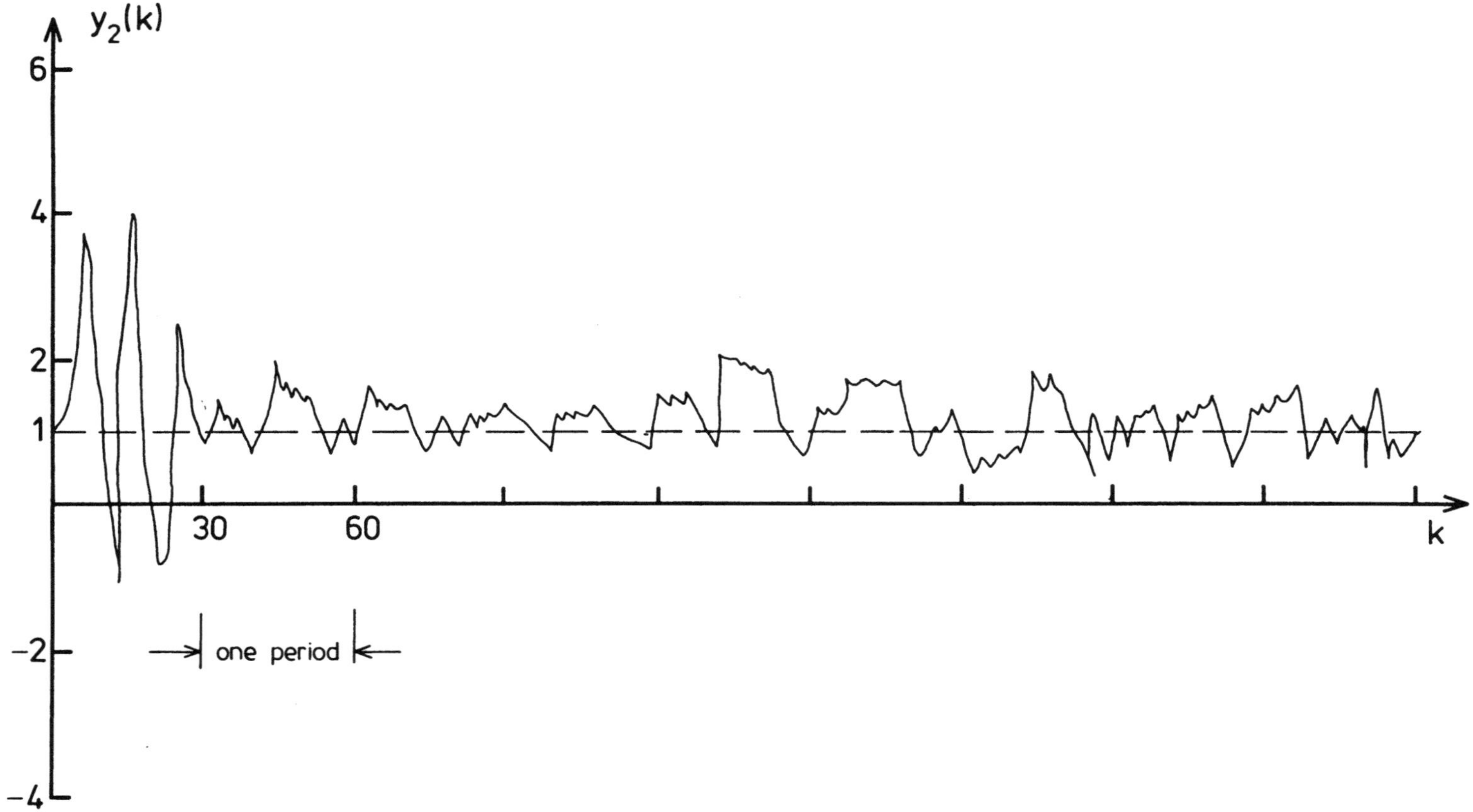

Fig. (7.19) Behaviour of output $y_2(k)$

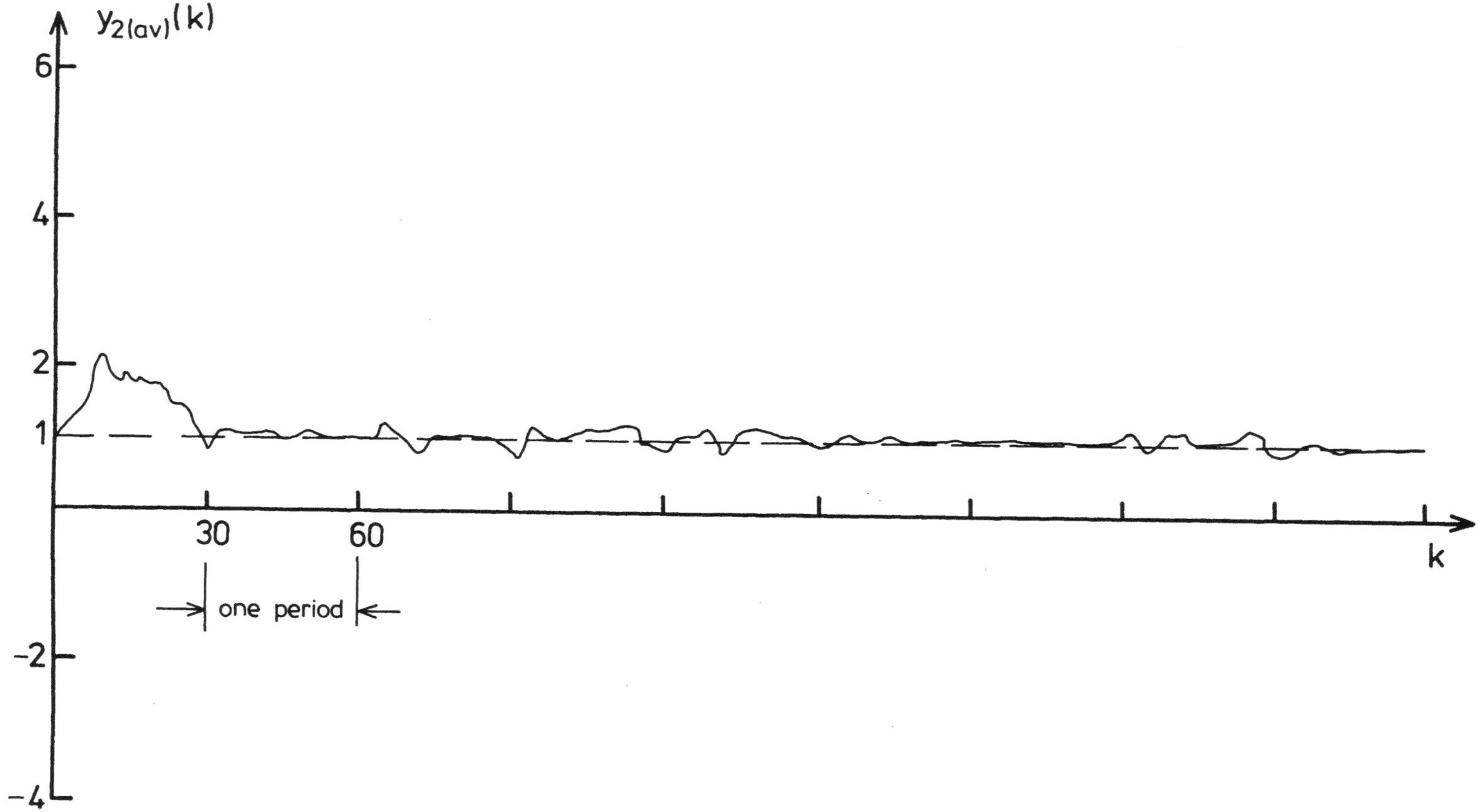

Fig. (7.20) Behaviour of average value of $y_2(k)$

outputs (y_1, y_2) and their average values are presented. We note that the performance of the algorithm is generally smooth.

Example 2

Consider the system whose output $y(k)$ evolves according to

$$\underline{y}(k) + A_1 \underline{y}(k-1) = \underline{u}(k-2) + \underline{\xi}(k) + C_1 \underline{\xi}(k-1)$$

where

$$A_1 = \begin{bmatrix} -.9 & .5 \\ .5 & -.2 \end{bmatrix} \quad ; \quad C_1 = \begin{bmatrix} -.2 & -.4 \\ .2 & -.8 \end{bmatrix}$$

$$E[\underline{\xi}(k)\underline{\xi}^t(k)] = R = \begin{bmatrix} .1 & .1 \\ .1 & .2 \end{bmatrix}$$

From (7.118) we calculate the roots of $A(z)$ and $C(z)$. We find that the zeros of $\det[C(z)]$ are $z_1 = 1.6667$ and $z_2 = 2.5$, and both are outside the unit circle. It is easy to see that $\det[A(z)]$ has the poles at $z_1 = -16.658$ and $z_2 = -.9434$, which means that the open-loop system is unstable. We shall use the analysis of part D, Section 7.4.5, with $d = 1$. The identity (7.191) gives:

$$F_1 = C_1 - A_1 = \begin{bmatrix} .7 & -.9 \\ -.3 & -.6 \end{bmatrix}$$

$$G_0 = -A_1 F_1 = \begin{bmatrix} .78 & -.51 \\ -.41 & .33 \end{bmatrix}$$

Using the quadratic criterion (7.194) with $Q = I_2$, the optimal control law (7.195) can be expressed as

$$[I_2 + \tilde{F}_1 z^{-1}]\underline{u}(k) = -\tilde{G}_0 \underline{y}(k)$$

where $\tilde{G}_0 = G_0$ and

$$\tilde{F}_1 = G_0 F_1 G_0^{-1}$$

$$= \begin{bmatrix} 1.4143 & .98571 \\ -1.1857 & -1.3143 \end{bmatrix}$$

That is

$$\underline{u}(k) = [-1/\Delta(z^{-1})] \begin{bmatrix} 1.1841-1.0252z^{-1} & -.8353-.6703z^{-1} \\ .5148-.5799z^{-1} & -.2747+.4667z^{-1} \end{bmatrix} \underline{y}(k)$$

with

$$\Delta(z^{-1}) = 2.1688 + .1z^{-1} - 1.8588z^{-2}$$

If we consider the parameters of the plant to be unknown. To tackle this situation, we use at each sampling interval a least squares estimation algorithm based on the model

$$\underline{y}(k) + A_0\, \underline{y}(k-2) = \underline{u}(k-2) + B_1\, \underline{u}(k-3) + \underline{\xi}(k)$$

together with a regulator of the form

$$[I_2 + B_1 z^{-1}]\underline{u}(k) = A_0\, \underline{y}(k)$$

The simulation results of the recursive estimation scheme (similar to (7.127)) are shown in Figs. 7.21 and 7.22 . It is readily seen that the estimated parameters approach the optimal values which have been computed from the previous case (known plant). We observe that the B-parameters converge slowly. For completeness, the loss function $f = \sum_{j=1}^{N} \underline{y}^t(j)\underline{y}(j)$ has been computed and plotted in Fig. 7.23 . It is clear that this

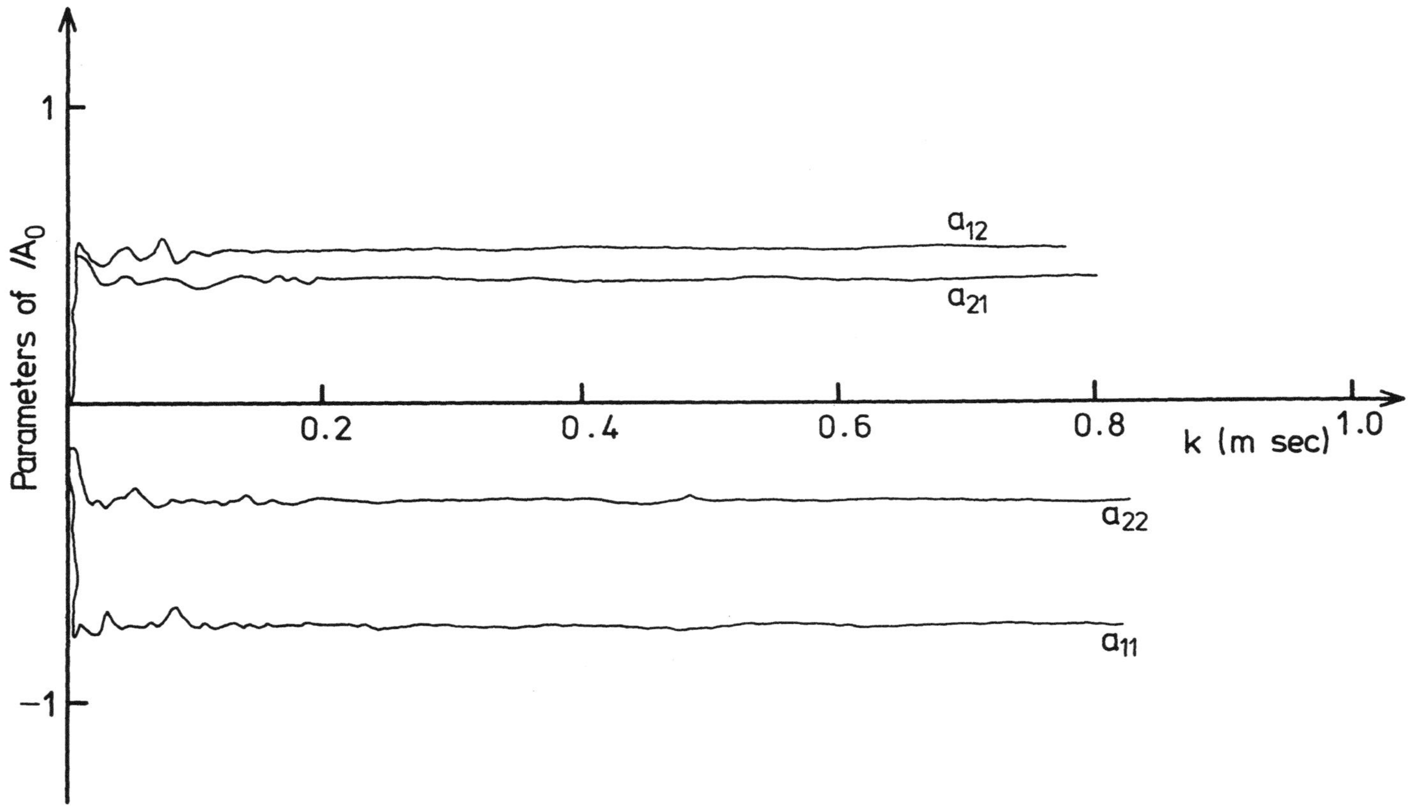

Fig. (7.21) Estimated parameter of $/A_0$

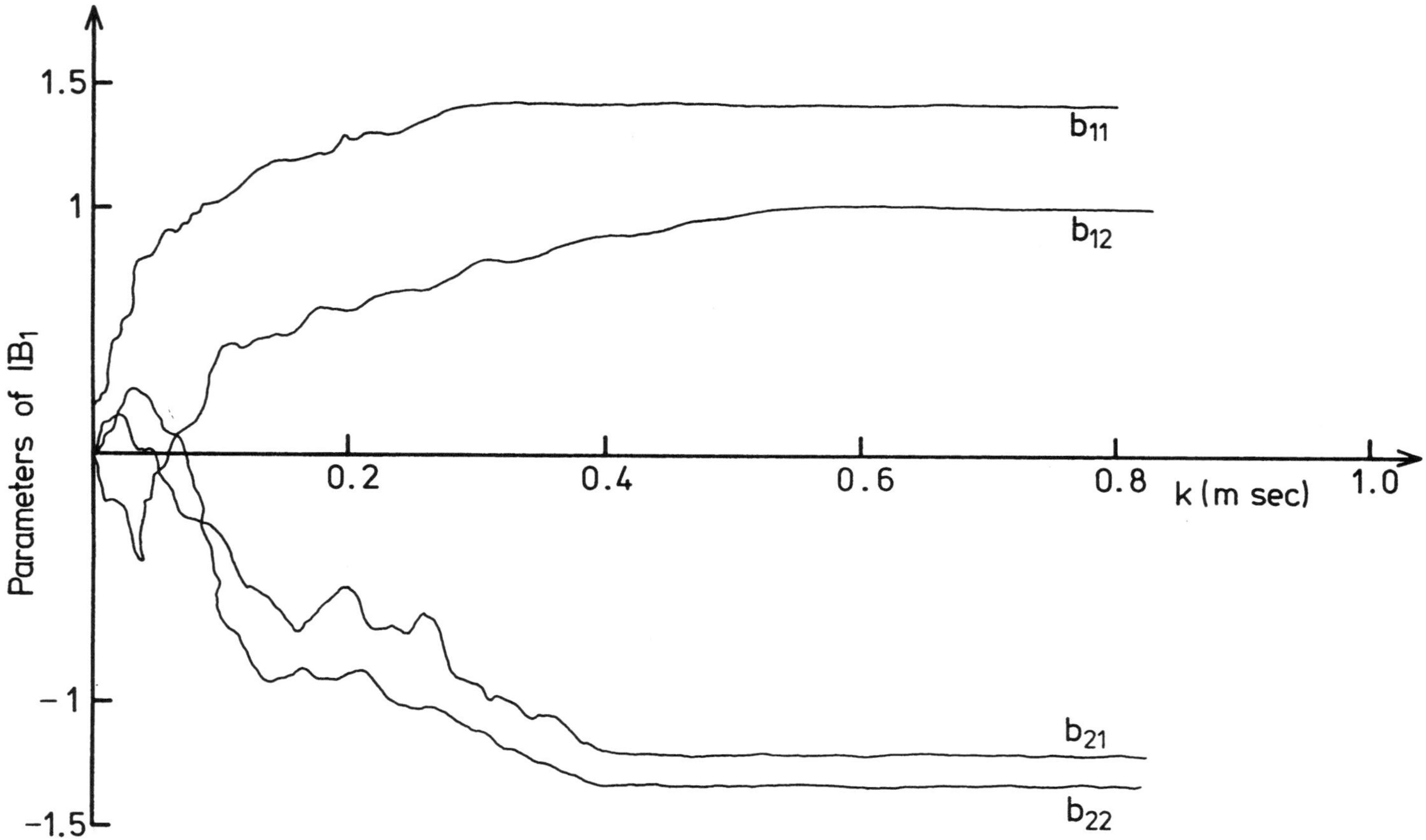

Fig. (7.22) Estimated parameter of $/B_1$

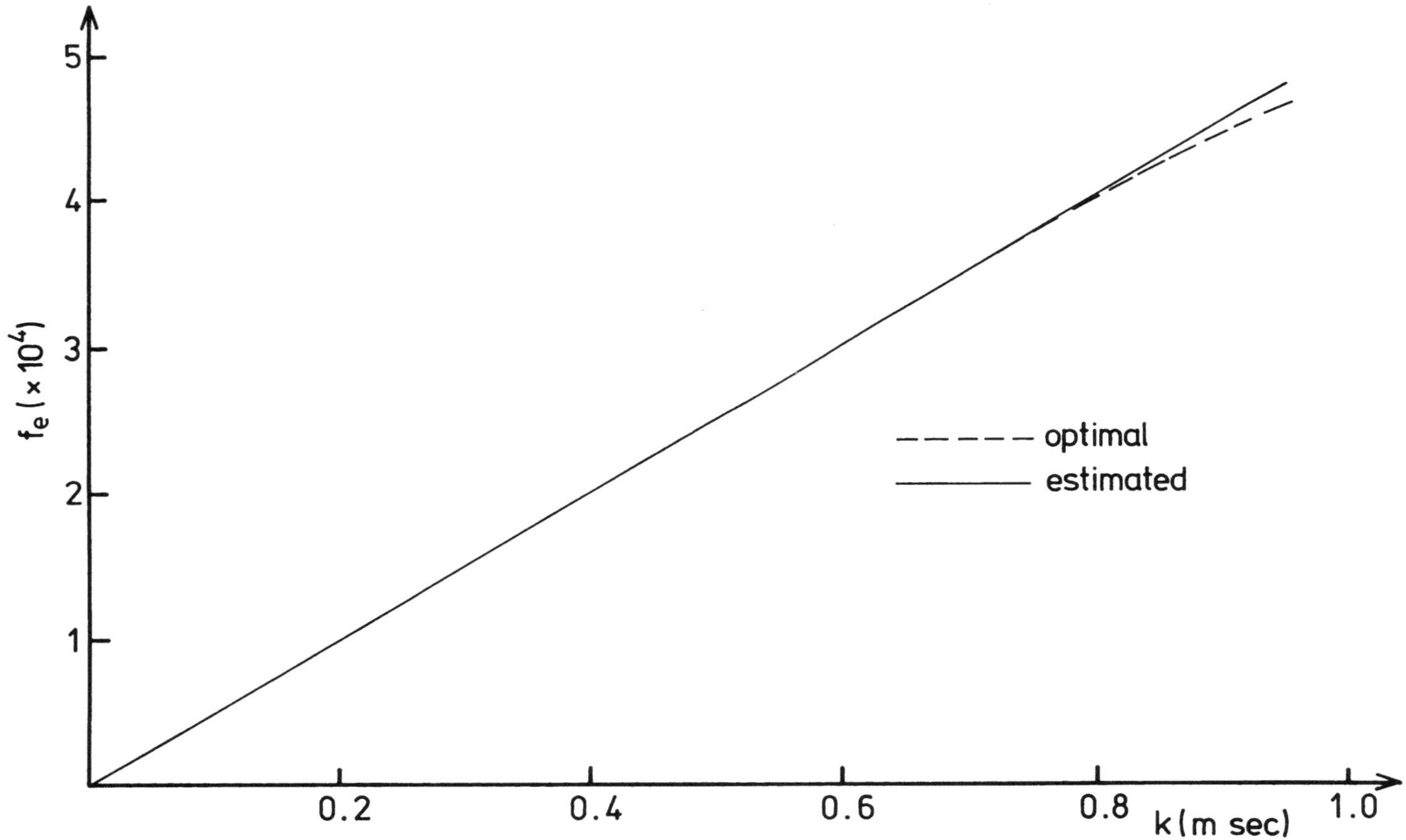

Fig. (7.23) Variation of loss function

function is almost always optimal over the simulation period.

Example 3

Consider the nonminimum-phase system defined by

$$[1-z^{-1}]y(k) = z^{-1}[z^{-1} + 1.5z^{-2}]u(k) + [1 - .2z^{-1}]\xi(k)$$

which is in the form (7.144) and (7.145) with $n_a = 1$, $d = 1$, $n_b = 2$ and $n_c = 1$. It can be argued that the MV regulation or its variants would probably go unstable. For this purpose we apply a self-tuning pole-assignment algorithm such that the closed-loop poles are given by

$$D(z^{-1}) = (1 + .4z^{-1})(1 - .4z^{-1})$$

$$= 1 - .16z^{-2}$$

From the pole assignment identity (7.152), the parameters of the controller (7.132) and (7.134) are given by:

$$f_1 = -5 \quad , \quad f_2 = .172$$

$$g_0 = .168$$

Let us now examine the regulation of the above plant, assuming its parameters are unknown, by a pole-assignment self-tuner. We use a prediction model of the form

$$y(k) = -q^{-1} A(q^{-1})y(k) + q^{-1} B(q^{-1})u(k) + \xi(k)$$

with

$$A(q^{-1}) = \alpha_1 q^{-1}$$

$$B(q^{-1}) = q^{-1} + \beta_2 q^{-2} + \beta_3 q^{-3}$$

Using this prediction model and the control law, just calculated, in (7.159) with $C(z^{-1}) = 1$, $\tilde{A}(z^{-1}) = A(z^{-1})$, $B(z^{-1}) = \tilde{B}(z^{-1})$, we arrive at:

$$[1+q^{-1}(\alpha_1 q^{-1})][1+.5q^{-1}+.172q^{-2}]+.168q^{-1}[q^{-1}+\beta_2 q^{-2}+\beta_3 q^{-3}]$$

$$= [1-.16q^{-2}][1+m_1 q^{-1}]$$

Equating coefficients of similar power, we obtain:

$$m_1 = .5 \quad ; \quad \alpha_1 = -.5$$

$$\beta_2 = 1.012 \quad ; \quad \beta_3 = .512$$

From (7.158), the closed-loop system output is described by

$$y(k) = [1+.5z^{-1}+.172z^{-2}]/[1-.16z^{-2}]\xi(k)$$

The results of the self-tuning simulation are shown in Fig. 7.24 for the plant parameters and in Fig. 7.25 for the controller parameters. It is clear from the figures that there is a biasing in parameter estimation as expected.

Example 4

Consider a multivariable system of the type (7.189) and (7.190)

$$[I_2+A_1 z^{-1}+A_2 z^{-2}]\underline{y}(k) = [B_1 z^{-1}+B_2 z^{-2}]\underline{u}(k)+[I_2+C_1 z^{-1}]\underline{\xi}(k)$$

where

$$A_1 = \begin{bmatrix} -1.4 & -.2 \\ -.1 & -.9 \end{bmatrix} \quad ; \quad A_2 = \begin{bmatrix} .48 & .1 \\ 0 & .2 \end{bmatrix}$$

$$B_1 = \begin{bmatrix} 1 & 0 \\ 0 & 0 \end{bmatrix} \quad ; \quad B_2 = \begin{bmatrix} 1.5 & 1 \\ 0 & 1 \end{bmatrix}$$

$$C_1 = \begin{bmatrix} -.5 & 0 \\ .1 & -.3 \end{bmatrix}$$

and $\underline{\xi}(k)$ is a white-noise vector sequence with zero-mean and

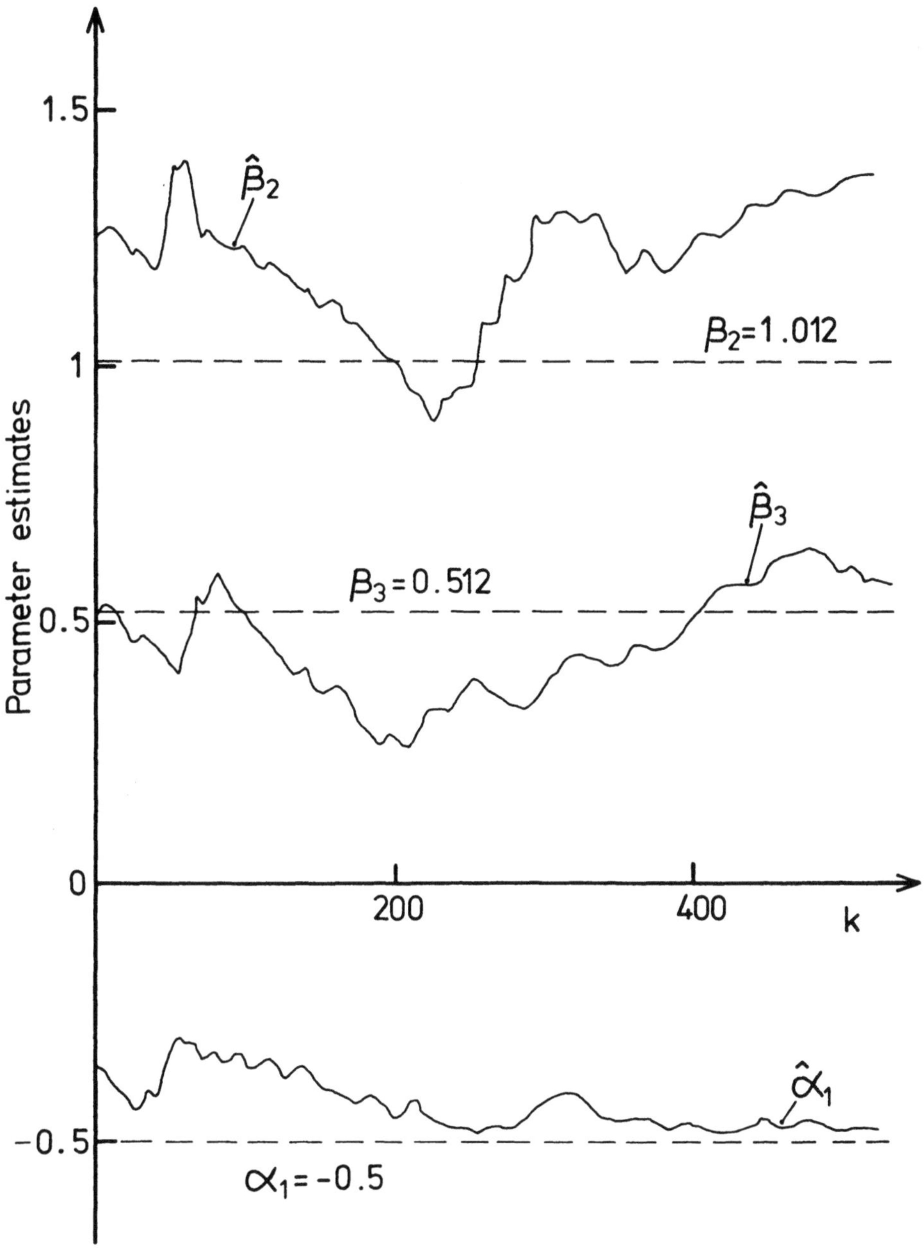

Fig. (7.24) Parameter estimates (Example 3)

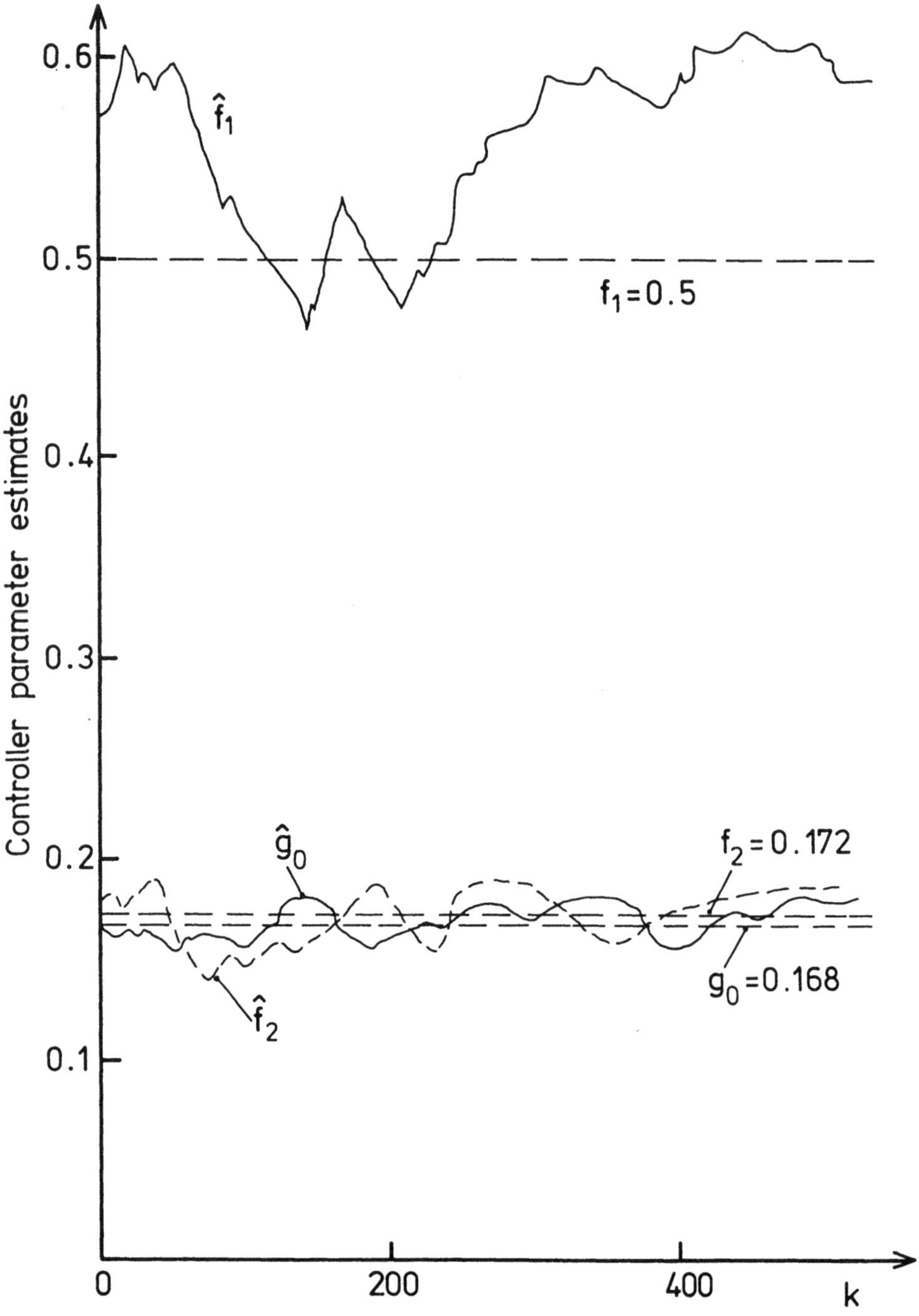

Fig. (7.25) Controller parameters estimate (Example 3)

variance

$$R = \begin{bmatrix} .1 & 0 \\ 0 & .1 \end{bmatrix}$$

It is desired to position the closed-loop poles at $z_1 = .5$ and $z_2 = .4$. A suitable choice of $[I_2+T(z^{-1})]$ would then be

$$I_2 + \begin{bmatrix} -.5 & 0 \\ 0 & -.4 \end{bmatrix} z^{-1}$$

We must emphasize that the system is nonminimum phase and B_1 is singular. (Note that this implies different time delays in the two input channels.) Therefore, direct application of self-tuning regulation would not work.

Proceeding to design a pole-placement regulator, we have to solve (7.201). To do this we employ a transformation of the type (7.193), that is

$$[I_2+\tilde{D}_1 z^{-1}]^{-1}[\tilde{N}_0+\tilde{N}_1 z^{-1}] = [N_0+N_1 z^{-1}][I_2+D_1 z^{-1}]^{-1}$$

so that the resulting control law takes the form (7.202) with

$$\tilde{D}_1 = \begin{bmatrix} .213 & .203 \\ .155 & .286 \end{bmatrix}$$

$$\tilde{N}_0 = \begin{bmatrix} -.1 & -.048 \\ -.16 & -.115 \end{bmatrix}$$

$$\tilde{N}_1 = \begin{bmatrix} .068 & .027 \\ .05 & .047 \end{bmatrix}$$

The self-tuning version requires the multivariable system to be modelled as

$$\underline{y}(k) = - A_1\, \underline{y}(k-1) - A_2\, \underline{y}(k-2)$$

$$+ B_1\, \underline{u}(k-1) + B_2\, \underline{u}(k-2) + \underline{\xi}(k)$$

where the coefficient matrices are estimated using recursive least squares and the control law is found by solving on-line

$$[I_2 + A_1z^{-1} + A_2z^{-2}][I_2 + F(z^{-1})] - [B_1z^{-1} + B_2z^{-2}][G(z^{-1})]$$

$$= [I_2 + T(z^{-1})]$$

under the transformation indicated earlier. The convergence of the control parameters are shown in Figs. 7.26 to 7.28, where the final result was

$$\tilde{D}_1 = \begin{bmatrix} .225 & .223 \\ .142 & .296 \end{bmatrix}$$

$$\tilde{N}_0 = \begin{bmatrix} -.103 & -.062 \\ -.161 & -.114 \end{bmatrix}$$

$$\tilde{N}_1 = \begin{bmatrix} .071 & .03 \\ .05 & .052 \end{bmatrix}$$

We note that the difference between the off-line design and the self-tuning simulation is manageable (maximum difference is only 10.6%).

Example 5

It is required to design a deadbeat self-tuning regulator for the system described by a state-space form

$$\underline{x}(k+1) = \begin{bmatrix} 1.8 & 1 \\ -.8 & 0 \end{bmatrix}\underline{x}(k) + \begin{bmatrix} 1 \\ -.4 \end{bmatrix}u(k) + \underline{w}(k)$$

$$y(k) = [1 \quad 0]\underline{x}(k) + v(k)$$

where $v(k) \sim N[0, \mu^2]$; $\mu^2 = .25$; $\underline{w}(k) \sim N[0,Q]$; $Q = \text{diag}[.25 \quad .25]$ and $\underline{x}(0) \sim [0,I_2]$.

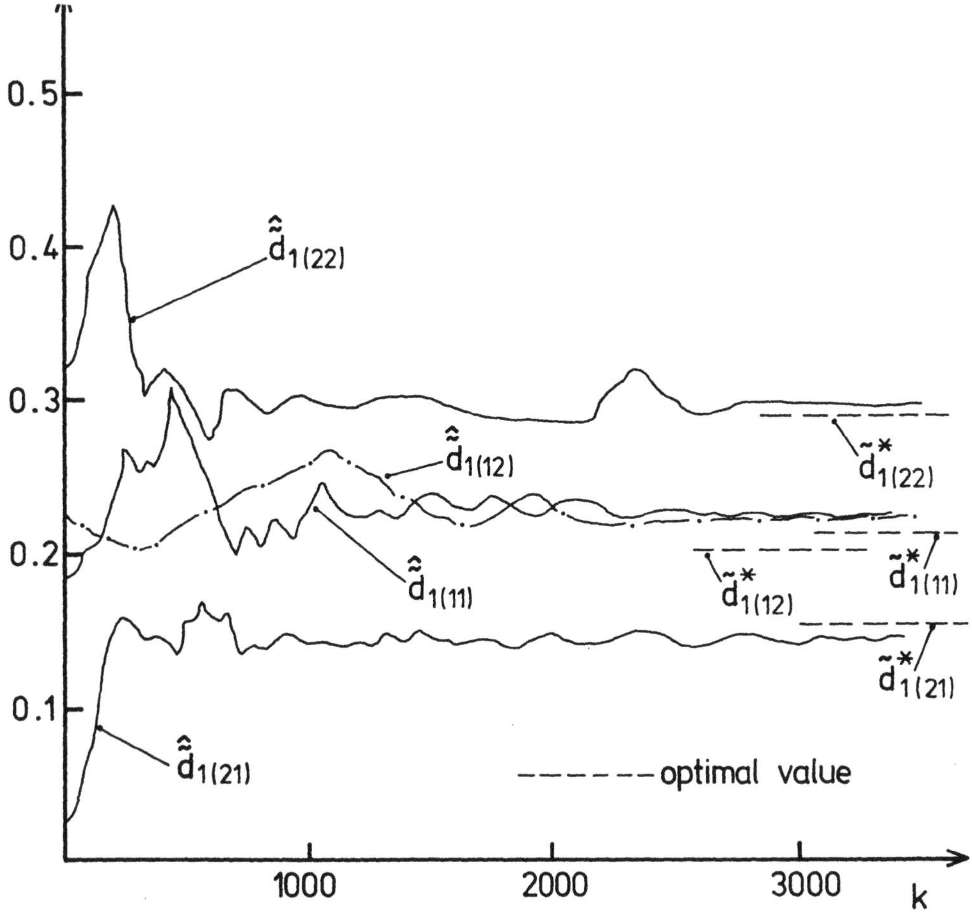

Fig. (7.26) Evolution of controller parameters $\tilde{D}_1$ (Example 4)

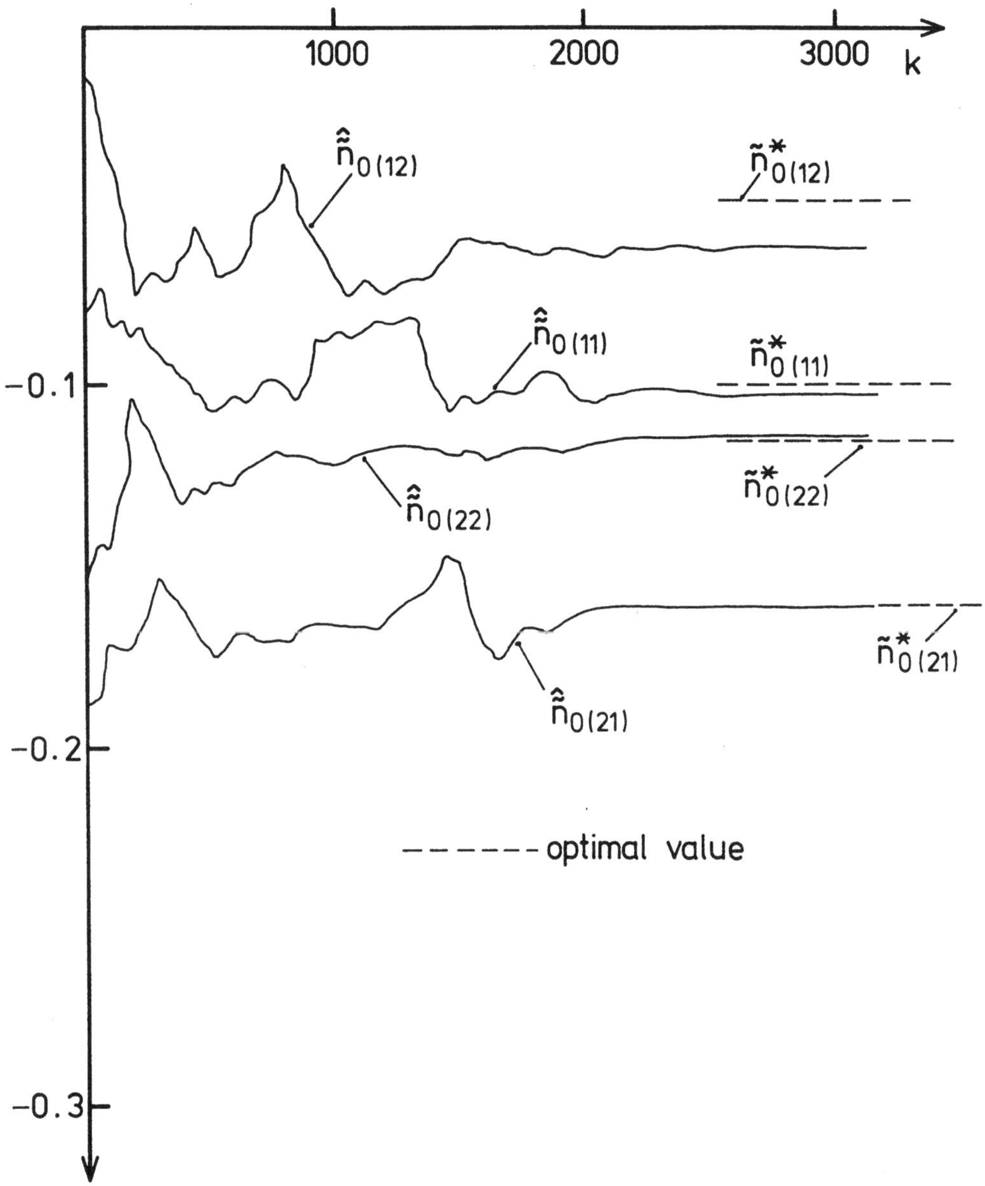

Fig. (7.27) Evolution of controller parameters $\tilde{N}_0$ (Example 4)

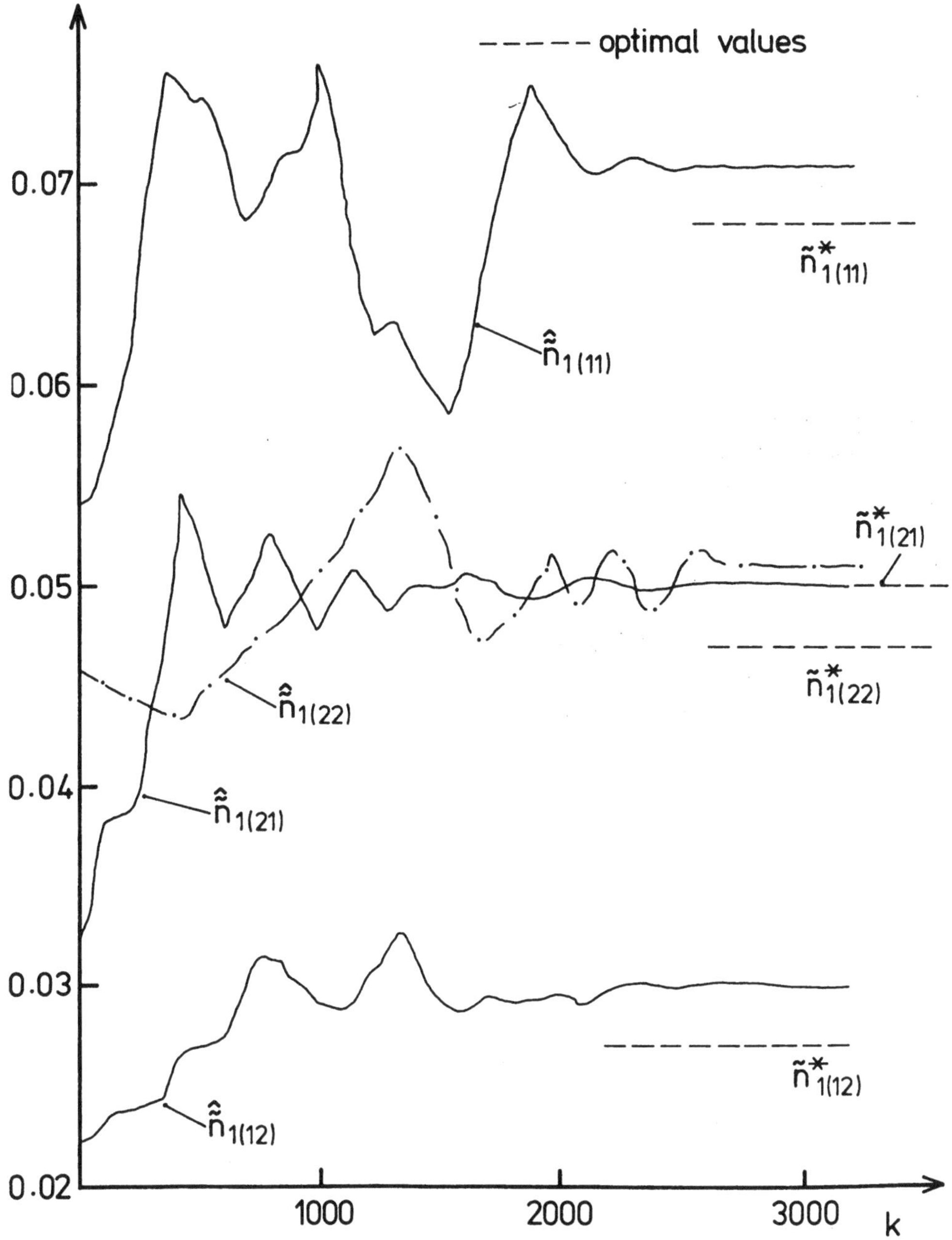

Fig. (7.28) Evolution of controller parameters $\tilde{N}_1$ (Example 4)

To put the above system into a state-space innovations form, we need to compute the stationary values of the error variance σ^2 and the Kalman gain $\underline{g}$. Since $\sigma^2(k) = \underline{c}^t P(k)\underline{c} + \mu^2$, then iterating on (7.171) using $P(0) = I_2$, we arrive at

$$P = \lim_{k\to\infty} P(k) = \begin{bmatrix} 1.0633 & -.253 \\ -.253 & .378 \end{bmatrix}$$

$$\sigma^2 = 1.3133$$

$$\underline{g} = \lim_{k\to\infty} \underline{g}(k) = \begin{bmatrix} 1.2647 \\ -.6477 \end{bmatrix}$$

It is significant to note that the problem at hand is in the observability canonical form (7.175) and there is no need to compute the transformation matrix H. More importantly, the parameter vector $\underline{\theta}$ of (7.179b) can be readily computed as

$$\underline{\theta} = [a_1 \quad a_2 \quad b_1 \quad b_2 \quad c_1 \quad c_2]^t$$

$$= [-1.8 \quad .8 \quad 1 \quad -.4 \quad -.5353 \quad .1523]^t$$

which in this case represents the exact value, (system with known parameters).

For deadbeat self-tuning design, the closed-loop eigenvalues of the state-feedback part are assigned at $z = 0, 0$. We now test the algorithm developed in part C, Section 7.4.5, using the information generated so far about the system plus the initial data

$$\underline{\theta}(0) = [-.9 \quad .4 \quad .5 \quad -.2 \quad -.3 \quad .1]^t$$

The result of simulating the recursive estimation algorithm (7.180a)-(7.180d) is displayed in Figures 7.29 - 7.31 . At $k = 1000$, it was recorded that

$$\underline{\theta} = [-1.8353 \quad .8329 \quad .9967 \quad -.431 \quad -.5361 \quad .1652]^t$$

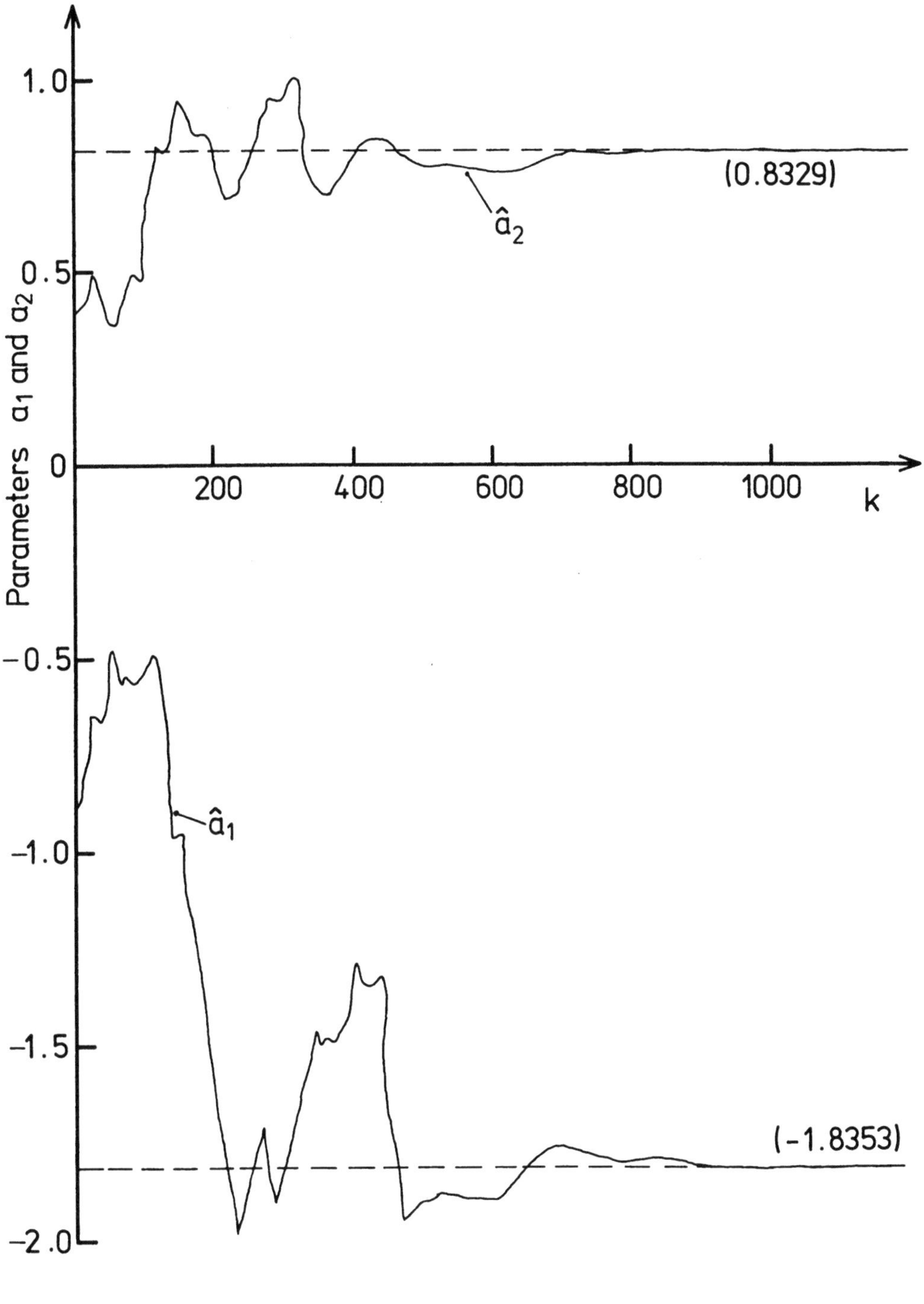

Fig. (7.29) Simulation of parameters a_1 and a_2

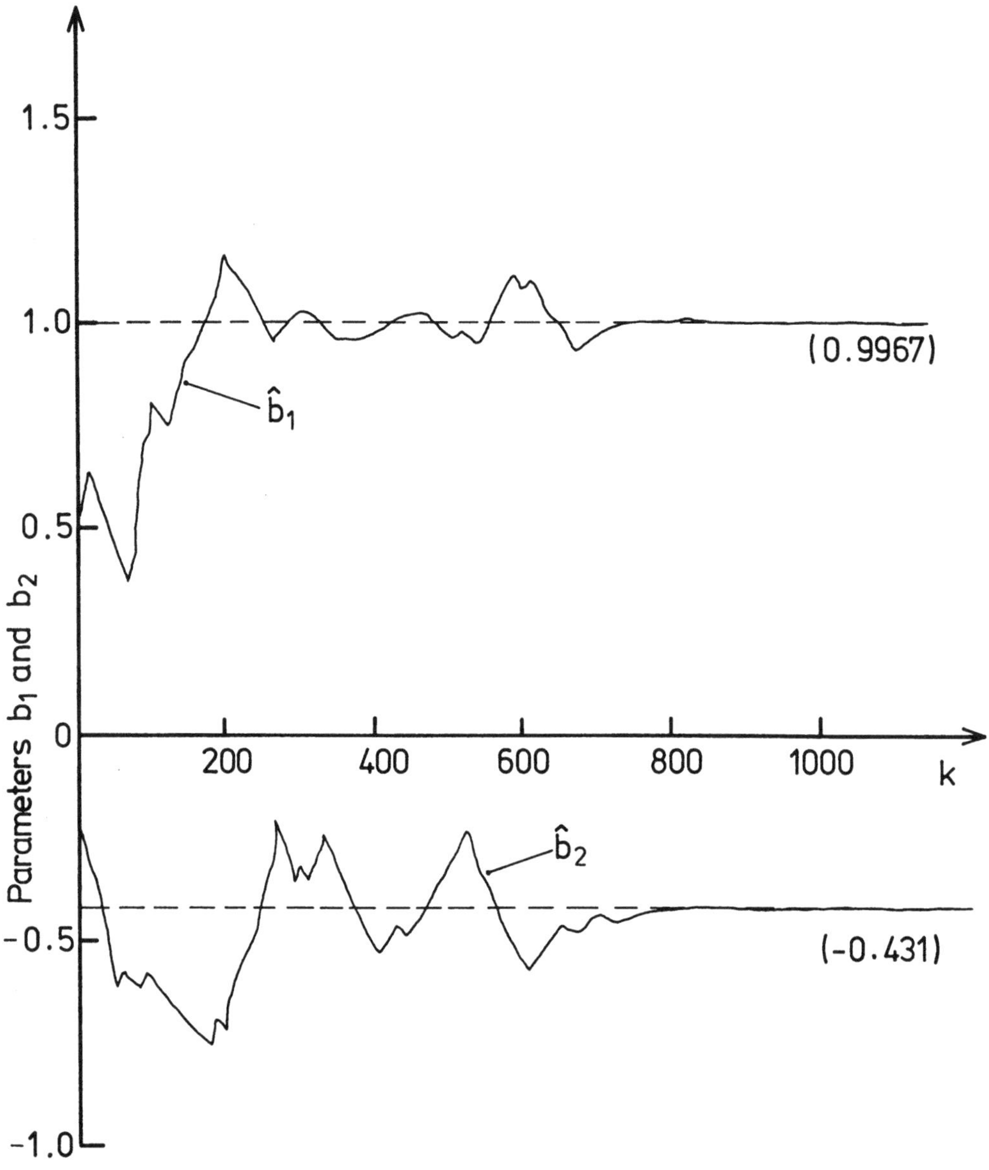

Fig. (7.30) Simulation of parameters b_1 and b_2

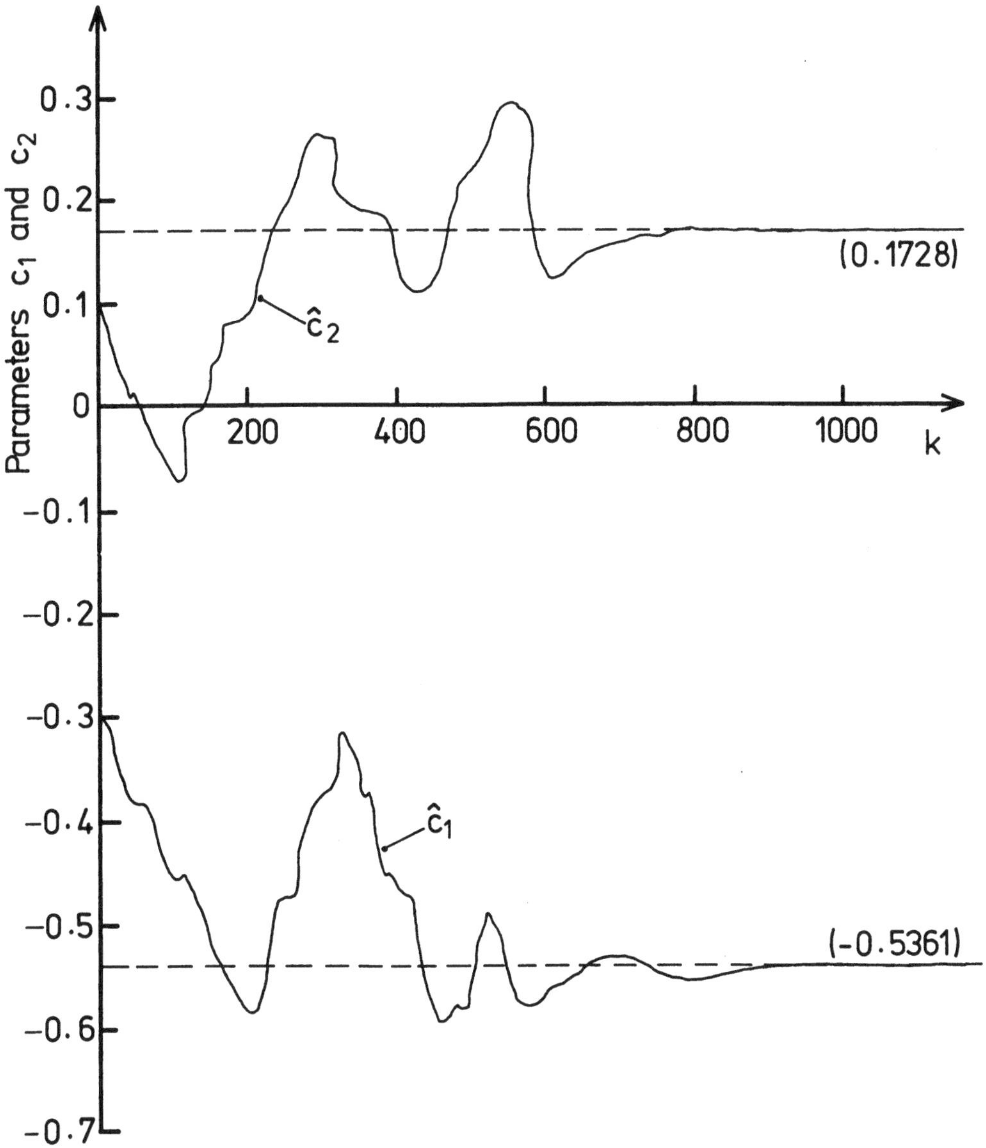

Fig. 7.31) Simulation of parameters c_1 and c_2

The different design methods have been demonstrated on typical examples. The solution of these examples relies on the availability of computer programs for the design algorithms. We must emphasize that the bulk of these programs are standard subroutines in every computer library.

7.5 Concluding Remarks

In this chapter we have examined the principal approaches to adaptive control for discrete dynamical systems. The two main classes which have been considered are (1) Model Reference Adaptive Systems (MRAS) and (2) Self Tuning Regulators. We have described the current state of the art in each of the two areas and we have examined the similarities and differences between the two approaches. The different design methods have been demonstrated on typical examples. The solution of these examples relies on the availability of computer programs for the design algorithms. We must emphasize that the bulk of these programs are standard subroutines in every computer library.

7.6 Problems

1. Consider the multivariable system

$$\underline{y}(k) = -A(z)\underline{y}(k) + B(z^{-1})\underline{u}(t-d) + C(z^{-1})\underline{\xi}(k)$$

where $d = 2$,

$$A(z^{-1}) = \begin{bmatrix} -1.5z^{-1}+.54z^{-2} & .3z^{-1}-.1z^{-2} \\ .2z^{-1}+.1z^{-2} & -1.5z^{-1}+.56z^{-2} \end{bmatrix}$$

$$B(z^{-1}) = \begin{bmatrix} 2-1.8z^{-1} & -.3+.2z^{-1} \\ .1-.1z^{-1} & 1-.2z^{-1} \end{bmatrix}$$

$$C(z^{-1}) = \begin{bmatrix} 1+.2z^{-1}-.48z^{-2} & .1z^{-1}-.2z^{-2} \\ -.1z^{-1}+.2z^{-2} & 1+.2z^{-1}-.24z^{-2} \end{bmatrix}$$

and $\underline{\xi}(k)$ is a white noise vector sequence with zero mean and variance $R = \text{diag}[.0025 \quad .0025]$. It is required to design a pole-assignment self-tuner such that the closed-loop system has eigenvalues placed at $z = .4$ and $z = .3$.

2. Design a self-tuning regulator for the non-minimum phase system

$$y(k) - .95y(k-1) = u(k-2) + 2u(k-3) + \xi(k) - .7\xi(k-1)$$

It is advisable to use a recursive maximum likelihood estimator coupled with a MV controller.

3. For the non-minimum phase system:

$$\underline{x}(k+1) = \begin{bmatrix} 2 & 1 & 0 \\ -1.5 & 0 & 1 \\ .5 & 0 & 0 \end{bmatrix} \underline{x}(k) + \begin{bmatrix} 1 \\ -3.5 \\ 1.5 \end{bmatrix} u(k) + \underline{w}(k)$$

$$y(k) = [1 \quad 0 \quad 0] \; \underline{x}(k) + v(k)$$

where $\sigma_v^2 = \mu^2 = .04, \quad E[\underline{x}(0)] = \underline{0}$

$$E\{[\underline{x}(0)-\hat{\underline{x}}(0)][\underline{x}(0)-\hat{\underline{x}}(0)]^t\} = I_3,$$

$$E[\underline{w}(k) \; \underline{w}^t(k)] = \mathrm{diag}[.04 \quad .04 \quad .04],$$

design a self-tuning control scheme to place the closed-loop eigenvalues at $z = .2$, $.4$ and $.6$. Evaluate your results.

4. Apply the analysis of Section 7.4.6 to a plant whose discrete transfer function (before a parameter change takes place) is characterised by:

$$H_p(z^{-1}) = \frac{z^{-2}(1+.4z^{-1})}{(1-.5z^{-1})(1-.8z^{-1})(1+.8z^{-1})}$$

The reference model is described by:

$$H_r(z^{-1}) = \frac{z^{-2}(.3+.2z^{-1})}{(1-.5z^{-1})(1-.7z^{-1})(1+.7z^{-1})}$$

When a change of the plant parameters occurs, the plant is characterised by:

$$\hat{H}_p(z^{-1}) = \frac{z^{-2}(.9+.5z^{-1})}{(1-.5z^{-1})(1-.9z^{-1})(1+.9z^{-1})}$$

In regulation, take the **R**-polynomial as (a) 1, (b) $(1-.4z^{-1})^3$, (c) $(1-.3z^{-1})(1-.5z^{-1})$. Give some interpretation of the results.

5. With reference to Table 7.1, consider the case when the

parallel adjustable system is described by

$$y(k) = \underline{p}^t(k)\ \underline{f}(k-1) + \underline{k}^t\ \underline{e}(k-1)$$

$$= \underline{p}^t(k)\ \underline{f}(k-1) + \sum_{j=1}^{n} k_j\ e_j(k-1)$$

Derive the adaptation algorithms with constant gain and with decreasing summation gain.

6. Apply the MRAC schemes of Section (7.3.2) to identify the parameters of a system whose simulated model is given by:

$$y(k) = -1.3y(k-1) + .2y(k-2) + .8y(k-3) + .3y(k-4) + u(k)$$

7. Experiment with the adaptive schemes in simulating the plant

$$y(k+2) + a_1y(k+1) + a_2y(k) = b_0\ u(k)$$

where the reference model is given by

$$y^m(k+2) + .5y^m(k+1) + .6y^m(k) = 2u(k)$$

7.7 References

[1] Donalson, D.D. and F.M. Kishi
"Review of Adaptive Control System Theories and Techniques", in Modern Control Systems Theory : edited by C.T. Leondes, vol. 2, McGraw-Hill, N.Y. pp.228-284, 1965.

[2] Tsypkin, Y.Z.
"Adaptation and Learning in Automatic Systems", Academic Press, N.Y., 1971.

[3] Davies, W.D.T.
"System Identification for Self-Adaptive Control", Wiley, N.Y., 1970.

[4] Landau, I.E.
"A Survey of Model Reference Adaptive Techniques: Theyory and Applications", Automatica, vol. 10, pp. 353-379, 1974.

[5] Astrom, K.J., U. Borisson, L. Ljung and B. Wittenmark
"Theory and Applications of Self-Tuning Regulators", Automatica, vol. 13, pp. 457-476, 1977.

[6] Landau, I.D.
"Adaptive Control : The Model Reference Approach", Marcel Dekker, Inc., N.Y., 1979.

[7] Astrom, K.J. and P. Eykoff
"System Identification : A Survey", Automatica, vol. 7, pp. 123-162, 1972.

[8] Landau, I.D.
"Design of Discrete Model Reference Adaptive Systems Using the Positivity Concept", Proceedings of the Third IFAC Symposium on Sensitivity, Adaptivity and Optimality Italy, pp. 307-314, 1973.

[9] Egardt, B.
"Unification of Some Discrete-Time Adaptive Control Schemes", IEEE Trans. Automat. Control, vol. AC-25, pp. 693-697, 1980.

[10] Popov, V.M.
"Hyperstability of Control Systems",
Springer-Verlag, N.Y., 1973.

[11] Serdyukov, V.A.
"Synthesis of a Generalized Adaptation Algorithm for a Non-search Self-adjusting System with Reference Model", Autom. Remote Control, vol. 31, pp. 1078-1084, 1970.

[12] Pearson, A.E. and V.R. Vanguir
"A Synthesis Procedure for Parameter Adaptive Control Systems", IEEE Trans. Autom. Control, vol. AC-16, pp. 440-449, 1971.

[13] Lindorff, D.P. and R.L. Caroll
"Survey of Adaptive Control Using Lyapunov Design", Int. J. Control, vol. 18, pp. 897-914, 1973.

[14] Narendra, K.S. and P. Kudva
"Stable Adaptive Schemes for System Identification and Control : Parts I and II", IEEE Trans. System, Man and Cybernetics, vol. SMC-4, pp. 542-560, 1974.

[15] Ionescu, T. and R. Monopoli
"Discrete Model Reference Adaptive Control with an Augmented Error Signal", Automatica, vol. 13, pp. 507-518, 1977.

[16] Landau, I.D.
"Unbiased Recursive Identification Using Model Reference Adaptive Techniques", IEEE Trans. Autom. Control, vol. AC-21, pp. 200-206, 1976.

[17] Lozano, R. and I.D. Landau
"Redesign of Explicit and Implicit Discrete Time Model Reference Adaptive Control Schemes", Int. J. Control, vol. 33, pp. 247-268, 1981.

[18] Landau, I.D. and R. Lozano
"Unification of Discrete Time Explicit Model Reference Adaptive Control Designs", Automatica, vol. 17, pp. 593-611, 1981.

[19] Subbayyan, R. and R. Nagarajan
"A Discrete-Time Model Reference Adaptive Control Scheme with Reduced On-line Computations", Int. J. Control, vol. 30, pp. 121-127, 1979.

[20] Bellman, R.E.
"Matrix Analysis", McGraw-Hill, N.Y., 1970.

[21] Anderson, B.D.O. and S. Vongpanitlerd
"Network Analysis and Synthesis : A Modern Systems Approach", Prentice-Hall, N.J., 1972.

[22] Hitz, L. and B.D.O. Anderson
"Discrete Positive Real Functions and Their Application to System Stability", Proc. IEEE, vol. 116, pp. 153-155, 1969.

[23] Ionescu, T.
"Hyperstability of Linear Time-varying Discrete Systems", IEEE Trans. Autom. Control, vol. AC-15, pp. 645-647, 1970.

[24] Erzberger, H.
"Analysis and Design of Model-Following Systems by State-Space Techniques", Proc. JACC, Ann Arbor, pp. 572-581, 1968.

[25] Chen, Y.T.
"Perfect Model Following with a Real Model", Proc. JACC, Paper 10-5, 1973.

[26] Morse, A.S.
"Structure and Design of Linear Model Following Control Systems", IEEE Trans. Autom. Control, vol. AC-18, pp. 346-354, 1973.

[27] Wolovich, W.A.
"The Use of State Feedback for Exact Model Matching", SIAM J. Control, vol. 10, pp. 512-523, 1972.

[28] Wang, S.H. and C.A. Desoer
"The Exact Model Matching of Linear Multivariable Systems", IEEE Trans. Autom. Control, vol. AC-17, pp. 347-349, 1972.

[29] Schneider, H. and G.P. Barner
"Matrices and Linear Algebra", Holt, Rienhart and Winston, N.Y., 1968.

[30] Landau, I.D. and B. Courtiol
"Design of Multivariable Adaptive Model-following Control Systems", Automatica, vol. 10, pp. 483-494, 1974.

[31] Bethoux, G. and B. Courtiol
"A Hyperstable Discrete Model Reference Adaptive Control System", Proceedings of the Third IFAC Symposium on Sensitivity, Adaptivity and Optimality, Italy, pp. 282-289, 1973.

[32] Borison, U.
"Self-Tuning Regulators for a Class of Multivariable Systems", Automatica, vol. 15, pp. 209-215, 1979.

[33] Clarke, D.W. and P.J. Gawthrop
"Self-Tuning Controller", Proc. IEE, vol. 122, pp. 929-934, 1975.

[34] Gawthrop, P.J.
"Some Interpretations of the Self-tuning Controller", Proc. IEE, vol. 124, pp. 889-894, 1977.

[35] Wellstead, P.E., J.M. Edmunds, D. Prager and P. Zanker
"Self-Tuning Pole/Zero Assignment Regulators", Int. J. Control, vol. 30, pp. 1-26, 1979.

[36] Wellstead, P.E., D. Prager and P. Zanker
"Pole Assignment Self-Tuning Regulator", Proc. IEE, vol. 126, pp. 781-787, 1979.

[37] Prager, D. and P.E. Wellstead
"Multivariable Pole-Assignment Self-Tuning Regulators", Proc. IEE, vol. 127, pp. 9-18, 1980.

[38] Tsay, Y.T. and L.S. Shieh
"State-Space Approach for Self-Tuning Feedback Control with Pole Assignment", Proc. IEE, vol. 128, pp. 93-101, 1981.

[39] Astrom, K.J. and B. Wittenmark
"On Self-Tuning Regulators", Automatica, vol. 9, pp. 185-199, 1973.

[40] Wittenmark, B.
"Stochastic Adaptive Control Methods : A Survey", Int. J. Control, vol. 21, pp. 705-731, 1975.

[41] Astrom, K.J.
"Introduction to Stochastic Control Theory", Academic Press, N.Y., 1970.

[42] Peterka, V.
"On Steady State Minimum Variance Control Strategy", Kybernetica, vol. 8, pp. 219-231, 1972.

[43] Ljung, L. and I.D. Landau
"Model Reference Adaptive Systems and Self-Tuning Regulators - Some Connections", Proc. 7th IFAC World Congress, Helsinki, pp. 1973-1980, 1978.

[44] Astrom, K.J., B. Westerberg and B. Wittenmark
"Self-Tuning Controllers Based on Pole-Placement Design", Dept. Automat. Contr., Lund Inst. Technol., Lund, Sweden, CODEN:LUTFD2/(TERT-3148)/1-52/(1978), 1978.

[45] Kailath, T.
"Linear Systems", Prentice-Hall, N.J., 1980.

[46] Ljung, L.
"Convergence Analysis of Parameter Identification Methods", IEEE Trans. Autom. Contr., vol. AC-25, pp. 770-783, 1978.

[47] Koivo, H.N.
"A Multivariable Self-Tuning Controller",
Automatica, vol. 16, pp. 351-362, 1980.

[48] Landau, I.D.
"Combining Model Reference Adaptive Controllers and Stochastic Self-Tuning Regulators", Automatica, vol. 18, pp. 77-84, 1982.

[49] Bayoumi, M.M., K.Y. Wong and M.A. El-Bagoury
"A Self-Tuning Regulator for Multivariable Systems",
Automatica, vol. 17, pp. 575-592, 1981.

[50] Allidina, A.Y. and F.M. Hughes
"Self-Tuning Controllers for Deterministic Systems",
Int. J. Control, vol. 37, pp. 831-841, 1983.

[51] Allidina, A.Y., F.M. Hughes and C. Tye
"Self-Tuning Control for Systems Employing Feedforward",
Proc. IEE, vol. 128, pp. 283-291, 1981.

Chapter 8
Dynamic Optimisation

8.1 Introduction

The main task of control engineering design is to ensure that the dynamic system under consideration behaves in some desirable way. In Chapters 4 and 5, we discussed how to design feedback controllers utilizing different schemes. The goal of this chapter is to solve this problem using an alternative technique which is based on optimal control theory *[1-6,23]*.

The problem of optimal control has received a great deal of attention during the past decade owing to increasing demand for systems of high performance and to the ready availability of digital computers. Optimal control methods are generally attractive because they handle multivariable systems easily and allow the designer to quickly determine many good candidate values for the feedback control matrix.

We will formulate the optimal control problem of discrete systems in a general setting first and then specialise it to linear system with quadratic performance criteria. Later we will extend the dynamic optimization results to deal with large-scale control problems.

8.2 The Dynamic Optimisation Problem

We shall study the dynamic optimization problem in which it is required to select a sequence of control variables which produce, via a set of difference equations, the sequence of state variables and at the same time to minimize (or maximise) a prescribed cost functional. The cost functional is a scalar quantity depending on the control and input sequences. It is formulated so as to reflect the design specifications on the performance of the system.

8.2.1 *FORMULATION OF THE PROBLEM*

Consider the dynamic system described by the nonlinear difference equations:

$$\underline{x}(k+1) = \underline{f}_k[\underline{x}(k), \underline{u}(k)] \quad ;$$
$$\underline{x}(0) = \underline{x}_0 \quad , \quad k = 0, \ldots, N-1 \tag{8.1}$$

This system, shown in Figure 8.1, can be viewed as a sequential set of equality constraints, where $\underline{x}(k)$, a sequence of n-vectors, is determined by $\underline{u}(k)$, a sequence of m-vectors.

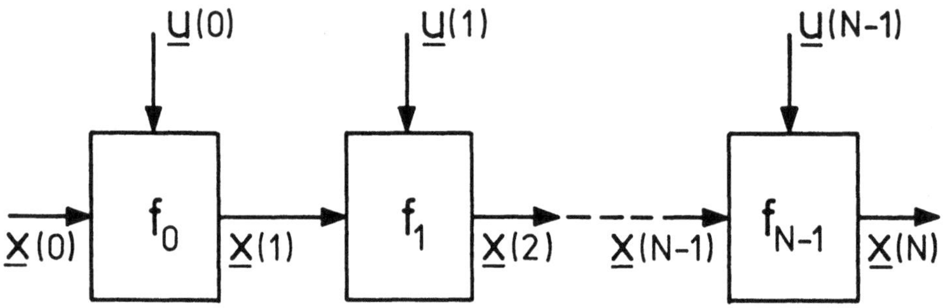

Fig. (8.1) Block-diagram of sequential discrete system

We note that 8.1 is an appropriate form of (3.81) where f_k .,. represents the kth value of $\underline{g}$.,.,. . For the system (8.1), we consider a cost functional of the form

$$J = L[\underline{x}(N)] + \sum_{j=0}^{N-1} G_j[\underline{x}(j), \underline{u}(j)] \tag{8.2}$$

where the first term, called final or terminal function, shows the dependence of the cost functional on the terminal state and the quantity under summation, $G_j[\cdot,\cdot]$, called intermediate function, shows the dependence of the functional on the control and state sequences.

The dynamic optimization problem is to find the sequence $\underline{u}(k)$ that minimizes J. We can write (8.1) as:

$$-\underline{x}(k+1) + \underline{f}_k[\underline{x}(k),\underline{u}(k)] = \underline{0} \quad ; \quad k = 0,\ldots,N-1 \qquad (8.3)$$

such that the dynamic optimization problem at hand reduces to a standard constrained-minimization problem *[5,6]*, which can be solved by different mathematical techniques, the most appealing of which is the method of Lagrange multipliers.

8.2.2 *CONDITIONS OF OPTIMALITY*

In the method of Lagrange multipliers *[5,6]*, there will be a Lagrange multiplier vector, $\underline{\pi}(k+1)$, for each value of k. The procedure is to adjoin (8.3) to J with the multiplier sequence as:

$$J_a = L[\underline{x}(N)] + \sum_{j=0}^{N-1} [G_j[\underline{x}(j),\underline{u}(j)] + \underline{\pi}^t(j+1)\{\underline{f}_j[\underline{x}(j),\underline{u}(j)] - \underline{x}(j+1)\}] \qquad (8.4)$$

and attempt to find the minimum of J_a which will correspond to that of J, with respect to $\underline{x}(k)$, $\underline{u}(k)$ and $\underline{\pi}(k)$. The argument of $\underline{\pi}$ is nominally arbitrary, but we will take it to be (j+1) because this choice will yield a particularly easy form of the equations later on. For convenience, we define a scalar sequence, often called the Hamiltonian, H_j :

$$H_j = G_j[\underline{x}(j),\underline{u}(j)] + \underline{\pi}^t(j+1)\underline{f}_j[\underline{x}(j),\underline{u}(j)] \quad ; \quad j = 0,\ldots,N-1 \qquad (8.5)$$

The use of (8.5) in (8.4) and changing the indices of summation yields:

$$J_a = L[\underline{x}(N)] - \underline{\pi}^t(N)\underline{x}(N) + \sum_{j=1}^{N-1} [H_j - \underline{\pi}^t(j)\underline{x}(j)] + H_0 \tag{8.6}$$

Now consider differential changes in J_a due to differential changes in $\underline{u}(j)$:

$$\begin{aligned} dJ_a = {} & [\partial L/\partial \underline{x}(N) - \underline{\pi}^t(N)]d\underline{x}(N) \\ & + \sum_{j=1}^{N-1} \{[\partial H_j/\partial \underline{x}(j) - \underline{\pi}^t(j)]d\underline{x}(j) \\ & + \partial H_j/\partial \underline{u}(j)\ d\underline{u}(j) + \partial H_0/\partial \underline{x}(0)\ d\underline{x}(0) \\ & + \partial H_0/\partial \underline{u}(0)\ d\underline{u}(0) \end{aligned} \tag{8.7}$$

We choose the multiplier sequence $\underline{\pi}(j)$ so as to obtain:

$$\underline{\pi}^t(j) - \partial H_j/\partial \underline{x}(j) = \underline{0}$$

which, with the aid of (8.5), reduces to

$$\underline{\pi}^t(j) = \partial G_j/\partial \underline{x}(j) + \underline{\pi}^t(j+1)\partial f_j/\partial \underline{x}(j) \ ; \quad j = 0,\dots,N-1 \tag{8.8a}$$

which are often called adjoint or costate equations. The boundary condition on (8.8a) is:

$$\underline{\pi}^t(N) = \partial L/\partial \underline{x}(N) \tag{8.8b}$$

The substitution of (8.8a) and (8.8b) into (8.7) results in

$$\begin{aligned} dJ_a & = \sum_{j=0}^{N-1} \partial H_j/\partial \underline{u}(j)d\underline{u}(j) + \underline{\pi}^t(0)d\underline{x}(0) \\ & \equiv \sum_{j=0}^{N-1} \partial H_j/\partial \underline{u}(j)d\underline{u}(j) \end{aligned} \tag{8.9}$$

since the second term vanishes because $\underline{x}(0)$ is given. From (8.9), the quantity $\partial H_j/\partial \underline{u}(j)$ is the gradient of J with respect to $\underline{u}(j)$ while satisfying (8.1). An extremum point requires that dJ_a be zero for arbitrary $d\underline{u}(j)$. This can only happen if we have:

$$\partial H_j/\partial \underline{u}(j) = \underline{0} \quad , \quad j = 0,\ldots,N-1 \qquad (8.10a)$$

Now, using (8.5) in (8.10a) it becomes:

$$\partial G_j/\partial \underline{u}(j) + \underline{\pi}^t(j+1)\partial f_j/\partial \underline{u}(j) = \underline{0} \quad ,$$

$$j = 0,\ldots,N-1 \qquad (8.10b)$$

In summary, to solve the dynamic optimization problem (8.1) and (8.2) we have to find a control sequence $\{\underline{u}(j)\}$ that produces a stationary value of the cost functional J. This, in turn, amounts to solving the following difference equations:

<u>for k = 0,...,N-1</u>

(a) $$\underline{x}(k+1) = f_k[\underline{x}(k),\underline{u}(k)] \quad ; \qquad \underline{x}(0) = \underline{x}_0 \qquad (8.11a)$$

(b) $$\underline{\pi}(k) = [\partial f_k/\partial \underline{x}(k)]^t \underline{\Pi}(k+1) + [\partial G_k/\partial \underline{x}(k)]^t \quad ; \qquad \underline{\pi}(N) = [\partial L/\partial \underline{x}(N)]^t \qquad (8.11b)$$

(c) $$[\partial G_k/\partial \underline{u}(k)] + \underline{\pi}^t(k+1)[\partial f_k/\partial \underline{u}(k)] = \underline{0} \qquad (8.11c)$$

It should be emphasized that the boundary conditions for (8.11a)-(8.11c) are split, that is, they are partly given for k = 0 and partly for k = N. For this reason, such problems are frequently called two-point boundary-value problems.

We note that (8.13) and (8.14) are coupled and in general they are rather difficult to solve. Finally, to ensure that a local minimum is attained, we require that the second-order expression for dJ w.r.t. its arguments must be nonnegative. This corresponds to the positive-semidefiniteness of the matrix

$$\left[\begin{array}{c|c} \partial^2 H_j/\partial \underline{x}^2(j) & \partial^2 H_j/\partial \underline{x}(j)\partial \underline{u}(j) \\ \hline \partial^2 H_j/\partial \underline{u}(j)\partial \underline{x}(j) & \partial^2 H_j/\partial \underline{u}^2(j) \end{array}\right]$$

Next, we present an alternative procedure to solve the dynamic optimization problem.

8.2.3 *THE OPTIMAL RETURN FUNCTION*

For the deterministic multistage system (8.1) suppose that the state $\underline{x}(m)$ at the mth stage is known. Then it follows directly from the structure of (8.1) that, given the sequence of controls $\underline{U}[m,r-1] = \{\underline{u}(m),\underline{u}(m+1),\dots,\underline{u}(r-1)\}$ together with $\underline{x}(m)$, the state of the rth stage $\underline{x}(r)$ is uniquely determined. This statement is often called *the Principle of Causality*, which is a fundamental property of multistage systems *[4,5,49]*. The importance of this principle is that it implies the existence of a function

$$F[\underline{x}(m),\underline{U}[m,r-1],m]$$

such that

$$\underline{x}(r) = F[\underline{x}(m),\underline{U}[m,r-1],m,r] \tag{8.12}$$

F is frequently called the transition function. It simply summarizes the effect of the initial state $\underline{x}(0)$ and the control sequences $\underline{U}[0,N-1]$ in uniquely determining the trajectory $\underline{X}[1,N] = \{\underline{x}(1),\underline{x}(2),\dots,\underline{x}(N)\}$. Consequently, the cost functional J in (8.2) can be written as a function of $\underline{x}(0)$ and $\underline{U}[0,n-1]$, that is, there exists a function $V\{\underline{x}(0),\underline{U}[0,N-1]\}$ such that

$$J = V\{\underline{x}(0),\underline{U}[0,N-1]\} \tag{8.13}$$

whose interpretation is simple; it is necessary only to

determine $\underline{U}[0,N-1]$ to minimize J provided $\underline{x}(0)$ is specified. V is usually called the return function.

Now consider the situation in which we have been able to determine the control profile $\underline{U}[0,k-1]$ in an optimal manner. Then from the above discussion we see that $\underline{X}[0,k]$ can also be determined. Using this observation, the cost functional J can be split into:

$$J_k = \sum_{j=0}^{k-1} G_j[\underline{x}(j),\underline{u}(j)] \tag{8.14a}$$

$$J_{N-k} = L[\underline{x}(N)] + \sum_{j=k}^{N-1} G_j[\underline{x}(j),\underline{u}(j)] \tag{8.14b}$$

such that $J = J_k + J_{N-k}$. This is an intuitively appealing presentation of Bellman's principle of optimality which can be stated as follows *[3-6]*:

> *An optimal sequence of controls in a multistage optimization problem has the property that whatever the initial stage, state and controls are, the remaining controls must constitute an optimal sequence of decisions for the remaining problem with stage and state resulting from the previous controls considered as initial conditions.*

We now demonstrate the use of the optimality and causality principles. First, from (8.2) and (8.13) in the light of the causality principle, we can write:

$$V\{\underline{x}(k),\underline{U}[k,N-1]\} = L[\underline{x}(N)] + \sum_{j=k}^{N-1} G_j[\underline{x}(j),\underline{u}(j)] \tag{8.15a}$$

in which, using the principle of optimality, $\underline{U}[k,N-1]$ must be chosen to minimize F. The optimal return function V^* is obtained when the optimal control sequence $\underline{U}^*[k,N-1]$ is substituted in the above relation. The method of constructing the optimal return function is called dynamic programming *[3-6]*. To explain its use in solving dynamic optimization problems consider that

(a) $\underline{u}[\underline{x}(j)]$; $j = k+1,\ldots,N-1$ is an arbitrary control law (control sequence as a function of the state).

(b) $\underline{X}[k,N]$ is chosen such that

$$\underline{x}(m+1) = f_m\{\underline{x}(m),\underline{u}[\underline{x}(m)]\} \quad ; \quad m = k,\ k+1,\ldots,N-1 \tag{8.15b}$$

It is readily evident from (8.15a) and (a) above that the return function must satisfy

$$V[\underline{x}(k)] = G_k\{\underline{x}(k),\underline{u}[\underline{x}(k)]\} + V[\underline{x}(k+1)] \tag{8.15c}$$

where $\underline{x}(k+1)$ is given by (8.15b). Note that (8.15c) is a backward difference equation with the final value $V[\underline{x}(N)] = L[\underline{x}(N)]$. Once an optimal control law is found, one can then use it in (8.15c) and by backward solution the optimal return function can be constructed. To elaborate on this point, recall from (8.15c) using (8.15b) that $V^*[\underline{x}(k+1)]$, the optimal return function, can be determined provided that the optimal control law $u^*[\underline{x}(j)]$, $j = k+1,\ldots,N-1$, has somehow been found. Let us define a return function, $E[\underline{x}(k),\underline{u}(k)]$, as the return obtained from an initial stage k and state $\underline{x}(k)$ by utilizing $\underline{u}(k)$ followed by $\underline{U}^*[k+1,N-1]$, that is:

$$E[\underline{x}(k),\underline{u}(k)] = G_k[\underline{x}(k),\underline{u}(k)] + V^*[\underline{x}(k+1)] \tag{8.15d}$$

Direct application of the principle of optimality entails

$$\underline{u}^*[\underline{x}(k)] = \arg\min_{\underline{u}(k)} \{E[\underline{x}(k),\underline{u}(k)]\} \tag{8.15e}$$

The procedure may now be summarized as follows:

1. Set
$$V^*[\underline{x}(N)] = L[\underline{x}(N)]$$

2. Given $V^*[\underline{x}(k+1)]$, select $\underline{u}^*[\underline{x}(k)]$ from (8.15e).

3. Given $\underline{u}^*[\underline{x}(k)]$ and $V^*[\underline{x}(k+1)]$, construct $V^*[\underline{x}(k)]$ from (8.15b) and (8.15c).

4. Set $k = k-1$. If $k > 0$ go to Step 2, otherwise terminate.

The continuous version of (8.15c) or (8.15d) is frequently called the Hamilton-Jacobi equation.

It should be emphasized that:

(1) The backward difference equation (8.15c) cannot be easily solved in general. However, when it can, the optimal control law $\underline{u}^*[\underline{x}(k)]$ is determined.

(2) The method can incorporate inequality constraints easily *[3-6]*.

For more detailed treatment of dynamic optimization the reader is referred to *[1-12]* where computational methods have also been discussed.

We next consider a special case, i.e. linear-quadratic discrete regulators.

8.3 Linear-Quadratic Discrete Regulators

We shall study the problem of choosing the control sequence $\{\underline{u}(k)\}$ in

$$\begin{aligned} \underline{x}(k+1) &= A\underline{x}(k) + B\underline{u}(k) \;; \quad \underline{x}(0) = \underline{x}_0 \\ \underline{y}(k) &= C\underline{x}(k) \end{aligned} \tag{8.16}$$

so as to minimize the cost functional

$$J_N = \frac{1}{2}\,\underline{x}^t(N)Q_f\underline{x}(N) + \frac{1}{2}\sum_{j=0}^{N-1}\{\underline{y}^t(j)Q\underline{y}(j) + \underline{u}^t(j)R\underline{u}(j)\} \tag{8.17}$$

where the symmetric weighting matrices Q and R are to be selected by the designer who bases his choice on the relative importance of the various states and controls. Some weight will almost always be placed on the control (det[R] $\neq$ 0), since otherwise the solution will include very large components in the control gains and the states would be driven to zero at a very fast rate. In general, Q, Q_f and R must be non-negative to yield a non-negative value of J for all possible $\underline{x}$ and $\underline{u}$. In fact, it is customary to assume that Q_f and Q are positive semi-definite, whereas R is posinive definite.

8.3.1 *DERIVATION OF THE OPTIMAL SEQUENCES*

A comparison between (8.1), (8.2) and (8.16), (8.17) reveals the following identifications:

$$\left.\begin{aligned} f_k[\underline{x}(k),\underline{y}(k)] &\longrightarrow A\underline{x}(k) + B\underline{u}(k) \\ L[\underline{x}(N)] &\longrightarrow \frac{1}{2}\,\underline{x}^t(N)Q_f\underline{x}(N) \\ G_j[\underline{x}(j),\underline{u}(j)] &\longrightarrow \frac{1}{2}\,\underline{x}^t(j)C^tQC\underline{x}(j) \\ &\qquad + \frac{1}{2}\,\underline{u}^t(j)R\underline{u}(j) \end{aligned}\right\} \tag{8.18}$$

Using (8.18) in (8.11a)-(8.11c) yields the necessary conditions of optimality:

for $k = 0,\ldots,N-1$

$$\underline{x}(k+1) = A\underline{x}(k) + B\underline{u}(k) \quad ; \quad \underline{x}(0) = \underline{x}_0 \tag{8.19}$$

$$\underline{\pi}(k) = A^t\underline{\pi}(k+1) + C^tQC\underline{x}(k) \quad ; \quad \underline{\pi}(N) = Q_f\underline{x}(N) \tag{8.20}$$

$$R\underline{u}(k) + B^t\underline{\pi}(k+1) = \underline{0} \tag{8.21}$$

From (8.21) we get

$$\underline{u}(k) = -R^{-1}B^t\underline{\pi}(k+1) \tag{8.22}$$

The solution to the dynamic optimization problem of the linear system (8.16) with quadratic cost functional (8.17) is now completely specified. It consists of (8.19), (8.20) and (8.22). One method to solve these equations, called the sweep method [5,6], is to assume

$$\underline{\pi}(k) = P(k)\,\underline{x}(k) \qquad (8.23)$$

then (8.22) becomes

$$\begin{aligned}\underline{u}(k) &= -R^{-1}B^tP(k+1)\underline{x}(k+1)\\ &= -R^{-1}B^tP(k+1)[A\underline{x}(k)+B\underline{u}(k)]\end{aligned}$$

Solving for $\underline{u}(k)$ we obtain:

$$\begin{aligned}\underline{u}(k) &= -S^{-1}B^tP(k+1)A\underline{x}(k)\\ &= -G(k)\underline{x}(k) \qquad (8.24)\end{aligned}$$

where

$$S = [R+B^tP(k+1)B] \qquad (8.25a)$$

$$G(k) = (R+B^tP(k+1)B)^{-1}B^tP(k+1)A \qquad (8.25b)$$

Proceeding further, we substitute (8.23) into (8.20) to obtain

$$P(k)\underline{x}(k) = A^tP(k+1)\underline{x}(k+1) + C^tQC\underline{x}(k)$$

which, upon using (8.19), reduces to:

$$P(k)\underline{x}(k) = A^tP(k+1)[A\underline{x}(k)+B\underline{u}(k)] + C^tQC\underline{x}(k) \qquad (8.26)$$

Next we use (8.24) for $\underline{u}(k)$ in (8.26),

$$\begin{aligned}P(k)\underline{x}(k) = {}& A^tP(k+1)A\underline{x}(k) - A^tP(k+1)BS^{-1}B^tP(k+1)A\underline{x}(k)\\ &+ C^tQC\underline{x}(k) \qquad (8.27)\end{aligned}$$

Since (8.27) must hold for arbitrary $\underline{x}(k)$, it follows that

$$P(k) = A^t[P(k+1)-P(k+1)BS^{-1}B^tP(k+1)]A + C^tQC \qquad (8.28)$$

which describes a backward difference equation in P(k). From (8.20) and (8.23), the boundary condition on the sequence {P(k)} is obtained as:

$$P(N) = Q_f \tag{8.29}$$

Fig. 8.2 shows a schematic diagram of the closed-loop system.

This completes the derivation of the optimal sequences for the linear-quadratic, dynamic optimization problem (8.16) and (8.17). Let us now summarize the entire procedure:

1. Set j (iteration index) = N,
   $P(j) = Q_f$ and $G(j) = 0$
2. Compute G(j-1) from (8.25) and store it.
3. Solve (8.28) for P(j-1). Then set j = j-1 and go to Step 2.

When the iteration index reaches the value 0, we stop and use the stored gains {G(k)} together with (8.24) in (8.19) to obtain the optimal state sequence. It should be noted that the optimal feedback gain {G(k)} is time-varying; however, it is independent of the initial state $\underline{x}_0$. Also, we note that the inverse of S exists since R > 0 by assumption and $B^tP(k+1)B$ is a non-negative definite matrix. P(k) is often called the discrete Riccati matrix. Perhaps it would be informative to evaluate the optimal value of the cost function J. To do this, we rewrite (8.17) as:

$$\begin{aligned} J_N = \frac{1}{2} \sum_{j=0}^{N-1} \{&\underline{x}^t(j)C^tQC\underline{x}(j) \\ &+ \underline{u}^t(j)R\underline{u}(j) - \underline{\pi}^t(j+1)\underline{x}(j+1) \\ &+ \underline{\pi}^t(j+1)[A\underline{x}(j) + B\underline{u}(j)]\} \\ &+ \frac{1}{2}\underline{x}^t(N)Q_f\underline{x}(N) \end{aligned} \tag{8.30}$$

where we have used (8.19).

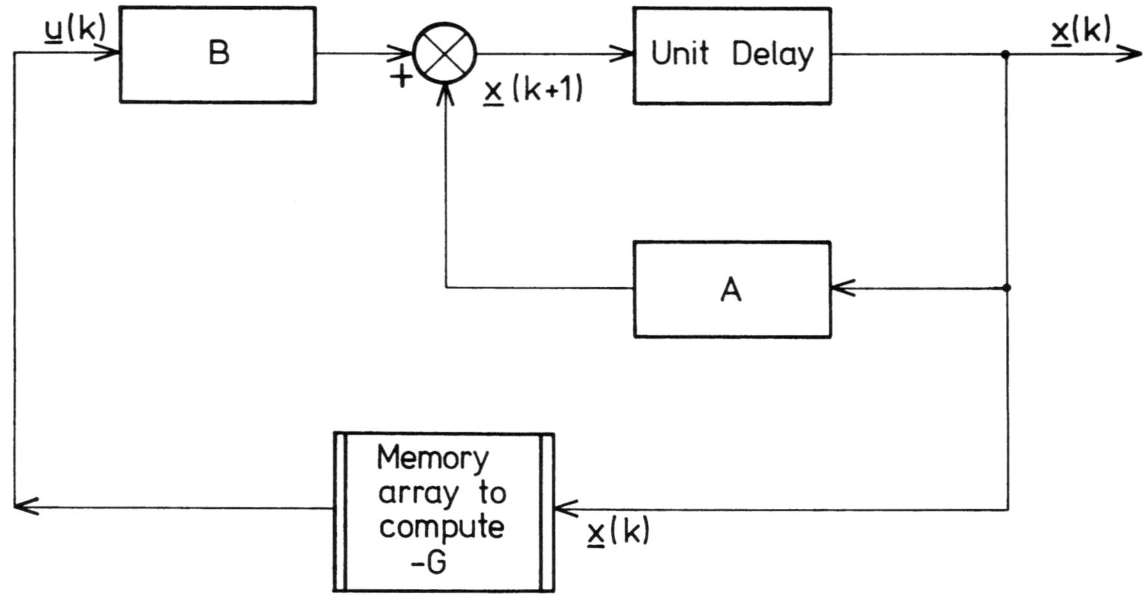

Fig. (8.2) A schematic diagram of the closed-loop regulator

By letting $\underline{\pi}^t(j+1)A = \underline{\pi}^t(j) - \underline{x}^t(j)C^tQC$ from (8.20) and $\underline{\pi}^t(j+1)B = -\underline{u}^t(j)R$ from (8.21), then (8.30) reduces to:

$$J_N = \frac{1}{2}\sum_{j=0}^{N-1} \{\underline{\pi}^t(j)\underline{x}(j) - \underline{\pi}^t(j+1)\underline{x}(j+1)\} + \frac{1}{2}\underline{x}^t(N)Q_f\underline{x}(N)$$

But $\underline{\pi}(N) = Q_f\underline{x}(N)$, hence

$$J_N = \frac{1}{2}\,\underline{\pi}^t(0)\underline{x}(0)$$

$$= \frac{1}{2}\,\underline{x}_0^t\,P(0)\underline{x}_0 \qquad (8.31)$$

which shows that the optimal cost functional depends only on the initial state and initial Riccati matrix. Finally, by transposing (8.28), it is easily verified that $P(k) = P^t(k)$ which means that the Riccati matrix is symmetric. This fact is usually exploited in the numerical computation to reduce the number of unknowns from n^2 to $\frac{1}{2}n(n+1)$.

8.3.2 *STEADY-STATE SOLUTION*

The results of the last section indicate that the optimal gain changes at each time step. This is undesirable in practice due to the difficulty encountered in the realization of this control. In this section we study the case where the control period extends from N to infinity, thus producing the steady-state optimal solution. Before deriving this solution, it is important to recall *[13-15]* that if the system (8.16) is both stabilizable and detectable (see Chapter 3), then the solution P of (8.28), for the infinite-horizon linear regulator, converges to a constant, unique and positive-semidefinite matrix.

Our approach will be based on manipulating (8.19)-(8.22) together and analyzing the result as $N \to \infty$. All the assumptions made in Section 8.3.1 are retained here. If we substitute (8.22) in (8.19) for $\underline{u}(k)$ and arrange the result, we find:

$$M\begin{bmatrix}\underline{x}(k)\\ \underline{\pi}(k)\end{bmatrix} = N\begin{bmatrix}\underline{x}(k+1)\\ \underline{\pi}(k+1)\end{bmatrix} \quad ; \qquad (8.32)$$

and $\underline{x}(0) = \underline{x}_0 \quad ; \quad \underline{\pi}(N) = P(N)\underline{x}(N)$ (8.33)

where

$$M = \begin{bmatrix} A & 0 \\ -C^tQC & I_n \end{bmatrix} \tag{8.34}$$

$$N = \begin{bmatrix} I_n & BR^{-1}B^t \\ 0 & A^t \end{bmatrix} \tag{8.35}$$

It is clear from (8.34) and (8.35) that M and N are invertible if and only if A is nonsingular. For the case when A^{-1} exists, the steady-state solution can be obtained by nonrecursive methods *[16,17]*. We shall consider the general case here and follow closely the approach of Gaalman *[18]*. For this purpose, the treatment starts by analyzing the generalized eigenvalue problem *[20]*:

$$M\underline{v} = \lambda N\underline{v} \tag{8.36}$$

and its reciprocal problem

$$\mu M^t\underline{W} = N^t\underline{w} \tag{8.37}$$

We must emphasize that (8.36) or (8.37) has 2n eigenvalues since M and N are of order (2nx2n). The characteristic polynomials of (8.36), (8.37) can be put in the form *[18]*:

$$p(\lambda) = \alpha_p\lambda^{m_0}(\lambda-\lambda_1)^{m_1}(\lambda-\lambda_2)^{m_2}\ldots(\lambda-\lambda_p)^{m_p} \quad ;$$

$$m_j > 0 \quad \text{for} \quad j = 0,\ldots,p \, , \quad \lambda_j \neq 0 \quad \text{for} \quad j \geq 1$$

$$\text{and} \quad \sum_{j=0}^{p} m_j \leq 2n \tag{8.38}$$

$$q(\mu) = \mu^{2n}p(1/\mu)$$

$$= \alpha_q\mu^{q_0}(\mu-{}^1/\lambda_1)^{m_1}(\mu-{}^1/\lambda_2)^{m_2}\ldots(\mu-{}^1/\lambda_p)^{m_p} \tag{8.39}$$

with $\alpha_q = \alpha_p \prod_{j=1}^{p} (-\lambda_j)^{p_j}$ and $q_0 = 2n - \sum_{j=0}^{p} p_j$

We note that:

(1) The stabilizability and detectability conditions prevent the scalar β with $|\beta| = 1$ to be one of the eigenvalues *[18]*.

(2) If a zero-eigenvalue appears with multiplicity p_0, then the number of finite eigenvalues is $2n-p_0$, since there will be p_0 infinite or "missing" *[21]* eigenvalues.

(3) Associated with (8.36) are eigenvectors and generalized eigenvectors formed by

$$M\underline{v}_j = \lambda_j N\underline{v}_j$$

$$M\underline{v}_k = \lambda_j N\underline{v}_k + M\underline{v}_{k-1} \; ; \quad k = j+1, \ldots .$$

such that there are $r = 2n - \sum_{j=0}^{p} m_j$ generalized eigenvectors.

(4) A complete set of eigenvalues and associated eigenvectors and generalized eigenvectors, which span $\mathbb{R}^{2n}$, can be formed from those of the original problem (8.36) or its reciprocal problem (8.37). The reason for dealing with both problems together is that either one of them would not provide complete information since either M or N could be singular.

(5) In view of the matrix identity

$$[M-\lambda N]^t = \begin{bmatrix} 0 & I_n \\ -\lambda I_n & 0 \end{bmatrix} [M - \tfrac{1}{2}N] \begin{bmatrix} 0 & I_n \\ -\lambda I_n & 0 \end{bmatrix}$$

it follows by evaluating the determinants of both sides that

$$p(\lambda) = \lambda^{2n} p(1/\lambda)$$

and consequently

$$\begin{aligned} p(\lambda) &= q(\lambda) \\ &= \alpha_p \lambda^{m_0} [(\lambda-\lambda_1)(\lambda-{}^1/\lambda_1)]^{s_1} \ldots [(\lambda-\lambda_r)(\lambda-{}^1/\lambda_r)]^{s_r} \end{aligned} \tag{8.40}$$

with

$$p = 2r, \quad 2\sum_{j=1}^{r} s_j = \sum_{j=1}^{p} m_j$$

and

$$p_0 = q_0 = n - \sum_{j=1}^{r} s_j$$

A particularly useful choice of the complete set of eigenvalues, based on the above consideration, is given by *[18,21]:*

$$\underbrace{0,\ldots,0}_{p_0},\ \underbrace{\lambda_{p_0+1},\ldots,\lambda_n}_{n-p_0},\ \underbrace{1/\lambda_{p_0+1},\ldots,1/\lambda_n}_{n-p_0},\ \underbrace{\infty,\ldots,\infty}_{p_0}$$

with $0 < |\lambda_j| < 1 \quad ; \quad j = p_0+1,\ldots,n$

Let V, W be the matrices of n generalized eigenvectors associated with (8.36), (8.37), resp'ctively, and define J as the Jordan canonical form (see Chapter 2); then,

$$MV = NVJ \tag{8.41a}$$

$$MWJ = NW \tag{8.41b}$$

Since each of the columns of V and W has dimension 2n and they are independent, thus the matrix

$$T = [V \mid W]$$

of dimension (2nx2n) is invertible. The vectors $(\underline{t}_1,\ldots,\underline{t}_{2n})$ constitute a basis for $\mathbb{R}^{2n}$. Define the transformation

$$\begin{bmatrix} \underline{z}(k) \\ \underline{h}(k) \end{bmatrix} = T^{-1}\begin{bmatrix} \underline{x}(k) \\ \_(k) \end{bmatrix} \tag{8.42}$$

such that (8.32) is converted to:

$$MT\begin{bmatrix} \underline{z}(k) \\ \underline{h}(k) \end{bmatrix} - NT\begin{bmatrix} \underline{z}(k+1) \\ \underline{h}(k+1) \end{bmatrix} = 0 \qquad (8.43)$$

Making use of (8.41) in (8.43), it becomes:

$$[NV \,\vdots\, MW]\left[\begin{pmatrix} J & 0 \\ 0 & I_n \end{pmatrix}\begin{bmatrix} \underline{z}(k) \\ \underline{h}(k) \end{bmatrix} - \begin{pmatrix} I_n & 0 \\ 0 & J \end{pmatrix}\begin{bmatrix} \underline{z}(k+1) \\ \underline{h}(k+1) \end{bmatrix}\right] = 0 \qquad (8.44)$$

It can be easily verified that the term $[NV \,\vdots\, MW]$ is invertible *[18]* and thus from (8.44) we obtain:

$$\begin{bmatrix} \underline{z}(N) \\ \underline{h}(N) \end{bmatrix} = \begin{bmatrix} J^N & 0 \\ 0 & J^{-N} \end{bmatrix}\begin{bmatrix} \underline{z}(0) \\ \underline{h}(0) \end{bmatrix} \qquad (8.45)$$

Note that since $|\lambda| < 1$, then the $\underline{h}$ components are unstable whereas the $\underline{z}$ components are stable. Let T be partitioned as

$$T = \begin{bmatrix} V_1 & W_1 \\ V_2 & W_2 \end{bmatrix} \qquad (8.46)$$

where V_1, V_2, W_1, W_2 are all of dimension (nxn) and V_1 is invertible.

To obtain the steady-state we simply let N go to infinity; therefore $\underline{z}(N)$ goes to zero and, in general, $\underline{h}(N)$ would grow indefinitely since each element of J^{-1} is greater than one. As a result, it is readily evident that the only realizable solution for the steady state $(N \longrightarrow \infty)$ case is for $\underline{h}(0) = \underline{0}$, implying that $\underline{h}(k) = \underline{0}$ for all k. From (8.42), (8.45) and (8.46), with $\underline{h}(k) \equiv \underline{0}$, we have:

$$\begin{aligned} \underline{x}(k) &= V_1\, \underline{z}(k) \\ &= V_1\, J^{-k}\, \underline{z}(0) \end{aligned} \qquad (8.47)$$

$$\begin{aligned}\underline{\pi}(k) &= V_2\ \underline{z}(k)\\ &= V_2\ J^{-k}\ \underline{z}(0)\end{aligned} \tag{8.48}$$

Substituting (8.47) for $\underline{z}(0)$ in (8.48) leads to:

$$\begin{aligned}\underline{\pi}(k) &= V_2 J^{-k} J^{k} V_1^{-1} \underline{x}(k)\\ &= V_2 V_1^{-1} \underline{x}(k)\\ &= P\ \underline{x}(x)\end{aligned} \tag{8.49}$$

We note that (8.49) is of the same form as our assumption (8.23), so we conclude that the steady state Riccati matrix given by:

$$P = V_2\ V_1^{-1} \tag{8.50}$$

is the solution to (8.28) when $N \longrightarrow \infty$, i.e. when we have an infinite-time horizon, linear optimization problem. From (8.24), the optimal control law for this system is:

$$\begin{aligned}\underline{u}(k) &= -\ G\underline{x}(k)\\ &= -\ [R+B^{t}PB]^{-1}B^{t}PA\underline{x}(k)\end{aligned} \tag{8.51}$$

where in (8.49)-(8.51) the subscripts of P and G are dropped to denote constant values at the steady state. By virtue of (8.50), the optimal cost functional associated with the control (8.51) is:

$$J = \frac{1}{2}\ \underline{x}_0^{t}\ P\ \underline{x}_0 \tag{8.52}$$

An alternative expression of (8.51) can be obtained using the matrix identity *[22]*:

$$R^{-1}B^{t}P[I_n+BR^{-1}B^{t}P]^{-1} = [R+B^{t}PB]^{-1}B^{t}P \tag{8.53}$$

whose proof is rather straightforward, so that (8.51) becomes:

$$\underline{u}(k) = - R^{-1}B^tP[I_n+BR^{-1}B^tP]^{-1}A\underline{x}(k) \tag{8.54}$$

Had we followed a direct route by setting $P(k) = P(k+1) = P$ in (8.28), we get the algebraic Riccati equation:

$$\begin{aligned} P &= A^tPA-A^tPBS^{-1}B^tPA+C^tQC \\ &= A^tP[I_n-B(R+B^tPB)^{-1}B^tP]A+C^tQC \end{aligned} \tag{8.55}$$

which can be simplified using the matrix identity *[22]*:

$$[I_n+BR^{-1}B^tP]^{-1} = I_n-B(R+B^tPB)^{-1}B^tP \tag{8.56}$$

into the form:

$$\begin{aligned} P &= A^tP[I_n+BR^{-1}B^tP]^{-1}A+C^tQC \\ &= A^t[P^{-1}+BR^{-1}B^t]^{-1}A+C^tQC \end{aligned} \tag{8.57}$$

Either (8.55) or (8.57) can be utilized and solved to yield the steady-state Riccati matrix. Despite the fact that the above analysis is non-rigorous, the real merit of the lengthy analysis leading to (8.50) is that it embodies an efficient computational technique to solve (8.55) or (8.57) in a non-recursive manner. We shall illustrate this point in Section 8.4 when discussing numerical algorithms.

8.3.3 *ASYMPTOTIC PROPERTIES OF OPTIMAL CONTROL*

In this section we present some of the properties of the optimal control (8.54) when applied to the sywtem (8.16). The first important property is that the optimal control sequence is a stabilizing one in the sense that the closed-loop system:

$$\underline{x}(k+1) = (A-BG)\underline{x}(k)$$

$$= (A-B[R+B^tPB]^{-1}B^tPA)\underline{x}(k)$$

$$= (I_n-B[R+B^tPB]^{-1}B^tP)A\underline{x}(k) \qquad (8{,}58)$$

has the eigenvalues within the unit circle. This property is quite obvious in the light of the preceding analysis. It can be demonstrated using Lyapunov analysis as follows. If we choose $\underline{x}^t(k)P\underline{x}(k)$ as a Lyapunov function, then direct application of (3.97) to (8.58) results in:

$$A^tPA-P = -Q+A^tPB(R+B^tPB)^{-1}B^tPA + A^tPB(R+BPB^t)^{-1}R(R+B^tPB)^{-1}B^tPA \qquad (8.59)$$

with Q negative definite. It is easy to see that the right hand side of (8.59) is positive definite. Thus, the closed-loop system (8.58) is asymptotically stable.

The quality of the optimal control sequence would, in general, be dependent on the weighting matrices Q and R. Their values represent the relative penalties on the state and control variables. Our interest here is to study the asymptotic behaviour of the steady-state optimal control when the control weighting matrix R is made variable, that is,

$$R = \rho R_0 \qquad (8.60)$$

where $\rho \to 0^+$ (ρ approaches zero from above). To do this we consider the optimal system (8.19)-(8.22) formulated in the Z-domain. Using the definitions of Section 2.2.1, we obtain:

$$\begin{aligned} zX(z)-zx_0 &= AX(z)-BR^{-1}B^t[z\Omega(z)-zp_0] \\ \Omega(z) &= A^t[z\Omega(z)-zp_0] + C^tQCX(z) \end{aligned} \qquad (8.61)$$

where $X(z)$ and $\Omega(z)$ are the Z-transforms of $\underline{x}$ and $\underline{\pi}$, respectively. Solving for $X(z)$ and $\Omega(z)$, we write:

$$\begin{bmatrix} X(z) \\ \Omega(z) \end{bmatrix} = \left[\begin{array}{c|c} zI_n - A & zBR^{-1}B^t \\ \hline -z^{-1}C^tQC & z^{-1}I_n - A^t \end{array}\right]^{-1} \left[\begin{array}{c} z\underline{x}_0 - zBR^{-1}B^t\underline{p}_0 \\ \hline -A^t\underline{p}_0 \end{array}\right] \quad (8.62)$$

Careful examination of (8.62) indicates that the components of $X(z)$ and $\Omega(z)$ are rational functions in z except at singular points given by:

$$\det \left[\begin{array}{c|c} zI_n - A & zBR^{-1}B^t \\ \hline -z^{-1}C^tQC & z^{-1}I_n - A^t \end{array}\right] = 0$$

which can equivalently be stated as *[23]*:

$$\det[zI_n - A].\det[z^{-1}I_n - A^t + C^tQC(zI_n - A)^{-1}BR^{-1}B^t] = 0 \quad (8.63)$$

It is readily evident from (8.63) that it is the product of two polynomials, one in z and the other in z^{-1}. This implies that the eigenvalues of (8.63) are formed by the pairs $(z_j, 1/z_j)$, $j = 1,2,\ldots n$. In view of the fact that the closed-loop system (8.58) is asymptotically stable, then $(n-p_0)$ eigenvalues of (8.63) having moduli strictly less than one are characteristic values of (8.58), and the remaining p_0 characteristic values are zero. This result corresponds to the analysis pursued in Section 8.3.2.

Now to study the effect of the control penalty factor ρ, we manipulate (8.63) to get:

$$\det[zI_n - A].\det[z^{-1}I_n - A^t].\det[I_n + (z^{-1}I_n - A^t)^{-1}C^tQC(zI_n - A)^{-1}BR^{-1}B^t] = 0$$

Using the determinant identity *[23]*, the above expression, together with (8.60), becomes:

$$\Psi(z)\Psi(z^{-1})\det[I_m + \frac{1}{\rho}R_0^{-1}H^t(z^{-1})QH(z)] = 0 \quad (8.64)$$

where $$\Psi(z) = \det[zI_n - A] \quad (8.65)$$

is the open-loop characteristic polynomial, and

$$H(z) = C[zI_n - A]^{-1}B \quad (8.66)$$

is the open-loop transfer matrix. Consideration of (8.64) leads us to:

(1) For $\rho = \infty$ (in which case the control effort is a heavily penalized-high cost control), the finite roots are those of

$$\Psi(z)\Psi(z^{-1}) \quad (8.67)$$

which correspond to the open-loop values since $\underline{u}(.) \to \underline{0}$. Let $\Psi(z)$ in (8.65) be expressed as:

$$\Psi(z) = z^{p_0} \prod_{j=1}^{n-p_0} (z-\alpha_j) \; ; \; \alpha_j \neq 0 \quad (8.68)$$

then
$$\Psi(z)\Psi(z^{-1}) = \prod_{j=1}^{n-p_0} (z-\alpha_j)(z^{-1}-\alpha_j) \quad (8.69)$$

From the method of root locus [17], it is readily seen from (8.69) that $(2(n-p_0)$ root loci of (8.63) originate for $\rho = \infty$ at the nonzero open-loop eigenvalues and their inverses.

(2) For $\rho \to 0^+$ (in which case the control effort is a lightly penalized-cheap control), it is clear that those eigenvalues that remain finite approach the zeroes of

$$\Psi(z)\Psi(z^{-1}) \det[H^t(z^{-1})QH(z)] \quad (8.70)$$

For simplicity let $H(z)$ be a square matrix (which corresponds to equal numbers of inputs and outputs) such that:

$$\det[H(z)] = \frac{\Phi(z)}{\Psi(z)}$$

$$= \frac{\omega z^{s-q} \prod_{j=1}^{q} (z-\beta_j)}{\Psi(z)} ; \quad \beta_j \neq 0 \qquad (8.71)$$

and consequently the zeros of (8.70) are the zeros of

$$\Phi(z)\Phi(z^{-1}) = \omega^2 \prod_{j=1}^{q} (z-\beta_j)(z^{-1}-\beta_j) \qquad (8.72)$$

This shows that the 2q root loci of (8.64) terminate for $\rho = 0$ at the finite zeros β_j, $j = 1,\ldots,p$ and their inverses $1/\beta_j$, $j = 1,\ldots,p$.

Depending on the number of nonzero poles and zeros, the resulting behaviour of the closed-loop poles is indicated in Table 8.1 . A little reflection reveals that as the control weighting matrix R approaches the zero matrix, the closed-loop poles remain finite, and subsequently the feedback gain matrix G remains finite. Other properties can be found in *[22,24,25]*. The preceding analysis is referred to in the literature as the reciprocal root locus, a recent account of which is presented in *[17]*.

8.4 Numerical Algorithms for the Discrete Riccati Equation

We now direct attention to the numerical solution (8.55) for P, the steady-state Riccati matrix. The available algorithms can be broadly categorized into:

(a) Successive approximation methods *[26-28]*.
(b) Hamiltonian methods *[16-19,21]*.
(c) Square-root approach *[32-38]*.

8.4.1 *SUCCESSIVE APPROXIMATION METHODS*

This class of methods determines P and G by direct iterative

TABLE 8.1 Behaviour of Closed-loop Poles

	Number of finite poles $(n-p_0 \geq$ Number of finite zeros q	Number of finite zeros $q >$ Number of finite poles $(n-p_0)$
No. of nonzero closed-loop poles	$(n-p_0)$	q
No. of zero closed-loop poles	p_0	$(n-q)$
Asymptotic values as $\rho \to 0^+$	(1) $(n-p_0-q) \to 0$ (2) $q \to \beta_j;\ \lvert\beta_j\rvert \leq 1$ $\searrow 1/\beta_j;\ \lvert\beta_j\rvert > 1$	$q \to \beta_j;\ \lvert\beta_j\rvert \leq 1$ $\searrow 1/\beta_j;\ \lvert\beta_j\rvert > 1$
Asymptotic values as $\rho \to \infty$	$(n-p_0) \to \alpha_j;\ \lvert\alpha_j\rvert \leq 1$ $\searrow 1/\alpha_j;\ \lvert\alpha_j\rvert > 1$	$(q-n-p) \to \alpha_j;\ \lvert\alpha_j\rvert \leq 1$ $\searrow 1/\alpha_j;\ \lvert\alpha_j\rvert > 1$

schemes. Recall that the algebraic Riccati equation can be put in the form

$$P = A^t P[I_n + DP]^{-1} A + C^t QC \tag{8.73}$$

and the associated feedback gain matrix is given by:

$$G = [R + B^t PB]^{-1} B^t PA \tag{8.74}$$

where $\quad D = BR^{-1}B^t$

The iterative scheme [26] is

$$P = \lim_{j\to\infty} Y_j \tag{8.75}$$

where, for $j = 0,1,2,\ldots,$

$$Y_j = E_j^t Y_j E_j + G_j^t R G_j + C^t QC \tag{8.76}$$

$$G_j = [R + B^t Y_{j-1} B]^{-1} B^t Y_{j-1} A \tag{8.77}$$

$$E_j = A - BG_j \tag{8.78}$$

Using the matrix identities (8.53) and (8.56), alternative expressions for (8.76), (8.78) are given by [27]:

$$Y_j = E_j^t [Y_j + Y_{j-1} D Y_{j-1}] E_j + C^t QC \tag{8.76a}$$

$$E_j = (I_n + DY_{j-1})^{-1} A \tag{8.78a}$$

We note that (8.76a), (8.78a) do not require the explicit use of G, which is a basic step in the procedure of (8.76)-(8.78). However, the matrix inversion in (8.77) involves matrices of order m, whereas (8.78a) requires the inversion of an (nxn) matrix. The convergence rate of (8.75) is almost quadratic [26,27]. Two remarks are in order here:

(1) The initialization of the above algorithm is given by:

$$Y_{-1} = (A^t)^N W_{N-1}^{-1} A^N \tag{8.79a}$$

$$W_N = \sum_{m=0}^{N} A^m D (A^t)^m \tag{8.79b}$$

and it makes the matrix

$$E_0 = (I_n + DY_{-1})^{-1} A \tag{8.79c}$$

a stable one. The integer N satisfies $N \geq n$, where n is the system order.

(2) The unique positive definite solution of (8.76) can be written as:

$$Y_j = \sum_{m=0}^{\infty} (E_j^t)^m [G_j^t R G_j + C^t Q C] E_j^m$$

It should be clear that the successive approximation scheme described above lends itself to computer programming, and basically requires few matrix operations at each cycle.

8.4.2 *HAMILTONIAN METHODS*

This class of methods determines P from the eigen-structure of the Hamiltonian equations

$$\begin{bmatrix} A & 0 \\ -C^t Q C & I_n \end{bmatrix} \begin{bmatrix} x(k) \\ \underline{\pi}(k) \end{bmatrix} = \begin{bmatrix} I_n & BR^{-1}B^t \\ 0 & A^t \end{bmatrix} \begin{bmatrix} \underline{x}(k+1) \\ \underline{\pi}(k+1) \end{bmatrix} \tag{8.80}$$

which can be obtained from (8.19)-(8.22) or equivalently from (8.32)-(8.35). We have seen in Section 8.3.2 that the steady state gain matrix P is given by (8.50). In what follows we summarize the implementation steps.

(a) Find a matrix V containing the generalized eigenvectors such that (8.41) is satisfied with J a stable matrix in

Jordan form. This operation amounts to computing the generalized eigenvalues of (8.36) and generalized eigenvectors corresponding to the stable eigenvalues. An efficient computer program to do this is available in *[29]*.

(b) Find P from the solution of the nth order linear matrix equation

$$PV_1 = V_2$$

Equivalently, since P is symmetric, we can solve

$$V_1^t P = V_2^t$$

to obtain $P = V_1^{-t} V_2^{-t} = V_2 V_1^{-1}$. Any good linear equation solver can be used for this purpose, including the Gaussian elimination and its variants *[31]*.

The above computation works well only if we do not have multiple eigenvalues. An alternative technique has been developed in *[21]* based on the real Schur vector approach. It is implemented in two steps similar to the generalized eigenvector approach described above. The basic difference lies in the first step which now becomes:

(a1) Compute the orthogonal matrices which transform the matrix M in (8.34) to the real Schur form and the matrix N to upper triangular form, in such a way that the diagonal blocks corresponding to the stable eigenvalues are in the upper left quarters of the matrices.

Again the subroutines included in *[29]* are recommended for use. In general, both methods are numerically stable and do not require inversion of the state transition matrix. Thus, they are directly applicable to problems with singular transition matrices and ill-conditioned matrices *[30]*.

8.4.3 *DISCUSSION*

In this section we have shown that the solution of linear dynamic problems with quadratic criteria results in a linear feedback configuration. The feedback gains are determined from the numerical computation of the Riccati difference equation. For the important case of infinite horizon, the difference equation becomes algebraic. We have presented two basic approaches to the solution of the algebraic Riccati equation: the successive approximation method and the generalized eigenvectors (or Schur eigenvectors) method.

For problems associated with linear recursive estimation and implementation of dynamic programming recurrence relations, a new approach, termed "square-root and related algorithms", has been recently developed *[32-37]* to compute the Riccati matrix. These algorithms are generally fast and require less computational storage and processing than other standard methods. The basic idea is to propagate not the Riccati matrix but its square root using certain orthogonal transformations. A brief exposition of this approach in the context of linear-quadratic discrete regulators is given in *[38]*. It must be emphasized that the potential of the approach lies in its suitability for time-varying problems or finite-horizon quadratic optimization.

One of the results that has been given in Chapter 3 is the lower and upper bounds on the Lyapunov matrix. Since (3.97) is a special case of (8.55), by setting $B = 0$ we expect some relationship to exist between the bounds on the Riccati and Lyapunov matrices. To demonstrate this point, we define in the manner of Section 3.4.5 the following quantities:

$\lambda_m(W), \lambda_M(W)$	are the minimum and maximum eigenvalues of W
$\lambda_m^+(W), \lambda_m^-(W)$	are the upper and lower bounds on the minimum eigenvalue of W, respectively

$\lambda_M^+(W), \lambda_M^-(W)$ are the upper and lower bounds on the maximum eigenvalue of W, respectively

The recent results developed in *[39]* indicate that

$$\left.\begin{aligned} \lambda_m^-(P) &= 2\lambda_m(C^tQC)/\Theta_1 \\ \lambda_m^+(P) &= 2\lambda_M(C^tQC)/\Theta_2 \quad ; \quad \Theta_2 \neq 0 \\ \lambda_M^-(P) &= 2\lambda_m(C^tQC)/\Theta_3 \\ \lambda_M^+(P) &= 2\lambda_M(C^tQC)/\Theta_4 \quad ; \quad \Theta_4 \neq 0 \end{aligned}\right\} \tag{8.81}$$

where

$$\left.\begin{aligned} \Delta_1 &= 1 - \lambda_m(A^tA) - \lambda_M(BB^t)\lambda_m(C^tQC) \\ \Theta_1 &= \Delta_1 + [\Delta_1^2 + 4\lambda_M(BB^t)\lambda_m(C^tQC)]^{1/2} \\ \Delta_2 &= 1 - \sigma_m^2 - \lambda_m(BB^t)\lambda_M(C^tQC) \\ \Theta_2 &= \Delta_2 + [\Delta_2^2 + 4\lambda_m(BB^t)\lambda_M(C^tQC)]^{1/2} \\ \Delta_3 &= 1 - \sigma_M^2 - \lambda_M(BB^t)\lambda_m(C^tQC) \\ \Theta_3 &= \Delta_3 + [\Delta_3^2 + 4\lambda_M(BB^t)\lambda_m(C^tQC)]^{1/2} \\ \Delta_4 &= 1 - \sigma_M(A^tA) - \lambda_m(BB^t)\lambda_M(C^tQC) \\ \Theta_4 &= \Delta_4 + [\Delta_4^2 + 4\lambda_m(BB^t)\lambda_M(C^tQC)]^{1/2} \\ \sigma_m &= \min_j \delta_j \quad ; \quad \delta_j = |\lambda_j(A)| \\ \sigma_M &= \max_j \delta_j \end{aligned}\right\} \tag{8.82}$$

It is readily evident that by setting $B = 0$ in (8.82) we obtain (3.115), but with C^tQC replacing Q.

A final point is related to the discrete-time Hamiltonian equations (8.80). An interesting interpretation of the Hamiltonian system in terms of flow graphs is shown *[32,33]* to lead to a scattering-theory framework.

Next, we illustrate the different concepts presented in this section by three examples.

8.4.4 *EXAMPLES*

Example 1

A discrete model of a steam power system is described by *[40]:*

$$A = \begin{bmatrix} .915 & .051 & .038 & .015 & .038 \\ -.03 & .889 & -.0005 & .046 & .111 \\ -.006 & .468 & .247 & .014 & .048 \\ -.715 & -.022 & -.021 & .24 & -.024 \\ -.148 & -.003 & -.004 & .09 & .026 \end{bmatrix}$$

$$B = \begin{bmatrix} .0098 \\ .122 \\ .036 \\ .562 \\ .115 \end{bmatrix}$$

The open-loop eigenvalues of this model are $\sigma(A) = \{.8928 \pm j.0937,\ .2506 \pm j.0252,\ .0295\}$. We solve the problem of determining the optimal control sequence to drive the above model with

$$R = 1 \ ; \qquad C^t Q C = .1 I_5$$

using the generalized eigenvectors method (programmed on the UNIVAC 1108 at Kuwait University). The results of computation are:

(1) The 2n eigenvalues of the generalized Hamiltonian are given by: $\{.02982,\ .24652 \pm j.0246,\ .88393 \pm j.08772,$

33.53454, 4.01645$\pm$j.40074, 1.12028$\pm$j.11118} from which it is readily seen that the first five eigenvalues are the reciprocal of the other five eigenvalues.

(2) The steady-state Riccati matrix P takes the form:

$$P = \begin{bmatrix} .95 & .066 & .044 & -.005 & .037 \\ .066 & .665 & .021 & .043 & .075 \\ .044 & .021 & .109 & .001 & .004 \\ -.005 & .043 & .001 & .11 & .006 \\ .037 & .075 & .004 & .006 & .111 \end{bmatrix}$$

(3) The feedback gain matrix is

$$G = [-.035 \quad .1 \quad .001 \quad .023 \quad .012]$$

and the eigenvalues of the closed-loop system are:

$$\{.02982,\ .24652 \pm j.0246,\ .88393 \pm j.08772\}$$

It is interesting to observe that these eigenvalues correspond to the stable eigenvalues of the Hamiltonian.

(4) The expected value of the optimal cost, taken as the trace of the Riccati matrix to avoid variation in the initial state, is

$$J^* = 1.945$$

Simulation of the problem using other values of Q, R were undertaken and the output is summarized below.

TABLE 8.2 Summary of Simulation Results (Ex. 1)

	$R = 1$, $C^tQC = I_5$	$R = .1$, $C^tQC = I_5$
Closed-loop eigenvalues	{.03013, .19363, .24324, .84628$\pm$j.03749}	{.04147$\pm$j.00527, .25096, .72624, .89301}
J^*	16.8495	15.03712
Gain matrix G	[-.13778, .55513, .01214, .14676, .07083]	[-.1556, 1.16358, .04098, .36318, .15498]

We can conclude the following:

(i) Higher ratios of the control to state penality matrices $(R{:}C^tQC)$ result in lower values of cost.

(ii) The feedback gain matrix G preserves its pattern in sign but changes in magnitude.

(iii) Although the closed-loop eigenvalues remain within the same numerical range, the number and value of complex eigenvalues change from one computer run to another.

Example 2

The state-space model of a boiler system *[41]* of order nine has been discretized and appropriately scaled *[40]* to yield:

$$A = \left[\begin{matrix} 1 & -.1489\times10^{-3} & .105\times10^{-3} & .1051\times10^{-3} & -.2894\times10^{-1} \\ 0 & .9866 & -.335\times10^{-3} & -.2745\times10^{-3} & .9544\times10^{-5} \\ 0 & -.1389\times10^{-2} & .9686 & .3156\times10^{-3} & -.0391 \\ 0 & .8048\times10^{-2} & .2856\times10^{-2} & .9057 & -.7275\times10^{-4} \\ 0 & -.2065\times10^{-2} & .3328\times10^{-2} & .7091\times10^{-3} & .8829 \\ 0 & .7152\times10^{-2} & .0259 & .0198 & -.8358\times10^{-3} \\ 0 & -6.016 & 312 & -1.336 & -231 \\ 0 & 249 & -.8749 & -.6724 & .0256 \\ 0 & -51.53 & 6.241 & 4.815 & -.1692 \end{matrix}\right.$$

$$\left.\begin{matrix} .3127\times10^{-3} & -.2667\times10^{-5} & -.5914\times10^{-6} & -.3823\times10^{-5} \\ -.0195 & -.1585\times10^{-7} & .4712\times10^{-2} & .503\times10^{-4} \\ .0257 & .8717\times10^{-4} & .9676\times10^{-5} & -.1144\times10^{-5} \\ .1951 & .1169\times10^{-6} & .3265\times10^{-5} & .1673\times10^{-4} \\ .0148 & -.1071\times10^{-4} & -.9028\times10^{-5} & .1334\times10^{-4} \\ .8705 & .1445\times10^{-5} & .1345\times10^{-4} & .1143\times10^{-3} \\ -100.6 & .2375 & .067 & -.0262 \\ -31.05 & -.4447\times10^{-4} & .1998 & .0828 \\ 329.1 & .2825\times10^{-3} & -.1018 & .149 \end{matrix}\right]$$

$$B = \begin{bmatrix} .1775\text{x}10^{-4} & .449\text{x}10^{-5} \\ -.3191\text{x}10^{-3} & .0116 \\ .2177\text{x}10^{-3} & .3889\text{x}10^{-4} \\ -.6494\text{x}10^{-4} & .1109 \\ -.1159\text{x}10^{-3} & .2689\text{x}10^{-4} \\ -.7698\text{x}10^{-3} & .1239\text{x}10^{-2} \\ 2.308 & .1651 \\ -.7292 & 1.81 \\ -.4393 & -.0509 \end{bmatrix}$$

For the dynamic optimization of the boiler system we choose:

$$R = \text{diag}[10^{-4} \quad 10^{-4}]$$

$$C^tQC = \text{diag}[10^{-5} \quad 10^{-5} \quad 10^{-5} \quad 10^{-5} \quad 10^{-5} \quad 10^{-5} \quad 10^{-5} \quad 10^{-5} \quad 10^{-5}]$$

Again we use the generalized eigenvalues method to yield the following results:

(1) The steady-state Riccati matrix can be put in the form (rounded to three significant decimals),

$$P = \begin{bmatrix} 1.188 & .021 & -.061 & -.002 & -.243 & -.03 & -.015 & 0 & .001 \\ .021 & 6.945 & -.039 & -.898 & 4.976 & -1.778 & -.008 & .033 & .003 \\ -.061 & -.039 & 1.98 & .019 & .672 & .152 & 0 & -.001 & 0 \\ -.002 & -.898 & .019 & .023 & .311 & .282 & 0 & -.001 & .001 \\ -.243 & 4.976 & .672 & .311 & -2.131 & 1.593 & 0 & -.003 & 0 \\ -.03 & -1.778 & .152 & .282 & 1.593 & 8.909 & .002 & -.018 & .021 \\ -.015 & -.008 & 0 & 0 & 0 & .002 & .251 & .03 & .019 \\ 0 & .033 & -.001 & -.001 & -.003 & -.018 & .03 & -.16 & .111 \\ .001 & .003 & 0 & .001 & 0 & .021 & .019 & .111 & .34 \end{bmatrix}$$

(2) The feedback gain matrix is:

$$G = \begin{bmatrix} 2.37 & -378.595 & 4865.777 & 33.825 & -2439.96 & 549.37 & .524 & -1.123 & -.043 \\ .442 & 98.205 & 449.914 & -1.383 & -237.23 & 5.853 & .052 & .288 & .016 \end{bmatrix}$$

which gives the closed-loop eigenvalues as:

$\{1, .967, .88, .668 \pm j.247, .478, .154, .111, -.546\}$

(3) The average value of the cost is:

$$J^* = tr[P] = 17.345$$

Similarly to Example 1, Table 8.3 contains a summary of the simulation results. A quick look at the simulation results leads us to conclude that the case of least expected value corresponds to higher ratio of $(R:C^tQC)$ and the pattern of closed-loop eigenvalues remains intact. This is in accordance with the results of Example 1.

Example 3

Consider a second-order system *[21]* of the type (8.16) and (8.17), with

$$A = \begin{bmatrix} 0 & 1 \\ 0 & 0 \end{bmatrix} ; \quad B = \begin{bmatrix} 0 \\ 1.4142 \end{bmatrix}$$

$$C = \begin{bmatrix} 1 & 0 \\ 0 & 1 \end{bmatrix} ; \quad Q_f = 0$$

$$R = 1 ; \quad Q = \begin{bmatrix} 1 & -1 \\ -1 & 1 \end{bmatrix}$$

and $N \longrightarrow \infty$

TABLE 8.3 Summary of Simulation Results (Ex. 2)

	$R = 10^{-4}I_2$ $C^tQC = 10^{-4}I_5$	$R = 10^{-5}I_2$ $C^tQC = 10^{-4}I_5$
Closed-loop eigenvalues	{1, .965, .876, .59 ±j.258, .269, .107, .038, -.516}	{1, .964, .875, .575 ±j251, .146, .036, .005, -.512}
J^*	172.027	126.268
Gain matrix G	[-2.762 -1335.028 -.075 45.492 178.132 103.656 8.072 3.009 -158.137 1187.085 -26.788 34.493 .129 -5.304 .039 .027 -.071 .031]	[.005 -651.241 2707.137 0 137.541 -193.1 62.342 -579.296 867.334 -2.231 - 7.057 -48.446 .34 -3.201 -.032 .037 -.059 .02]

Simple calculation shows that the generalized eigenvalues of (8.36), that is,

$$\det\left[\begin{pmatrix} 0 & 1 & 0 & 0 \\ 0 & 0 & 0 & 0 \\ \hline -1 & 1 & 1 & 0 \\ 1 & -1 & 0 & 1 \end{pmatrix} - \lambda \begin{pmatrix} 1 & 0 & 0 & 0 \\ 0 & 1 & 0 & 2 \\ \hline 0 & 0 & 0 & 0 \\ 0 & 0 & 1 & 0 \end{pmatrix}\right] = 0$$

are $0, .5, \infty, 2$. The eigenvectors associated with the stable eigenvalues can be put in the form:

$$V = \begin{bmatrix} V_1 \\ \\ V_2 \end{bmatrix} = \begin{bmatrix} 1 & 4 \\ 0 & 2 \\ \hline 1 & 2 \\ -1 & -1 \end{bmatrix}$$

from which we find the steady state Riccati matrix:

$$P = V_2 V_1^{-1}$$

$$= \begin{bmatrix} 1 & -1 \\ -1 & 1.5 \end{bmatrix}$$

which has the eigenvalues $\{.219224, 2.280776\}$.

To examine the bounds on the extremal eigenvalues of the P matrix we calculate

$$\lambda[A^tA] = \{0,1\}$$

$$\lambda[C^tQC] = \{0,2\}$$

$$\lambda[BB^t] = \{0,2\}$$

$$\sigma_m = \sigma_M = 0$$

From (8.82) we have:

$$\Delta_1 = 1, \quad \Theta_1 = 2, \quad \Delta_2 = 1$$
$$\Theta_2 = 2, \quad \Delta_3 = 1, \quad \Theta_3 = 2$$
$$\Delta_4 = 1, \quad \Theta_4 = 2$$

and the substitution in (8.81) gives:

$$\lambda_m^-(P) = 0, \quad \lambda_m^+(P) = 2$$
$$\lambda_M^-(P) = 0, \quad \lambda_M^+(P) = 2$$

It is easy to see that

$$\lambda_m^-(P) < \lambda_m(P) < \lambda_m^+(P)$$

$$\lambda_M^-(P) < \lambda_M(P)$$

but $\lambda_M^+(P) < \lambda_M(P)$. This result is to be expected in view of the fact that the system has two zero eigenvalues.

Next, let us consider large scale systems within a discrete time framework.

8.5 Hierarchical Optimisation Methodology

In this section we shall consider the dynamic optimization of linear systems consisting of an interconnection of subsystems. The division of a system into subsystems may be done purely mathematically. It can also be done along physical boundaries in the sense that each subsystem may represent a physical entity.

8.5.1 *PROBLEM DECOMPOSITION*

We begin by formulating the problem of optimisation and control of interconnected dynamical systems. We assume that the overall system, of the type (8.16), comprises N_s subsystems which are interconnected together as shown, for instance, in Fig. 8.3 . For any subsystem j, let

$\underline{x}_j$ be the n_j-dimensional state vector

$\underline{u}_j$ be the m_j-dimensional control vector

$\underline{z}_j$ be the r_j-dimensional vector of inputs which are generated by the states of the other subsystems.

The dynamics of the jth subsystem are assumed to be of the form:

$$\underline{x}_j(k+1) = A_j \underline{x}_j(k) + B_j \underline{u}_j(k) + \underline{z}_j(k) \tag{8.83a}$$

$$\underline{z}_j(k) = \sum_{i \neq j}^{N_s} A_{ji} \, \underline{x}_i(k) \tag{8.83b}$$

and

$$\underline{x}_j(0) = \underline{x}_{j0} \tag{8.83c}$$

$$\underline{y}_j(k) = C_j \, \underline{x}_j(k) \tag{8.83d}$$

In the light of (8.16) and (8.83), we have:

$$A = \begin{bmatrix} A_1 & A_{12} & \cdots & A_{1N_s} \\ A_{21} & \ddots & & \vdots \\ \vdots & & A_j & \vdots \\ \vdots & & & \ddots \\ A_{N_s 1} & \cdots & \cdots & A_{N_s} \end{bmatrix} \tag{8.84a}$$

$$B = \text{diag}\{B_1 \; \ldots \; B_j \; \ldots \; B_{N_s}\} \tag{8.84b}$$

$$C = \text{diag}\{C_1 \; \ldots \; C_j \; \ldots \; C_{N_s}\} \tag{8.84c}$$

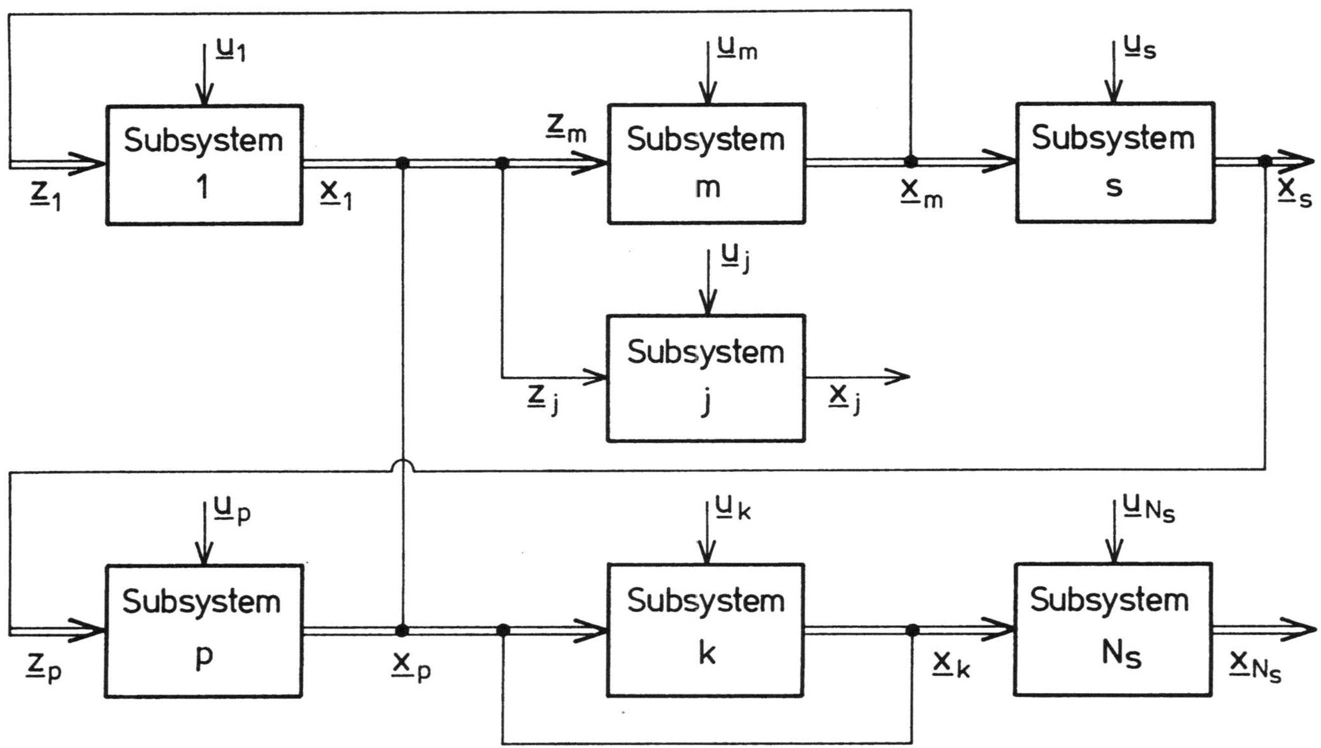

Fig. (8.3) Interconnected dynamical system

in which the decomposition of the overall system into smaller subsystems is obvious. It should be emphasized that the interaction among the subsystems is assumed to come from the states and not from the controls.

In a similar way, we associate with the subsystem (8.83) a part of the overall cost as given by (8.17), that is:

$$J_j = \frac{1}{2} \underline{x}_j^t(N)\, Q_{f_j}\, \underline{x}_j(N)$$

$$+ \frac{1}{2} \sum_{r=0}^{N-1} \{\underline{y}_j^t(r) Q_j \underline{y}_j(r) + \underline{u}_j^t(r) R_j \underline{u}_j(r)$$

$$+ \underline{z}_j^t(r) W_j \underline{z}_j(r)\} \tag{8.85}$$

where

$$J_N = \sum_{j=1}^{N_s} J_j \tag{8.86a}$$

$$Q_f = \text{diag}\{Q_{f1} \ldots Q_{fj} \ldots Q_{fN_s}\} \tag{8.86b}$$

$$Q = \text{diag}\{Q_1 \ldots Q_j \ldots Q_{N_s}\} \tag{8.86c}$$

$$R = \text{diag}\{R_1 \ldots R_j \ldots R_{N_s}\} \tag{8.86d}$$

$$W = \text{diag}\{W_1 \ldots W_j \ldots W_{N_s}\} \tag{8.86e}$$

The first term of (8.85) represents the subsystem terminal cost and the terms within the inner summation denote the cost over the rest of the optimization sequence $[0,N-1]$.

Now the problem of interest is that of minimizing (8.85) over the N_s subsystems and subject to the equality constraints (8.83). It has been established that the solution of this problem is identical to the overall solution obtained by minimizing (8.17) subject to (8.16); for reference see *[41]*. In this regard, we should mention that the term $\underline{z}_j^t(r) W_j \underline{z}_j(r)$ has been included in (8.85) to avoid singular solutions, although it has no physical interpretation in the overall case.

8.5.2 *OPEN-LOOP COMPUTATION STRUCTURES*

We now proceed to solve the dynamic optimization problem of the jth subsystem using hierarchical optimization techniques.

A. *The Goal Coordination Method*

The first technique is often called "the Goal Coordination" or "the Interaction Balance" method, and this was developed by Pearson *[42]*. According to this approach, the original minimization problem at hand is converted into a simpler maximization problem and then solved using a two-level iterative calculation structure. To do this, define

$$\Phi_d(\underline{\pi}) = \underset{\underline{x},\underline{u},\underline{z}}{\text{Min}} \{L(\underline{x},\underline{u},\underline{z},\underline{\pi}) \text{ subject to } (8.83)\} \qquad (8.87)$$

as a dual function to the Lagrangian $L(\underline{x},\underline{u},\underline{z},\underline{\pi})$ given by:

$$\begin{aligned} L(\underline{x},\underline{u},\underline{z},\underline{\pi}) &= \sum_{j=1}^{N_s} \{\tfrac{1}{2}\, \underline{x}_j^t(N) Q_{f_j} \underline{x}_j(N) \\ &\quad + \tfrac{1}{2} \sum_{r=0}^{N-1} [\underline{y}_j^t(r) Q_j \underline{y}_j(r) + \underline{u}_j^t(r) R_j \underline{u}_j(r) \\ &\quad + \underline{z}_j^t(r) W_j \underline{z}_j(r) + \underline{\pi}_j^t(r) \underline{z}_j(r) \\ &\quad - \sum_{j \neq i}^{N_s} \underline{\pi}_i^t(r) A_{ij} \underline{x}_j(r)]\} \\ &= \sum_{j=1}^{N_s} L_j \end{aligned} \qquad (8.88)$$

where L_j designates the part of the Lagrangian associated with the jth subsystem. It can be obtained from (8.88) directly. The Lagrange multiplier $\underline{\pi}_j$ is an r_j-dimensional vector. By the theorem of strong duality *[43]* the minimization of the quadratic cost function J w.r.t. $\underline{u}$ subject to the linear dynamic equality constraint (minimization of convex problems) is equivalent to the maximization of the dual function w.r.t. $\underline{\pi}$; that is,

$$\underset{\underline{\pi}}{\text{Max}}\ \Phi_d(\underline{\pi}) \quad = \quad \underset{\underline{u}}{\text{Min}}\ J_N \tag{8.89}$$

The maximization can be done within a two-level structure where for given sequences $\underline{\pi}(k) = \underline{\pi}^*(k)$, supplied by a second level, the Lagrangian can be separated into N_s independent subproblems constituting the first level. A description of the two-level structure is as follows:

<u>Level 2</u> Guess the sequences $\underline{\pi}^L(k)$, set $L = 1$ and convey these to Level 1.

<u>Level 1</u> For $j = 1,\ldots,N_s$, minimize the subLagrangian $\mathcal{L}_j$ w.r.t. $\underline{x}_j$, $\underline{z}_j$ and $\underline{u}_j$, given $\underline{\pi}_j(k) = \underline{\pi}_j^L(k)$. Send these values to Level 2.

<u>Level 2</u> Compute the gradient of the dual function

$$\nabla\Phi_d(\underline{\pi})\Big|_{\underline{\pi}=\underline{\pi}^L} = \underline{e}^L$$

$$= \begin{bmatrix} \underline{e}_1^L \\ \cdot \\ \cdot \\ \cdot \\ \underline{e}_j^L \\ \cdot \\ \cdot \\ \cdot \\ \underline{e}_{N_s}^L \end{bmatrix} \tag{8.90a}$$

where $$\underline{e}_j^L = [\underline{z}_j(k) - \sum_{i\neq j}^{N_s} A_{ji}\ \underline{x}_i(k)] \tag{8.90b}$$

and use this information to improve the Lagrange multipliers iteratively. A suitable updating mechanism has the form:

$$\underline{\pi}^{L+1}(k) = \underline{\pi}^L(k) + \alpha_\pi\ \underline{f}(\underline{e}^L) \tag{8.91a}$$

when $\underline{f}(\underline{e}^L) = \underline{e}^L$ (8.91b)

we have the standard steepest ascent routine. On the other hand, when

$$\left.\begin{aligned} \underline{f}(\underline{e}^L) &= \underline{d}^L(k) \\ \underline{d}^L(k) &= \underline{e}^L(k) + \beta^L \underline{d}^{L-1}(k) \\ \beta^L &= \sum_{r=1}^{N} \underline{e}^{L^t}(r)\underline{e}^L(r) / \sum_{r=1}^{N} \underline{e}^{(L-1)^t}(r)\underline{e}^{(L-1)}(r) \\ \underline{d}^0(k) &= \underline{e}^0(k) \end{aligned}\right\} \quad (8.91c)$$

we get the conjugate gradient method *[41]*. In (8.91a), α_π is an appropriate step size that can be determined by one-dimensional search methods like Golden-section or quadratic interpolation *[41]*. The overall optimum is achieved when $||\underline{e}^L||$ becomes sufficiently small. The two-level structure is shown in Fig. 8.4 . It should be emphasized that minimization of L_j w.r.t. $\underline{u}$ yields a set of necessary conditions similar to (8.13)-(8.15).

B. The Method of Tamura

We note the first level of the Goal Coordination structure performs a functional minimization. Tamura *[41]* suggested treating the Lagrangian by duality and decomposition. Thus, instead of decomposing the Lagrangian into the sub-Lagrangians for each subsystem, the subsystem Lagrangian itself can be decomposed by the discrete index k leading at the lowest level to a parametric optimization. Here we will consider a decomposition in discrete time as opposed to the decomposition by subsystems that we considered earlier.

We start by defining the dual problem of minimizing L_j in (8.88) subject to (8.83a) as

$$\underset{\underline{\rho}}{\text{Maximize}} \quad M(\underline{\rho})$$

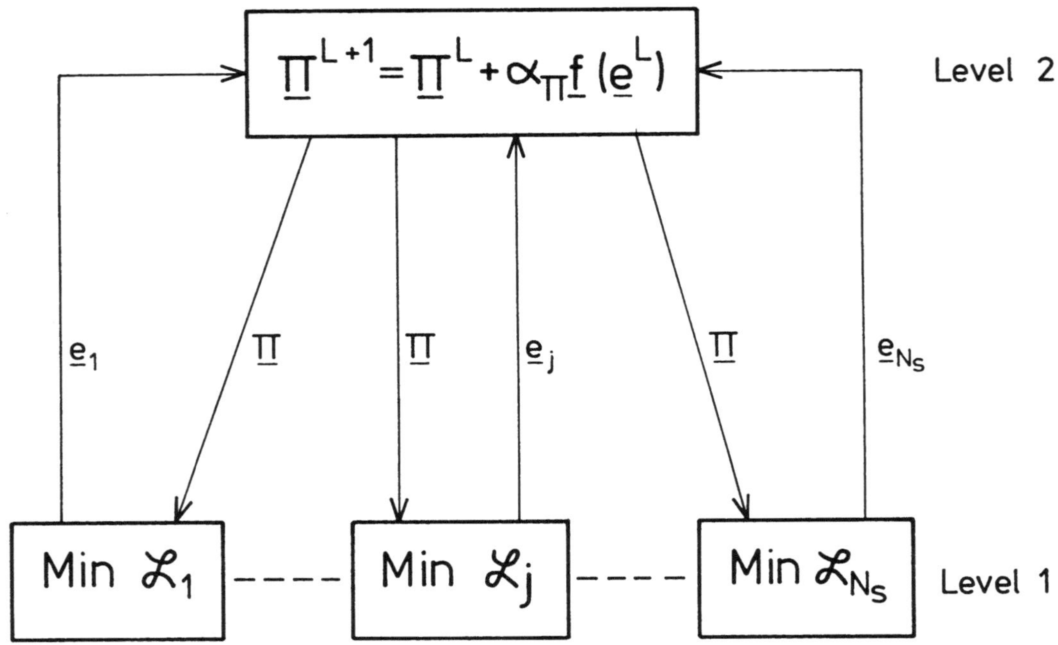

Fig. (8.4) The two-level goal coordination structure

where

$$M(\rho) = \underset{\underline{x},\underline{u}}{\text{Min}}\{\frac{1}{2}\, \underline{x}_j^t(N) Q_{f_j} \underline{x}_j(N)$$

$$+ \frac{1}{2} \sum_{r=0}^{N-1} [\underline{y}_j^t(r) Q_j \underline{y}_j(r) + \underline{u}_j^t(r) R_j \underline{u}_j(r)$$

$$+ \underline{z}_j(r) W_j \underline{z}_j(r) + \underline{\pi}_j^t(r) \underline{z}_j(r)$$

$$+ \underline{\rho}_j^t(r) \{A_j \underline{x}_j(r) + B_j \underline{u}_j(r) + \underline{z}_j(r)$$

$$- \underline{x}_j(r+1)\} - \sum_{i \neq j}^{N_s} \underline{\pi}_i^t(r) A_{ij} \underline{x}_j(r)]\} \qquad (8.92)$$

subject to (8.83c) and (8.83d). It is interesting to note that the gradient of $M(\underline{\rho})$ is given by

$$\nabla M(\underline{\rho})\Big|_{\underline{\rho}=\underline{\rho}^*} = A_j \underline{x}_j(r) + B_j \underline{u}_j(r) + \underline{z}_j(r) - \underline{x}_j(r+1)$$

$$r = 0,\ldots,N-1; \quad j = 1,\ldots,N_s \qquad (8.93a)$$

where $\underline{\rho} = \underline{\rho}^*$ is a known sequence, and $\underline{x}_j$, $\underline{u}_j$ are the solutions obtained after minimizing L_j subject to (8.83a). To solve the dual problem numerically, it is necessary to compute the value of $M(\underline{\rho})$ for a given sequence $\underline{\rho} = \underline{\rho}^*$ and then to maximize $M(\underline{\rho})$ using the gradient in (8.93a). The computation of $M(\underline{\rho})$ for fixed sequences $\{\underline{\rho} = \underline{\rho}^*, \underline{\pi} = \underline{\pi}^*\}$ can be performed by minimizing the function independently for each time index r. A three-level computation structure can be constructed to implement the dynamic optimization and can be summarized as follows:

<u>Level 3</u> Guess sequences $\underline{\pi}^L(k)$, set $L = 1$ and transmit these to Level 1.

<u>Level 2</u> Guess sequences $\underline{\rho}^M(k)$, set $M = 1$ and send these to Level 1.

Level 1 Using the sequences $\{\underline{\pi}^L(k), \underline{\rho}^M(k)\}$, perform the minimization of the Lagrangian over the discrete instants to yield:

for k = 0

$$\underline{x}_j(0) = \underline{x}_{j0}$$

$$\underline{u}_j(0) = -R_j^{-1} B_j^t \underline{\rho}_j^M(0)$$

$$\underline{z}_j(0) = -W_j^{-1} [\underline{\rho}_j^M(0) + \underline{\pi}_j^L(0)]$$

for k = 1,2,...,N-1

$$\underline{x}_j(k) = -(C_j^t Q_j C_j)^{-1} [A_j^t \underline{\rho}_j^M(k) - \rho_j^M(k-1)] - \sum_{i \neq j}^{N_s} \{A_{ij}^t \; \underline{\pi}_i^L\}$$

$$\underline{u}_j(k) = -R_j^{-1} B_j^t \underline{\rho}_j^M(k)$$

$$\underline{z}_j(k) = -W_j^{-1} [\underline{\rho}_j^M(k) + \underline{\pi}_j^L(k)]$$

for k = N

$$\underline{x}_j(N) = Q_{f_j}^{-1} \underline{\rho}_j^M(N-1)$$

The sequences $\underline{x}$, $\underline{u}$, $\underline{z}$ are conveyed to Level 2.

Level 2 Compute the gradient vector (8.93) and use it to improve $\underline{\rho}$ as

$$\underline{\rho}^{M+1} = \underline{\rho}^M + \alpha_\rho \; \underline{g}[\nabla M(\underline{\rho})] \qquad (8.93b)$$

where $[\underline{g}]$. is a correction term that can be obtained by steepest ascent or conjugate gradient.

This iterative procedure continues until $||\underline{\rho}^{M+1} - \underline{\rho}^M||$ becomes sufficiently small.

Level 3 Update the $\underline{\pi}$ sequences using the rule (8.91).

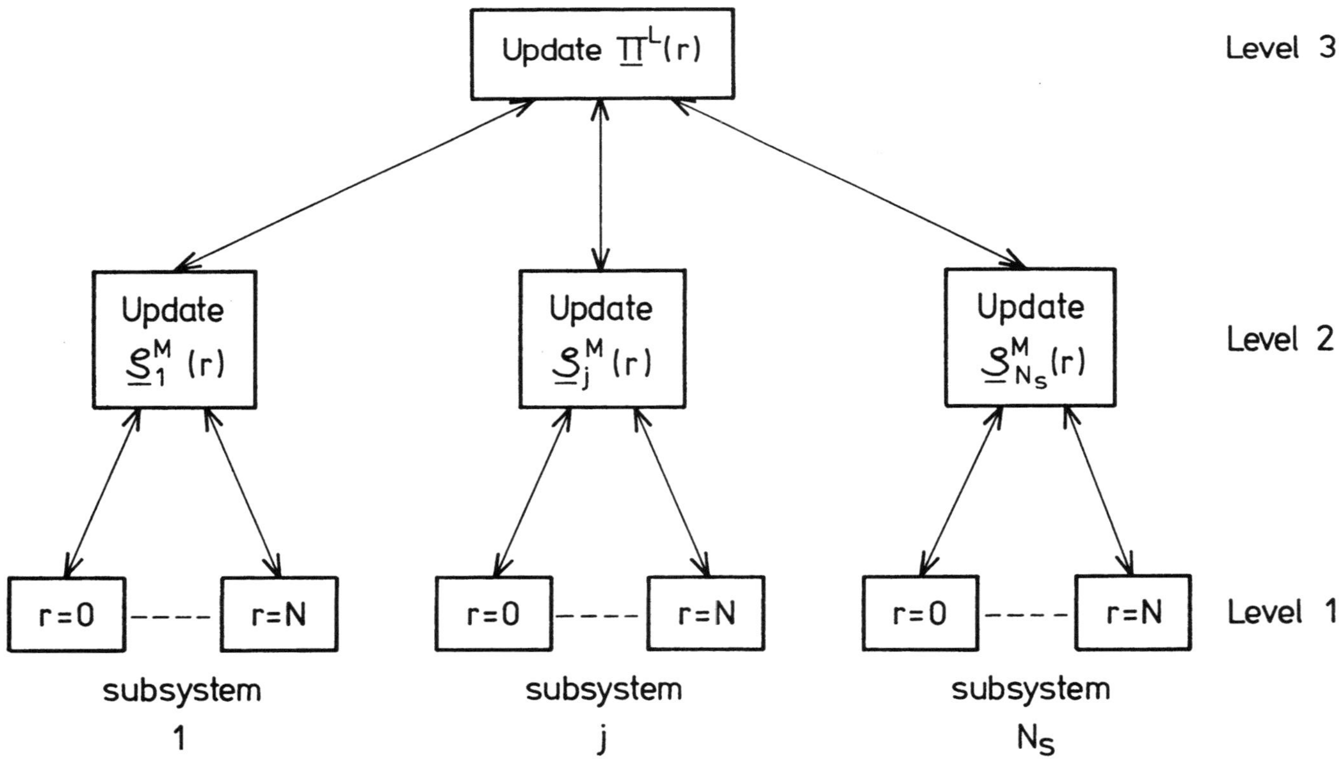

Fig. (8.5) The three-level hierarchical structure

Figure 8.5 shows a block-diagram of the proposed three-level structure.

Experience indicates that the three-level computation structure is attractive because an explicit solution is obtained at the first level and simple updating mechanisms are used for the higher levels. Looked at in this light, it has generally proved to be superior to the standard Goal Coordination structure *[41]*.

C. The Interaction Prediction Method

We now turn to hierarchical techniques which do not require the use of penality terms on the interaction vector.

We begin by deleting this term in (8.85), appropriately appending (8.83) and fixing the sequences $\underline{z}_j = \underline{z}_j^*$ and $\underline{\pi}_j = \underline{\pi}_j^*$ to yield:

$$\begin{aligned} J_j^a &= \frac{1}{2}\,\underline{x}_j^t(N)Q_{f_j}\underline{x}_j(N) \\ &+ \frac{1}{2}\sum_{r=0}^{N-1}\{\underline{y}_j^t(r)Q_j\underline{y}_j(r) + \underline{u}_j^t(r)R_j\underline{u}_j(r) \\ &+ \underline{\pi}_j^{*t}(r)\underline{z}_j^*(r) - \sum_{i\neq j}^{N_s}\underline{\pi}_i^{*t}(r)A_{ij}\underline{x}_j(r) \\ &- \underline{\rho}_j^t(r+1)[A_j\underline{x}_j(r) + B_j\underline{u}_j(r) + \underline{z}_j^*(r)]\} \end{aligned} \qquad (8.94)$$

where we retained the notation of Section 8.3 for convenience. The necessary conditions of optimality are given by:

$\underline{r = 0,\ldots,N-1}$

$$\underline{x}_j(r+1) = A_j\underline{x}_j(r) + B_j\underline{u}_j(r) + \underline{z}_j^*(r) \; ; \quad \underline{x}_j(0) = \underline{x}_j(\;) \qquad (8.95a)$$

$$\underline{u}_j(r) = -R_j^{-1}B_j^t\underline{\rho}_j(r+1) \qquad (8.95b)$$

$$\underline{\rho}_j(r) = C_j^t Q_j C_j \underline{x}_j(r) + A_j^t \underline{\rho}_j(r+1)$$

$$- \sum_{i \neq j}^{N_s} A_{ij}^t \underline{\pi}_i^*(r) \quad ;$$

$$\underline{\rho}_j(N) = Q_{f_j} \underline{x}_j(N) \tag{8.95c}$$

and they constitute the first level. At the second level, the sequences $\underline{\pi}$ and $\underline{z}$ are improved by the prediction rule

$$\begin{bmatrix} \underline{\pi}^*(r) \\ \\ \underline{z}^*(r) \end{bmatrix}_{L+1} = \begin{bmatrix} \underline{\rho}_j(r) \\ \\ \sum_{i \neq j}^{N_s} A_{ji} \underline{x}_i(r) \end{bmatrix}_L \tag{8.96}$$

where L is the iteration index. This method is called the Interaction Prediction method and it is summarized by the following steps:

<u>Level 2</u> Supply the sequences $\underline{\pi}^*(r)$ and $\underline{z}^*(r)$ to the first level.

<u>Level 1</u> Solve the optimality conditions (8.95) together for each subsystem to produce $\underline{x}_j(r)$, $\underline{u}_j(r)$ and $\underline{\rho}_j(r)$. Convey the results to Level 2.

<u>Level 2</u> The second level computes the r.h.s. of (8.96) to obtain new predictions of $\underline{\pi}^*$ and $\underline{z}^*$ and sends these back to Level 1. The process continues until the error in prediction, i.e. the difference in norm between two successive iterates, becomes sufficiently small.

Figure 8.6 gives a schematic diagram of the computation structructure. It should be clear that the updating routine (8.96) is much simpler than (8.91) for the standard Goal Coordination method and (8.91), (8.93) for the three-level structure. Computer simulation studies *[41]* have demonstrated that the convergence behaviour of the Interaction Prediction structure

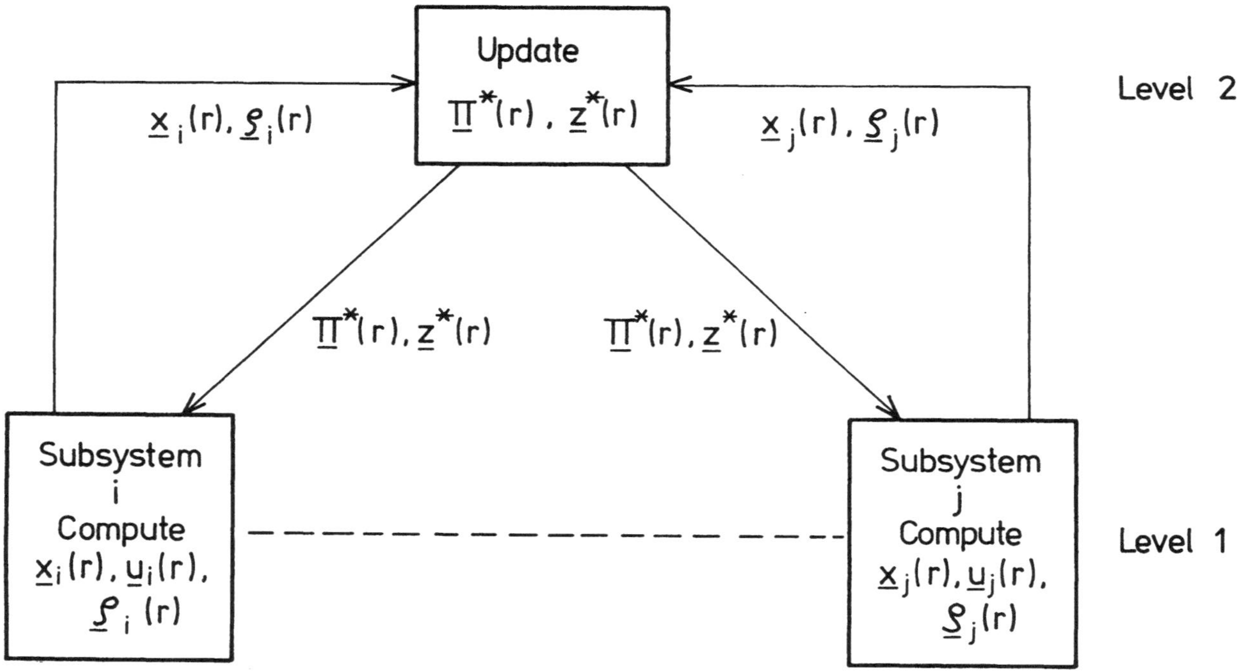

Fig. (8.6) The two-level interaction prediction structure

is much better than that of other structures. It can also be adapted to compute the feedback control. This is the subject of the next section.

8.5.3 *CLOSED-LOOP CONTROL STRUCTURE*

In order to derive a feedback control scheme, we consider the two-level hierarchical structure based on the Interaction Prediction approach. At the subsystem level we use a Riccati-type transformation similar to (8.23) which now takes the form:

$$\underline{\rho}_j(r) = P_j(r)\underline{x}_j(r) + \underline{s}_j(r) \tag{8.97}$$

where $\underline{s}_j(r)$ is the open-loop tracking vector, introduced to account for the coupling between the subsystems. The substitution of (8.97) into (8.95b) leads to

$$\underline{u}_j(r) = -R_j^{-1}B_j^t[P_j(r+1)\underline{x}_j(r+1) + \underline{s}_j(r+1)] \tag{8.98}$$

which, when used in (8.95a), yields:

$$\underline{x}_j(r+1) = [I_j+B_jR_j^{-1}B_j^tP_j(r+1)]^{-1}[A_j\underline{x}_j(r) + \underline{z}_j^*(r) - B_jR_j^{-1}B_j^t\underline{s}_j(r+1)] \tag{8.99}$$

where I_j is the identity matrix of order n_j . Manipulating (8.95c), using (8.97) and (8.99), results in:

$$\begin{aligned}&\{P_j(r) - C_j^tQ_jC_j - A_j^tP_j(r+1)[I_j+B_jR_j^{-1}B_j^tP_j(r+1)]^{-1}A_j\}\underline{x}_j(r)\\&+ \{\underline{s}_j(r) - A_j^tP_j(r+1)[I_j+B_jR_j^{-1}B_j^tP_j(r+1)]^{-1}\underline{z}_j^*(r)\\&- A_j^t\underline{s}_j(r+1) + \sum_{i\neq j}^{N} A_{ij}^t\ \underline{\pi}_i^*(r)\\&+ A_j^tP_j(r+1)[I_j+B_jR_j^{-1}B_j^tP_j(r+1)]^{-1}B_jR_j^{-1}B_j^t\underline{s}_j(r+1)\\&= \underline{0}\end{aligned}$$

which is valid for arbitrary $\underline{x}_j(r)$. Thus,

$$P_j(r) = C_j^t Q_j C_j + A_j^t P_j(r+1)[I_j+B_j R_j^{-1} B_j^t P_j(r+1)]^{-1} A_j \; ;$$

$$P_j(N) = Q_{f_j} \tag{8.100a}$$

and

$$\underline{s}_j(r) = A_j^t P_j(r+1)[I_j+B_j R_j^{-1} B_j^t P_j(r+1)]^{-1} \underline{z}_j^*(r)$$
$$+ A_j^t\{I_j - P_j(r+1)[I_j+B_j R_j^{-1} B_j^t P_j(r+1)]^{-1} B_j R_j^{-1} B_j^t\} \underline{s}_j(r+1)$$
$$- \sum_{i\neq j}^{N_s} A_{ij}^t \, \underline{\pi}_j^*(r) \quad ; \qquad \underline{s}_j(N) = \underline{O}$$

which can be simplified using (8.56) to:

$$\underline{s}_j(r) = A_j^t P_j(r+1)[I_j+B_j R_j^{-1} B_j^t P_j(r+1)]^{-1} \underline{z}_j^*(r)$$
$$+ A_j^t [I_j + P_j(r+1) B_j R_j^{-1} B_j^t]^{-1} \underline{s}_j(r+1)$$
$$- \sum_{i\neq j}^{N_s} A_{ij}^t \, \underline{\pi}_j^*(r) \tag{8.100b}$$

We note that:

(1) The P_j in (8.100a) is independent of the initial state $\underline{x}_j(0)$. Thus, the N_s matrix Riccati-equations, each involving $n_j(n_j+1)/2$ equations, can be solved independently and recursively from the final condition $P_j(N) = Q_{f_j}$. This gives a partial feedback control. It can be argued that this feedback around each subsystem does provide some degree of stabilization against small disturbances and moreover, allows one to correct the control based on the available state as opposed to the initial condition.

(2) The $\underline{s}_j$ in (8.100b) is not independent of the initial state $\underline{x}_j(0)$. It provides, however, open-loop compensation since at the optimum it can be written as:

$$\underline{s}_j(r) = A_j^t[I_j+P_j(r+1)B_jR_j^{-1}B_j^t]^{-1}\underline{s}_j(r+1) - \sum_{i\neq j}^{N_s} A_{ij}^t[P_j(r)\underline{x}_j(r) + \underline{s}_j(r)] + A_j^tP_j(r+1)[I_j+B_jR_j^{-1}B_j^tP_j(r+1)]^{-1}[\sum_{i\neq j}^{N_s} A_{ji}\underline{x}_i(r)]$$

which shows that $\underline{s}_j(r)$ depends on the initial state $\underline{x}(0)$ of the overall system.

For the infinite-horizon linear regulators, N , P_j reaches its steady-state value anc can be computed in the manner of Section 8.3.2 . With $P_j(r) = P_j(r+1) = P_j$, it is readily evident that

$$\underline{s} = E\,\underline{x} \tag{8.101}$$

where E is an (nxn) matrix [41].

Define G_d as a block diagonal matrix with $[R_j^{-1}B_j^tP_j]$ as the block elements. Then from (8.98) and (8.101), we get the overall closed-loop control as:

$$\underline{u} = -G_d\,\underline{x} + G_0\,\underline{x} \tag{8.102}$$

where G_0 depends on R, B and E. The interpretation of (8.102) is that G_0 represents the off-diagonal elements in the feedback gain matrix.

Recall that G_d can be calculated from the decoupled Riccati matrices of the subsystems. The computation of G_0 is not straightforward.

It has been suggested [41] that since around $r = 0$ the E matrix is constant whereas $\underline{x}$ and $\underline{s}$ are not, then the values of $\underline{x}$ and $\underline{s}$ can be recorded at the first $\sum_{j=1}^{N_s} n_j$ discrete instants close to $r = 0$. Then from the matrices

$$S = [\underline{s}(0),\underline{s}(r_1),\ldots,\underline{s}(r_n)]$$

$$X = [\underline{x}(0), \underline{x}(r_1), \ldots, \underline{s}(r_n)]$$

and (8.101) we obtain:

$$E = S X^{-1} \tag{8.103}$$

The inversion of X should not pose much of a problem in numerical computation. An alternative procedure would be [44] to solve the problem off-line n times successively from the initial conditions:

$$\underline{x}(0) = \begin{pmatrix} 1 \\ 0 \\ \cdot \\ \cdot \\ \cdot \\ 0 \end{pmatrix}, \begin{pmatrix} 0 \\ 1 \\ 0 \\ \cdot \\ \cdot \\ 0 \end{pmatrix}, \ldots, \begin{pmatrix} 0 \\ 0 \\ \cdot \\ \cdot \\ \cdot \\ 1 \end{pmatrix}$$

then $E = S$. Here, the computation is off-line and in a decentralized way.

To summarize, the solution of the linear-quadratic regulator problem can be obtained within a decentralized calculation structure. The result will be the feedback control scheme (8.102) which is independent of the initial state. Thus, this scheme can bring the discrete system back to steady state optimally from any initial disturbance. Fig. 8.7 shows the hierarchical implementation of the feedback control scheme.

8.5.4 *EXAMPLES*

Example 1

The purpose of this example is to provide a comparison between the hierarchical computation structures given in the last two sections. We consider the pollution control problem of a two-reach "no delay" river model [45]. The state vector represents the B.O.D. and D.O. concentration in the two reaches, whereas the maximum fraction of B.O.D. removed from the effluent in the

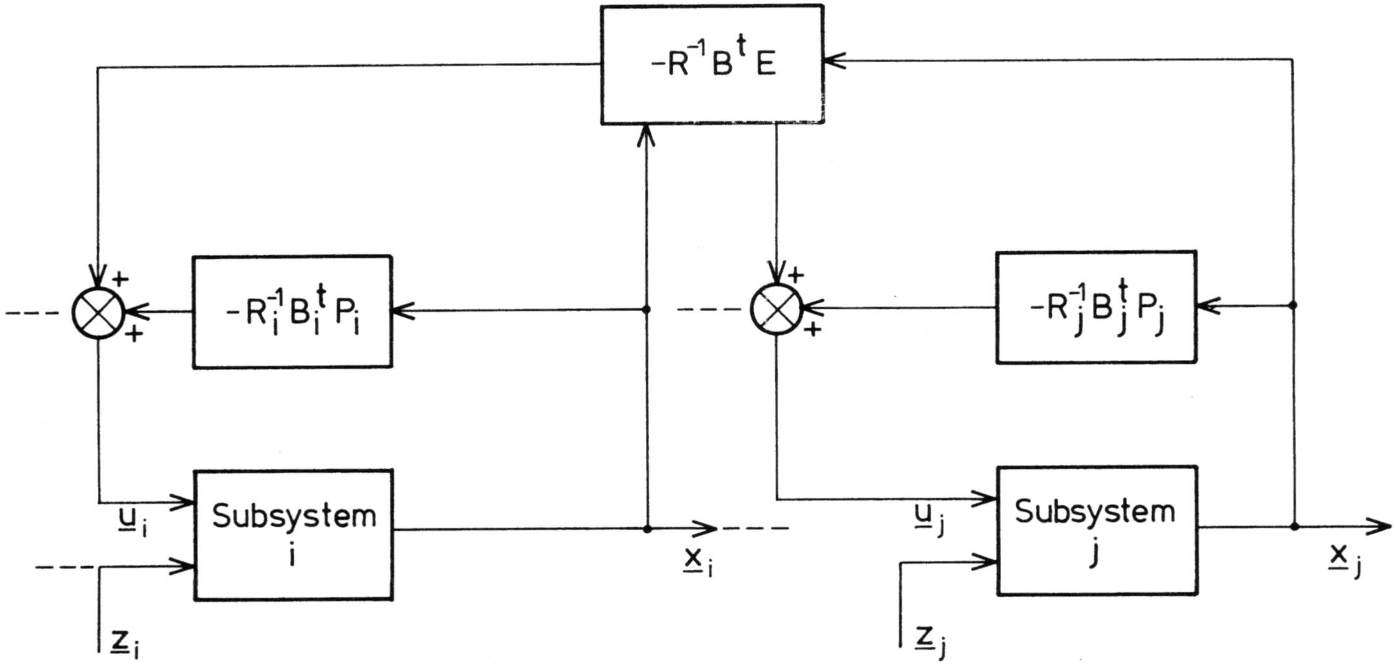

Fig. (8.7) Hierarchical implementation of feedback control scheme

reaches is the control vector. The model takes the form [45]:

$$\underline{x}(k+1) = A\underline{x}(k) + B\underline{u}(k) + \underline{d}$$

where

$$A = \begin{bmatrix} .18 & 0 & 0 & 0 \\ -.27 & .27 & 0 & 0 \\ .55 & 0 & .18 & 0 \\ 0 & .55 & -.25 & .27 \end{bmatrix}$$

$$B = \begin{bmatrix} -2 & 0 \\ 0 & 0 \\ 0 & -2 \\ 0 & 0 \end{bmatrix} ; \quad \underline{d} = \begin{bmatrix} 4.5 \\ 6.15 \\ 2 \\ 2.65 \end{bmatrix}$$

A suitable cost function for this system is

$$J = \frac{1}{2}||\underline{x}(N)||^2_{I_4} + \sum_{k=0}^{N-1} \frac{1}{2}(||\underline{x}(k) - \underline{x}^d||^2_{I_4} + ||\underline{u}(k)||^2_{100I_2})$$

where $||\underline{a}||_Q = \underline{a}^t Q \underline{a}$ and the desired values $\underline{x}^d$ are:

$$\underline{x}^d = [5 \quad 7 \quad 5 \quad 7]^t$$

which implies that it is desired to maintain the stream near the B.O.D., D.O. values of 5 and 7 units respectively while minimizing the treatment at the sewage works.

In simulation studies, N was chosen to be 23 which is certainly sufficiently long for the system to settle to a steady state since the sampling interval is .5 day. The initial state vector is:

$$\underline{x}(0) = [0 \quad 0 \quad 0 \quad 1]^t$$

A summary of the simulation results of the three hierarchical techniques is given in Table 8.4 .

TABLE 8.4 Summary of the Simulation Results*

Method	Number of Iterations at		CPU time sec.	Accuracy
	Level 2	Level 3		
1. Goal Coordination Method	117	-	49.88	10^{-5}
2. Method of Tamura	29	8	32.65	10^{-5}
3. Interaction Prediction Method	43	-	14.78	10^{-5}

*All simulations were performed on the UNIVAC 1108 facilities at Kuwait University.

The optimum cost is recorded to be $J^* = 1607$. Figures 8.8 - 8.11 show the optimal profile of B.O.D. and D.O. in reaches 1 and 2 and the corresponding controls.

Due to the effect of the $\underline{d}$ vector, the closed-loop control can now be written as:

$$\underline{u} = -G\,\underline{x} + \underline{u}_0$$

where G can be computed according to the analysis of Section 8.5.3. It takes the value

$$G = \begin{bmatrix} -.0074 & .0011 & -.0006 & .0001 \\ -.0126 & .0015 & -.0042 & .0004 \end{bmatrix}$$

and

$$\underline{u}_0 = \underline{u}(0) + G\,\underline{x}(0)$$

$$= \begin{bmatrix} .05449 \\ .00668 \end{bmatrix}$$

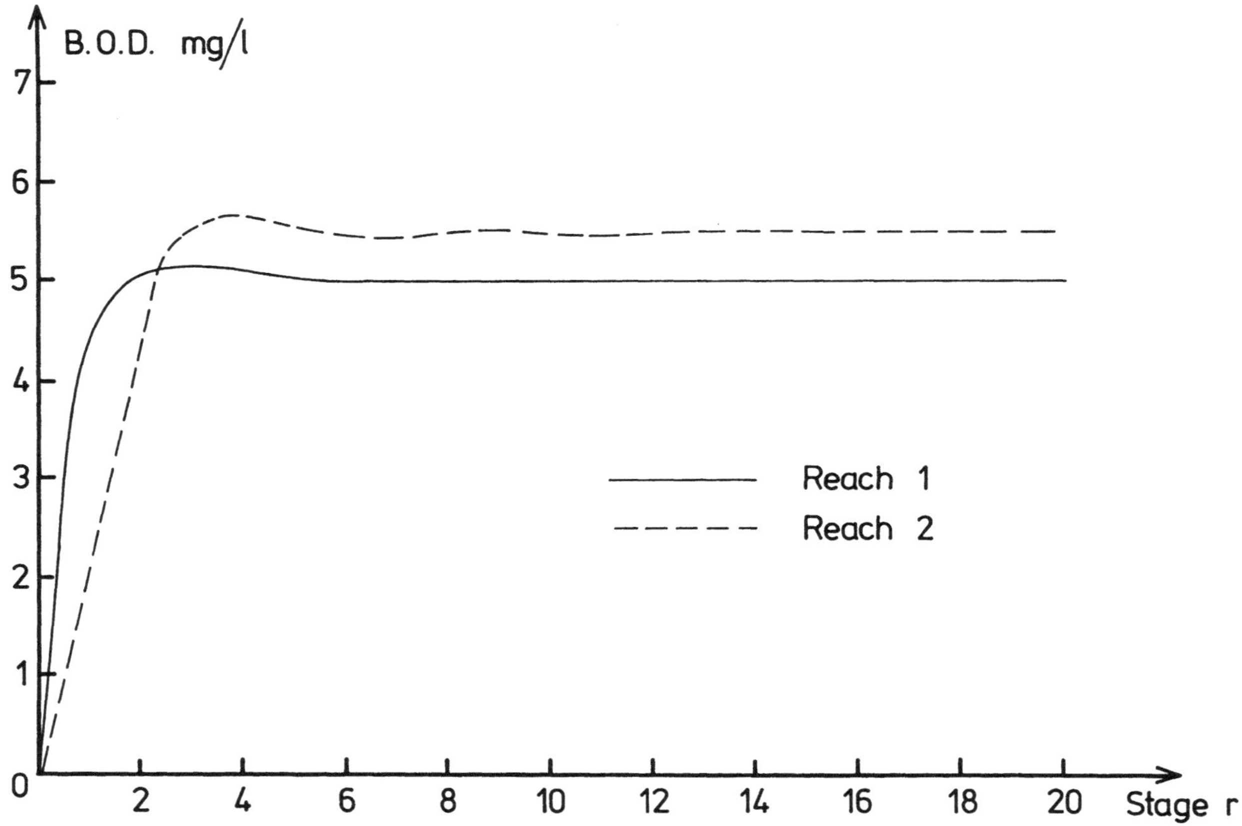

Fig. (8.8) Optimal B.O.D. sequences

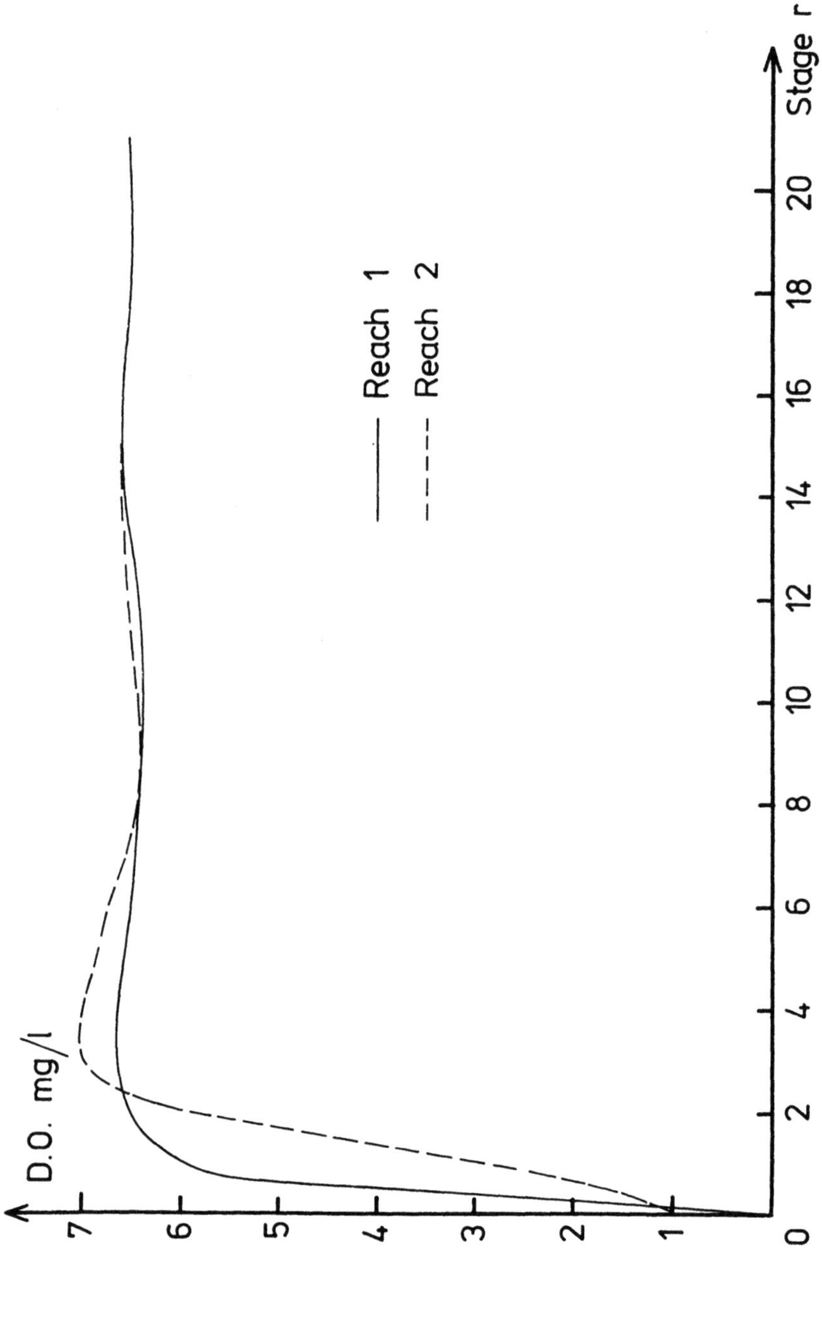

Fig. (8.9) Optimal D.O. sequences

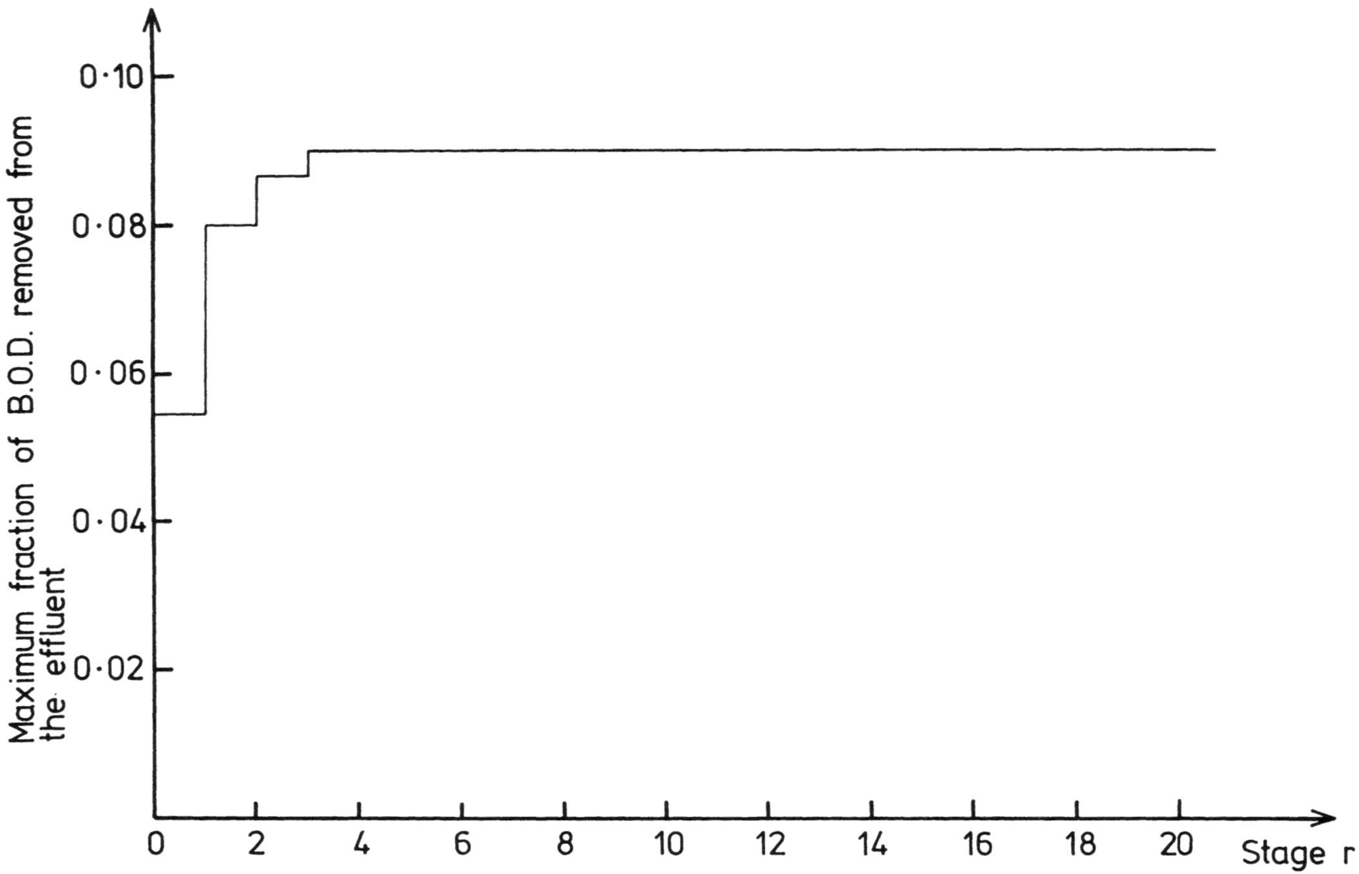

Fig. (8.10) Optimal control sequence for Reach 1

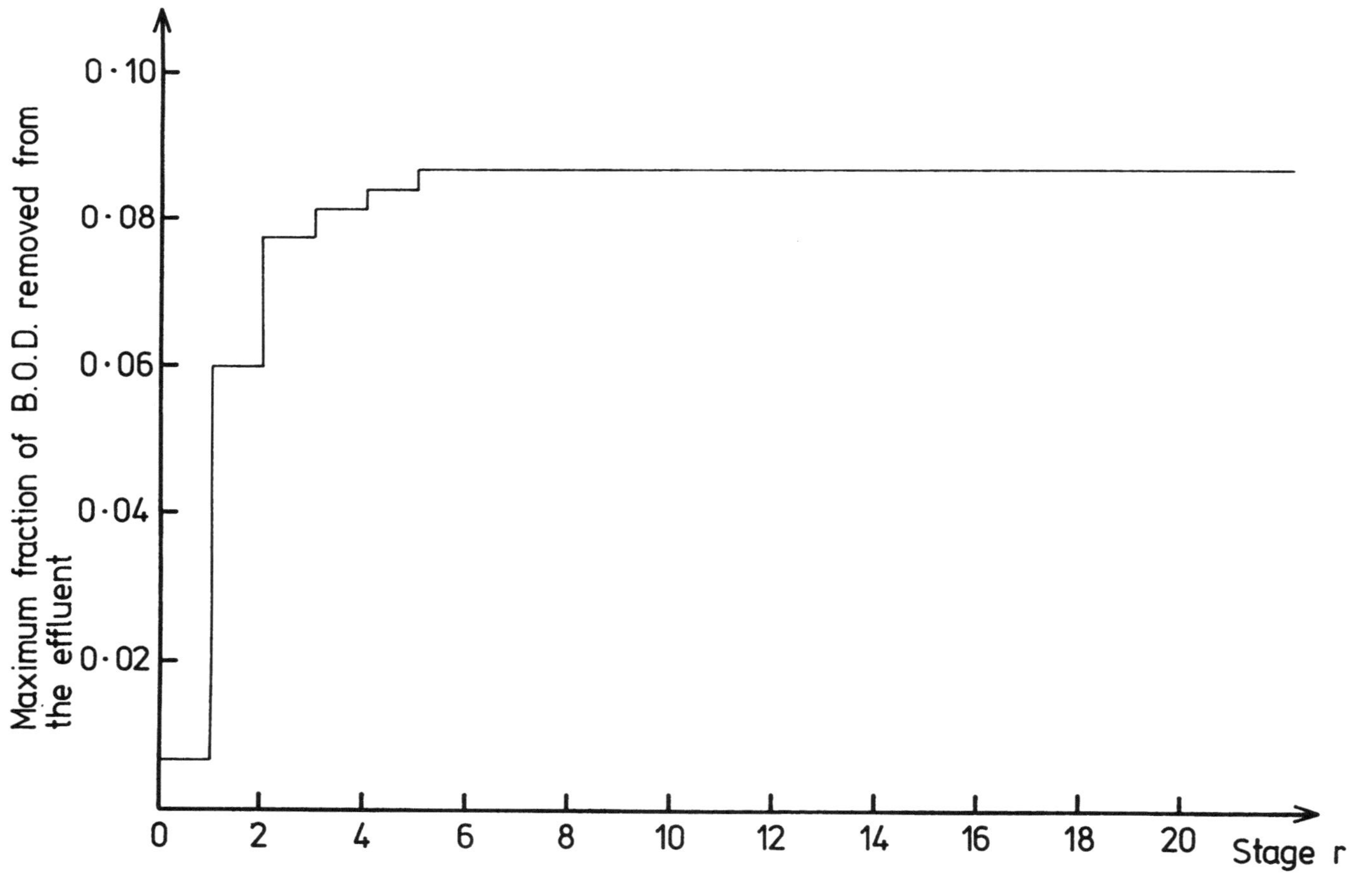

Fig. (8.11) Optimal control sequence for Reach 2

Example 2

A steam power system model is described by:

$$A = \left[\begin{array}{cccc|cccc} .835 & 0 & 0 & 0 & 0 & 0 & 0 & 0 \\ .096 & .861 & 0 & 0 & 0 & 0 & 0 & 0 \\ -.002 & -.005 & .882 & -.253 & .041 & -.003 & -.025 & -.001 \\ .007 & .014 & -.029 & .928 & 0 & .006 & .059 & .002 \\ \hline -.03 & -.061 & 2.028 & -2.303 & .088 & -.021 & -.224 & -.008 \\ .048 & .758 & 0 & 0 & 0 & .165 & 0 & .023 \\ -.012 & -.027 & 1.209 & -1.4 & .161 & -.013 & .156 & .006 \\ .815 & 0 & 0 & 0 & 0 & 0 & 0 & .011 \end{array}\right]$$

$$B = \left[\begin{array}{c|c} .003 & 0 \\ .001 & 0 \\ .294 & 0 \\ -.038 & 0 \\ \hline 0 & -.051 \\ 0 & .056 \\ 0 & -.015 \\ 0 & 2.477 \end{array}\right]$$

$$C = \left[\begin{array}{cccc|cccc} 0 & 1 & 0 & 0 & 0 & 0 & 0 & 0 \\ 0 & 0 & 0 & 1 & 0 & 0 & 0 & 0 \\ \hline 0 & 0 & 0 & 0 & 0 & 1 & 0 & 0 \end{array}\right]$$

This model is of the form (8.84). It can be decomposed into two subsystems, $N_s = 2$, along the dotted lines to yield the form (8.83). Our purpose is to derive the closed-loop optimal control sequence by applying the results of Section 8.5.3, which minimizes a cost functional of the form (8.85) with

$$Q_1 = \begin{bmatrix} 1 & 0 \\ 0 & 1 \end{bmatrix}, \qquad Q_2 = [1]$$

$$R_1 = R_2 = [1]$$

$$Q_{f_1} = Q_{f_2} = W_1 = W_2 = [0]$$

$$N \to \infty \quad \text{(infinite time-horizon)}$$

The solution of this problem should take the form (8.102), and hence our objective is to compute the gains G_d and G_0. We first consider the computation of the gain matrix G_d when the two subsystems are decoupled. The solutions of the independent Riccati equations, each of the type (8.100a), truncated to four decimals, are given by:

$$P_1 = \begin{bmatrix} .0652 & .3037 & -.0004 & .0184 \\ .3037 & 3.1997 & -.0135 & -.0201 \\ -.0004 & -.0135 & .2680 & .0479 \\ .0184 & -.0201 & .0479 & 3.3763 \end{bmatrix}$$

$$P_2 = \begin{bmatrix} .001 & .0001 & .0018 & 0 \\ .0001 & 1.0297 & .0013 & .0039 \\ .0018 & .0013 & .0149 & .0005 \\ 0 & .0039 & .0005 & .0006 \end{bmatrix}$$

From (8.98) and (8.102) we get:

$$G_d = \begin{bmatrix} .0004 & -.008 & .1113 & .1381 & 0 & 0 & 0 & 0 \\ 0 & 0 & 0 & 0 & .0004 & .0093 & .0034 & .0012 \end{bmatrix}$$

Following the procedure in *[44]* and by simulating the system model from the initial conditions,

$$\underline{x}(0) = \begin{bmatrix} 1 \\ 0 \\ \cdot \\ \cdot \\ \cdot \\ 0 \end{bmatrix}, \begin{bmatrix} 0 \\ 1 \\ \cdot \\ \cdot \\ \cdot \\ 0 \end{bmatrix}, \ldots \begin{bmatrix} 0 \\ 0 \\ \cdot \\ \cdot \\ \cdot \\ 1 \end{bmatrix}$$

it was found that the gain matrix G_0 has the value:

$$G_0 = \begin{bmatrix} -.0003 & .0013 & -.022 & .0254 & -.0084 & -.0006 & -.0103 & -.0004 \\ -.179 & -.6633 & -.0012 & -.0401 & -.0001 & 0 & 0 & -.0002 \end{bmatrix}$$

The overall optimal control is thus given by:

$$\underline{u} = \begin{bmatrix} -.0007 & .0093 & -.1334 & -.1127 & -.0084 & -.0006 & -.0103 & -.0004 \\ -.179 & -.6633 & -.0012 & -.0401 & -.0005 & -.0093 & -.0034 & -.0014 \end{bmatrix} \underline{x}$$

Now to interpret the simulation results, we note that the open loop system has the eigenvalues $\{.8745 \pm j.1696, .861, .835, .2866, .165, .0184, .011\}$ which, in the light of Chapter 4, is a two-time-scale system. In fact the first subsystem is approximately the slow subsystem and the second subsystem is the fast subsystem. The closed-loop eigenvalues are given by $\{.8356 \pm j.1744, .5004 \pm j.2942, .286, .1663, .0185, .0109\}$. A simple comparison shows that the fast eigenvalues remain nearly undisturbed whereas two real slow eigenvalues have formed a complex conjugate pair. This is because the two subsystems are weakly coupled, the input matrix B is small in magnitude and the penalty matrices are of equal weight. The reader is encouraged to work out the example with different weighting matrices. Another point to note is that G has entries with small values, which again is due to the effect of weak coupling.

8.6 Decomposition-Decentralisation Approach

We have seen in the last section that the hierarchical optimization methodology provides decomposition and multilevel control schemes to deal with linear interconnected dynamical systems. Such schemes are iterative in nature and they consider that the source of coupling among subsystems is only due to the off-diagonal blocks in the A matrix, (see (8.83b)). Although almost any computational technique embodies some iterative procedures, yet it would be desirable in control design to have explicit feedback gains that can be computed with minimal off-line calculations. The case in which the B matrix contains nonzero off-diagonal blocks, implying that the individual controls of the subsystems are interacting, can be dealt with in the same way at the expense of having more coordination variables. This, in turn, will complicate the updating mechanisms at higher levels.

In this section we examine the linear discrete systems in which A and B are full matrices. Our purpose is to develop a multilevel structure that computes the feedback control scheme with the least amount of iterative computation.

8.6.1 *STATEMENT OF THE PROBLEM*

Let us consider that the linear discrete model be composed of N_s interconnected subsystems and described by:

$$\underline{i = 1,\ldots,N_s}$$

$$\underline{x}_i(k+1) = A_i\underline{x}_i(k) + B_i\underline{u}_i(k) + \underline{h}_i(k) \; ; \qquad \underline{x}_i(0) = \underline{x}_{i0}$$

$$\underline{y}_i(k) = C_i\,\underline{x}_i(k) \tag{8.104}$$

$$\underline{h}_i(k) = \sum_{j=1}^{N_s} [D_{ij}\underline{x}_j(k) + F_{ij}\underline{u}_j(k)]$$

where $\underline{x}_i(k)$ is an n_i dimensional state vector, $u_i(k)$ is an m_i dimensional control vector, $\underline{y}_i(k)$ is a p_i dimensional output vector and $\underline{h}_i(k)$ is an n_i dimensional vector representing the coupling between the ith subsystem and the remaining subsystems. The block-matrices D_{ij}, F_{ij}, $i,j = 1,\ldots,N_s$, denote the off-diagonal blocks in the overall A, B matrices, respectively. The cost function to be minimized is a quadratic form of the outputs and controls, that is:

$$J_i = \frac{1}{2}\sum_{k=0}^{\infty} \{||\underline{y}_i(k)||^2_{Q_i} + ||\underline{u}_i(k)||^2_{R_i} \qquad (8.105)$$

where $||\underline{a}||^2_G = \underline{a}^t G \underline{a}$, Q_i is $(p_i \times p_i)$ symmetric nonnegative-definite and R_i is $(m_i \times m_i)$ symmetric positive-definite. To ensure a finite solution, the usual assumptions that the N_s-pairs (A_i, B_i) are completely reachable and the N_s-pairs (A_i, C_i) are completely observable are made. We note that:

(1) The overall model of (8.104) can be put in the form (8.16) such that:

$$A = \begin{bmatrix} A_1+D_{11} & D_{12} & \cdots & D_{1N_s} \\ D_{21} & A_2+D_{22} & & \vdots \\ \vdots & & \ddots & \vdots \\ D_{N_s1} & \cdots & \cdots & A_{N_s}+D_{N_sN_s} \end{bmatrix}$$

$$B = \begin{bmatrix} B_1+F_{11} & F_{12} & \cdots & F_{1N_s} \\ F_{21} & B_2+F_{11} & & \vdots \\ \vdots & & \ddots & \vdots \\ F_{N_s1} & \cdots & \cdots & B_{N_s}+F_{N_sN_s} \end{bmatrix}$$

$$C = [C_1 \quad C_2 \quad \cdots \quad C_{N_s}]$$

$$\underline{x} = \begin{bmatrix} \underline{x}_1 \\ \vdots \\ \underline{x}_{N_s} \end{bmatrix}; \quad \underline{u} = \begin{bmatrix} \underline{u}_1 \\ \vdots \\ \underline{u}_{N_s} \end{bmatrix}; \quad \underline{y} = \begin{bmatrix} \underline{y}_1 \\ \vdots \\ \underline{y}_{N_s} \end{bmatrix}$$

(2) In a similar way, the overall cost can be put in the form (8.17) with $N \to \infty$, $Q_f = 0$ and

$$Q = \text{diag}[Q_1 \ \dots \ Q_{N_s}]$$

$$R = \text{diag}[R_1 \ \dots \ R_{N_s}]$$

such that

$$J \ (= J_N \text{ as } N \to \infty) = \sum_{i=1}^{N_s} J_i$$

which means that the integrated cost function is separable.

Our objective is to determine the optimal sequence $\{\underline{u}_j(k)\}$, $j = 1,\dots,N_s$ which minimizes (8.105) for the N_s subsystems such that the equality constraints (8.104) are satisfied.

8.6.2 *THE DECOUPLED SUBSYSTEMS*

We start by considering the case in which the coupling term $\underline{h}_j(k)$ is identically zero. This implies that the N_s subsystems are completely decoupled, hence from (8.104) we get their dynamics:

$$\underline{i = 1,\dots,N_s}$$

$$\underline{x}_i(k+1) = A_i \underline{x}_i(k) + B_i \underline{u}_i(k) \ ; \qquad \underline{x}_i(0) = \underline{x}_{i0}$$

$$y_i(k) = C_i \ \underline{x}_i \tag{8.106}$$

Under the decoupling condition, the problem at hand reduces to the derivation of the optimal sequences $\{\underline{u}_i(k)\}$ which can be applied to (8.106) while minimizing (8.105). It is easy to see that each subsystem optimization problem is a particular version of the problem addressed in Section 8.3.2 . Consequently, the optimal control of the ith subsystem is given by:

$$\underline{u}_i^*(k) = -R_i^{-1} B_i^t P_i [I_i + B_i B_i^{-1} B_i^t P_i]^{-1} A_i \underline{x}_i(k)$$

$$= -G_i^* \ \underline{x}_i(k) \tag{8.107}$$

where P_i is the positive semi-definite solution of the discrete algebraic Riccati equation

$$P_i = C_i^t Q_i C_i + A_i^t P_i [I_i + B_i R_i^{-1} B_i^t P_i]^{-1} A_i \tag{8.108}$$

where I_i is the $(n_i x n_i)$ identity matrix. The associated optimal cost has the value:

$$J^* = \frac{1}{2} \underline{x}_{i0}^t P_i \underline{x}_{i0} \tag{8.109}$$

It should be emphasized that the optimal feedback control (8.107) provides a basic regulation task of stabilizing the decoupled system (8.106).

8.6.3 *MULTI-CONTROLLER STRUCTURE*

We now consider the coupled system (8.104). In view of the fact that the difference between (8.104) and (8.106) is the coupling pattern, we define

$$\underline{u}_i(k) = \underline{u}_i^*(k) + \underline{u}_i^c(k) \tag{8.110}$$

where $\underline{u}_i(k)$ is the control input to the system (8.104) and $\underline{u}_i^c(k)$ is a corrective control component to handle the effect of interactions. It is interesting to compare (8.102) and (8.110) and realize that $\underline{u}_i^c(k)$ has the role of $G_0\underline{x}(k)$. However, our analysis of the component $\underline{u}_i^c(k)$ will be fundamentally different.

Let us fix the coupling term at known sequences, that is:

$$\underline{h}_i(k) \quad \sum_{j=1}^{N_s} [D_{ij}\underline{x}_j^f(k) + F_{ij}\underline{u}_j^f(k)] \tag{8.111}$$

The substitution of (8.110), (8.111) into (8.104), using (8.107) leads us to:

$$\begin{aligned}\underline{x}_i(k+1) = \; & [A_i - B_i G_i]\underline{x}_i(k) + B_i \underline{u}_i^c(k) \\ & +\{\sum_{j=1}^{N_s} [D_{ij}\underline{x}_j^f(k) + F_{ij}\underline{u}_j^f(k)]\}\end{aligned} \tag{8.112}$$

Depending on the objective of the design problem and the nature of interaction, one may have different ways of determining $\underline{u}_i^c(k)$, $\underline{x}_j^f(k)$, $\underline{u}_j^f(k)$. Here we consider the effect of $\underline{x}_j^f(k)$, $j = 1,\ldots,N_s$, on the state $\underline{x}_i(k)$ as harmful causing undesirable deviations. In order to reduce these undesirable deviations, we choose:

$$\underline{u}_i^c(k) = -\sum_{j=1}^{N_s} L_{ij}\, \underline{x}_j^f(k)$$

or

$$\underline{u}^c(k) = -L\underline{x}^f(k) \tag{8.113}$$

Summing up (8.112) over the subsystems and making use of (8.113), we arrive at:

$$\underline{x}(k+1) = (A-BG)\underline{x}(k)+[D-BL]\underline{x}^f(k)+F\underline{u}^f(k) \tag{8.114}$$

where $A = \text{diag}\{A_i\}$, $B = \text{diag}\{B_i\}$, $G = \text{diag}\{G_i\}$.

It is important to note that the matrix $[D-BL]$ depends on L only. This matrix reduces to the null matrix if and only if [45]:

$$\text{Rank}[B] = \text{Rank}[B \quad D] = m \tag{8.115}$$

in which case the unknown gains L are given by:

$$L = [B^tB]^{-1}B^tD \tag{8.116}$$

The rationale behind nullifying the interaction matrix $[D-BL]$ is quite intuitive; it simply leaves the original system (8.104) partially decoupled except for the term $F\underline{u}^f(k)$.

In practice, the rank condition (8.115) is rarely satisfied. Thus, in most cases a residual interaction term will arise and take the form:

$$\begin{aligned} E &= D-BL \\ &= D-B[B^tB]^{-1}B^tD \\ &= [I-B[B^tB]^{-1}B^t]D \end{aligned} \tag{8.117}$$

Note that the effect of the E matrix is against the autonomy of the individual subsystems and the feedback gain L has a global nature, that is obtained from the overall state vector, see (8.113).

We now turn to the term $F\underline{u}^f(k)$. In order to provide for some improvement in the overall feedback scheme, we select:

$$F\underline{u}^f(k) = -SG^*\underline{x}(k) \tag{8.118}$$

to emphasize that the interaction among the control signals will strengthen the overall feedback scheme. Here S is an arbitrary gain matrix of proper dimension. The combined use of (8.117), (8.118) in (8.114) yields:

$$\underline{x}(k+1) = [A-BG]\underline{x}(k) + E\underline{x}^f(k) - SG^*\underline{x}(k)$$

which at the optimum, $\underline{x}^f(k) = \underline{x}(k)$, becomes:

$$\underline{x}(k+1) = (A+E)\underline{x}(k) - (B+S)G^*\underline{x}(k) \tag{8.119}$$

The interpretation of (8.119) is that the result of the control actions (8.113) and (8.118) is to perturb the decoupled matrices in (8.106) by the terms E and S, that is the system matrix becomes $\text{diag}\{A_i\} + E$ and the input matrix becomes $\text{diag}\{B_i\}+S$. We are now left with the way of evaluating the matrix S. Since the L matrix will eventually cancel out the undesirable deviations, some loss in performance index will be incurred. To partially eliminate this loss and to provide for some improvement in the performance of the overall system, we require that both the decoupled system (8.106) and the new structured system (8.119) have the same gain. This will be explained in the following analysis.

Consider the minimization of the cost function J, formed by summing up (8.105), subject to the dynamic constraint:

$$\underline{x}(k+1) = (A+E)\underline{x}(k) + (B+S)\underline{u}(k)$$
$$\underline{y}(k) = C\underline{x}(k) \tag{8.120}$$

which is quite similar to (8.119) by replacing $-G^{*}\underline{x}(k)$ by $\underline{u}(k)$.

In the light of the analysis of Section 8.3.2, the result is:

$$\underline{u}^{0}(k) = -R^{-1}(B+S)^{t}Y[I_{n}+(B+S)R^{-1}(B+S)^{t}Y]^{-1}(A+E)\underline{x} \tag{8.121}$$

where Y is the symmetric, positive semidefinite solution of

$$Y = C^{t}QC + (A+E)^{t}Y[I_{n}+(B+S)R^{-1}(B+S)^{t}Y]^{-1}(A+E) \tag{8.122}$$

and the associated cost is

$$J^{0} = \frac{1}{2}\,\underline{x}_{0}^{t}\,Y\,\underline{x}_{0} \tag{8.123}$$

From (8.107) and (8.108), the feedback gain can be expressed as:

$$G^{*} = R^{-1}B^{t}A^{-t}[P-C^{t}QC] \tag{8.124}$$

and from (8.121), (8.122) the corresponding gain is given by:

$$G^{0} = R^{-1}(B+S)^{t}(A+E)^{-t}[Y-C^{t}QC] \tag{8.125}$$

By setting $G^{*} = G^{0}$, we get:

$$S = (A+E)[Y-C^{t}QC]^{-1}[P-C^{t}QC]A^{-1}B-B \tag{8.126}$$

which defines the input perturbation matrix. Had we followed another route based on Section 8.3.1, we could arrive at an alternative form of S ussng the equality

$$R^{-1}(B+S)^{t}Y = R^{-1}B^{t}P \tag{8.127a}$$

which implies that

$$S = Y^{-1}PB - B \tag{8.127b}$$

Define

$$\left.\begin{aligned} X &= P - Y \\ P_1 &= P - C^tQC \\ Y_1 &= Y - C^tQC \end{aligned}\right\} \tag{8.128}$$

Manipulating (8.107), (8.122), (8.127) and (8.128) together, we arrive at:

$$Y_1(A+E)^{-1} - P_1A^{-1} = (A+E)^t(X+P) - A^tP \tag{8.129}$$

We can further simplify (8.129) with the aid of (8.128) to get explicit expression of X:

$$X = (A+E)^tX(A+E) - E^tP(A+E) + (C^tQC-P)A^{-1}E \tag{8.130}$$

Note that the above expression is independent of Y and thus avoids the solution of (8.122). The real benefit of (8.130) is the improvement in the performance index as given by:

$$\begin{aligned} \tilde{J} &= J^* - J^0 \\ &= \frac{1}{2}[\underline{x}_0^tP\underline{x}_0 - \underline{x}_0^tY\underline{x}_0] \\ &= \frac{1}{2}\,\underline{x}_0^t[P-Y]\underline{x}_0 \\ &= \frac{1}{2}\,\underline{x}_0^t\,X\,\underline{x}_0 \end{aligned} \tag{8.131}$$

It is easy to see that the solution of (8.130) would be positive semidefinite if the residual interaction term E satisfies the inequality

$$||C^tQCA^{-1}E|| \geq ||PA^{-1}E + E^tP(A+E)||$$

Under this condition, (8.131) yields $\tilde{J} \geq 0$, and more importantly we have:

$$J^+ < J^0 < J^*$$

which stresses the fact that the optimal cost of the new structured system J^0 is better than the decoupled optimal cost J^*; and both of them are generally less than the global cost value J^+.

To summarize, the design procedure is given by the following steps:

(1) Solve (8.108) for the N_s decoupled Riccati matrices $\{P_j\}$ and use them in (8.107) to compute the N_s decoupled feedback gains $\{G_j^*\}$.

(2) Use (8.116) to compute the global gain L and from (8.117) obtain the interaction matrix E.

(3) From (8.122) and (8.127a) we obtain:

$$Y = C^tQC+(A+E)^tY[I_n+Y^{-1}PBR^{-1}B^tP]^{-1}(A+E) \tag{8.132}$$

which can be solved to yield the Y matrix.

(4) The S matrix is computed from (8.127b).

(5) Solve (8.130) to obtain the cost perturbation matrix X and then from (8.131) we get the improvement in the performance index.

Figure 8.12 shows a block-diagram of the multicontroller structure.

Next, we illustrate the above analysis by two examples.

8.6.4 *EXAMPLES*

Example 1

A ninth-order model of a tubular ammonia reactor *[46]* is written in the format (8.104) with $N_s = 1$, $n_1 = 5$, $n_2 = 4$, $m_1 = 1$, $m_2 = 2$, and the subsystem matrices are:

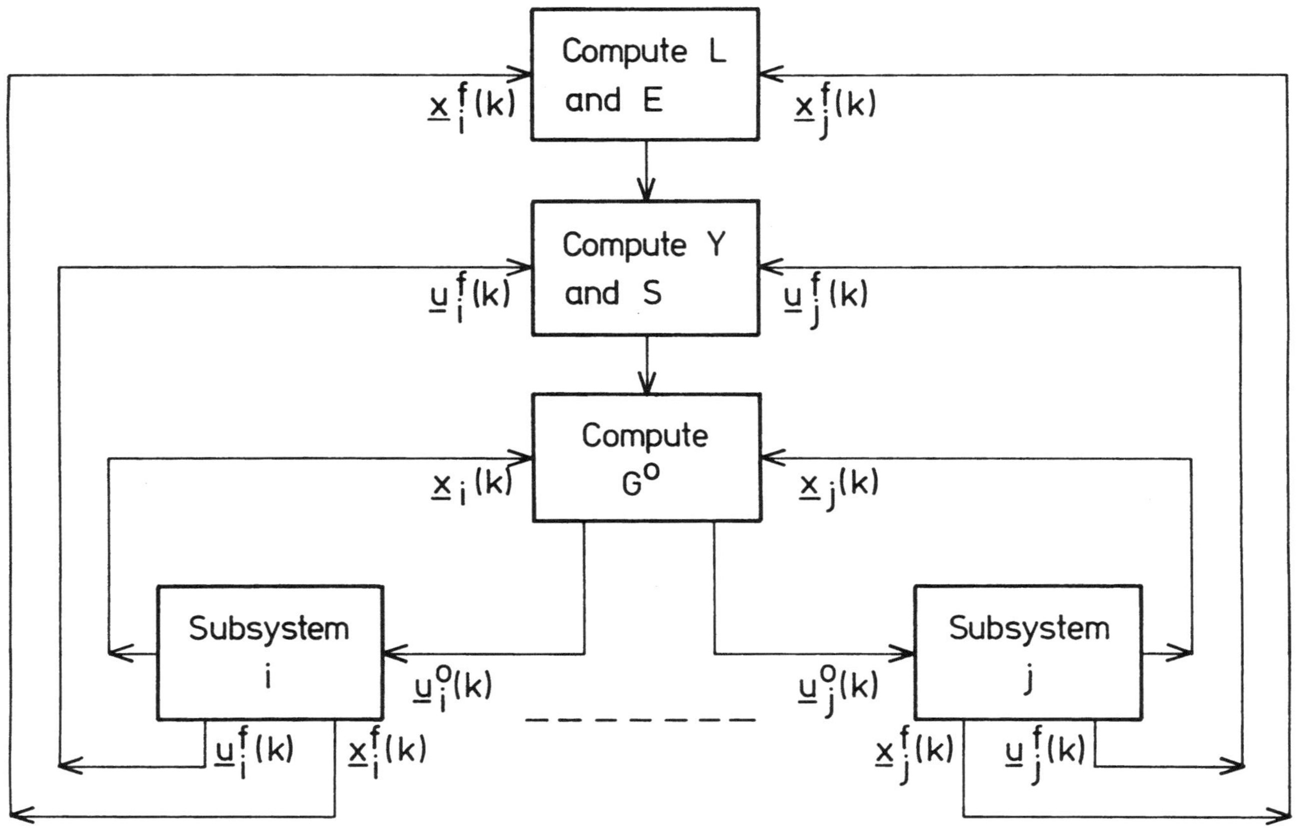

Fig. (8.12) Multi-controller structure

$$A_1 = \begin{bmatrix} .8701 & .135 & .1159\times10^{-1} & .5014\times10^{-3} & -.3722\times10^{-1} \\ .7665\times10^{-1} & .8974 & .1272\times10^{-1} & .5504\times10^{-3} & -.4016\times10^{-1} \\ -.1272 & .3975 & .817 & .1455\times10^{-2} & -.1028 \\ -.3635 & .6339 & .7491\times10^{-1} & .7966 & -.2735 \\ -.96 & .1646\times10^{-1} & -.1289 & -.5597\times10^{-2} & .7142\times10^{-1} \end{bmatrix}$$

$$A_2 = \begin{bmatrix} .136\times10^{-1} & 0 & .1443 & .1061 \\ .1249\times10^{-1} & .1063\times10^{-3} & .9997\times10^{-1} & .6967\times10^{-1} \\ .2216\times10^{-1} & 0 & .2139 & .3554\times10^{-1} \\ .1986\times10^{-1} & 0 & .2191 & .2152 \end{bmatrix}$$

$$B_1 = \begin{bmatrix} .476\times10^{-3} \\ .879\times10^{-4} \\ .1482\times10^{-4} \\ .3892\times10^{-3} \\ .1034\times10^{-2} \end{bmatrix} ; \quad B_2 = \begin{bmatrix} -.6159\times10^{-2} & .3834\times10^{-2} \\ -.3683\times10^{-2} & .2029\times10^{-2} \\ -.1554\times10^{-2} & .6937\times10^{-3} \\ -.302\times10^{-2} & .1469\times10^{-2} \end{bmatrix}$$

$$D_{11} = D_{22} = 0$$

$$D_{12} = \begin{bmatrix} .3484\times10^{-3} & 0 & .4282\times10^{-2} & .7249\times10^{-2} \\ .3743\times10^{-3} & 0 & .453\times10^{-2} & .7499\times10^{-2} \\ .987\times10^{-3} & 0 & .1185\times10^{-1} & .1872\times10^{-1} \\ .2653\times10^{-2} & 0 & .317\times10^{-1} & .4882\times10^{-1} \\ .7108\times10^{-2} & 0 & .8452\times10^{-1} & .1259 \end{bmatrix}$$

$$D_{22} = \begin{bmatrix} -.6644 & .1129\times10^{1} & -.8889\times10^{-1} & -.3854\times10^{-2} & .8447\times10^{-1} \\ -.4102 & .693 & -.5471\times10^{-1} & -.2371\times10^{-1} & .6649\times10^{-1} \\ -.1799 & .3017 & -.2393\times10^{-1} & -.1035\times10^{-2} & .6059\times10^{-1} \\ -.3451 & .5804 & -.4596\times10^{-1} & -.1989\times10^{-2} & .1056 \end{bmatrix}$$

$$F_{11} = F_{22} = 0$$

$$F_{12} = \begin{bmatrix} -.571\times10^{-4} & -.8368\times10^{-2} \\ -.4773\times10^{-3} & -.273\times10^{-3} \\ -.1312\times10^{-2} & .8876\times10^{-3} \\ -.3513\times10^{-2} & .248\times10^{-2} \\ -.9275\times10^{-2} & .668\times10^{-2} \end{bmatrix}$$

$$F_{21} = \begin{bmatrix} .7203x10^{-3} \\ .4454x10^{-3} \\ .1971x10^{-3} \\ .3773x10^{-3} \end{bmatrix}$$

The weighting matrices are given by:

$$Q_1 = 50\ I_5,\quad Q_2 = 50\ I_4,\quad R_1 = I_1,$$

$$R_2 = I_2,\qquad C_1 = I_5,\qquad C_2 = I_4$$

and the initial state is:

$$\underline{x}_0 = [.1 \quad .001 \quad 1 \quad .9 \quad .5 \quad .001 \quad 0 \quad .5 \quad .1]$$

Solution of (8.108) results in the local Riccati matrices:

$$P_1 = \begin{bmatrix} 1456.4 & 638.6 & 255.61 & -18.04 & -112.26 \\ 638.6 & 1323.4 & 296.06 & 105.77 & -145.22 \\ 255.61 & 296.06 & 251.14 & 40.48 & -62.24 \\ -18.04 & 105.77 & 40.48 & 138.27 & -38.91 \\ -112.26 & -145.22 & -62.24 & -38.91 & 81.97 \end{bmatrix}$$

$$P_2 = \begin{bmatrix} 50.07 & .661x10^{-1} & .7085 & .4156 \\ .661x10^{-4} & 50 & .5282x10^{-3} & .3681x10^{-3} \\ .7085 & .5282x10^{-3} & 57.18 & 4.331 \\ .4156 & .3681x10^{-3} & 4.331 & 53.43 \end{bmatrix}$$

From (8.117) the residual interaction matrix E, in the form:

$$E = \begin{array}{c} \\ 5 \\ 4 \end{array}\begin{array}{c} \begin{array}{cc} 5 & 4 \end{array} \\ \left[\begin{array}{c|c} E_{11} & E_{12} \\ \hline E_{21} & E_{22} \end{array}\right] \end{array}$$

has the value

$$E_{11} = E_{22} = 0$$

$$E_{12} = \begin{vmatrix} -.2464\times10^{-2} & 0 & -.292\times10^{-1} & -.4305\times10^{-1} \\ -.1451\times10^{-3} & 0 & -.1655\times10^{-2} & -.179\times10^{-2} \\ .1113\times10^{-3} & 0 & .1422\times10^{-2} & .306\times10^{-2} \\ .3532\times10^{-3} & 0 & .4313\times10^{-2} & .7693\times10^{-2} \\ .9979\times10^{-3} & 0 & .1176\times10^{-1} & .1664\times10^{-1} \end{vmatrix}$$

$$E_{21} = \begin{vmatrix} .1861\times10^{-3} & -.2322\times10^{-3} & .1693\times10^{-4} & .1031\times10^{-5} & .5174\times10^{-2} \\ -.4876\times10^{-3} & .5637\times10^{-3} & -.3929\times10^{-4} & -.2667\times10^{-5} & -.1759\times10^{-1} \\ .557\times10^{-3} & -.8344\times10^{-3} & .661\times10^{-4} & .3189\times10^{-5} & .2535\times10^{-2} \\ -.5233\times10^{-4} & .1857\times10^{-3} & -.1808\times10^{-4} & -.3814\times10^{-6} & .9585\times10^{-2} \end{vmatrix}$$

Solving (8.132) for Y and then substituting in (8.127b) we get:

$$S = \begin{vmatrix} .6902\times10^{-4} & -.2793\times10^{-3} & .1366\times10^{-3} \\ .114\times10^{-4} & .215\times10^{-4} & .1053\times10^{-4} \\ -.5614\times10^{-5} & .1217\times10^{-4} & -.5947\times10^{-5} \\ -.2417\times10^{-5} & .632\times10^{-5} & -.3068\times10^{-5} \\ .8486\times10^{-4} & .2231\times10^{-5} & -.1064\times10^{-5} \\ .5071\times10^{-4} & .218\times10^{-5} & -.1069\times10^{-5} \\ -.9109\times10^{-8} & -.4657\times10^{-9} & -.4657\times10^{-9} \\ .5749\times10^{-3} & .2433\times10^{-4} & -.119\times10^{-4} \\ .671\times10^{-3} & .2981\times10^{-4} & .1873\times10^{-4} \end{vmatrix}$$

Computation of the cost functionals for the integrated, decoupled and new-structured systems give the values:

$$J^{+} = 388.2, \quad J^{*} = 428.5, \quad J^{0} = 405.5$$

which clearly justify the inequality $J^{+} < J^{0} < J^{*}$. The amount of performance improvement $\tilde{J}$ is 5.3676% .

Example 2

A mathematical model of an a.c. turbogenerator system comprises a synchronous generator tied, via a step-up transformer and transmission line, to an infinite busbar developed in *[47]*, has been linearized about an operating point of a 37.5 MW generator *[48]* and then sampled every .05 seconds. The resulting model is of order 6 and has the rotor angle, rotor velocity, field flux linkage, field voltage, steam power and mechanical torque input as the state variables. The model explicitly consists of two third-order subsystems as follows:

$$A_1 = \begin{bmatrix} .958 & 0 & -.038 \\ 0 & .607 & 0 \\ -.017 & .007 & .973 \end{bmatrix}$$

$$A_2 = \begin{bmatrix} .888 & .059 & 1.338 \\ 0 & .607 & 0 \\ 0 & .075 & .905 \end{bmatrix}$$

$$B_1 = \begin{bmatrix} 0 \\ .393 \\ .002 \end{bmatrix} ; \quad B_2 = \begin{bmatrix} .015 \\ .559 \\ .029 \end{bmatrix}$$

$$D_{11} = D_{22} = 0$$

$$D_{12} = \begin{bmatrix} .048 & .001 & .035 \\ 0 & 0 & 0 \\ 0 & 0 & 0 \end{bmatrix}$$

$$D_{21} = \begin{bmatrix} -1.628 & -.006 & -1.468 \\ 0 & 0 & 0 \\ 0 & 0 & 0 \end{bmatrix}$$

$$F_{11} = F_{12} = F_{22} \quad 0$$

$$F_{21} = \begin{bmatrix} -.001 \\ 0 \\ 0 \end{bmatrix}$$

The weighting matrices are: $R_1 = .5$, $R_2 = 1$, $C_1 = C_2 = I_3$ and

$$Q_1 = \begin{bmatrix} 200 & 0 & 1.684 \\ 0 & 0 & 0 \\ 1.684 & 0 & 124 \end{bmatrix}$$

$$Q_2 = \begin{bmatrix} 0 & 0 & 0 \\ 0 & 10 & 0 \\ 0 & 0 & 200 \end{bmatrix}$$

Computer simulation of the turbogenerator model yields the local Riccati matrices:

$$P_1 = \begin{bmatrix} 2283 & -7.8591 & -594.25 \\ -7.8591 & .2366 & 14.751 \\ -594.25 & 14.751 & 1107 \end{bmatrix}$$

$$P_2 = \begin{bmatrix} .6538\text{x}10^{-2} & .201\text{x}10^{-2} & .3549\text{x}10^{-1} \\ .201\text{x}10^{-2} & 12.198 & 20.816 \\ .3549\text{x}10^{-1} & 20.816 & 539.12 \end{bmatrix}$$

The residual interaction matrix E has the values:

$$E = \begin{bmatrix} 0 & 0 & 0 & -.048 & .01 & .035 \\ 0 & 0 & 0 & 0 & 0 & 0 \\ 0 & 0 & 0 & 0 & 0 & 0 \\ -1.6268 & -.5996\text{x}10^{-2} & -1.4669 & 0 & 0 & 0 \\ .4354\text{x}10^{-1} & .1605\text{x}10^{-5} & .3926\text{x}10^{-1} & 0 & 0 & 0 \\ .2259\text{x}10^{-2} & .8324\text{x}10^{-5} & .2037\text{x}10^{-2} & 0 & 0 & 0 \end{bmatrix}$$

and the input perturbation matrix S is given by:

$$S = \begin{bmatrix} -.718\text{x}10^{-2} & .3174\text{x}10^{-1} \\ -.4008 & 3.3465 \\ .6008\text{x}10^{-2} & -.9888\text{x}10^{-2} \\ .1988\text{x}10^{-1} & -.2286 \\ .2447\text{x}10^{-1} & -.9631\text{x}10^{-1} \\ .4868\text{x}10^{-3} & -.1344\text{x}10^{-1} \end{bmatrix}$$

For the initial state $\underline{x}(0) = [.2 \quad .001 \quad .002 \quad 10^{-5} \quad .1 \quad 0]$ the cost functions for the integrated, decoupled and new-structured system have the values:

$$J^{+} = 35.4, \quad J^{*} = 91, \quad J^{0} = 38.4$$

which shows that $J^{+} < J^{0} < J^{*}$ as expected. In addition, it is interesting to observe that the overall improvement in performance index is 57.69%.

8.7 Concluding Remarks

The purpose of this chapter has been to present approaches and computational algorithms for the dynamic optimization of linear shift-invariant models with quadratic cost functionals. Both single and interconnected systems have been considered.

In Section 8.2, the optimality conditions have been derived in a general setting and then specialized in Section 8.3 to linear, shift-invariant models. Properties of the optimal control law and the associated Riccati matrix are examined in the steady-state. In particular, the behaviour of the closed-loop poles for limiting values of the control penality matrix, is illustrated. Computational algorithms to solve the algebraic Riccati equation are discussed in Section 8.4, with the primary focus on the generalized eigenvalue approach.

Sections 8.5 and 8.6 have been devoted to interconnected systems. The design of hierarchical structures for both open-loop and closed-loop control schemes has been treated in Section 8.5 . A decomposition-decentralization approach has been introduced in Section 8.6 to build up a multi-controller structure capable of computing the feedback control scheme in a non-iterative manner. It should be emphasized that the material covered in this chapter is only a part of the literature. The interested reader is referred to the references cited here and at this stage we terminate our discussions.

8.8 Problems

1. Find the optimal control for the system

$$x_1(k+1) = x_1 + 2u(k) ; \qquad x_1(0) = 1 ;$$

$$x_2(k+1) = .5x_1(k) + 2x_2(k) ; \qquad x_2(0) = 0$$

where the cost function to be minimized is:

$$J = \sum_{j=0}^{\infty} [x_1^2(j) + 2x_2^2(j) + u^2(j)]$$

2. Consider the minimization of the cost functional:

$$j = \frac{1}{2} \underline{x}^t(N) Q_f \underline{x}(N)$$

$$+ \frac{1}{2} \sum_{j=0}^{N-1} \{\underline{x}^t(j) Q_1 \underline{x}(j) + \underline{u}^t(j) R_1 \underline{u}(j)$$

$$+ \underline{x}^t(j) Q_2 \underline{u}(j) + \underline{u}^t(j) R_2 \underline{x}(j)$$

$$+ Q_3 \underline{x}(j) + R_3 \underline{u}(j)\}$$

subject to:

$$\underline{x}(k+1) = A\underline{x}(k) + B\underline{u}(k)$$

Show that the optimal control law has the form:

$$\underline{u}^*[\underline{x}(k)] = D_1(k)\underline{x}(k) + D_2\underline{u}(k)$$

Derive recursive relations for $D_1(k)$ and $D_2(k)$.

3. A linear dynamical system is described by:

$$A = \begin{bmatrix} 1 & .01 & 0 \\ 0 & 1 & 7.561 \\ 0 & -.35 & 1 \end{bmatrix} \; ; \quad B = \begin{bmatrix} 0 \\ 0 \\ .04 \end{bmatrix}$$

It is required to find the Riccati matrix when

$$C = 1 \quad 0 \quad 0 \quad , \quad Q =$$
$$R = 1 \qquad , \quad N \quad \text{large}$$
$$\text{for} \quad = 10^{-2}, 1, 10^{2}, 10^{4} .$$

Comment on the results.

4. A linearized model of a chemical process is described by:

$$A = \begin{bmatrix} .989 & 9.5 & 0 & -560 & 0 & 4.14 & 0 & 0 \\ 0 & 1 & .25 & 0 & 0 & 0 & 0 & 0 \\ -.0007 & 0 & .92 & .39 & -.99 & -249 & 0 & 0 \\ .0002 & 0 & 0 & 1.03 & -.159 & -.158 & 0 & 0 \\ 0 & 0 & 0 & 0 & -1.5 & 0 & 0 & 0 \\ 0 & 0 & 0 & 0 & 0 & -1.5 & 0 & 0 \\ 0 & 0 & 0 & 0 & 0 & 0 & -.05 & -1.25 \\ 0 & 0 & 0 & 0 & 0 & 0 & 1.25 & 1 \end{bmatrix}$$

$$B = \begin{bmatrix} .01 & 0 \\ 0 & 0 \\ .001 & 0 \\ .25 & 0 \\ 0 & .2 \\ 0 & 0 \\ 0 & .25 \\ 0 & .03 \end{bmatrix} \; ; \qquad C = I_8$$

$$Q = \text{diag}\{10^{-5}, 1, 1, 10, 10^{-2}, 1, 1, 10^{2}\}$$

$$R = \text{diag}\{.05, .02\}$$

Find the optimal state and control trajectories using hierarchical computation methods for N_s = 2 and 4. Evaluate the computational effort for all the methods used.

5. An interconnected system is described by:

$$\begin{bmatrix} \underline{x}_1(k+1) \\ \underline{x}_2(k+1) \end{bmatrix} = \begin{bmatrix} A_1 & \varepsilon_1 A_2 \\ \varepsilon_2 A_3 & A_4 \end{bmatrix} \begin{bmatrix} \underline{x}_1(k) \\ \underline{x}_2(k) \end{bmatrix} + \begin{bmatrix} B_1 & 0 \\ 0 & B_2 \end{bmatrix} \begin{bmatrix} \underline{u}_1(k) \\ \underline{u}_2(k) \end{bmatrix}$$

$$\min_{\underline{u}_1,\underline{u}_2} J = \sum_{j=0}^{\infty} \{\underline{x}_1^t(j)Q_1\underline{x}_1(j) + \underline{x}_2^t(j)Q_2\underline{x}_2(j) + \underline{u}_1^t(j)R_1\underline{u}_1(j) + \underline{u}_2^t(j)R_2\underline{u}_2(j)\}$$

where $0 \le \varepsilon_1, \varepsilon_2 \le 1$.

(a) Find the optimal control law and cost function in the general case.

(b) Derive the following special cases:

1. $\varepsilon_1 = \varepsilon_2 = 0$
2. $\varepsilon_1 = 0$, $\varepsilon_2 \neq 0$
3. $\varepsilon_1 \neq 0$, $\varepsilon_2 = 0$
4. $\varepsilon_1 = 0$, $\varepsilon_2 = 1$
5. $\varepsilon_1 = 1$, $\varepsilon_2 = 0$
6. $\varepsilon_1 = \varepsilon_2 = 1$

Comment on the results of these cases, and point out the relationship between them (if it exists).

8.9 References

[1] Kalman, R.E.
"Contributions to the Theory of Optimal Control".
Bol. Soc. Mat. Mexicana, vol. 5, pp. 102-119, 1960.

[2] Kalman, R.E.
"When is a Linear Control System Optimal?".
J. Basic Engineering, Trans. ASME, vol. 86, pp. 51-60, 1964.

[3] Bellman, R. and S. Dreyfus
"Applied Dynamic Programming".
Princeton University Press, N.J., 1962.

[4] White, D.J.
"Dynamic Programming",
Holden-Day, Inc., San Francisco, 1969.

[5] Bryson, A.E. and Y.C. Ho
"Applied Optimal Control".
Hemisphere Publishing Co., N.Y., 1975.

[6] Sage, A.P. and C.C. Whitte,
"Optimum Systems Control", Prentice-Hall, N.J., 1977.

[7] Bellman, R., R. Kalaba and B. Kotkin
"Polynomial Approximation - A New Computational Technique in Dynamic Programming - I, Allocation Processes".
Mathematics of Computation, vol. 17, pp. 155-161, 1963.

[8] Larson, R.E.
"Dynamic Programming with Reduced Computational Requirements", IEEE Trans. Automatic Control, vol. AC-10, pp. 135-143, 1965.

[9] Bryson, A.E. and W.F. Denham
"A Steepest Ascent Method for Solving Optimum Programming Problems". J. Applied Mechanics, vol. 29, pp. 247-257, 1962.

[10] Breakwell, J.V., J.L. Speyer and A.E. Bryson
"Optimization and Control of Nonlinear Systems Using the Second Variation". SIAM J. Control, vol. 1, pp. 193-223, 1963.

[11] Bellman, R. and R. Kalaba
"Quasilinearization and Nonlinear Boundary-Value Problems".
Elsevier Press, N.Y., 1965.

[12] Tabak, D. and B.C. Kuo
"Optimal Control by Mathematical Programming",
Prentice-Hall, N.J., 1971.

[13] Kwakernaak, H. and R. Sivan
"Linear Optimal Control Systems", Wiley Interscience, N.Y. 1971.

[14] Caines, P.E. and D.Q. Mayne
"On the Discrete-Time Matrix Equation of Optimal Control". Int. J. Control, vol. 12, pp. 785-794, 1970.

[15] Caines, P.E. and D.Q. Mayne
"On the Discrete-Time Matrix Equation of Optimal Control - A Correction". Int. J. Control, vol. 14, pp. 205-207, 1971.

[16] Vaughan, D.R.
"A nonrecursive Algebraic Solution for the Discrete Riccati Equation". IEEE Trans. Automatic Control, vol. AC-15, pp. 597-599, 1970.

[17] Franklin, G.F. and J.D. Powell
"Digital Control of Dynamic Systems". Addison-Wesley Publishing Co., Mass., 1980.

[18] Gaalman, G.J.
"Comments on A Nonrecursive Algebraic Solution for the Discrete Riccati Equation". IEEE Trans. Automatic Control, vol. AC-25, pp. 610-612, 1980.

[19] Michelsen, M.L.
"On the Eigenvalue-Eigenvector Method for Solution of the Stationary Discrete Matrix Riccati Equation", IEEE Trans. Automatic Control, vol. AC-24, pp. 480-481, 1979.

[20] Stewart, G.W.
"Introduction to Matrix Computations", Academic Press, N.Y., 1973.

[21] Pappas, T., A.J. Lamb and N.R. Sandell, Jr.
"On the Numerical Solution of the Discrete-Time Algebraic Riccati Equation". IEEE Trans. Automatic Control, vol. AC-25, pp. 631-641, 1980.

[22] Bar-ness, Y.
"Solution of the Discrete Infinite-Time, Time-Invariant Regulator by the Euler Equation". Int. J. Control, vol. 22, pp. 49-56, 1975.

[23] Aoki, M.
"Optimal Control and System Theory in Dynamic Economic Analysis". North-Holland, 1976.

[24] Payne, H.J. and L.M. Silverman
"On the Discrete-Time Algebraic Riccati Equation". IEEE Trans. Automatic Control, vol. AC-18, pp. 226-234, 1973.

[25] Rappaport, D. and L.M. Silverman
"Structure and Stability of Discrete-Time Optimal Systems". IEEE Trans. Automatic Control, vol. AC-16,

pp. 227-233, 1971.

[26] Hewer, G.A.
"An Iterative Technique for the Computation of the Steady State Gains for the Discrete Optimal Regulator". IEEE Trans. Automatic Control, vol. AC-16, pp. 382-384, 1971.

[27] Kleinman, D.L.
"Stabilizing a Discrete, Constant, Linear System with Application to Iterative Methods for Solving the Riccati Equation". IEEE Trans. Automatic Control, vol. AC-19, pp. 252-254, 1974.

[28] Sandell, Jr., N.R.
"On Newton's Method for Riccati Equation Solution". IEEE Trans. Automatic Control, vol. AC-19, pp. 254-255, 1974.

[29] Garbow, B.S., et. al.
"Matrix Eigensystem Routines - EISPACK Guide Extension". Lecture Notes in Computer Science, vol. 51, Springer-Verlag, N.Y., 1977.

[30] Laub, A.J.
"A Schur Method for Solving Algebraic Riccati Equations". IEEE Trans. Automat. Control, vol. AC-24, pp. 913-921, 1979.

[31] Dahlquist, G. and A. Bjorck
"Numerical Methods". Prentice-Hall, N.J., 1974.

[32] Kailath, T.
"Redheffer Scattering Theory and Linear State-Space Estimation". Richerche di Automatica, Special Issue on System Theory and Physics, January 1979.

[33] Kailath, T.
"Some New Algorithms for Recursive Estimation in Constant Linear Systems". IEEE Trans. Inform. Thy., vol. IT-19, pp. 750-760, 1973.

[34] Morf, M. and T. Kailath
"Square-Root Algorithms for Least-Squares Estimation". IEEE Trans. Automatic Control, vol. AC-20, pp. 487-497, 1975.

[35] Silverman, L.M.
"Discrete Riccati Equations : Alternative Algorithms, Asymptotic Properties and System Theory Interpretations" in Control and Dynamic Systems, vol. 12 (C.T. Leondes, Ed.), Academic Press, pp. 313-386, 1976.

[36] Morf, M., J. Dobbins, B. Friedlander and T. Kailath
"Square-Root Algorithms for Parallel Processing in Optimal Estimation". Automatica, vol. 15, 1979.

[37] Morf, M., G.S. Sidhu and T. Kailath
"Some New Algorithms for Recursive Estimation in Constant, Linear, Discrete-Time Systems". IEEE Trans. Automatic Control, vol. AC-19, pp. 315-323, 1974.

[38] Kailath, T.
"Linear Systems", Prentice-Hall, N.J., 1980.

[39] Yasuda, K. and K. Hirai
"Upper and Lower Bounds on the Solution of the Algebraic Riccati Equation". IEEE Trans. Automatic Control, vol. AC-24, pp. 483-487, 1979.

[40] Mahmoud, M.S. and M.G. Singh
"Large Scale Systems Modelling". Pergamon Press, Oxford, 1981.

[41] Singh, M.G. and A. Titli
"Systems : Decomposition, Optimization and Control". Pergamon Press, Oxford, 1978.

[42] Pearson, J.D.
"Dynamic Optimization Techniques" in Optimization Methods for Large Scale Systems, edited by D.A. Wismer, McGraw-Hill, N.Y., 1971.

[43] Geoffrion, A.M.
"Duality in Nonlinear Programming". SIAM Review, vol. 13, pp. 1-37, 1971.

[44] Singh, M.G., M.F. Hassan and A. Titli
"Multi-Level Feedback Control for Interconnected Dynamical Systems Using the Prediction Principle". IEEE Trans. Systems, Man and Cybernetics, vol. SMC-6, pp. 233-239, 1976.

[45] Albert, A.
"Regression and the Moore-Penrose Pseudo Inverse". Academic Press, N.Y., 1972.

[46] Patnaik, L.M., N. Viswanadham and I.G. Sarma
"Computer Control Algorithms for a Tubular Ammonia Reactor". IEEE Trans. Automatic Control, vol. AC-25, pp. 642-650, 1980.

[47] Shackshaft, G.
"General-Purpose Turbo-Alternator Model." Proc. IEE, vol. 110, pp. 703-713, 1963.

[48] Walker, P.A. and O.H. Abdalla
"Discrete Control of An A.C. Turbogenerator by Output Feedback". Proc. IEE, vol. 125, pp. 1031-1038, 1978.

[49] Dyer, P. and S.R. McReynolds
"The Computation and Theory of Optimal Control". Academic Press, N.Y., 1970.

Author Index

Subject Index

MIX
Papier aus verantwortungsvollen Quellen
Paper from responsible sources
FSC® C105338

If you have any concerns about our products,
you can contact us on
ProductSafety@springernature.com

In case Publisher is established outside the EU,
the EU authorized representative is:
Springer Nature Customer Service Center GmbH
Europaplatz 3, 69115 Heidelberg, Germany

Printed by Libri Plureos GmbH
in Hamburg, Germany